ABOUT THE AUTHOR

J. WESLEY BARNES, Cullen Trust for Higher Education Endowed Professor in Engineering at the University of Texas at Austin, is a past winner of the Institute of Industrial Engineers Book-of-the-Year Award. Professor Barnes has gained extensive engineering experience while employed at Tracor and the Bell Telephone Laboratories and in his consulting activities with numerous major manufacturing firms. Reviewers, including practicing engineers and engineering professors, have consistently praised the sharp and entertaining style of the award-winning author of this book.

McGraw-Hill Series in Industrial Engineering and Management Science

Barnes: *Statistical Analysis for Engineers and Scientists: A Computer-Based Approach*
Bedworth, Henderson, and Wolfe: *Computer-Integrated Design and Manufacturing*
Black: *The Design of the Factory with a Future*
Blank: *Statistical Procedures for Engineering, Management, and Science*
Denton: *Safety Management: Improving Performance*
Dervitsiotis: *Operations Management*
Hicks: *Industrial Engineering and Management: A New Perspective*
Juran and Gryna: *Quality Planning and Analysis: From Product Development through Use*
Khoshnevis: *Discrete Systems Simulation*
Law and Kelton: *Simulation Modeling and Analysis*
Lehrer: *White-Collar Productivity*
Moen, Nolan, and Provost: *Improving Quality through Planned Experimentation*
Niebel, Draper, and Wysk: *Modern Manufacturing Process Engineering*
Polk: *Methods Analysis and Work Measurement*
Riggs and West: *Engineering Economics*
Riggs and West: *Essentials of Engineering Economics*
Taguchi, Elsayed, and Hsiang: *Quality Engineering in Production Systems*
Wu and Coppins: *Linear Programming and Extensions*

STATISTICAL ANALYSIS FOR ENGINEERS AND SCIENTISTS

A COMPUTER-BASED APPROACH

Also available from McGraw-Hill

Schaum's Outline Series in Engineering

Most outlines include basic theory, definitions, and hundreds of example problems solved in step-by-step detail, and supplementary problems with answers.

Related titles on the current list include:

Acoustics
Computer Graphics
Continuum Mechanics
Descriptive Geometry
Dynamic Structural Analysis
Elementary Statics & Strength of Materials
Engineering Economics
Engineering Mechanics
Fluid Dynamics
Fluid Mechanics & Hydraulics
Heat Transfer
Introductory Surveying
Lagrangian Dynamics
Linear Algebra
Machine Design
Mathematical Handbook of Formulas & Tables
Mechanical Vibrations
Operations Research
Programming with C
Programming with Fortran
Programming with Pascal
Reinforced Concrete Design
Statics & Mechanics of Materials
Strength of Materials
Structural Analysis
Structural Steel Design (LRFD Method)
Theoretical Mechanics
Thermodynamics with Chemical Applications
Vector Analysis

Schaum's Solved Problem Books

Each title in this series is a complete and expert source of solved problems with solutions worked out in step-by-step detail.

Related titles on the current list include:

3000 Solved Problems in Calculus
2500 Solved Problems in Differential Equations
3000 Solved Problems in Electric Circuits
2500 Solved Problems in Fluid Mechanics & Hydraulics
1000 Solved Problems in Heat Transfer
3000 Solved Problems in Linear Algebra
2000 Solved Problems in Mechanical Engineering Thermodynamics
2000 Solved Problems in Numerical Analysis
700 Solved Problems in Vector Mechanics for Engineers: Dynamics
800 Solved Problems in Vector Mechanics for Engineers: Statics

STATISTICAL ANALYSIS FOR ENGINEERS AND SCIENTISTS

A COMPUTER-BASED APPROACH

J. Wesley Barnes

The University of Texas at Austin

McGRAW-HILL, INC.

New York St. Louis San Francisco Auckland Bogotá
Caracas Lisbon London Madrid Mexico City Milan
Montreal New Delhi San Juan Singapore Sydney Tokyo Toronto

This book was set in Times Roman by Publication Services.
The editors were Eric M. Munson and John M. Morriss;
the production supervisor was Denise L. Puryear.
The cover was designed by Initial Graphics Systems, Inc.
Project supervision was done by Publication Services.
R. R. Donnelley & Sons Company was printer and binder.

STATISTICAL ANALYSIS FOR ENGINEERS AND SCIENTISTS

A Computer-Based Approach

This book is printed on recycled, acid-free paper containing 10% postconsumer waste.

3 4 5 6 7 8 9 0 DOH DOH 9 0 9 8 7 6 5

P/N 005093-7

Library of Congress Cataloging-in-Publication Data

Barnes, J. Wesley.
Statistical analysis for engineers and scientists: a computer-based approach / J. Wesley Barnes.
p. cm.
Rev. ed. of: Statistical analysis for engineers. c1988.
Includes index.
ISBN 0-07-839608-5 (IBM)—ISBN 0-07-839605-0 (Mac)
1. Engineering—Statistical methods—Data processing. I. Barnes, J. Wesley. Statistical analysis for engineers. II. Title.
TA340.B35 1994 93-43834
519.5′ 02462—dc20

To Susan, Kelley, and Erin

whose support and patience made the writing of this book possible

CONTENTS

LIST OF FIGURES

LIST OF TABLES

PREFACE

Why another book on probability and statistics for engineers? After all, excellent books are available that are more than sufficient for teaching the fundamentals of probability theory and statistical analysis to engineering and science students. The answer to this question is that, without exception, currently available books lack a critical element for students aspiring to be engineers and, indeed, for practicing engineers in today's ever more demanding professional environment.

To the best of my knowledge, no engineering statistics text currently available addresses statistical analysis from a modern computer-based viewpoint. Anyone with a background in "real-world" uses of statistical analysis knows that the availability of reliable, user-friendly statistical computer software is absolutely essential for all but the most trivial applications. Clearly, any book that ignores this fact is presenting the material from a dated, if not truly obsolete, viewpoint.

There are two primary reasons for these statements. First, the size of practical data sets is usually so large that the execution of even the simplest analysis methods becomes a cumbersome and heavy burden if not performed by a computer. For example, the computation of a sample mean of 1000 values requires that all 1000 values be summed, necessitating a marked clerical effort; yet a data set of 1000 values would be considered small in many practical environments. This dimensionality also carries a significant chance of human error both in mathematical computation and data entry. Repeated analysis using the same data set only magnifies these difficulties. On the other hand, use of computer software not only requires that the data set be entered only a single time, with appropriate verification procedures to minimize input errors, but also removes both the computational burden and the chance of computational error.

Second, the computational complexity of many very useful and conceptually understandable analysis methods presents a formidable obstacle to many analysts even when the data set is not large (and all the more when the data set is large). The availability of appropriate software removes this difficulty.

An equally important fact is that the author who ignores computer methods is severely limiting himself or herself in the pedagogical tools that might be used. In contrast, by using the computer as in this text, the author can construct various Monte Carlo simulations and numerous dynamic graphical illustrations to be used both in the text itself and in the exercises that accompany the text.

I have written this text to enable the instructor to teach the student how statistical analyses are usually performed in practical situations while still assuring the student a sound understanding of the basic mathematical concepts underpinning the methods to be learned.

The book is directed at junior and senior undergraduate students in engineering and the sciences. As such, it is applications-oriented, and most theoretical arguments are approached from an intuitive level rather than from the viewpoint of rigorous mathematical proofs. Anyone possessing a mathematical background through integral calculus is adequately prepared to study all the topics presented in this book. Persons with less mathematical preparation will still find the great majority of the material well within their reach and understanding. Some familiarity with computers and computer programming is also helpful but not essential.

Two companions to the text exist. The first is a set of microcomputer programs accompanied by an extensive User's manual. Both the programs and the User's manual were written especially for use with this book. The second companion to the book is the comprehensive *Instructor's Manual*. This manual contains not only detailed answers to the exercises at the end of the chapters but also suggestions, keyed to each chapter, for using the aforementioned statistical computer software. The computer software and the User's manual are packaged with the book. The *Instructor's Manual* is available through McGraw-Hill.

The primary difference between books that are already in use and this book is the computer-oriented viewpoint of this book. Thus it is appropriate, at this point, to discuss exactly how the computer-oriented viewpoint is melded with the text.

In Chapter 1, in addition to some historical background, the rationale of the text is explained and an overview of the software designed to accompany the text is presented. In Chapter 2 the primary uses of the computer are to illustrate the frequency interpretation of the probability of the events in a finite discrete sample space by means of a Monte Carlo simulation model and to evaluate factorials, permutations, and combinations.

In Chapters 3, 4, and 5 the uses of the computer are as follows:

1. Direct evaluation of the probability distribution function and the cumulative distribution function, for the normal, t, χ^2, F, binomial, uniform, exponential, gamma, lognormal, beta, Weilbull, Poisson, hypergeometric, geometric, and negative binomial distributions
2. Direct evaluation of the inverse cumulative distribution for these same distributions given the desired cumulative probability; i.e., the student specifies the cumulative probability and the program returns the value of the random variable that is associated with that cumulative probability. The availability of this software essentially negates the need for the statistical tables of these

distributions. However, in deference to tradition, I have elected to include the classical tables in the Appendix. The capabilities cited in (1) and (2) prove quite useful in the chapters associated with statistical inference that appear later in the book.

3. A generalized plotting package plots any of the named probability distributions presented in Tables 3.2 and 4.1. Also, the values of the plotted distribution's mean, variance (standard deviation), skewness, and kurtosis are presented.
4. A plotting package that accepts any user-specified probability distribution, both discrete and continuous, and plots it. In addition, the specified distribution's mean, variance, skewness, and kurtosis are given.
5. A generalized discrete convolution program that enables the student to study the results of adding, subtracting, dividing, and multiplying independent discrete random variables. This program can be used in a number of different ways, not the least of which is the empirical verification of central limit theorem. The results from this program can easily be passed to the plotting programs described above for graphical realization.

In association with Chapter 6, software is available to take a raw data set and produce the following items or any appropriate subset of them, as desired:

1. A frequency table. This is constructed using either automatically generated class limits or a user-specified lowest class limit and class width.
2. Plots of the frequency polygon associated with the frequency table in (1).
3. Plots of the histogram for the frequency table of (1).
4. Computed values, for the data set, of the sample mean, the sample variance, the sample standard deviation, the coefficient of variation, the maximum and minimum sample values, the sample range, the sample median, the sample mode, the sample skewness, and the sample kurtosis.

In association with Chapter 7, the software of Chapter 2 is expanded to allow Monte Carlo sampling from any of the named distributions included in Tables 3.2 and 4.1. These capabilities provide data sets that can then easily be used with the software of Chapter 6, making it easy for the student to conduct his or her own empirical study of such things as the distribution of the sample mean or of the sample variance.

The software associated with Chapter 8 allows the student easily and automatically to access the confidence intervals for the mean and variance based on a specific sample data set. Explicit use of direct access to the normal, t, and χ^2 distribution evaluator programs, as developed in Chapter 3, is made there. The software for this chapter is very user-friendly and is designed to assist the student in learning the specific circumstances under which each of the normal, t, and χ^2 distributions should be used.

Chapters 9 and 10 have associated programming to perform each of the classical tests of inference for one or two means, one or two variances, or one or two proportions. Here direct access to the F distribution evaluator program is used in addition to that of the normal, t, and χ^2 distributions. Note that with this direct access to the distribution evaluators, the student is no longer limited to the usual error levels of 0.1, 0.05, and 0.01. Programs that compute and plot the operating characteristic curves for any of the hypothesis tests just mentioned are also provided. These programs make it easy for the student to compute the sample size required for specified Type I and II error levels.

A bivariate linear regression program is associated with Chapter 11, and a program is available for performing the data transformations necessary for curvilinear regression in a bivariate environment. Output from the regression program provides all of the standard information, such as data input listing, ranges of each variable input, correlation between the input variables, parameter estimates and their variances, univariate confidence intervals on the model parameters, the correlation between the parameter estimators, a plot of the confidence region for β_0 and β_1, and the mean square for error. In addition to the standard output, a confidence band on the predicted line throughout the range of the data can be superimposed on the plot of the fitted line and the scatterplot of the data. The final output options consist of residual plots against the independent variable, the order of data entry, the predicted values of the dependent variable, and a plot of the empirical cumulative distribution function of the residuals on a normal probability scale.

The software for Chapter 12 consists of a multivariate regression analysis program that can fit a specified model and any subset of that model, one at a time, to a multivariate data set. With the exception of the scatterplot and confidence band, the output from this program is a higher-dimension analog to the output from the program of Chapter 11. Also in this chapter, the text discusses other widely available computer software packages, such as SPSS and SAS. Software to perform the more sophisticated regression procedures, such as stepwise regression or the "all regressions" approach, are not included among the programs of this book. I believe that students who progress this far in the text will be able to access the great store of sophisticated software in packages such as SPSS. Of course, several of the more sophisticated approaches are described in Chapter 12.

The programs for Chapters 13 and 14 consist of software to perform

1. A completely randomized single-factor analysis of variance
2. A single-factor randomized block analysis of variance
3. A two-factor factorial analysis of variance
4. Construction of o.c. curves for the above designs
5. A three-factor factorial analysis of variance

The programs associated with Chapter 15 are formulated to construct $\bar{x}$-R charts, p charts and o.c. curves for single sampling plans. Each of these programs present the associated results in both a tabular and graphical form. The programs associated with Chapter 16 perform the χ^2 goodness of fit test and three nonparametric analysis techniques, the Wilcoxon signed rank test, the Wilcoxon rank sum test, and the Kruskal-Wallis test. The last three analysis methods form an introduction to distribution-free alternatives to the more classical methods of Chapters 9, 10, and 13.

It has been our experience at the University of Texas at Austin, in over eight years of use, that this book with its supporting cast significantly enhances the student's learning experience and that these tools also enhance the teaching of the material contained within the book.

My sincere thanks go to the following friends and colleagues who used and critiqued the first edition of this book. Their extremely helpful comments are reflected in this revision: Mica Endsley, Texas Technological University; Earnest W. Fant, University of Arkansas; Chuanching Ho, Oklahoma State University; Ali A. Houshmand, University of Cincinnati; Norma Hubele, Arizona State University; Laura Raiman, Pensylvania State University; and Alice E. Smith, University of Pittsburgh. I also sincerely thank the graduate and undergraduate students whose numerous comments have provided unique insights and improvements that could not have been obtained without their diligent study and use of the book and software.

In addition to providing a careful reworking of the entire text, incorporating clearer nomenclature, to improve clarity and completeness, this revised version of *Statistical Analysis for Engineers and Scientists* incorporates many new explanations, examples, and exercises. The revised text is joined by an updated version of the IBM-compatible software, complete with a full-screen data editor, and by a completely new Macintosh version of the software.

It is my belief that old and new users of the *Statistical Analysis for Engineers and Scientists* text and software package will find the revised version to be a powerful and effective tool for learning, teaching, and applying statistical methods.

J. Wesley Barnes

STATISTICAL ANALYSIS FOR ENGINEERS AND SCIENTISTS

A COMPUTER-BASED APPROACH

CHAPTER 1

INTRODUCTION

Statistical analysis, at its most basic level, is a collection of methods that have been developed (1) to help describe some aspect or characteristic of a phenomenon under study and (2) to use our description of that characteristic. These two branches of statistical analysis are commonly known as *descriptive statistics* and *inferential statistics,* respectively.

For example, one application of statistical analysis could be to study the starting incomes of new engineering college graduates in the United States. Characteristics of interest might be the average income of such individuals, the maximum income, the minimum income, or the income level that only 10% of them exceed. In order to describe such characteristics, we would have to obtain *data* about that particular segment of the population, perhaps from federal government data banks or by conducting our own data gathering in some fashion.

We could easily arrive at the appropriate values of the above characteristics for those engineers whose starting incomes were included in our *sample*. However, since we will not have all graduating engineers in the United States in our sample, it becomes reasonable to ask if the average income in our sample is close to the average income for all such persons in the United States. In other words, we might want to answer a question like, "How likely is the average from our sample to be within \$3000 dollars of the *true* average starting income of *all* graduating engineers in the United States?" In this book you are shown how to answer questions just like this.

An argument can be made for the proposition that the use of basic descriptive statistical ideas has been present in peoples' endeavors from the very earliest civilizations. We have evidence of the tabulations of peoples and possessions in biblical accounts and in records found through the excavation of sites from even earlier times. These kinds of descriptive activities have become even more intense in the last two or three hundred years. Indeed, nowadays one can hardly turn on the television or radio or pick up a newspaper without being bombarded with all sorts of *averages, ranges, tables, graphs,* and *frequencies*.

It is well known that the basic tenets of *probability theory,* with which we will measure "likelihood," were developed as a by-product of the study of gambling by Pascal and Fermat during the middle of the seventeenth century. Many persons since that time *have used* the knowledge gained by the two French mathematicians to prevail in games of chance over their less well informed acquaintances. For this reason, we can say that the same time period saw the infancy of inferential statistics.

The history of the development of statistical analysis from that point to the science that it has become today and a recounting of the intellectual giants that have contributed to that growth is inappropriate in a text like this. It is sufficient to state that statistical analysis has grown and diversified to the point at which virtually none of the activities of the human race are free from the impact of modern statistical analysis. Nowhere is the influence more important than in the various branches of engineering. A great number of examples of the application of statistical analysis to engineering problems are given throughout this text both in the chapters and in the exercises.

Until comparatively recent times, it had been uncommon for an undergraduate engineering curriculum to have even a single course in probability and statistics among the required courses. However, there is now a trend that appears to be correcting that situation. The primary reason for this turn of events is the engineering community's realization that probabilistic phenomena must be considered in practical uses of engineering if the best job is to be done. Indeed, in many cases, serious errors can occur if probabilistic content is ignored.

Thus one must move from a strictly *deterministic* viewpoint in the practice of engineering to one in which the probabilistic content of the "real world" is considered. As you might suspect, such a move is not without cost. The added cost, for the most part, appears in two ways. The first is in the augmented complexity of the theoretical concerns that must be considered in the engineering analysis. The second is in the added clerical effort that is required to assess correctly the probabilistic components that bear directly on that analysis.

Happily, there is a product of modern technology that can lift most of the additional burden from the engineer. This labor-saving device is the microcomputer or its predecessor, the mainframe. You will have access to a set of computer programs that have been developed especially for use with this book.

If you use these programs, not only will you be relieved from hours of repetitive pencil-pushing toil but you will also learn how statistical analysis is practiced by engineers working in today's professions. In addition, you will be able to conduct exercises and learn concepts that, before this time, were impossible for students in a class like yours simply because of the clerical effort that was required.

Each of the chapters that follow begins with a list of the concepts contained in the chapter that you are expected to understand completely. Following that list, the material is presented in a fashion that includes written definitions, theorems, rules, and numerous examples that illustrate *how to use* the ideas that have been described. There are no *formal* proofs or derivations in this book.

Following each chapter, a set of exercises is given. Some of the exercises are conceptual in nature and may not involve any numbers. Other exercises involve some hand or calculator computation. The third and final kind of exercise involves the use of the computer software (i.e., programs) to arrive at the correct answer.

Chapter 2 begins the presentation of the content of this book with a discussion of *probability*—its meaning and how it is applied.

CHAPTER 2

PROBABILITY: FUNDAMENTAL CONCEPTS AND OPERATIONAL RULES

The following list summarizes the essential content of this chapter. When you have completed your study of this chapter, you should understand the concepts and be able to perform the operations named in this list.

- Understand what is meant by a *repeatable experiment*
- Understand what is meant by an *outcome* of an experiment
- Be able to list all of the possible outcomes of simple experiments similar to those described in the chapter
- Know the definition of a *sample space* and understand the differences between *discrete* and *continuous* sample spaces
- Understand what is meant by the *frequency interpretation of probability*
- Know which numbers can be assigned as probabilities of the events of a sample space
- Understand the meanings of *simple* and *composite* events
- Be able to assign probabilities to the outcomes of various experiments and then use these probabilities to calculate the probabilities of certain composite events
- Understand what is meant by the concepts of *mathematical symmetry* and *equally likely probabilities* and how they apply to certain experiments
- Know what is meant by $A_1 \cup A_2, A_1 \cap A_2$, and A_1', where A_1 and A_2 are events

- Know what is meant by *mutually exclusive events*
- When dealing with the events of a sample space, be able to use the *addition rule*
- Understand what is meant by $P(A_2 \mid A_1)$, where A_1 and A_2 are events
- Be able to use the general form of the *multiplication rule* to solve probability problems similar to those problems found in this chapter
- Know the definition of *statistically independent events*
- When given relevant facts concerning $P(A), P(B), P(A \cap B)$, or $P(A \mid B)$, be able to determine when two events, A and B, are statistically independent
- Be able to appropriately apply *Bayes' formula*
- Know what is meant by a *permutation* of a set of objects
- Know what is meant by the *factorial* of zero or any positive integer
- Know what is meant by a *combination of r objects*
- Know what is meant by the number of possible *combinations of n things taken r at a time*
- Be able to compute and use both ${}_nP_r$ and ${}_nC_r$ to solve counting and probability problems

2.1 REPEATABLE EXPERIMENTS AND SAMPLE SPACES

Few physical processes can be analyzed using mathematical tools without some simplifying assumptions. As an example, consider the inherent complexity of something so simple as flipping a coin, say, a 1992 United States quarter, or 25¢ piece. What we have is a laminated, asymmetric metallic disk with an embossed image of George Washington on one side and the American Eagle on the other side. When we toss the coin, the chemical energy in the arm muscles pushes the disk through the viscous fluid medium (which is the air we breathe) and, with luck, we catch the coin when it returns from its flight. Professors in other classes in engineering might see this as a marvelous opportunity to characterize the equations of motion governing the coin's travel.

To simplify our view of the experiment of flipping a coin, we ignore all of that rich inherent physical complexity. We also ignore other possibilities such as the coin landing on edge or the coin hitting the floor and disappearing into a crack. All we want to know is whether we get a "heads" or a "tails." If anything else happens, we ignore it and perform the experiment again until we get a heads or tails.

In statistical analysis, these simplifications and assumptions lead to a *sample space*, the set of all possible unique outcomes of a repeatable experiment, i.e., an experiment that, from the viewpoint of the experimenter, may be repeated as many times as desired. Only one of the outcomes in the sample space can occur as the result of any single performance of a repeatable experiment. For the coin-flipping experiment, the sample space consists of the two unique outcomes—heads or tails. Only one can occur as the result of any experiment, a flip of the coin.

There are two basic kinds of sample spaces, *discrete* sample spaces and *continuous* sample spaces. Discrete sample spaces may be divided into those that are *finite* and those that are *countably infinite*. The coin-tossing experiment has a discrete finite sample space. It is discrete because each of the two outcomes may be uniquely identified by a label—heads or tails. It is finite because the number of outcomes is 2, a finite number. The outcomes of a countably infinite sample space may be placed in one-to-one correspondence with the set of positive integers. For example, consider the sample space describing the possible *number of heads* that can occur in repeated flips of a coin *before the first tails* occurs. What might happen if we perform that experiment?

Possibly, the first toss would be tails. If that happens, the number of heads that occurs before the first tails would be 0. It is also possible that the first toss would be a heads and the second toss, tails. In that case, the number of heads would be 1. Continuing with this reasoning, we might get 2 or 3 or more heads before the first tails. Indeed, if we carry this argument to its logical conclusion, it is conceptually possible that we might continue to toss the coin an infinite number of times and still not get tails.

Nevertheless, since there is no upper bound on the number of possible heads that can occur before the first tails, the sample space for this experiment is discrete and countably infinite. It is discrete because each of the outcomes may be uniquely identified by a label, which would be either the number 0 or one of the set of positive integers. The sample space is countably infinite, because the number of possible outcomes is infinite.

The second kind of sample space, the continuous sample space, is always *uncountably infinite* because it is defined on a *continuum*. Suppose we are measuring the amount of time between the emission of two different particles from a radioactive substance. Because time is one of the few things, if not the only thing, in our physical environment that is a truly continuous phenomenon, the number of possible intervals of time that might occur is infinite. Indeed, if we change the experiment so that we record the interval only if it is less than 2 minutes, there would still be an uncountably infinite number of possible outcomes.

Herein lies the difference between countably and uncountably infinite sample spaces. No matter how short an interval you select on a continuum, an infinite number of possibilities still exist on the interval. Obviously, it is impossible to label each and every element of a continuous sample space.

2.2 EVENTS AND THE VENN DIAGRAM

An *event* is defined as any subset of outcomes from a sample space. For example, consider the sample space describing the roll of a die (small cube marked on each face with 1 to 6 dots and used in various games). The outcomes in that sample space are 1, 2, 3, 4, 5, and 6. Each outcome is, by our definition, an event. Indeed, an outcome is often called a *simple event*. The event "getting a three or greater" is made up of the subset of outcomes 3, 4, 5, and 6. Often, events made up of two or more outcomes are called *composite, or compound, events*. Indeed, the sample space itself is the largest of the possible composite events.

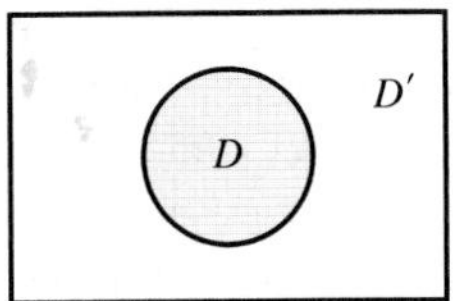

FIGURE 2-1
A Venn diagram.

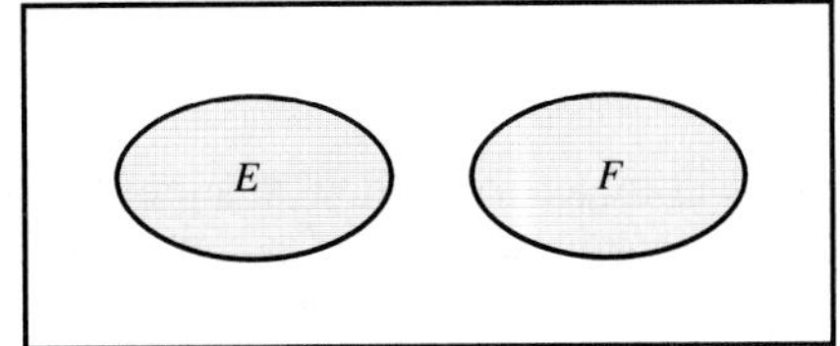

FIGURE 2-2
Mutually exclusive events.

The *union* of two or more events is defined as the event consisting of all outcomes present in one or more of the events contributing to the union. The *intersection* of two or more events is the set of outcomes that is present in each and every one of the contributing events. The *complement* of an event consists of the outcomes in the sample space that are not in that event. Two events are *mutually exclusive* if their intersection is *null*, i.e., if no single outcome is present in both events.

To assist in understanding these ideas, consider the events $A = \{1, 2, 4\}$, $B = \{2, 4, 6\}$, and $C = \{3, 5\}$, defined on our sample space, $S = \{1, 2, 3, 4, 5, 6\}$. The union of A and B is $A \cup B = \{1, 2, 4, 6\}$. Note that the simple events 2 and 4 appear in both A and B but are counted only once in $A \cup B$. The intersection of A and B is $A \cap B = \{2, 4\}$, and the complement of B is $B' = \{1, 3, 5\}$. Since they share no common outcomes, the events A and C are mutually exclusive and their intersection is null; i.e., $A \cap C = \varnothing$.

A *Venn diagram* is a useful pictorial tool to describe and analyze such things as unions and intersections. An example Venn diagram is given in Figure 2.1, where the sample space S is represented by the large rectangle and the events D and D *complement*, D', are represented as shown. (Note that Venn diagrams are rarely drawn to scale. Thus the size of an event pictured on a Venn diagram does not necessarily imply the proportion of the sample space contained in that event.)

Events E and F, as shown in Figure 2.2, are mutually exclusive. $A \cup B$ and $A \cap B$ are shown in Figure 2.3, where $A \cup B$ is represented by the entire shaded region and $A \cap B$ is the doubly shaded region.

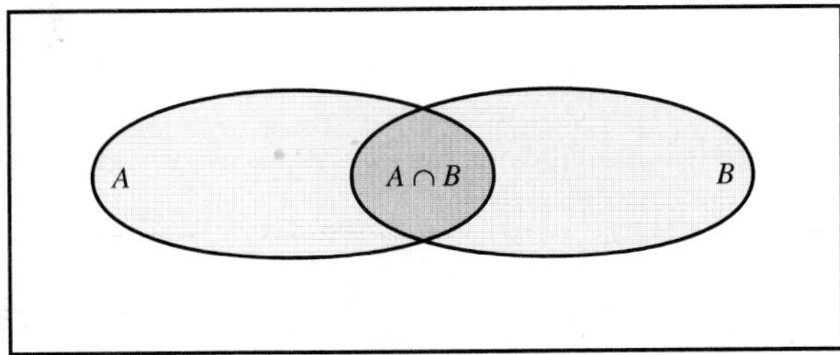

FIGURE 2-3
The union and intersection of two events.

Here are two examples that use the ideas just presented.

Example 2.1. Suppose that the testing laboratory of a particular electronics firm has been asked to determine how many of 5 components of type A and 3 components of type B function properly. One representation of the sample space is given on the following graph, where the vertical axis is associated with type A and the horizontal axis with type B.

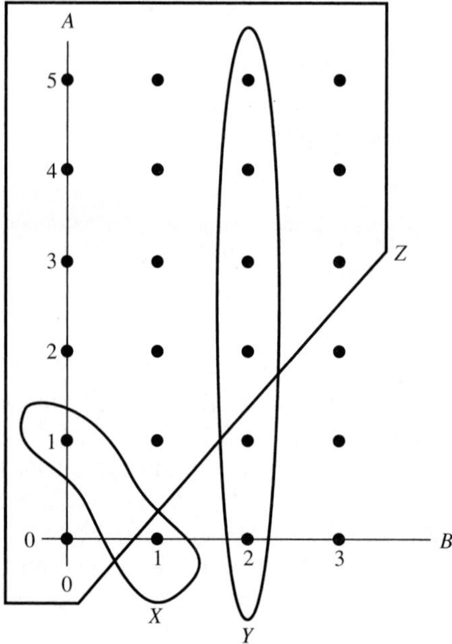

The sample space for this situation has 24 outcomes. Events X, Y, and Z are defined as follows.

X = only 1 of the 8 components functions = $\{(1, 0), (0, 1)\}$

Y = exactly 2 of the type B components function

= $\{(2, 0), (2, 1), (2, 2), (2, 3), (2, 4), (2, 5)\}$

Z = the same amount or more of the type A components function than do the type B

Events X, Y, and Z are graphically indicated directly on the sample space. Note that X and Y are mutually exclusive; i.e., $X \cap Y = \varnothing$.

$$X \cup Y = \{(1, 0), (0, 1), (2, 0), (2, 1), (2, 2), (2, 3), (2, 4), (2, 5)\}$$

$$Y \cap Z = \{(2, 2), (2, 3), (2, 4), (2, 5)\}$$

$$Z = \{(0, 0), (0, 1), (0, 2), (0, 3), (0, 4), (0, 5), (1, 1), (1, 2), (1, 3), (1, 4), (1, 5), (2, 2), (2, 3), (2, 4), (2, 5), (3, 3), (3, 4), (3, 5)\}$$

Example 2.2. A newly manufactured switching gear may be used in any one of the three modes, in any two of three modes, or in all three modes: manual (M), semiautomatic (S), and automatic (A). In tests on 100 gears, the following uses were found.

Number	Use
15	$M \cap S \cap A$
20	$M \cap S$
10	A-only
20	$S \cap A$
Remainder	M-only

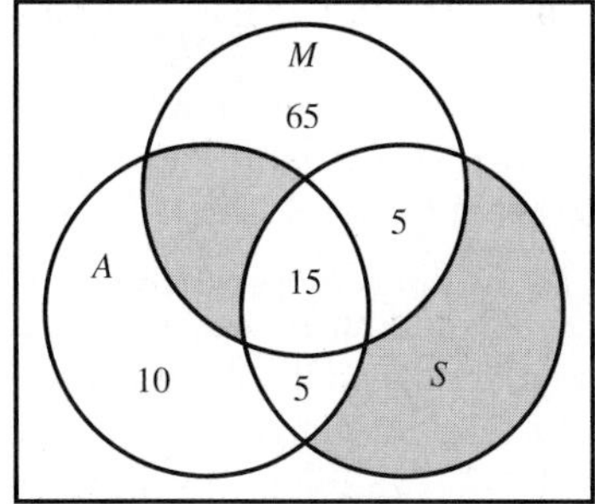

(a) Define M, S, and A as sets of gears used. Draw a Venn diagram, and show the number of gears in each use.

(b) How many gears are in M? in S? in (M ∪ S)? in (M ∪ S)′? in M ∪ S′?

Solution

(a) The preceding Venn diagram shows the number of gears in each set. First, the given information states that 15 of the gears must be in the triple intersection, $M \cap S \cap A$, and 10 of the gears must be in the set A-only, i.e., the set that has type A gears that are not used with either of the M or S modes. The intersection $M \cap S$ has 20 gears, but 15 are in the intersection $M \cap S \cap A$. This leaves 5 gears in $M \cap S$-only. In like manner, 5 gears are in $S \cap A$-only.

The considerations just discussed account for 35 of the gears. Since there is a total of $M \cup S \cup A = 100$ gears, there must be 65 gears in M-only. The sets S-only and $M \cap A \cap S'$ (shaded) are empty sets because there are no gears being used in these modes.

(b) Mode M has $65 + 15 + 5 = 85$ gears. Therefore, only 15 gears are never used in the manual mode. The S mode has 25 gears. The union $M \cup S$ has 90, so $(M \cup S)'$ has 10 gears. The union $M \cup S'$ has 95 gears, which are the gears contained in M added to the gears in A-only. (Because they are contained in M, the 20 gears in $M \cap S$ are counted as part of $M \cup S'$.)

2.3 PROBABILITY AND OPERATIONAL RULES

Now that you have seen how to deal with various events, let us discuss some methods that are employed to assess and communicate the likelihood of certain events occurring. The concept of probability is essential to these methods.

The *probability* of an event A, expressed as $P(A)$, enjoys the following properties.*

*Equivalent notation used by other authors for the probability of an event A is Pr(A) or PR(A).

$$\begin{aligned} 0 \le P(A) &\le 1 \quad & P(A) &= \text{sum of the probabilities} \\ P(S) &= 1 & &\text{of the outcomes of the} \\ P(\varnothing) &= 0 & &\text{sample space making up} \\ P(A) + P(A') &= 1 & &\text{the event } A. \end{aligned} \tag{2.1}$$

where S is the sample space. Thus an event that is certain to occur has a probability equal to 1 and an event that cannot occur has a probability equal to 0.

Often the probability of an event can be determined by *mathematical symmetry*. For example, symmetry indicates that the probability of a heads occurring on a single toss of a *fair* coin is 0.5; i.e., there are only 2 *equally likely* events that can occur. Since they equally share the total probability of 1 unit, each event has a probability of 0.5. Similarly, each face of a fair die has a probability $\frac{1}{6}$ of occurring. The following example illustrates the use of mathematical symmetry with a continuous sample space.

Example 2.3. As pictured, a 1-inch needle is pinned exactly at its midpoint, precisely in the middle of two parallel lines.

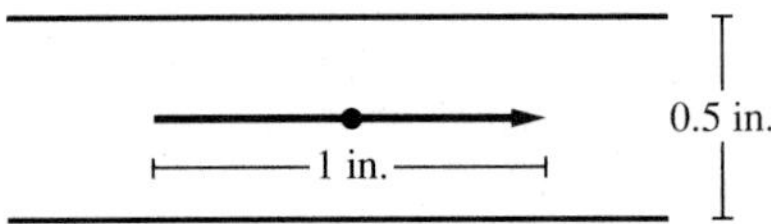

The needle may spin freely on its pin. When the needle is spun, it stops in a random position. What is the probability that the needle will stop in a position that crosses the parallel lines?

Solution. The needle is *equally likely* to stop at any position in the uncountably infinite set of possible stopping positions. As pictured here, suppose we turn the needle until it is just touching the lines.

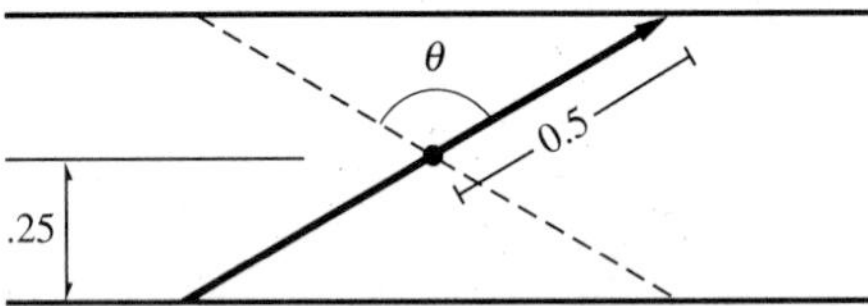

The angle θ is easily shown to be 120°. Since the needle will touch or cross the lines for 240° out of the 360° total, the required probability is $\frac{2}{3}$.

If symmetry arguments are not possible for some event, we can use the *frequency interpretation* of probability. Loosely speaking, this interpretation states that the proportion of times that A occurs during a large number of repeated experiments approaches the probability of A, that is, $P(A)$. Stated mathematically, this implies that

$$P(A) = \lim_{N\to\infty} \frac{N(A)}{N} \tag{2.2}$$

where $N(A)$ is the number of times A occurred in the N repetitions of the experiment.

Thus we expect that heads will occur about one-half of the time in a large number of fair coin flips. Notice that this applies only to a large number of experiments. No reasonable person would expect every face of a fair die to appear in only six rolls.

A natural question that could be asked at this point is, How many rolls should one expect to make before the appearance of each face of a fair die would approach the expected $\frac{1}{6}$ (0.166) of the total rolls? How to answer this question depends on how closely you want your experimental results to agree with theory. One way of answering it would be to obtain a fair die, begin to roll it, and record the results. Fortunately, because we have access to the microcomputer software package provided with this book, there is an easier way. Using the program for simulating finite discrete sample spaces, Table 2.1 was formed.

We can observe that the simulation results get closer to the theoretical values as the number of rolls increases. In particular, the maximum deviation of any outcome's proportion, from the theoretical value of $\frac{1}{6}$, experiences a steady decline with only one minor exception, at 288 rolls. The random variability present in any experiment like this will often yield such "reversals" for smaller sample sizes. However, as the sample size gets larger and larger, this will become more and more rare. We return to this discussion in much greater detail in Chapter 8.

Once the sample space of an experiment is defined and a probability has been assigned to each outcome, we can compute the probability of an event by adding the probabilities of the outcomes making up that event. For example, the probability of rolling an even number with a fair die is $P(\text{Even}) = P(2)+P(4)+P(6) = \frac{1}{6}+\frac{1}{6}+\frac{1}{6} = \frac{1}{2}$. A direct extension of this rule is applicable

TABLE 2.1
Simulation results for the proportion of rolls of a fair die

	Number of Rolls								
Outcome	**6**	**12**	**24**	**48**	**96**	**144**	**192**	**288**	**384**
1	0.16	0.16	0.21	0.25	0.115	0.131	0.133	0.166	0.151
2	0.00	0.08	0.25	0.08	0.145	0.174	0.133	0.184	0.179
3	0.66	0.08	0.16	0.15	0.188	0.183	0.179	0.128	0.141
4	0.00	0.25	0.12	0.19	0.177	0.188	0.199	0.197	0.169
5	0.00	0.33	0.16	0.14	0.188	0.132	0.173	0.170	0.169
6	0.16	0.08	0.08	0.19	0.188	0.194	0.184	0.153	0.190
Maximum absolute deviation	0.50	0.17	0.09	0.09	0.051	0.035	0.033	0.038	0.025

when two or more events are mutually exclusive. The probability associated with their union is equal to the sum of the probabilities of all the events; i.e., if A, B, and C are mutually exclusive, $P(A \cup B \cup C) = P(A) + P(B) + P(C)$.

When events are not mutually exclusive, a somewhat more complicated rule applies. Suppose events C and D are not mutually exclusive, as in Figure 2.4. The *general addition rule* states that

$$P(C \cup D) = P(C) + P(D) - P(C \cap D) \tag{2.3}$$

Subtracting $P(C \cap D)$ is required because $P(C) + P(D)$ adds $P(C \cap D)$ into the sum *two times*. Similar relations apply in the case of three or more events. For example, it is easily shown that

$$P(A \cup B \cup C) = P(A) + P(B) + P(C) - P(A \cap B) - P(A \cap C)$$

$$-P(B \cap C) + P(A \cap B \cap C) \tag{2.4}$$

When all events are mutually exclusive, all intersections such as $C \cap D$ have a probability of zero—$P(C \cap D) = 0$—and the formula, given in the preceding paragraph, for mutually exclusive events emerges as a special case of the general addition rule.

The following example provides an additional illustration of the preceding concepts.

Example 2.4. A vendor's experience has shown that, in units of a particular product, 4 out of 100 electronic components have fabrication errors and 3 out of 100 have both fabrication errors and the presence of impurities. If a particular bin contains components of which 8% have impurities, what is the probability of finding, in a unit from that bin, (a) fabrication errors or impurities, (b) neither fabrication errors nor impurities, and (c) only impurities?

Solution. Let I be the presence of impurities and E be the presence of fabrication errors.

(a) $P(I \cup E) = P(I) + P(E) - P(E \cap I) = 0.08 + 0.04 - 0.03 = 0.09$

(b) $P\{(I \cup E)'\} = 1 - P(I \cup E) = 1 - 0.09 = 0.91$

(c) $P(I\text{-only}) = P(I) - P(I \cap E) = 0.08 - 0.03 = 0.05$

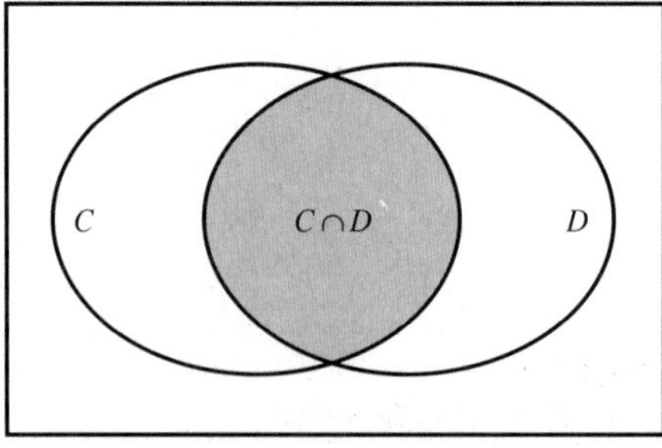

FIGURE 2-4
Two events that are not mutually exclusive.

2.4 CONDITIONAL PROBABILITY AND STATISTICAL INDEPENDENCE

The concept of *conditional probability* is a very important one in the practical use of probability theory and statistical analysis. Consider the Venn diagram given in Figure 2.5. Now assume that we are given knowledge that B has occurred. This causes a dramatic change in the sample space that needs to be considered. The reduced sample space is shown in Figure 2.6. Clearly, the *conditional probability* of A occurring given B has occurred is

$$P(A \mid B) = \frac{P(A \cap B)}{P(B)} \qquad P(B) \neq 0 \tag{2.5}$$

This definition of conditional probability directly yields the following statement of the *general rule of multiplication*.

$$P(A \cap B) = P(B)P(A \mid B) \quad \text{or, equivalently,} \quad P(A \cap B) = P(A)P(B \mid A) \tag{2.6}$$

Note that there are two ways of computing the intersection probability. The appropriate formula to use depends on what information is available about events A and B.

The concept of *statistical independence* is another important idea in probability and statistics. An event A is statistically independent of another event B if and only if $P(A \mid B) = P(A)$—which implies that $P(B \mid A) = P(B)$. Clearly, mutually exclusive events are *not* independent; i.e., $P(A \mid B) = 0$ if A is mutually exclusive of B.

If events A and B are statistically independent, then, by Equation 2.6, $P(A \cap B) = P(A)P(B)$; i.e., the intersection probability of two statistically independent events is simply the product of their unconditional probabilities. To extend this idea to three or more events, let us suppose that we have n statistically independent events, $A_1, A_2, A_3, \ldots, A_{n-1}, A_n$. Using Equation 2.6 repeatedly, it is easily shown that the probability of the intersection of all of the A_i, $i = 1, \ldots, n$, is

$$P(\cap_{i=1}^{n} A_i) = P(A_1 \cap A_2 \cap \cdots \cap A_{n-1} \cap A_n) = \prod_{i=1}^{n} P(A_i) \tag{2.7}$$

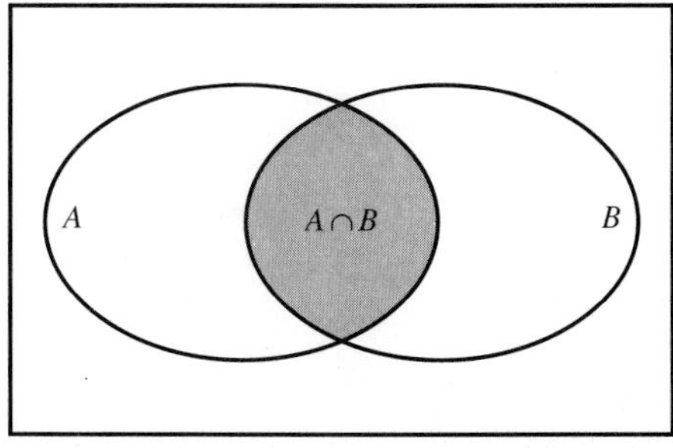

FIGURE 2-5
The original sample space.

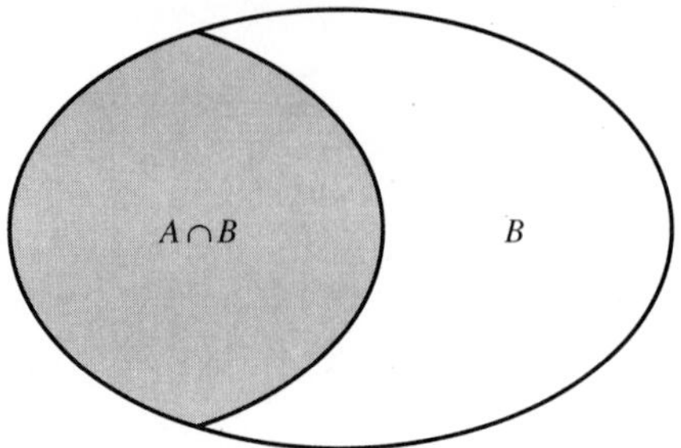

FIGURE 2-6
The reduced sample space.

Example 2.5. In order to clarify these concepts, consider the discrete finite sample space associated with the pistons in an 8-cylinder tractor used in heavy hauling. Occasionally, a piston fails, requiring replacement and associated repairs. Numbering the cylinders, we may represent the sample space as $S = (1, 2, 3, 4, 5, 6, 7, 8)$. Let $A = (2, 3, 4, 5)$, i.e., the event that one of the pistons numbered 2, 3, 4, or 5 fails $[P(A) = \frac{1}{2}]$ and let $B = (3, 4, 5, 6, 7, 8)[P(B) = \frac{3}{4}]$. Consider the following representation of these events.

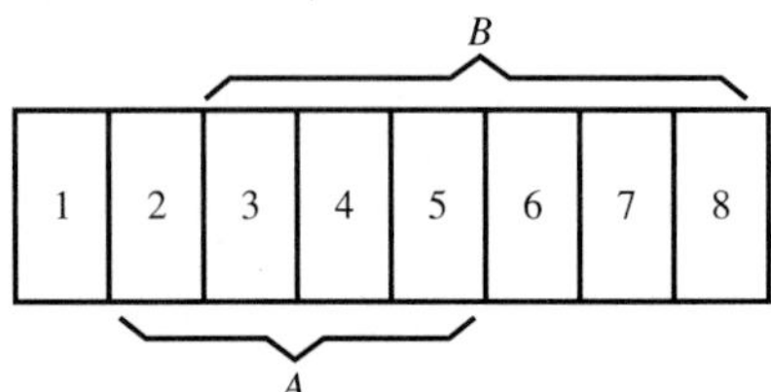

The reduced sample space, given B occurred, is represented by

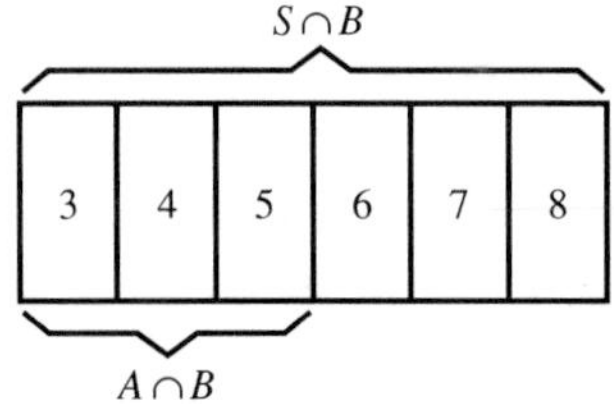

and we observe that

$$P(A \mid B) = \frac{P(A \cap B)}{P(B)} = \frac{3/8}{6/8} = \frac{1}{2} = P(A)$$

Therefore, the events A and B are statistically independent. Note that when the sample space was reduced, the sample space S and the event A were *proportionally reduced* by the same amount. In this specific case, S and A were both reduced by $\frac{1}{4}$.

Example 2.6. An electrical device requires that two linked subsystems function. The following schematic shows that A must function and that at least one of the two B's must function.

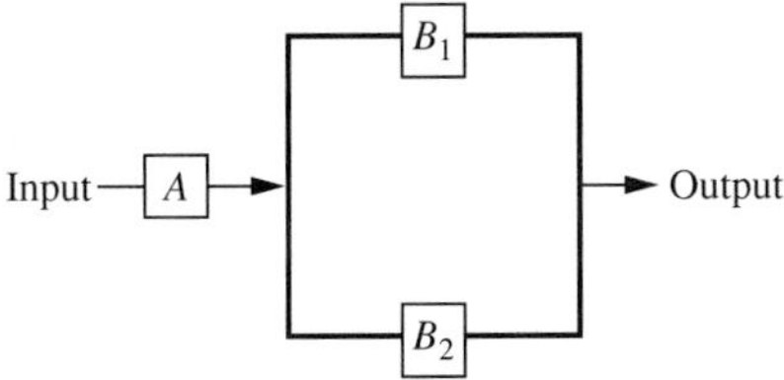

Assume that the B components function independently of A and of each other. The *reliability* (probability of functioning) of A is 0.9 and that of each B is 0.8. Calculate the reliability of the device.

Solution. There are several ways to compute the device reliability. The most straightforward approach is to list the outcomes in the sample space and add the probabilities of the outcomes that allow the system to function. Let $\overline{A}(\overline{B})$ designate that component $A(B)$ has failed. The sample space is made up of eight outcomes as follows.

	Outcome	Probability	Outcome	Probability
·	AB_1B_2	$(0.9)(0.8)(0.8) = 0.576$	$\overline{A}B_1B_2$	$(0.1)(0.8)(0.8) = 0.064$
·	$AB_1\overline{B}_2$	$(0.9)(0.8)(0.2) = 0.144$	$\overline{A}B_1\overline{B}_2$	$(0.1)(0.8)(0.2) = 0.016$
·	$A\overline{B}_1B_2$	$(0.9)(0.2)(0.8) = 0.144$	$\overline{A}\overline{B}_1B_2$	$(0.1)(0.2)(0.8) = 0.016$
	$A\overline{B}_1\overline{B}_2$	$(0.9)(0.2)(0.2) = 0.036$	$\overline{A}\overline{B}_1\overline{B}_2$	$(0.1)(0.2)(0.2) = 0.004$

The three marked outcomes (·) are the only ones that allow the device to function, and the sum of their probabilities is 0.864.

Because of the independence assumptions, the desired probability is also given by the formula

$$P(\text{Device}) = P(A)P[(B_1 \cup B_2) \mid A)] = P(A)P(B_1 \cup B_2)$$

With regard to components B_1 and B_2, four mutually exclusive things can occur: Both components can fail, both components can succeed, B_1 can succeed and B_2 can fail, or B_2 can succeed and B_1 can fail. Hence, an easy way to calculate $P(B_1 \cup B_2)$ is first to calculate the probability that both B_1 and B_2 will fail. Because of independence, that probability is given by $P(\overline{B}_1 \cap \overline{B}_2) = (0.2)(0.2) = 0.04$. Recalling that $P(B_1 \cup B_2)$ is the probability that either B_1 or B_2 or both B_1 and B_2 will succeed, $P(B_1 \cup B_2) = 1 - 0.04 = 0.96$. Alternatively, $P(B_1 \cup B_2) = P(B_1) + P(B_2) - P(B_1 \cap B_2) = 0.8 + 0.8 - 0.64 = 0.96$

Since $P(A) = 0.9$, the reliability of the device is given by $(0.9)(0.96) = 0.864$.

Example 2.7. Eight chemical solutions were checked for impurity types S and D. Four had S, six had D, and two had both S and D. Unfortunately, the data sheets have been mixed up and, therefore, the solutions must be rechecked. Compute the probability that a solution

(a) has S given that it has D on recheck

(b) has D given that it has S on recheck

Solution. From the information provided, the events and probabilities are

Event	Description	Probability
S	Impurity S	0.50
D	Impurity D	0.75
$S \cap D$	Both S and D	0.25

(a) $P(S \mid D) = \dfrac{P(S \cap D)}{P(D)} = \dfrac{0.25}{0.75} = \dfrac{1}{3}$ (b) $P(D \mid S) = \dfrac{P(S \cap D)}{P(S)} = \dfrac{0.25}{0.5} = \dfrac{1}{2}$

2.5 BAYES' FORMULA

Bayes' formula is a natural outgrowth of conditional probability and the general rule of multiplication. Suppose you have a group of events, $B_1, B_2, \ldots, B_n$, that are *mutually exclusive and exhaustive*; i.e., the B_i's have no outcomes in common and, together, contain all the outcomes in the sample space. In mathematical terms, the mutually exclusive B_i's are exhaustive if and only if $\sum_{i=1}^{n} P(B_i) = 1$. Further, suppose that another event A is defined on the same sample space. Since the B_i's are exhaustive, A must intersect with one or more of the B_i's. Therefore, one way to obtain the probability of A would be to sum the probabilities of $A \cap B_i$ over all values of i: $\sum_{i=1}^{n} P(A \cap B_i) = P(A)$.

Let us consider an example in which $n = 4$. This may be pictured as in Figure 2.7. Notice that A consists wholly of $A \cap B_1, A \cap B_2, A \cap B_3$, and $A \cap B_4$. Thus $P(A) = P(A \cap B_1) + P(A \cap B_2) + P(A \cap B_3) + P(A \cap B_4)$. This implies that $P(A) = \sum_{i=1}^{4} P(A \cap B_i) = \sum_{i=1}^{4} P(B)_i)P(A \mid B_i)$. Now, suppose we turn the problem around a bit and assume that A is known to have occurred. How does one obtain the probability that each of the B_i's also occurred? If A occurred, the sample space is reduced. The reduced sample space is pictured in Figure 2.8.

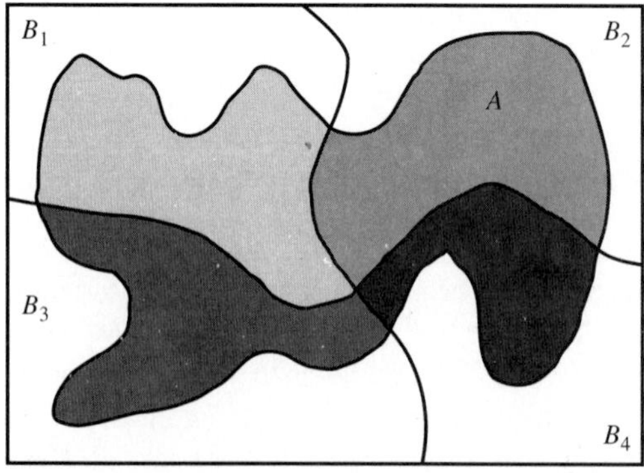

FIGURE 2-7
An illustration of Bayes' formula.

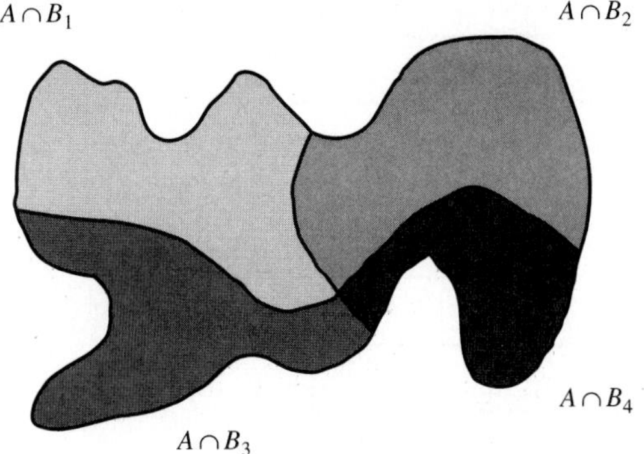

FIGURE 2-8
The reduced sample space when A has occurred.

From Figure 2.8, we see that

$$P(B_i \mid A) = \frac{P(B_i \cap A)}{P(A)} = \frac{P(B_i)P(A \mid B_i)}{\sum_{j=1}^{n} P(B_j)P(A \mid B_j)} = \frac{P(B_i)P(A \mid B_i)}{\sum_{j=1}^{n} P(B_j \cap A)} \tag{2.8}$$

Equation 2.8 is known as *Bayes' formula.*

Example 2.8. Vendors I, II, III, and IV provide all the bushings that the Acme Axle Co. purchases in the amounts 25%, 35%, 10%, 30%, respectively. It is known from long experience that vendors I, II, III, and IV provide 80%, 95%, 70%, and 90% good components. What is the probability that a randomly selected bushing is bad? Given that a bushing is bad, what is the probability it came from vendor III?

Solution. Let A represent the selection of a bad bushing, and let B_1, B_2, B_3, and B_4 represent the selection of a bushing from vendors I, II, III, and IV, respectively.

$$\begin{aligned} P(A) = \sum_{i=1}^{4} P(A \cap B_i) &= \sum_{i=1}^{4} P(B_i)P(A \mid B_i) \\ &= 0.25(0.2) + 0.35(0.05) + 0.1(0.3) + 0.3(0.1) \\ &= 0.1275 \end{aligned}$$

Therefore,

$$P(B_3 \mid A) = \frac{P(B_3 \cap A)}{P(A)} = \frac{0.03}{0.1275} = 0.2353$$

2.6 COUNTING TECHNIQUES: TREES, COMBINATIONS, AND PERMUTATIONS

Often, in physical applications, repeatable experiments can be quite complex, sometimes consisting of several stages. In evaluating probabilities associated with such experiments, the *probability tree* is a convenient pictorial tool. Suppose that, in stage 1, one of two events can occur, A_1 or A_2. Further, in stage 2, one of three events can occur, B_1, B_2, or B_3. A probability tree depicting the sample space for this two-stage experiment is given in Figure 2.9. Note that the probabilities

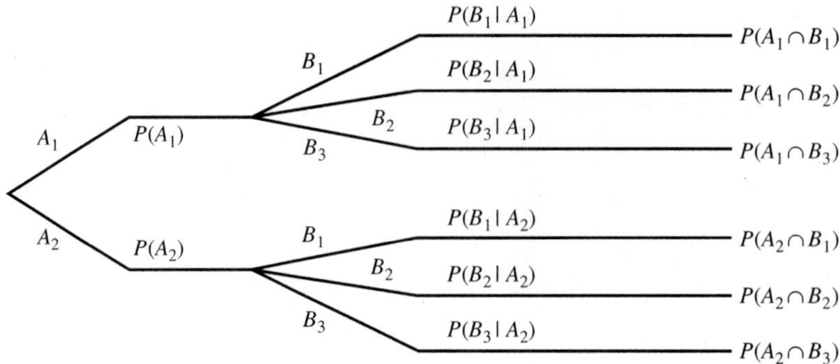

FIGURE 2-9
A probability tree.

along each branch may be multiplied to obtain the probabilities at the far right; i.e., $P(A_1 \cap B_1) = P(A_1)P(B_1 \mid A_1)$. Trees with more stages simply require more multiplications; i.e., $P(A_1 \cap B_1 \cap C_1) = P(A_1)P(B_1 \mid A_1)P(C_1 \mid A_1 \cap B_1)$.

Let us consider two example applications of probability trees.

Example 2.9. A sensitive electronic component is part of a guidance system on a military vehicle. Because it fails on the average about once every hundred flights, it is important to have a reliable performance test for it. Quality engineers have devised a test that indicates that a component is defective 90% of the time when the component *is defective*. Further, the test will indicate that a component *is good* 99% of the time when the component is good.

The guidance system on one of the vehicles failed. The component in that system was tested, and the test indicated that the component was defective. What is the probability that the component was indeed defective?

Solution. In this problem we have two stages, each with two possible events, as pictured here.

A_1 0.01 — B_1 0.9 — $P(A_1 \cap B_1) = 0.0090$

A_1 0.01 — B_2 0.1 — $P(A_1 \cap B_2) = 0.0010$

A_2 0.99 — B_1 0.01 — $P(A_2 \cap B_1) = 0.0099$

A_2 0.99 — B_2 0.99 — $P(A_2 \cap B_2) = 0.9801$

where A_1 = component is defective, A_2 = component is good, B_1 = test says component is defective, and B_2 = test says component is good. We *know* that the test said the component was defective. Therefore, only outcomes containing B_1 may form part of the reduced sample space.

$$P(A_1 \mid B_1) = \frac{P(A_1 \cap B_1)}{P(B_1)} = \frac{P(A_1 \cap B_1)}{P(A_1 \cap B_1) + P(A_2 \cap B_1)} = 0.476$$

This implies that when the test says that the component is defective the component is more likely to be good! This apparently nonsensical result stems from the relatively high probability (0.01) of the test saying the component is defective when it is indeed good.

Example 2.10. Two electronic components are in parallel and one is in series.

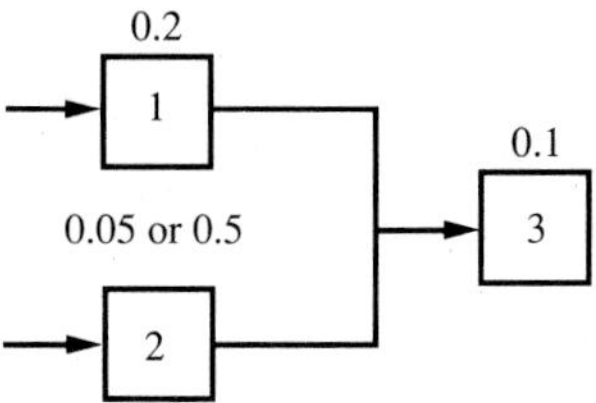

Event	Description	Probability
I	component 1 fails	0.20
II|I′	component 2 fails|1 succeeds	0.05
III	component 3 fails	0.10
II|I	component 2 fails|1 fails	0.50

The probability of failure is given for components 1 and 3. In addition, because component 1 will give off considerable heat when it fails, it is estimated that if component 1 fails there is a 0.5 probability that component 2 will also fail. If component 1 does not fail, component 2 will fail with a probability of only 0.05. The parallel components 1 and 2 are redundant, so they must both fail before current cannot flow to component 3. Compute the probability that

(a) the device does not fail

(b) the device fails

(c) either component 1 or 2 fails and component 3 fails

(d) component 2 failed given that the device fails

Solution. In the accompanying probability tree, the events I′, II′, and III′ indicate component success.

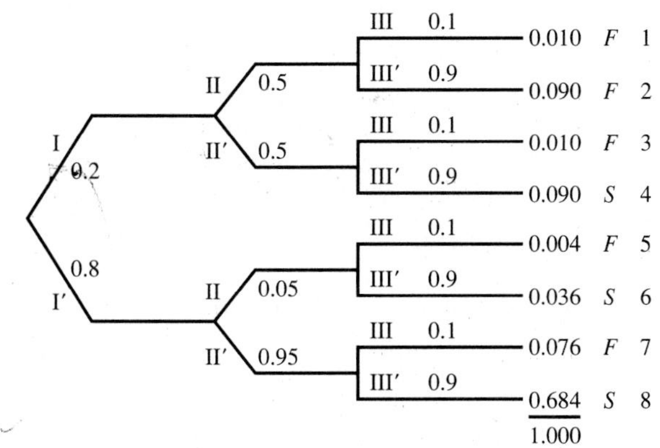

The sum of all probability values is 1.0, as required.

(a) The probability of success, $P(S)$, is the sum of results 4, 6, and 8, $P(S) = 0.090 + 0.036 + 0.684 = 0.810$.

(b) The probability of failure, $P(F)$, is found by $P(F) = 1 - P(S) = 1 - 0.810 = 0.190$.

(c) If the event (I ∪ II) ∩ III occurs, the probability tree results 1, 3, and 5 apply. Therefore, $P[(\text{I} \cup \text{II}) \cap \text{III}] = 0.01 + 0.01 + 0.004 = 0.024$. Another way to reach this result is to notice that events I and II are not mutually exclusive. First we compute $P(\text{I} \cup \text{II}) = P(\text{I}) + P(\text{II}) - P(\text{II}|\text{I})P(\text{I}) = 0.20 + 0.14 - (0.5)0.2 = 0.24$. To obtain $P(\text{II}) = 0.14$, we add the probabilities of the events where II occurs in results 1, 2, 5, and 6. Since the events I, II, and III are independent, $P[(\text{I} \cup \text{II}) \cap \text{III}] = P(\text{I} \cup \text{II})P(\text{III}) = 0.24(0.10) = 0.024$.

(d) If failure results, $P(\text{II}|\text{F})$ is a probability that II was, in part, a cause of the failure. Recall that $P(\text{II}|\text{F}) = P(\text{II} \cap F)/P(F)$. Failure and II occur in results 1, 2, and 5; $P(F)$ was computed in (b). Therefore,

$$P(\text{II} \mid F) = \frac{0.010 + 0.090 + 0.004}{0.190} = \frac{0.104}{0.190} = 0.547$$

There is only a 19% chance that the device will fail, but there is a better than 50% chance that II will have failed once device failure is observed.

It is possible to compute a similar probability for all events, given device failure. You should verify the following values.

$$P(\text{I} \mid F) = 0.579 \qquad P(\text{II} \mid F) = 0.547 \qquad P(\text{III} \mid F) = 0.526$$

$$P(\text{I} \mid S) = 0.111 \qquad P(\text{II} \mid S) = 0.044 \qquad P(\text{III} \mid S) = 0$$

The probability tree approach will quickly become cumbersome in the presence of a large number of stages and events at each stage. Therefore, probability trees must be viewed as useful conceptual tools and not ends in themselves. Clearly, an algebraic and systematic method of counting elements of the sample space is needed.

Consider a general, 2-stage system with r and s possibilities at stages 1 and 2, respectively. The experiment would have $(r)(s)$ outcomes. A third stage with t possibilities would yield $(r)(s)(t)$ outcomes for the experiment. Such a general tree may be used to assist in answering the question, How many ways can r things be selected from n distinct things?

To clarify this, consider how many three-letter "words" can be made from the letters a, b, c, and d, if each letter may be used only once. Thinking of this as a 3-stage decision problem, the answer is easily seen to be (4) (3) (2) = 24, since the first stage has 4 selectable letters whereas the second and third stages have only 3 and 2 selectable letters, respectively. The specific tree for this problem is presented in Figure 2.10.

The answer is that r things may be selected from n things in

$$_nP_r = n(n-1)(n-2)\cdots(n-r+1) = \frac{n!}{(n-r)!} \tag{2.9}$$

ways.

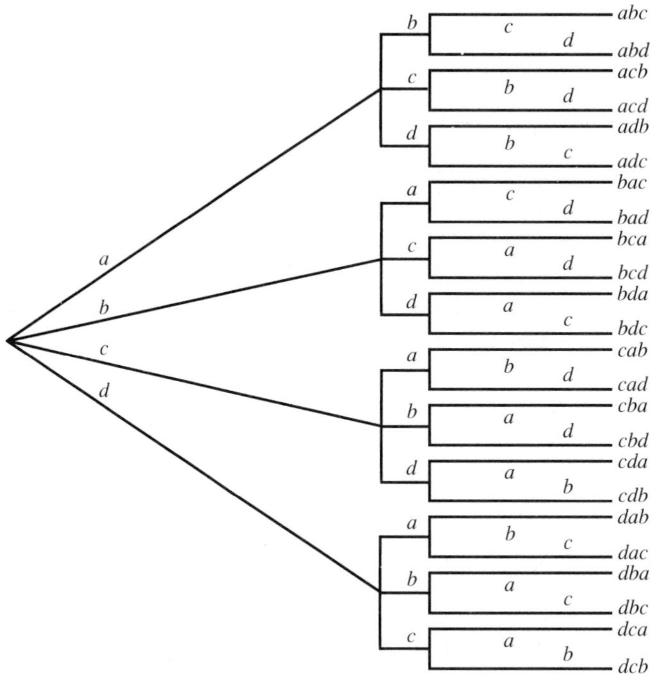

FIGURE 2-10
A tree illustrating $_4P_3$.

The notation $_nP_r$ symbolizes the number of *permutations* of n things taken r at a time. Formally speaking, a permutation is an ordered arrangement of all or some of the elements of a set. The notation $n!$ is read *n factorial* and is defined to mean the product of all integers from 1 up to n. For example,

$$4! = (1)(2)(3)(4) = 24 \quad \text{and} \quad {}_4P_3 = \frac{4!}{(4-3)!} = \frac{4!}{1!} = \frac{(4)(3)(2)(1)}{1} = 24$$

Zero factorial (0!) is a special case and is defined to have a value of 1.

From Figure 2.10, we can note a very significant fact: The use of permutations implies that the *order* of the elements is important. For example, in Figure 2.10, *abc* is different from *acb, bac, bca, cab*, and *cba*. If order is important, such as selecting positions on a baseball team, then you should find the number of permutations. However, order is often not important. Consider the selection of a committee made up of specific individuals. In this case, a committee of John, Mary, and Joe is identical to one of Mary, Joe, and John or to any of the other permutations of the three names. *If order is not important*, then the quantity of interest is the number of *combinations* of n things taken r at a time. As an example, if order is not important, the 24 permutations in Figure 2.10 would reduce to 4 combinations—*abc, abd, acd*, and *bcd*. In general, the number of combinations may be obtained from the number of permutations by dividing $_nP_r$

by $r(r-1)(r-2)\cdots(3)(2)(1) = r!$; i.e.,

$$_nC_r = \frac{_nP_r}{r!} = \frac{n!}{r!\,(n-r)\,!} = \binom{n}{r} \tag{2.10}$$

The notation $_nC_r$ is read *n combinatorial r* and symbolizes the number of combinations of n things taken r at a time. (The final expression in Equation 2.10 is a somewhat older but equivalent symbol that is still in wide use.)

To assist you in solving problems that require evaluations of this kind, your computer software package includes a program that will quickly and accurately evaluate any factorial, permutation, or combination. Let us consider some applications of these counting techniques.

Example 2.11. Employees in a certain firm are given an aptitude test when first employed. Experience has shown that of the 60% who passed the test, 80% of them were good workers. Of the 40% who failed, only 30% were rated as good workers. What is the probability that an employee selected at random will be a good worker?

Solution

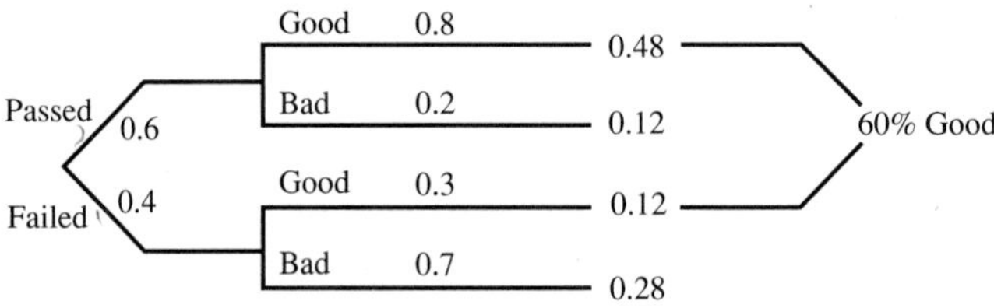

Example 2.12. Ten electronic components are available. However, it is known that 3 are defective. A set of 3 of the 10 components is selected at random.

(a) What is the probability that at least 2 of the components are defective?
(b) What is the probability that at least 1 component is defective?

Solution

(a) We are interested in the probability of exactly 2 or exactly 3 components being bad. Order is not important. Random selection implies that all components are equally likely to be selected; therefore any combination of three components is equally likely. The total number of possible sets of 3 is $_{10}C_3$. The number of ways that 2 of the 3 bad components can be chosen *and* 1 of 7 good components can be chosen is $(_3C_2)(_7C_1)$. In like manner, the number of ways that 3 components can be bad and zero components can be good is $(_3C_3)(_7C_0)$. Therefore, the probability of at least 2 bad components is

$$\frac{(_3C_2)(_7C_1) + (_3C_3)(_7C_0)}{_{10}C_3} = \frac{(3!/2!1!)(7!/6!1!) + (3!/3!0!)(7!/7!0!)}{10!/7!3!}$$

$$= \frac{(3)(7)+1}{[(10)(9)(8)]/[(3)(2)(1)]} = \frac{22}{120} = \frac{11}{60}$$

(b) The fact that P(no components bad) is

$$\frac{({}_3C_0)({}_7C_3)}{{}_{10}C_3} = \frac{35}{120}$$

yields

$$P(\text{at least 1 bad}) = 1 - \frac{35}{120} = \frac{17}{24}$$

(Example 2.12 is an example of the use of the hypergeometric distribution, one of the discrete probability distributions we study in Chapter 3.)

EXERCISES

2.1. State the definition of a sample space. Give two examples of each of the three different kinds of sample spaces: discrete finite, discrete countably infinite, and continuous.

2.2. Give one example of a repeatable experiment and one example of a nonrepeatable experiment.

2.3. The accompanying table shows a frequency distribution of the lifetime of 200 light bulbs tested at the Everdim bulb factory.

Lifetime (hours)	Number of Bulbs
75–99	24
100–124	36
125–149	80
150–174	35
175–200	25

(a) Using the frequency interpretation of probability, what probability is associated with the class of bulbs whose lifetimes are between 100 and 124 hours, inclusive?

(b) What probability is associated with bulbs whose lifetimes fall between 100 and 174 hours, inclusive?

2.4. Suppose an experiment consists of rolling three dice a single time. What is the probability that the three faces that turn up are not all the same number?

2.5. Suppose an experiment consists of the drawing of a single card at random from an ordinary deck of playing cards.

(a) What is the probability that the card is an ace?

(b) a king?

(c) either an ace or a king?

2.6. A bin contains 10 times as many good parts as bad parts. One part is drawn at random from this bin. What is the probability that it is a bad part?

2.7. Let A, B, and C be events in a sample space S defined by $S = (1, 2, 3, 4, 5, 6, 7, 8)$, where $A = (2, 4, 6, 8)$, $B = (1, 3, 5, 7)$, and $C = (2, 3, 5, 8)$.

(a) Find the following events: (1) $(A \cup C)$, (2) $(A \cup B)$, (3) $(B \cup C)$, (4) $(A \cap C)$, (5) $(A \cap B)$, and (6) $(B \cap C)$.

(b) Are A and B mutually exclusive events?

2.8. In Exercise 2.7, suppose that each of the simple events is equally likely; i.e., $P(1) = 0.125, P(2) = 0.125, \ldots, P(8) = 0.125$. Find the following probabilities: (a) $P(A)$, (b) $P(B)$, (c) $P(C)$, (d) $P(A \cup C)$, (e) $P(A \cup B)$, (f) $P(B \cup C)$, (g) $P(A \cap C)$, (h) $P(A \cap B)$, and (i) $P(B \cap C)$.

2.9. Write the formula for $P(A \mid B)$. What is the effective sample space for finding this probability? Also give the formula for $P(B \mid A)$. What advantage does each formula possess?

2.10. Is there a time order implied for the two events A and B in the symbol $P(B \mid A)$?

2.11. When A_1 and A_2 are independent events, what can we say about $P(A_2 \mid A_1)$, $P(A_1 \mid A_2)$, and $P(A_1 \cap A_2)$? [Assume that $P(A_1) \neq 0$ or $P(A_2) \neq 0$ when considering the conditional probabilities.]

2.12. There are 7 red Brewick Trylarks on a dealer's lot. The lot contains 60 cars. Let A = a car is red, and let B = a car is a Brewick Trylark. Given that there are 40 red cars on the lot, find $P(B \mid A)$, i.e., the probability that a car is a Brewick Trylark given the fact that it is a red car.

2.13. A bin contains three types of ropes. Each type of rope looks like the other types, but they differ in strength. Type A can support weights up to 100 pounds. Type B can support weights up to 200 pounds, and Type C can support weights up to 300 pounds. The bin contains 25 ropes of Type A, 25 ropes of Type B, and 50 ropes of Type C. One rope is chosen at random from this bin, and it is then tested by using it to suspend a 150-pound weight. The rope does not break.
(a) What is the probability that this rope is a Type A rope?
(b) a Type B rope?
(c) a Type C rope?

2.14. A box contains 12 GM bearings, 3 Ford bearings, and 5 Renault bearings. Three bearings are drawn successively from this box. Find the probability that they are drawn in the order Ford, GM, and Renault if each bearing is (a) replaced and (b) not replaced.

2.15. A box contains 4 brass, 2 copper, and 4 cast iron bushings. Three bushings are drawn successively from this box without replacement. What is the probability that 2 bushings drawn are brass and 1 bushing is cast iron?

2.16. Three cards are successively drawn from a well-shuffled, ordinary deck of cards without replacement. What is the probability that 2 kings and 1 ace will be drawn?

2.17. One container contains 3 good circuit boards and 2 defective circuit boards; another container contains 2 good circuit boards and 8 defective circuit boards. If 1 circuit board is drawn from each container, find the probability that
(a) both circuit boards are defective
(b) both circuit boards are good
(c) one circuit board is good and one circuit board is defective

2.18. Suppose A and B are events of the same sample space S. Suppose also that $P(A \cap B) = \frac{1}{4}$, $P(A) = \frac{1}{3}$, and $P(B) = \frac{3}{4}$.
(a) Are A and B independent events?
(b) What is $P(B \mid A)$?
(c) Are A and B mutually exclusive events?

2.19. Suppose that A and B are events of the same sample space S. Suppose also that $P(A \mid B) = \frac{1}{3}$, $P(A) = \frac{1}{3}$, and $P(B) = \frac{1}{2}$.
(a) Are A and B independent events?
(b) What is $P(A \cap B)$?
(c) Are A and B mutually exclusive events?

2.20. R distinct letters can be arranged along a line to produce how many distinct words? (Assume that any permutation of the R letters is a word.)

2.21. Define a combination of r objects. How many ways can you take r things from n distinct things?

2.22. What does the factorial symbol indicate? What does 0! equal?

2.23. Suppose A and B are events of the same sample space S. Suppose also that $P(A \cap B) = 0.25$, $P(A) = 0.5$, and $P(B) = 0.5$.
(a) Are A and B independent events?
(b) What is $P(B \mid A)$?

2.24. A fair die is rolled until a 4 appears. What is the probability that this die must be rolled more than 10 times?

2.25. A box contains 10 floppy disks, 3 of which are defective. Three disks are drawn at random from this box without replacement. Events A and B are defined as follows.

A = at least 2 of the floppy disks that are drawn are defective

B = at least 1 of the floppy disks that are drawn is defective

(a) What is $P(A)$?
(b) What is $P(B)$?
(c) What is $P(A \cap B)$?
(d) Are A and B independent events?

2.26. Suppose that 1 male student out of 10 and 1 female student out of 40 are engineering majors. Suppose that the student population has twice as many males as females. A student is selected at random from this population and is found to be an engineering major. What is the probability that this student is male?

2.27. Box I contains 3 white and 1 black balls. Box II contains 2 white and 3 black balls. A box is selected at random, and 2 balls are drawn from it. Given that 1 ball is white and 1 is black, what is the probability that Box I is chosen?

2.28. Evaluate the following expressions: (a) 8! (b) ${}_8C_2$ (c) ${}_8C_8$ (d) ${}_7C_3$ (e) ${}_8C_0$.

2.29. If a number may not begin with a 0, how many 4-digit numbers can be formed using the digits 0, 2, 4, 6, 8 if (a) repetitions are allowed and (b) repetitions are not allowed.

2.30. There are 6 roads from A to B and 3 roads from B to C. In how many ways can one go from A to C via B?

2.31. In how many ways can a committee of 5 people be chosen from a group of 9 people?

2.32. In how many ways can 6 questions be selected out of 10?

2.33. How many different committees composed of 2 boys and 3 girls can be formed from a group of 6 girls and 4 boys?

2.34. How many diagonals are there in a 12-sided convex polygon?

2.35. What is the probability of obtaining three 4s in five rolls of a fair die?

2.36. Three dice are thrown. What is the probability that the same number appears on exactly two of the dice?

2.37. Consider 2 boxes, 1 containing 1 black and 1 white marble, and the other, 2 black and 1 white marble. A box is selected at random, and a marble is drawn at random from the selected box. What is the probability that the marble is black ? Given that the selected marble is black, what is the probability that the first box was selected?

2.38. A manufacturer has 9 distinct motors in stock, 2 of which come from a particular supplier. The motors must be divided among 3 production lines, with 3 motors going to each line. If the assignment of motors to lines is random, find the probability that both motors from the particular supplier go to the first line.

2.39. While dressing in the dark you select 2 socks from a drawer containing 5 differently colored pairs. What is the probability that your socks match?

2.40. A police radar gun is 98% accurate; in other words, it indicates that a car is speeding when the car actually is with probability 0.98 and indicates that a car is not speeding when it is not with probability 0.98. Your teenager speeds 75% of the time. If she comes home and tells you that she got a ticket, what is the probability that she was speeding?

2.41. Joe attends school at The University and is graded on the 4.0 scale. It is finals time, and he is trying to determine the probability of making at least a 3.0. He knows he will make an A in probability and statistics. In thermodynamics there is a 30% possibility he'll make a C, a 60% possibility he'll make a B, and a 10% possibility, an A. In mechanics there is a 40% chance he'll make a C and a 60% chance for a B.
(a) What is the probability he'll make at least a 3.0?
(b) With what combination of grades is Joe most likely to get at least a 3.0?

2.42. Five democrats are running in the presidential primaries. Suppose the order of their competence could be ranked 1, 2, 3, 4, and 5, with 1 denoting the best-qualified candidate. Two of these candidates end up on the democratic ticket, with the first choice as president and the second choice as vice president. Determine the probability that the ticket
(a) carries the 2 most competent candidates
(b) does not carry the most competent candidate

Assume voters are ignorant of the candidates' competence, and therefore each candidate has the same chance of being elected. The voter votes for a president and a vice president (2 votes, for 2 different candidates).

2.43. In the ME building, the coffee vending machine has a probability of 0.8 of serving burnt coffee, a probability of 0.4 of being out of sugar, and a probability of 0.2 of both. What is the probability of the coffee being
(a) burnt or sugarless?
(b) neither burnt nor sugarless?

2.44. An engineering student applies for 4 different scholarships. The probabilities of obtaining each of them is 0.7, 0.5, 0.6, and 0.3, respectively. If receiving any scholarship has no effect on receiving any other scholarship, what is the probability that
(a) he will obtain all of the scholarships?
(b) he will receive no scholarship?
(c) he will receive 2 scholarships?

2.45. A company has two machines to produce a particular product. Machine A produces 45% of the product, and Machine B produces 55%. The defective rate for Machine A is 8% and is 10% for Machine B. If a defective item is observed, what is the probability that it was from Machine A?

2.46. Three cards will be drawn without replacement from a well-shuffled, ordinary deck of cards. What is the probability that all of them will be from different suits?

2.47. A survey of customers in an auto repair shop showed that 20% were not satisfied with the service they received. Half the complaints dealt with mechanic A. If mechanic A performs 40% of the jobs in that repair shop, find the probability that a customer will obtain unsatisfactory service given that the mechanic was A.

2.48. What is the probability that the sum will be at least 4 if 3 fair dice are thrown?

2.49. What is the probability that 2 or more students will have the same birthday in a statistics class of 25 students?

2.50. Five students will be randomly selected to make a study committee from a group of 6 undergraduate and 4 graduate students. Find the probability that exactly 3 undergraduates will be selected.

COMPUTER-BASED EXERCISES

C2.1. Duplicate the simulation of rolls of a fair die that was performed in Section 2.3. How do your results compare with those in this book? Do things "smooth out" in the same general way? Determine, approximately, how many tosses must be made before

(a) the maximum absolute deviation is in the neighborhood of 0.01

(b) the average absolute deviation is in the neighborhood of 0.01

C2.2. Your 12-year-old cousin is an avid Dungeons and Dragons fan. One of his friends has suggested that his 12-sided die is unfair. In an attempt to disprove this suggestion, your cousin has rolled the die 250 times and has obtained the following results.

Side	1	2	3	4	5	6	7	8	9	10	11	12
Frequency	20	24	20	19	25	19	19	17	24	20	19	24

What do you think about the friend's suggestion?

C2.3. Use the computer program for factorials, permutations, and combinations to evaluate

(a) 77!

(b) ${}_{88}P_{33}$

(c) ${}_{169}C_{63}$

CHAPTER 3

DISCRETE RANDOM VARIABLES

The following procedures and concepts outline the primary things you should learn from this chapter.

- Understand what is meant by the term *random variable*
- Understand what is meant by the *probability distribution* and *cumulative probability distribution of a random variable*, and know the *two basic properties* of any probability distribution function
- Be able to determine the probability distributions of random variables like those found in the exercises and examples of this chapter
- Thoroughly understand what is meant by a *binomial distribution*. Understand what n, p, q, and x represent in the general formula for a binomial distribution
- Be familiar with the conditions that must be satisfied for the proper application of the binomial distribution (i.e., What properties must the experiment and trials possess in order that the binomial formula can be properly applied?)
- Be able to use the binomial distribution to solve problems similar to those found in this chapter
- Be able to discuss, at the *conceptual level*, the *relationships* that join the various discrete random variables discussed in this chapter
- Know when the Poisson distribution can be applied to a problem that *satisfies the assumptions* governing the binomial distribution
- Know when to apply the geometric, negative binomial, and hypergeometric distributions to solve problems like those of the chapter
- Understand what is meant by a *discrete bivariate probability distribution function*, $p(x_1, x_2)$
- When given $p(x_1, x_2)$, be able to obtain the *cumulative* bivariate probability distribution function, $F(x_1, x_2)$, and the *marginal* probability distribution function for either X_1 or X_2

- When given $p(x_1, x_2)$ and additional information about one of the random variables, be able to obtain the *conditional* probability distribution function for the other random variable
- When given $p(x_1, x_2)$, be able to determine whether X_1 and X_2 are statistically independent

3.1 RANDOM VARIABLES AND GENERAL PROPERTIES OF PROBABILITY DISTRIBUTIONS

The formal definition of a *random variable* is

> A random variable is a numerically valued function defined on a sample space.

Like many formal definitions, this one is useful only after you have a practical mastery of its content. For simplicity, let us consider a single die. The die has 6 sides, and each side has some probability of being up at the end of any particular roll. However, on any particular roll of a die, any side may come up, and the side that does appear is *governed strictly by chance*.

In the foregoing statements, we have defined a discrete finite sample space consisting of 6 outcomes, the 6 sides of the die. All that remains is to explain the meaning of a *numerically valued function*, as contained in our formal definition. We will use a very popular function and will assign one of the numbers 1 through 6 uniquely to each side. With this function defined, we can speak of "rolling a 6," "rolling a 3," and so on. Thus, we see that a random variable is simply a function that assigns a number (numeric value) to each and every outcome of the sample space.

For the discussions in this book, we use uppercase letters, such as X, to denote random variables and lowercase letters, such as x, to denote values that the random variable may assume. The letter X symbolizes any of the possible values that could be observed when the experiment is performed; x symbolizes the value that is actually observed, often called the *realization* of X, after the experiment has been performed. The expression $p(x) = P(X = x)$ is read as the "probability that X will assume the value x."

Having only a numerically valued function defined on a sample space, a random variable, is not very useful. The next step is to *assign probabilities* to each outcome. In our current example, if the die is fair, then by arguments of symmetry, we can say that the following *probability distribution function* (pdf) governs the random variable, X, just defined.

x	1	2	3	4	5	6
$p(x)$	$\frac{1}{6}$	$\frac{1}{6}$	$\frac{1}{6}$	$\frac{1}{6}$	$\frac{1}{6}$	$\frac{1}{6}$

Note that $p(x)$ is *defined to be zero* for any value x that has not been assigned to an outcome in the sample space; i.e., $p(7) = 0$ for this $p(x)$.

Probability distribution functions for discrete random variables are also known as probability *mass* functions. This equivalent name arises from the physical analogy of thinking of each outcome's probability as an amount of "probability mass." We make use of this analogy in Chapters 5 and 6.

Like *every* probability distribution function, this one is essentially a theoretical model. However, we expect that it is close enough to reality to be used for drawing practical conclusions. All probability distribution functions satisfy *two basic properties*:

$$\begin{aligned} &1. \quad p(x) \geq 0 \quad \text{for each value } x \\ &2. \quad \sum_{\text{all } x} p(x) = 1 \end{aligned} \tag{3.1}$$

Associated with each probability distribution function, $p(x)$, is a *cumulative distribution function* (cdf), $F(x) = P(X \leq x)$, which gives the probability that the random variable X will assume a value less than or equal to a stipulated value x. The cdf for the example of a fair die is

x	1	2	3	4	5	6
$F(x)$	$\frac{1}{6}$	$\frac{1}{3}$	$\frac{1}{2}$	$\frac{2}{3}$	$\frac{5}{6}$	1

Any discrete mathematical function that satisfies the properties of Equation 3.1 is a discrete probability distribution function. There are several well-known discrete distributions that have received wide attention and practical use. They are discussed in detail later in this chapter. Let us now consider four examples of probability distributions to clarify this discussion.

Example 3.1. Formulate and state the probability distribution of the random variable, Y, the sum of the faces of two fair dice in a single roll of the dice.

Solution. Let X_1 and X_2 be the random variables associated with the outcomes of dice 1 and 2, respectively. We know that the roll of each die has 6 possible outcomes; i.e., X_1 and X_2 can both assume any of the values 1, 2, 3, 4, 5, or 6. Further, each outcome of the first die could occur with each outcome of the second die.

Therefore, there are (6)(6) = 36 outcomes possible when 2 dice are thrown. Because the dice are *fair* and because each die's possible result is *statistically independent* of the other's, the probability of any outcome is, by Equation 2.6, $(\frac{1}{6})(\frac{1}{6}) = \frac{1}{36}$. The sample space and the values of our new random variable, $Y = X_1 + X_2$, are illustrated in the following table.

x_2 \ x_1	1	2	3	4	5	6
1	2	3	4	5	6	7
2	3	4	5	6	7	8
3	4	5	6	7	8	9
4	5	6	7	8	9	10
5	6	7	8	9	10	11
6	7	8	9	10	11	12

Value of $y = x_1 + x_2$

Using this table, it is a simple matter to count along the diagonals that move upward and to the right and obtain the numbers of outcomes yielding the same unique value of the sum, y. For example, there are exactly 5 outcomes that have a sum of 6. Hence, $P(Y = 6) = p(6) = \frac{1}{36} + \frac{1}{36} + \frac{1}{36} + \frac{1}{36} + \frac{1}{36} = \frac{5}{36}$. Proceeding as before, the probability distribution of Y is

y	2	3	4	5	6	7	8	9	10	11	12
$p(y)$	$\frac{1}{36}$	$\frac{2}{36}$	$\frac{3}{36}$	$\frac{4}{36}$	$\frac{5}{36}$	$\frac{6}{36}$	$\frac{5}{36}$	$\frac{4}{36}$	$\frac{3}{36}$	$\frac{2}{36}$	$\frac{1}{36}$

This problem and its solution are an illustration of a fundamental fact:

> Anything that is a function of one or more random variables is a random variable itself.

Here, Y is a function of two random variables, X_1 and X_2. As we have shown, Y is a random variable possessing its own probability distribution. This is a very special case of a general operation that mathematicians call a *summation convolution*. Your software package contains a general program to perform convolutions of the sort contained in this example. We discuss this in general in Chapter 5.

Example 3.2. An engineer is analyzing a new microcircuit device. Each of 191 of the devices has been tested 1000 times. The observed number of times that the devices failed to perform up to specification was

Number of failures per 1000 tests, x	**Number of devices having x failures**
2	125
3	37
4	16
5	8
6	5
	191

The engineer observed that the number of devices having x failures per 1000 tests appeared to approximately follow the equation

$$\text{Devices having } x \text{ failures} = \frac{1000}{x^3}$$

For example, $1000/6^3 = 4.63$, which is approximately the observed value 5. However, since the engineer wanted to compute probabilities that are bounded by 0 and 1, 4.63 was not an allowable number. In order to bring the numbers closer to what probabilities must be, the engineer decided to divide the results from the foregoing equation by 100. Doing so yielded the following equation for an approximate probability distribution function.

$$\hat{p}(x) = \frac{10}{x^3} \qquad x = 2, 3, 4, 5, 6$$

(a) Is $\hat{p}(x)$ a true probability distribution function?

(b) If not, correct it.

Solution

(a) The properties of a discrete probability distribution function must be present. The function $\hat{p}(x)$ is always positive. However, the sum of $\hat{p}(x)$ over all x is not 1; i.e.,

$$\sum_{x=2}^{6} \hat{p}(x) = \sum_{x=2}^{6} \frac{10}{x^3} = 10(\tfrac{1}{8} + \tfrac{1}{27} + \tfrac{1}{64} + \tfrac{1}{125} + \tfrac{1}{216}) = 1.9029167$$

Therefore, $\hat{p}(x)$ is *not* a legitimate probability distribution function. The engineer is using an incorrect formula to explain mathematically the probabilities associated with the number of failures that a device might have in 1000 trials.

(b) If we wish to form a probability distribution function, we must force the sum in part (a) to be 1. Suppose that we multiply $\hat{p}(x)$ by $c = 1/\sum_{x=2}^{6}(10/x^3) = 1/1.9029167 = 0.52551$. This yields the probability distribution function $p(x) = c\hat{p}(x) = 5.2551/x^3$ for $x = 2, 3, 4, 5, 6$:

x	$p(x)$
2	0.6569
3	0.1946
4	0.0821
5	0.0421
6	0.0243
	1.0000

The operation just demonstrated is an example of the process of normalization in which the *normalizing factor,* c, causes the new sum to be equal to 1.

Example 3.3 Suppose that a primary device has failed as a result of a high-temperature environment. The probability that an electronic switch will successfully activate a backup device is 0.6. If switch failures are statistically independent and the switches are tried one at a time, how many parallel switches are required to achieve at least a 95% probability of successful switching?

Solution. One way to answer this question is to determine the probability distribution function of the random variable, X, the number of switch failures prior to the first success. The cumulative distribution function of X may then be used to determine the number of switches required. By statistical independence,

$$\begin{aligned} p(0) &= P(\text{1st parallel switch activates backup switch}) \\ &= 0.6 \\ p(1) &= P(\text{1st parallel switch fails and 2nd switch activates backup switch}) \\ &= (0.4)(0.6) \\ p(2) &= P(\text{1st and 2nd parallel switches fail, and 3rd switch activates backup switch}) \\ &= (0.4)^2(0.6) \end{aligned}$$

In general, $p(x) = (0.4)^x(0.6)$. This countably infinite sample space does have probabilities that constitute a probability distribution function; i.e., all of the $p(x) \geq 0$, and the infinite sum of the probabilities of the outcomes in the sample space does equal 1:

$$\sum_{x=0}^{\infty}(0.4)^x 0.6 = 0.6\sum_{x=1}^{\infty}(0.4)^{x-1} = 0.6\frac{1}{1-0.4} = 1.$$

Thus, to answer our question, we must find the smallest x such that $F(x) = \sum_{k=0}^{x}(0.4)^k 0.6 > 0.95$. Since the cumulative distribution function is

x	0	1	2	3	...
$F(x)$	0.6	0.84	0.936	0.9744	

the probability of three or fewer switch failures is 0.9744. Therefore, four parallel switches will provide the required probability in excess of 95%.

Example 3.4. A carton contains 6 fuses: 3 are 10-amp fuses, 1 is a 15-amp fuse, and 2 are 20-amp fuses. Two fuses are drawn at random, and without replacement, from the carton. Let X be the random variable defined by the sum of the amperages of the 2 fuses that are drawn. What is the probability distribution of X?

Solution. Suppose that each fuse is uniquely identifiable and the order in which the fuses are drawn is important. As shown in the following table, there are ${}_6P_2 = 30$ equally likely outcomes.

Sample space for drawing 2 fuses without replacement

	2nd fuse drawn					
1st fuse drawn	10	10	10	15	20	20
10	—	10, 10	10, 10	10, 15	10, 20	10, 20
10	10, 10	—	10, 10	10, 15	10, 20	10, 20
10	10, 10	10, 10	—	10, 15	10, 20	10, 20
15	15, 10	15, 10	15, 10	—	15, 20	15, 20
20	20, 10	20, 10	20, 10	20, 15	—	20, 20
20	20, 10	20, 10	20, 10	20, 15	20, 20	—

The dashes along the major diagonal indicate that the same fuse cannot be drawn twice. Note that, if order is considered unimportant, there would be only ${}_6C_2 = 15$ equally likely outcomes, enumerated by the nonempty cells either above or below the diagonal in the table.

Counting the number of outcomes that yields the same value x and summing their probabilities [i.e., 6 mutually exclusive outcomes generate a sum of 20; $P(X = 20) = \frac{6}{30}$] yield the following probability distribution function for X.

x	20	25	30	35	40
$p(x)$	0.2	0.2	0.4	0.133	0.067

3.2 THE BINOMIAL DISTRIBUTION

The binomial distribution is one of the most commonly used discrete probability distributions in modern statistical analysis. It is also closely related to other popular distributions, as we see later. Its use is appropriate if the following conditions are satisfied.

1. Each repetition of the experiment, commonly called a *trial*, can result in only one of two possible outcomes, a *success* or *failure*.
2. The probability of a success, p, (and likewise the probability of a failure, $1 - p$) is *constant* from trial to trial.
3. All trials are statistically independent; i.e., no trial outcome has any effect on any other trial outcome.
4. The number of trials, n, is a *specified constant* (stated before the experiment begins).

If these conditions are satisfied, the binomial distribution is the distribution that governs the random variable, X, *the number of successes that occur in the n trials*. There need not be a preference, in ethical, moral, or practical terms, for the outcome labeled "success." For example, "the axle failed" could very well be labeled a success in an experiment involving the strengths of truck axles.

Suppose the probability of a success is $p = 0.1$ and $n = 5$. What is the probability of getting 2 successes and then 3 failures in that *specific order*? Schematically, we have

$$\underbrace{SS}_{2} \quad \underbrace{FFF}_{3}$$

with $P(S) = 0.1$ and $P(F) = 0.9$. Since the trials are statistically independent, the answer to the question is

$$P(SSFFF) = (0.1)(0.1)(0.9)(0.9)(0.9) = (0.1)^2(0.9)^3$$

What is the probability of a failure, a success, a failure, a success, and finally a failure? Using the same ideas, $P(FSFSF) = (0.9)(0.1)(0.9)(0.1)(0.9) = (0.1)^2(0.9)^3$. Under the conditions just stated, *any sequence* with 2 successes and 3 failures has a probability of $(0.1)^2(0.9)^3$. Since all such sequences are mutually exclusive, we need count only the number of ways that 2 successes and 3 failures can occur. Summing their identical probabilities yields the probability of obtaining 2 successes and 3 failures in *any order* in the 5 trials.

How many ways can 2 successes occur in 5 trials? Asked another way, How many ways can 5 things be taken 2 at a time? Naturally, the answer is ${}_5C_2$. So the answer to the preceding question is

$$P(X = 2) = p(2) = {}_5C_2(0.1)^2(0.9)^3 = \frac{5!}{2!\,(5-2)!}(0.1)^2(0.9)^{5-2}$$

Generalizing from this example, we see that the *basic mathematical form* of the binomial distribution is

$$p(x) = {}_nC_x\, p^x(1-p)^{n-x} = {}_nC_x p^x q^{n-x} \tag{3.2}$$

where $q = 1 - p$. Note that this is a *two-parameter* (n and p) finite discrete probability distribution, with a cumulative distribution defined by

$$F(x) = \sum_{k=0}^{x} {}_nC_k\, p^k q^{n-k} \tag{3.3}$$

where it is easy to show that $\sum_{k=0}^{n} {}_nC_k\, p^k q^{n-k} = 1$.

In general, a *parameter* of a probability distribution function is any symbol defined in the function's basic mathematical form such that the value of the parameter may be specified by the user of that function. For example, in the binomial distribution, n may be set to any positive integer value and p may be set to any value from 0 to 1, inclusive; i.e., $0 \le p \le 1$.

If the value of n is small, it is relatively easy to evaluate the binomial and cumulative binomial distributions from their mathematical definitions. However, these evaluations quickly become difficult as n increases to a larger number. For this reason, you have been provided both with *a computer program* and with *a table*, Table A.1, to assist you in desired evaluations. (All statistical tables referred to in this book are given at the end of the book.) Table A.1 presents the values of the cumulative binomial distributions for $n = 2$ through $n = 20$ and for $p = 0.01$, $p = 0.05$ through $p = 0.95$ in increments of 0.05, and $p = 0.99$. For parameter values that fall within these ranges but not precisely on one of the values of p, linear interpolation may be performed to achieve approximations to the desired cumulative probability.

If, as described in the *User's Manual* for the computer programs, you had selected the program for evaluating the binomial distribution and had stipulated $n = 12$ and $p = 0.4$, the information given in Table 3.1 would have been presented.

Notice in Table 3.1 that the first column gives the number of successes, x; the second column gives the probability of achieving exactly x successes, $p(x)$; the third column gives the probability of achieving x or fewer successes, $F(x)$; and the fourth column gives the probability of $x + 1$ or more successes, $1 - F(x)$.

TABLE 3.1
Output from the binomial distribution program

Binomial Distribution
n = 12 p = .4

x	PR(x)	PR(≤ x)	PR(≥ x + 1)
0	.00218	.00218	.99782
1	.01741	.01959	.98041
2	.06385	.08344	.91656
3	.14189	.22534	.77466
4	.21284	.43818	.56182
5	.22703	.66521	.33479
6	.17658	.84179	.15821
7	.10090	.94269	.05731
8	.04204	.98473	.01527
9	.01246	.99719	.00281
10	.00249	.99968	.00032
11	.00030	.99998	.00002
12	.00002	1	0

The third and fourth columns, within the limits of roundoff and truncation error, will always add to 1. In Table 3.1, the maximal error is of the order of 10^{-6}.

There are two advantages to using the program that created Table 3.1. First, the program provides you with much more *immediate* information than is provided by Table A.1. Second, there are no restrictions, except those on the possible values of n and p; i.e., a table for $n = 52$ and $p = 0.2347$ would be just as easily achieved by the program as for $n = 12$ and $p = 0.4$. In contrast to using Table A.1, when using the computer program you will never need to interpolate to obtain required results.

Figure 3.1 presents several selected plots of the binomial probability distribution function that have been created by one of the programs in your software

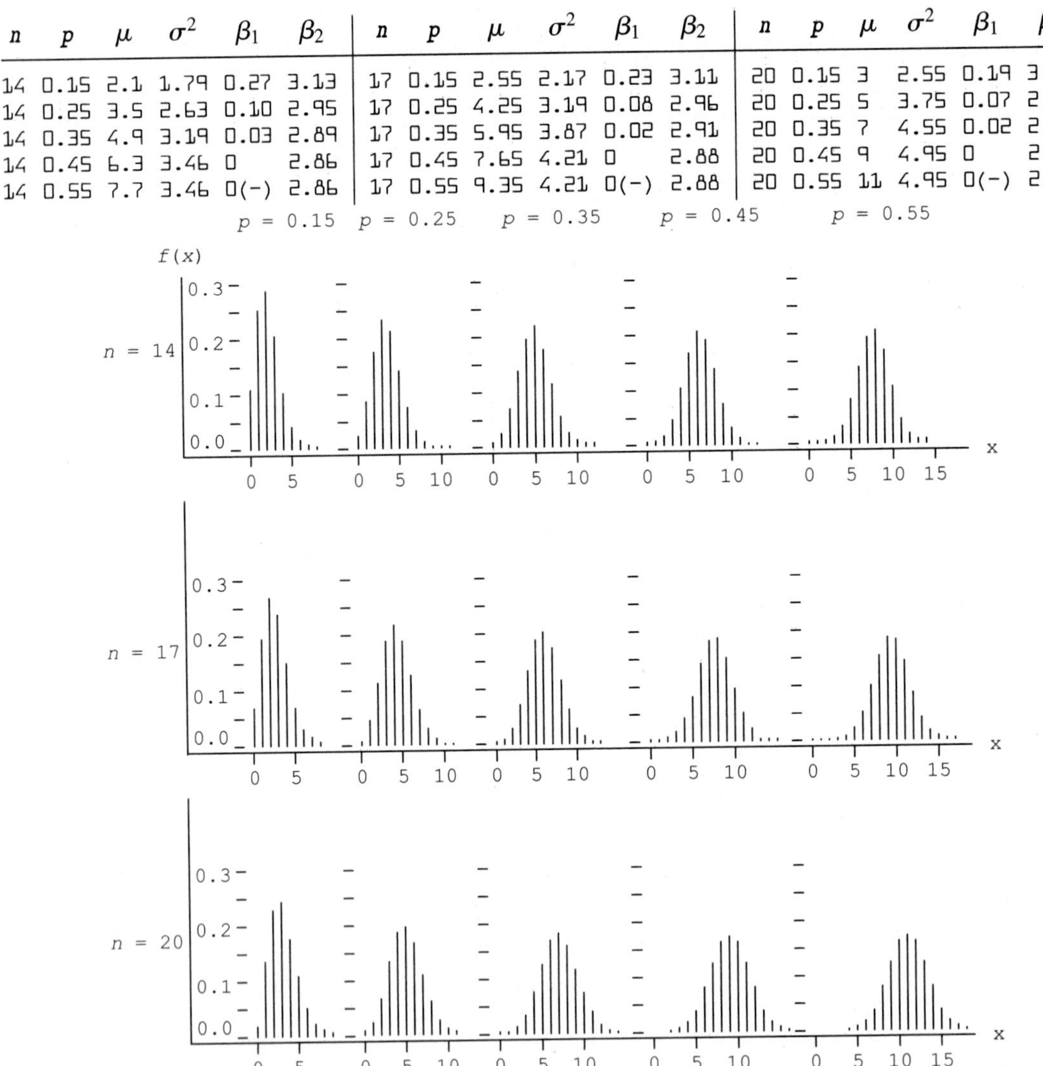

n	p	μ	σ^2	β_1	β_2	n	p	μ	σ^2	β_1	β_2	n	p	μ	σ^2	β_1	β_2
14	0.15	2.1	1.79	0.27	3.13	17	0.15	2.55	2.17	0.23	3.11	20	0.15	3	2.55	0.19	3.09
14	0.25	3.5	2.63	0.10	2.95	17	0.25	4.25	3.19	0.08	2.96	20	0.25	5	3.75	0.07	2.97
14	0.35	4.9	3.19	0.03	2.89	17	0.35	5.95	3.87	0.02	2.91	20	0.35	7	4.55	0.02	2.92
14	0.45	6.3	3.46	0	2.86	17	0.45	7.65	4.21	0	2.88	20	0.45	9	4.95	0	2.9
14	0.55	7.7	3.46	0(-)	2.86	17	0.55	9.35	4.21	0(-)	2.88	20	0.55	11	4.95	0(-)	2.9

FIGURE 3-1
The binomial distribution for selected values of n and p.

package. Pay particular attention to how the *shape* and *location* of the distribution changes in response to changes in the values of the parameters n and p. The values of μ, σ^2, and β_1 and β_2 are included here for completeness. The meanings of these quantities are discussed in detail in Chapter 5, where these plots are used again.

In addition to plotting the binomial distribution, you can use the program that created the plots of Figure 3.1 to plot 15 other popular probability distribution functions and to study the effects that changes in their parameter values have on the shapes of their distributions. You can also plot *any* probability distribution or mathematical function that you care to specify. Let's consider some examples of the concepts just presented.

Example 3.5. Suppose that in a very large shipment of housings, the probability of a cosmetic defect for any 1 housing is 0.35. Suppose that the assumptions underlying the binomial distribution are met.

(a) What is the probability that *at most* 3 housings have defects in the sample of 17?

(b) What is the probability that 5 *or more* housings have defects?

(c) What is the probability that exactly 12 housings do *not* have defects?

(d) What is the probability that 3, 4, or 5 housings have defects?

Solution. Using the distribution evaluation program, the following table has been generated.

Binomial Distribution
n = 17 p = .35

X	PR(X)	PR(≤ X)	PR(≥ X+1)
0	.00066	.00066	.99934
1	.00604	.00670	.99330
2	.02602	.03273 (d)	.96727
3	.07006	.10279 (a)	.89721
4	.13205	.23484	.76516 (b)
5	.18486 (c)	.41970 (d)	.58030
6	.19908	.61878	.38122
7	.16846	.78724	.21276
8	.11338	.90062	.09938
9	.06105	.96167	.03833
10	.02630	.98797	.01203
11	.00901	.99698	.00302
12	.00243	.99941	.00059
13	.00050	.99991	.00009
14	.00008	.99999	.00001

(a) "At most 3" is the same as "3 or less" and is found in the PR(≤x) column as $F(3) = 0.10279$.

(b) The probability that 5 or more housings have defects is found to be 0.76516 in the PR(≥x+1) column at $x = 4$. This is the same as 1 minus the probability that 4 or fewer have defects. Therefore, the probability that 5 or more have defects is also given by $1 - F(4) = 1 - 0.23484 = 0.76516$.

(c) Since there are 17 housings in our sample, the probability that exactly 12 housings *do not* have defects is the same as the probability that exactly 5 housings *do* have defects: $p(5) = 0.18486$.

(d) The probability that 3, 4, or 5 housings have defects is given by $p(3) + p(4) + p(5) = 0.07006 + 0.13205 + 0.18486 = 0.38697$. This is equivalent to $F(5) - F(2) = 0.41970 - 0.03273 = 0.38697$.

Example 3.6. A quality control engineer wants to check whether (in accordance with specifications) 95% of the components shipped by her company are in good working condition. Prior to shipment, she randomly selects 15 components from each very large lot. She allows the lot to be shipped only if all 15 components are in good working condition. Otherwise, each of the components in the lot is tested, and any bad components are replaced with good components.

(a) What is the probability that the engineer will commit the error of holding a lot for further inspection even though 95% of the components are in good working condition?

(b) What is the probability that the engineer will commit the error of shipping a lot without further inspection when only 90% of the components are in good working condition?

Solution

(a) We want the probabiiity that there are one or more bad components in the sample of 15 components when the probability of a good component is 0.95. This is exactly the same as 1 minus the probability that all 15 of the components are good; i.e., 1 minus the probability of 15 successes in 15 trials. If we define a success as a good component, $p = 0.95$. Therefore, p(one or more bad components) is equal to

$$1 - P(\text{no failures}) = 1 - p(15) = 1 - {}_{15}C_{15}(0.95)^{15}(0.05)^{0} = 1 - (0.95)^{15} = 0.5367$$

If we define a "success" as a *bad* component, $p = 0.05$; and p(one or more bad components) is equal to

$$1 - P(\text{no successes}) = 1 - {}_{15}C_{0}(0.05)^{0}(0.95)^{15} = 1 - (0.95)^{15} = 0.5367$$

(b) We want the probability that no bad components are present in the sample of 15 components when the probability of a good component is 0.90. That probability is equal to the probability that all of the components are good, $p(15) = {}_{15}C_{15}(0.9)^{15}(0.1)^{0} = (0.9)^{15} = 0.205891$.

From these results, we conclude that the engineer's plan is not a good one. In Chapter 15, techniques for designing good sampling plans for this kind of application are presented.

In Section 3.3 we discuss the relationships between the binomial distribution and five other discrete distributions. In that discussion, we see that the assumption of a "very large lot" is important to the correct use of the binomial distribution in solving the problem presented in Example 3.6.

Example 3.7. One production line produces 2000 parts of type A.B2-4 each day. On a particular day, a 1% sample was taken. The probability that a part is defective is 5%. Determine the probability that the sample of 20 parts is no more than 10% defective.

Solution. Let the random variable X denote the number defective in the sample; $n = 2000(0.01) = 20$ and $p = 0.05$. Therefore, the problem requires that we determine the probability that no more than $0.1(20) = 2$ parts are defective. Thus, we must compute the sum of the probabilities that $X = 0, 1$, or 2. Using the evaluator program, we have obtained

```
                 Binomial Distribution
                  n = 20    p = .05
x     PR(x)              PR(≤x)            PR(≥x+1)
---   ---------          ---------         ---------
0     .35849             .35849            .64151
1     .37735             .73584            .26416
2     .18868             .92452            .07548
3     .05958             .98410            .01590
...   ...                ...               ...
```

The probability that the sample will be 10% or less defective is 0.92452.

3.3 SOME POPULAR DISCRETE DISTRIBUTIONS AND THEIR RELATIONSHIPS

In addition to the binomial distribution, Table 3.2 presents the discrete probability distributions that are discussed in the remainder of this chapter. The columns labeled mean, variance, and β_1 and β_2 are included for completeness and are discussed in detail in Chapter 5. With the exception of the Bernoulli distribution, all of the distributions given in Table 3.2 have evaluator programs in the computer software package very similar to the evaluator program for the binomial distribution. Let us consider the discrete probability distributions contained in Table 3.2.

The binomial distribution, as discussed in Section 3.2, characterizes the number of successes in n trials under four specific conditions. If n is *set equal to* 1, we have a special case of the binomial distribution—the *Bernoulli distribution*.

By Equation 3.2, the binomial distribution has $p(x) = {}_nC_x\, p^x(1-p)^{n-x}$, where $x = 0, 1, 2, \ldots, n$. If $n = 1$,

$$p(x)\begin{cases} = {}_1C_x\, p^x(1-p)^{1-x} & x = 0, 1 \\ = p^x(1-p)^{1-x} & x = 0, 1 \end{cases} \quad \text{which yields} \quad \begin{aligned} p(0) &= 1 - p \\ p(1) &= p \end{aligned} \tag{3.4}$$

In verbal terms, we see from Equation 3.4 that the Bernoulli distribution governs the situation in which a single experiment, or *Bernoulli trial*, is to be performed and only two outcomes are possible. In this single trial, the probability of a success is p and, therefore, the probability of a failure is $1 - p$. An example plot of the Bernoulli distribution is given in Figure 3.2.

From another viewpoint, the binomial distribution is a generalization of the Bernoulli distribution in which the binomial distribution gives $P(X = x)$ in $n > 1$ Bernoulli trials. To explain further, suppose that we have n random variables, X_i; $i = 1, 2, \ldots, n$, that are all distributed according to the same Bernoulli distribution with probability of success, p. If we form their sum, $Y = \sum_{i=1}^{n} X_i$, Y will be distributed according to a binomial distribution with parameters n and p.

Indeed, the Bernoulli distribution "generates" all five of the other distributions in Table 3.2. Figure 3.3 shows the relationships schematically. In the remainder of this section we discuss all of the concepts contained in Figure 3.3.

TABLE 3.2
Some Frequently Encountered Discrete Probability Distribution Functions

	Probability distribution function p(.)	Mean	Variance	β_1	β_2
Bernoulli	$p(x) = p \quad x = 1 \quad$ or $p(x) = p^x q^{1-x}$ $= p \quad x = 0$ $0 \le p \le 1$	p	pq	$\frac{q}{p} + \frac{p}{q} - 2$	$\frac{1 - 6pq}{pq} + 3$
Binomial	$p(x) = {}_nC_x\, p^x q^{n-x} \quad x = 0, 1, 2, \ldots, n$ $0 \le p \le 1 \quad n = 1, 2, \ldots$	np	npq	$\frac{q}{np} + \frac{p}{nq} - \frac{2}{n}$	$\frac{1 - 6pq}{npq} + 3$
Poisson	$p(x) = e^{-\lambda}\frac{\lambda^x}{x!} \quad x = 0, 1, 2, \ldots$ $\lambda > 0$	λ	λ	$\frac{1}{\lambda}$	$\frac{1}{\lambda} + 3$
Geometric	$p(x) = pq^x \quad x = 0, 1, 2, \ldots$ $0 \le p \le 1$	$\frac{1 - p}{p}$	$\frac{1 - p}{p^2}$	$\frac{(2 - p)^2}{1 - p}$	$\frac{p^2}{1 - p} + 9$
Negative binomial	$p(x) = {}_{r+x-1}C_x\, p^r q^x \quad x = 0, 1, 2, \ldots$ $r > 0 \quad 0 \le p \le 1, \quad q = 1 - p$	$\frac{rq}{p}$	$\frac{rq}{p^2}$	$\frac{(2 - p)^2}{r(1 - p)}$	$\frac{3r(1 - p) + p^2 + 6(1 - p)}{r(1 - p)}$
Hyper-geometric	$p(x) = \frac{{}_{Np}C_x\, {}_{Nq}C_{n-x}}{{}_NC_n} \quad x = 0, 1, \ldots, n$ $N = 1, 2, \ldots$ $n = 1, 2, \ldots N$ $p = 0, \frac{1}{N}, \frac{2}{N}, \ldots, 1$	np	$npq\left(\frac{N - n}{N - 1}\right)$	$\frac{(q - p)^2(N - 2n)^2(N - 1)}{npq(N - n)(N - 2)^2}$	$\frac{(N - 1)N(N + 1)}{np(N - n)(n - 2)(N - 3)}$ $- \frac{(N - 1)6n(N - n)}{np(N - n)(n - 2)(N - 3)}$ $+ \frac{3(N - 1)p\{N^2(n - 2) - Nn^2 + 6n(N - n)\}}{np(N - n)(n - 2)(N - 3)}$

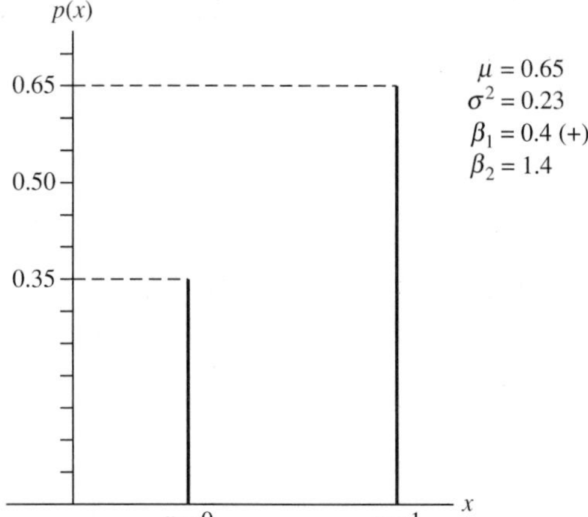

FIGURE 3-2
The Bernoulli distribution with $p = 0.65$.

Instead of the number of successes in n Bernoulli trials, let us consider a *different random variable*: the number of Bernoulli trial failures that would have to be performed prior to achieving the first success. Physical examples of this are the number of coin flips that are tails before a heads is achieved or the number of microwave signals sent unsuccessfully through a thunderstorm until a signal is successfully transmitted. The distribution that governs this situation is the *geometric distribution*, where *X is the random variable representing the number of failures before the first success*. Thus we see that $X + 1$ statistically independent Bernoulli trials will be performed, i.e., X failures followed by a single success. The mathematical form of the geometric probability distribution is

$$p(x) = p(1 - p)^x \qquad x = 0, 1, 2, \ldots \tag{3.5}$$

and an example plot of the geometric distribution is given in Figure 3.4.

An example of the application of the geometric distribution has been given in Example 3.3. A part of the table that has been generated using the evaluator program to solve Example 3.3 is given on page 42 (the answer to Example 3.3 is underlined).

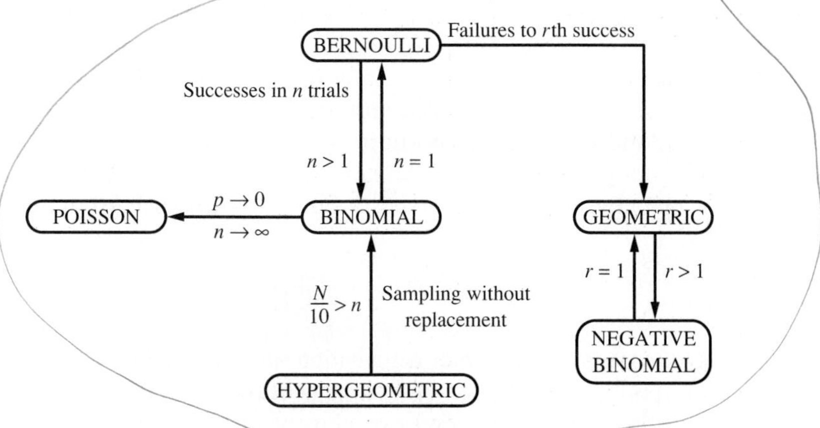

FIGURE 3-3
Relationships between the distributions of Chapter 3.

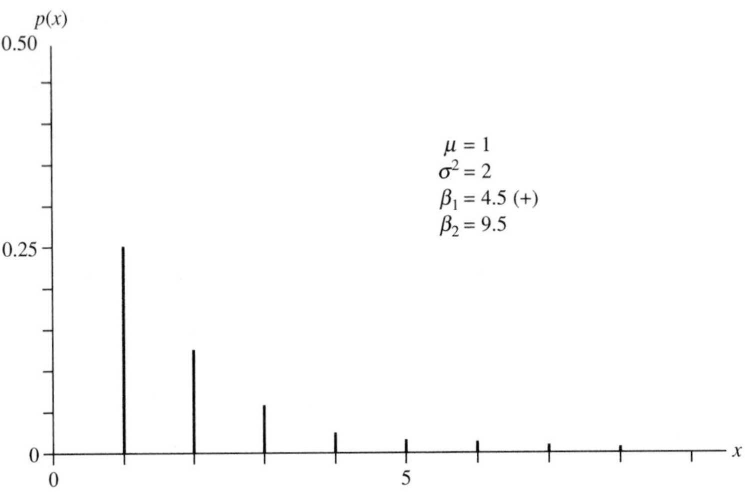

FIGURE 3-4
The geometric distribution with $p = 0.5$.

Geometric Distribution
p = .6

X	PR(X)	PR(≤ X)	PR(≥ X+1)
0	.60000	.60000	.40000
1	.24000	.84000	.16000
2	.09600	.93600	.06400
3	.03840	.97440	.02560
4	.01536	.98976	.01024
5	.00614	.99590	.00410

A direct generalization of the geometric distribution is the *negative binomial distribution*, which governs another random variable: the *number of failures, X, that occur before the rth success* ($r = 1, 2, \ldots$) *occurs*. (As indicated in Figure 3.3 the geometric distribution is a special case of the negative binomial distribution where $r = 1$.) Thus, the total number of statistically independent trials that will be performed at the rth success is $X + r$. The *final* trial will be a success. The other $r - 1$ successes and the X failures can occur in any order during the previous $X + r - 1$ trials. Once this is understood, it is easy to derive the mathematical form of the negative binomial distribution given in Equation 3.6. (See Exercise 3.11.)

$$p(x) = {}_{r+x-1}C_x\, p^r (1 - p)^x \qquad x = 0, 1, 2, \ldots \tag{3.6}$$

Physical examples of the negative binomial distribution are the number of coin flips until 3 heads occurs or the number of tailpipe checks until 2 cars pass EPA emission requirements. Figure 3.5 shows two plots of the negative binomial distribution.

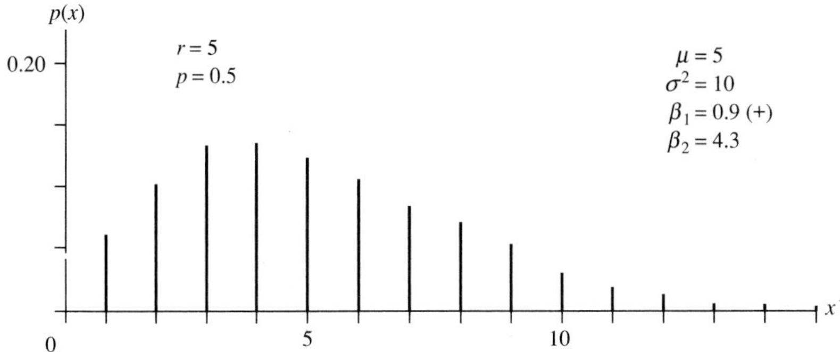

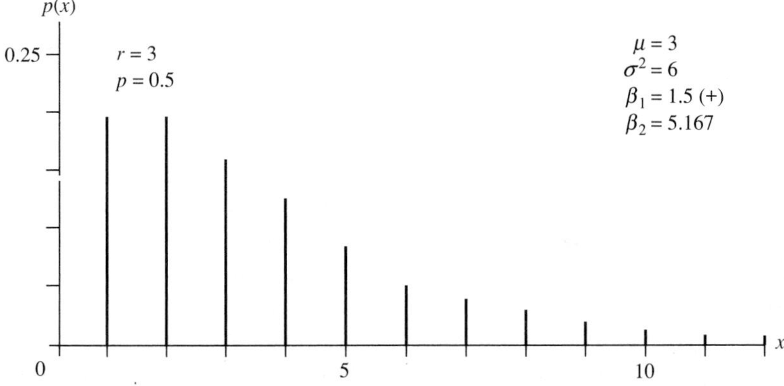

FIGURE 3-5
Two examples of the negative binomial distribution.

Example 3.8. Historical records associated with privately owned automobiles in Gotham City indicate that the probability that such an automobile will pass the Federal EPA standards is 0.45. How many such cars will a Gotham City inspector have to check before the probability is greater than 0.95 that he will have found 3 cars that pass the EPA standards?

Solution. We can use the negative binomial distribution with $r = 3$ and $p = 0.45$ and can determine the value x that will cause the cumulative negative binomial distribution to exceed 0.95, i.e., the value x at which

$$\sum_{i=0}^{x} {}_{r+i-1}C_i \, p^r (1 - p)^i > 0.95$$

Evaluation of this expression, using the negative binomial distribution evaluator program, has yielded the following table.

Negative Binomial Distribution
r = 3 p = .45

X	PR(X)	PR(≤ X)	PR(≥ X+1)
0	.09112	.09112	.90888
1	.15036	.24148	.75852
2	.16539	.40687	.59313
3	.15161	.55848	.44152
4	.12508	.68356	.31644
5	.09631	.77987	.22013
6	.07063	.85050	.14950
7	.04994	.90044	.09956
8	.03434	.93478	.06522
9	.02308	.95786	.04214
10	.01523	.97309	.02691
...	...	...	...

The cdf yields $F(9) = 0.95786$. Therefore, the total number of cars that the inspector will have to check (to ensure a probability of at least 0.95 that 3 cars will be found that pass the EPA standards) is $x + r = 9 + 3 = 12$.

The *Poisson distribution*

$$p(x) = \frac{\lambda^x e^{-\lambda}}{x!} \qquad x = 0, 1, 2, \ldots \tag{3.7}$$

is often used to characterize physical situations in which the number of events during a specific period of time is of interest, like the number of customers arriving at a bank during one hour. The random variable governed by Equation 3.7 is the number of occurrences, X, and $\lambda > 0$ is the rate parameter—the average number of occurrences in a specified time period; e.g., λ might have a value of 10 customers per hour. In using the Poisson distribution, we assume that occurrences are equally likely to happen during any time interval and that one occurrence has no effect on the probability of another. Readers interested in detail about this application of the Poisson distribution are referred to Hillier and Lieberman [1974]. A table of the cumulative Poisson distribution function is given in Table A.2. Example 3.9 illustrates the use of the Poisson distribution.

Example 3.9. The number of customer orders for 21-inch ("two-page") microcomputer monitors at a large mail-order firm averages 20 monitors per week. The average level of customer demand is constant, and customers do not affect one another in their buying habits.

(a) Determine the probability that more than 20 monitors will be purchased in a particular week.

(b) Find the probability that exactly 17 monitors will be purchased in a week.

(c) There are currently 22 monitors in stock, and no more will arrive from the manufacturer until the start of the next week. What is the probability that no monitor will have to be backordered?

Solution. From the information in the problem statement, this is an application of the Poisson distribution with $\lambda = 20$. Using the distribution evaluator program, the following table has been generated:

Poisson Distribution
$\lambda = 20$

X	PR(X)	PR(≤ X)	PR(≥ X+1)
0	.00000	.00000	1
1	.00000	.00000	1
2	.00000	.00000	1
3	.00000	.00000	1
4	.00001	.00002	.99998
5	.00005	.00007	.99993
6	.00018	.00026	.99974
7	.00052	.00078	.99922
8	.00131	.00209	.99791
9	.00291	.00500	.99500
10	.00582	.01081	.98919
11	.01058	.02139	.97861
12	.01763	.03901	.96099
13	.02712	.06613	.93387
14	.03874	.10486	.89514
15	.05165	.15651	.84349
16	.06456	.22107	.77893
17	.07595 (b)	.29703	.70297
18	.08439	.38142	.61858
19	.08884	.47026	.52974
20	.08884	.55909	.44091 (a)
21	.08461	.64370	.35630
22	.07691	.72061 (c)	.27939
23	.06688	.78749	.21251
24	.05573	.84323	.15677
...	...	...	...

The answer to part (a) is 0.44091, the probability that 21 or more monitors will be purchased (i.e., more than 20 monitors). The answer to part (b) is 0.07595, the probability that exactly 17 monitors will be purchased, and the answer to part (c) is 0.72061, the probability that 22 or fewer monitors will be purchased during this week.

The Poisson distribution has another popular use. Suppose that you want to know the probability of 58 or fewer successes in 1000 trials, in which the probability of success is $p = 0.008$ and the assumptions associated with the binomial distribution are satisfied. Table A.1 goes only to $n = 25$, so you cannot use it to evaluate such a probability. Even your binomial distribution evaluator program would have to work for a while to evaluate the probability just described, i.e., the program would have to evaluate explicitly

$$F(58) = P(X \leq 58) = \frac{1000!}{0!\,1000!}0.008^0(1-0.008)^{1000} + \frac{1000!}{1!\,999!}0.008^1(1-0.008)^{999} + \cdots + \frac{1000!}{57!\,943!}0.008^{57}(1-0.008)^{943} + \frac{1000!}{58!\,942!}0.008^{58}(1-0.008)^{942}$$

Such an evaluation is a cumbersome task, even for a computer. Fortunately, in the case of $p = 0.008$ and $n = 1000$, you can use the Poisson distribution to give a *good approximate answer*.

As indicated in Figure 3.3 the Poisson distribution is the *asymptotic* form of the binomial distribution when λ is set equal to np and is held constant, p is driven small, and, simultaneously, n is driven large. The mathematician would state that

$$\lim_{\substack{n\to\infty \\ p\to 0}} {}_nC_x\, p^x(1-p)^{n-x} = \frac{\lambda^x e^{-\lambda}}{x!} \tag{3.8}$$

where $\lambda = np$.

Example 3.10. Use the Poisson distribution to approximate the probability of 9 or fewer successes in 1000 trials, in which the probability of success is $p = 0.006$ and the assumptions associated with the binomial distribution are satisfied.

Solution. Let $\lambda = np = 1000(0.006) = 6$. Using the Poisson distribution evaluator program, we obtain

$$\sum_{x=0}^{9} \frac{\lambda^x e^{-\lambda}}{x!} = 0.9161$$

which compares favorably with the value of 0.9167 obtained by using the binomial distribution evaluator.

A good *guideline* for deciding between the binomial and the Poisson distributions is to use the Poisson distribution to approximate the binomial distribution when either ($n > 20$ and $p < 0.05$) or when ($n > 100$ and $np < 10$). Figure 3.6 presents two plots of the Poisson distribution, first with $\lambda = 5.95$ and then with $\lambda = 2.55$. These plots may be compared with the plot of the binomial distribution with $n = 17$ and $p = 0.35$ ($np = 5.95$) and with the plot of the binomial distribution with $n = 17$ and $p = 0.15$ ($np = 2.55$), both found in Figure 3.1. The two comparisons show that neither of the approximations are good. This, of course, was to be expected, since the guideline had not been satisfied in either case.

One of the assumptions in the use of the binomial distribution is that the probability of success is constant from trial to trial. There are several physical examples very similar to the binomial situation in which this assumption is not satisfied. The classical case of *sampling without replacement* is a particularly important instance in which the probability of success is not constant from trial

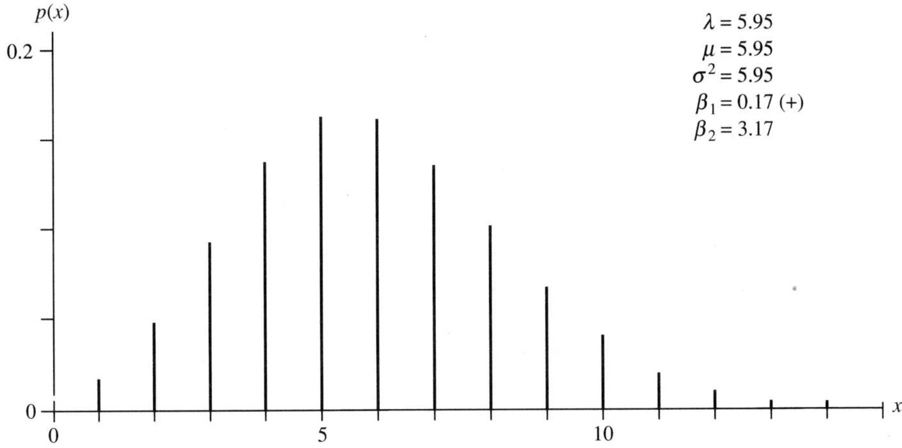

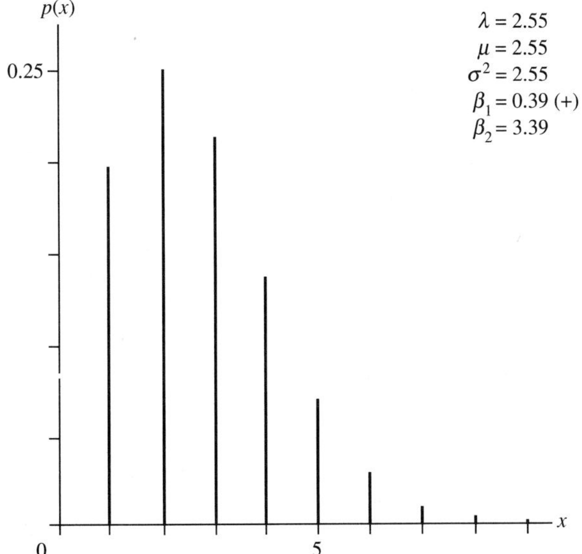

FIGURE 3-6
Two example plots of the Poisson distribution.

to trial. Suppose that you have a finite set of N unique things and a of them possess a property of interest; i.e., there are *a successes* in the *N things*. Let X be the number of successes that occur in a sample, without replacement, of n things from the total of N things. What is $P(X = x)$?

First, there are ${}_NC_n$ equally likely samples, ${}_aC_x$ ways to get x successes, and ${}_{N-a}C_{n-x}$ ways to get $(n - x)$ failures. Thus there are $({}_aC_x)({}_{N-a}C_{n-x})$ ways to get x successes and $(n - x)$ failures, and the required probability is

$$p(x) = \frac{({}_aC_x)({}_{N-a}C_{n-x})}{{}_NC_n} \qquad x = 0, 1, 2, \ldots, n \tag{3.9}$$

which follows directly from the frequency interpretation of probability as discussed in Chapter 1. If p is the initial proportion of successes, $p = \frac{a}{N}$, and if we define $a = Np$ and $q = 1 - p$, the preceding formula for $p(x)$ becomes identical to the formula for the *hypergeometric distribution*, given in Table 3.2.

Let us suppose that N is *large*. If this is so, sampling without replacement may not have a significant effect on the value of p from trial to trial. For example, suppose that $N = 150$ and $a = 50$. If we sample $n = 10$ things from $N = 150$, we would be unlikely to effect a major change in the probability of getting a success. For example, if the first item sampled is a success, p would change very little (from an initial value of $\frac{50}{150} = \frac{1}{3} = 0.333333$ to a new value of $\frac{49}{149} = 0.328859$). Now suppose $N = 15$, $a = 5$, and $n = 10$. If the first item sampled is a success, p would change from $\frac{5}{15} = 0.333333$ to $\frac{4}{14} = 0.2857$. This is a *marked* change, and the binomial distribution may not be appropriately applied because the assumption of a constant probability of success from trial to trial has been *significantly* violated.

Example 3.11. Suppose that in a lot of 15 items there are 5 defective items. If 10 items are sampled without replacement, determine the probability distribution of the number of defectives in the sample.

Solution. Using the hypergeometric distribution evaluator program with $N = 15$, $a = 5$, and $n = 10$, we obtain the following probability distribution of the number of defectives in the sample, $p(x)$. The parenthetical probabilities in the last column are the analogous values obtained from the binomial distribution evaluator with $n = 10$ and $p = \frac{a}{N} = \frac{1}{3}$. These probabilities vividly illustrate that the binomial approximation is not valid for this example.

x	$p(x)$	
0	0.00033	(0.01734)
1	0.01665	(0.08671)
2	0.14985	(0.19509)
3	0.39960	(0.26012)
4	0.34965	(0.22761)
5	0.08392	(0.13656)
6	—	(0.05690)
7	—	(0.01626)
8	—	(0.00305)
9	—	(0.00034)
10	—	(0.00002)

As indicated in Figure 3.3, a good *guideline* in deciding whether to use the binomial or the hypergeometric distribution is that one may use the binomial distribution as an approximation of the hypergeometric distribution when $\frac{N}{10} > n$. (Of course, when the evaluator program is available there is less motivation to do so.) The two plots in Figure 3.7 show the hypergeometric distribution with

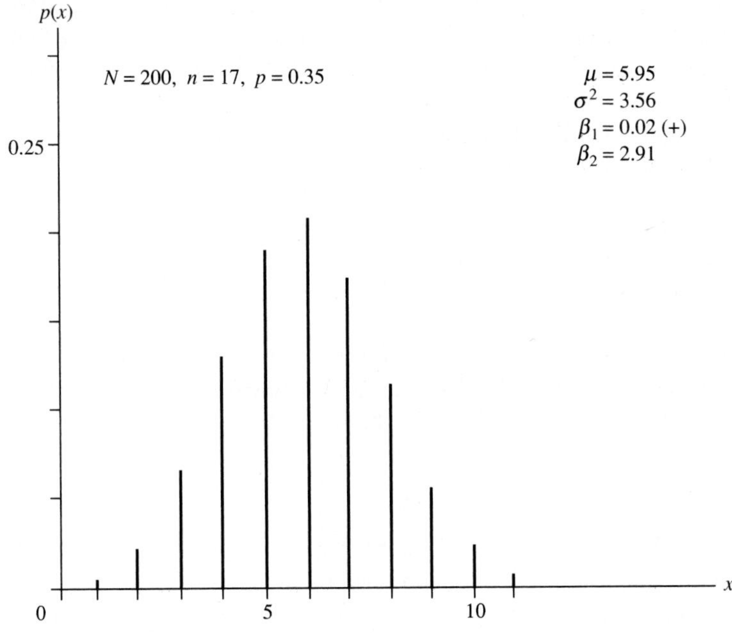

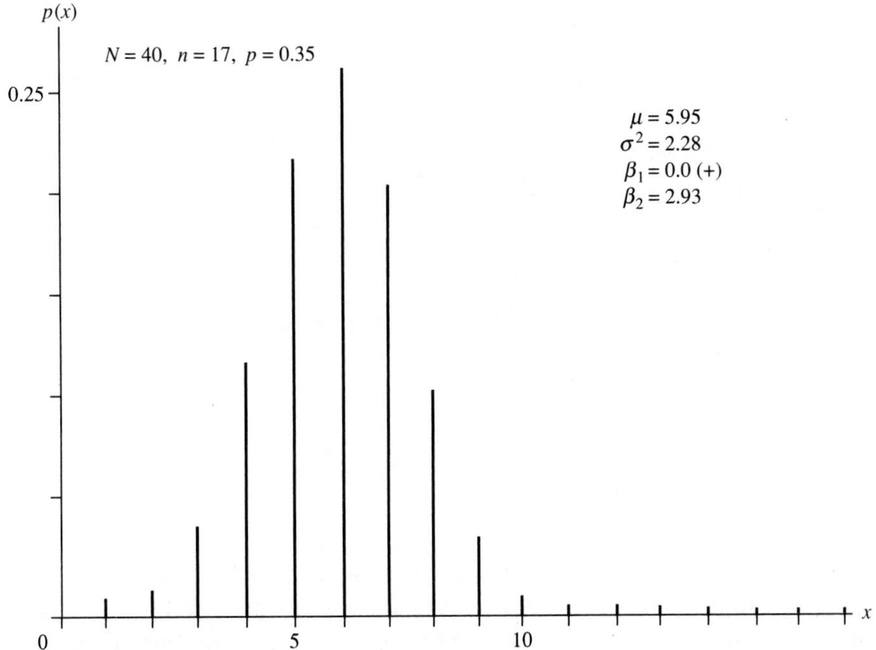

FIGURE 3-7
Two sample plots of the hypergeometric distribution.

$N = 200, n = 17$, and $p = 0.35$ and with $N = 40, n = 17$, and $p = 0.35$. There is very close agreement between the first plot in Figure 3.7, for which $N/10 > n$, and the plot of the binomial distribution with $n = 17$ and $p = 0.35$. Conversely, in the second plot, $N/10 < n$; the agreement is not good between the hypergeometric and binomial distributions.

3.4 A SUGGESTION FOR SOLVING PROBLEMS INVOLVING DISCRETE RANDOM VARIABLES

In Section 3.1, we have seen that any problem involving a discrete random variable may be solved by constructing a detailed and complete specification of the sample space, with all of the probabilities of the outcomes in the sample space. Like many "brute force" methods, the sample space approach pays for its general ability to approach any problem by often being cumbersome and difficult to use.

As discussed in Sections 3.2 and 3.3, certain types of problems occur very commonly in practical applications and streamlined methods have been developed to solve such problems quickly. To efficiently approach a discrete random variable problem,

1. Understand the random variable under consideration, and determine if the random variable fits the description and satisfies the assumptions associated with any of the six random variables presented in Table 3.3 (and described in detail in Sections 3.2 and 3.3).
2. If you find a match in Table 3.3, use the evaluator program for that distribution to solve your problem.
3. If none of the six random variables in Table 3.3 match the random variable associated with your problem, you can proceed with confidence in your use of the sample space method.

One way to gain proficiency with this approach is to review Examples 3.1 through 3.11 in the light of these ideas.

3.5 DISCRETE BIVARIATE PROBABILITY DISTRIBUTION FUNCTIONS

In the preceding discussion, we have considered only *univariate* probability distribution functions. However, on occasion, we need to deal with two or more random variables that occur together simultaneously. Because the ideas presented next are readily extended to three or more variables, we limit our attention to *bivariate* probability distribution functions.

TABLE 3.3

Descriptions and assumptions for the random variables of Table 3.2

Distribution	Random Variable, X	Assumptions
Bernoulli	The number of successes in one Bernoulli trial	1. The probability of success is known. 2. The trial can result in only one of two possible outcomes—success or failure.
Binomial	The number of successes in n Bernoulli trials (may be approximated by the Poisson distribution, letting $\lambda = np$, when either $n > 20 \cap p < 0.05$ or $n > 100 \cap np < 10$)	1. Each trial can result in only one of two possible outcomes—a success or failure. 2. The probability of a success, p, is constant from trial to trial. 3. All trials are statistically independent. 4. The number of trials, n, is a specified constant.
Geometric	The number of failures prior to the first success in a sequence of Bernoulli trials	1. Each trial can result in only one of two possible outcomes—success or failure. 2. The probability of a success, p, is constant from trial to trial. 3. All trials are statistically independent. 4. The sequence of trials terminates after the first success.
Negative binomial	The number of failures prior to the rth success in a sequence of Bernoulli trials	1. Each trial can result in only one of two possible outcomes—success or failure. 2. The probability of a success, p, is constant from trial to trial. 3. All trials are statistically independent. 4. The sequence of trials terminates after the rth success.
Hypergeometric	The number of successes in a sample of size n (may be approximated by binomial distribution when $N > 10n$)	1. Sampling is performed without replacement from a finite set of size N containing a successes. 2. Each member of the sample can result in only one of two possible outcomes—a success or failure. 3. The sample size, n, is a specified constant.
Poisson	The number of event occurrences during a specified period of time	1. The average rate of occurrences ($\lambda > 0$) is known. 2. Occurrences are equally likely to occur during any time interval. 3. Occurrences are statistically independent.

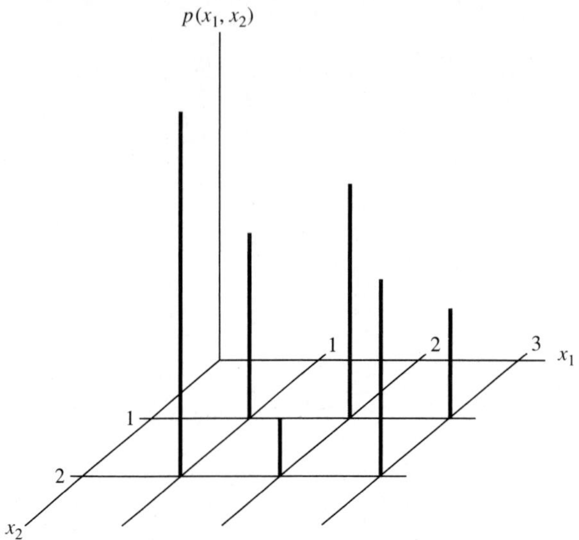

FIGURE 3-8
The bivariate pdf $p(x_1, x_2)$ for Example 3.12.

Example 3.12. Consider the following *joint* probability distribution function of X_1 and X_2, $p(x_1, x_2)$.

		x_1 = 1	2	3
x_2	1	0.15	0.20	0.10
	2	0.35	0.05	0.15

Here we see that $P(X_1 = 1 \cap X_2 = 1) = p(1, 1) = 0.15, P(X_1 = 1 \cap X_2 = 2) = p(1, 2) = 0.35, \ldots,$ and $P(X_1 = 3 \cap X_2 = 2) = p(3, 2) = 0.15$. A pictorial representation of p(x_1, x_2) is given in the perspective drawing of Figure 3.8.

The definitions of sample space, numerically valued function, and random variable that apply to any univariate probability distribution function, $p(x)$, apply equally as well to any $p(x_1, x_2)$. The basic properties of Equation 3.1 also apply to any bivariate probability distribution function; i.e.,

$$p(x_1, x_2) \geq 0 \qquad \text{and} \qquad \sum_{\text{all } x_1, x_2}\sum p(x_1, x_2) = 1$$

The *cumulative bivariate probability distribution function*, $F(x_1, x_2)$, is defined as

$$F(x_1, x_2) = P(X_1 \leq x_1 \cap X_2 \leq x_2) \tag{3.10}$$

This gives the probability that simultaneously X_1 is less than or equal to x_1 and X_2 is less than or equal to x_2.

Example 3.13. Find $F(2, 1)$ for $p(x_1, x_2)$ of Example 3.12.

Solution. The boxed area of $p(x_1, x_2)$, as pictured here, satisfies the requirements of $F(2, 1)$. Hence, $F(2, 1) = 0.35$.

		x_1: 1	2	3
x_2	1	0.15	0.20	0.10
	2	0.35	0.05	0.15

The *marginal probability distribution function* of X_1, $p_{X_1}(x_1)$, is a univariate probability distribution function derived from $p(x_1, x_2)$ by summing the probabilities over all x_2 values.

$$p_{X_1}(x_1) = \sum_{\text{all } x_2} p(x_1, x_2)$$

Similarly,

$$p_{X_2}(x_2) = \sum_{\text{all } x_1} p(x_1, x_2) \tag{3.11}$$

Example 3.14. Obtain the marginal probability distributions of X_1 and X_2 for $p(x_1, x_2)$ of Example 3.12.

Solution. To obtain $p_{X_1}(x_1)$, we simply sum the probabilities in each column of $p(x_1, x_2)$. This yields

x_1	1	2	3
$p_1(x_1)$	0.50	0.25	0.25

To obtain $p_{X_2}(x_2)$, we sum the probabilities in each row, yielding

x_2	1	2
$p_2(x_2)$	0.45	0.55

Note that both of these marginal distributions obey all rules associated with univariate probability distribution functions.

We say that X_1 and X_2 are *statistically independent random variables* if and only if

$$p(x_1, x_2) = p_{X_1}(x_1)p_{X_2}(x_2) \tag{3.12}$$

This means that for *each and every* cell (x_1, x_2), the product of the marginal probabilities must equal the joint probability. If this is *not* true for any cell, we must conclude that X_1 and X_2 are not statistically independent.

Example 3.15. Determine whether X_1 and X_2 for the joint density of Example 3.12 are statistically independent.

Solution. Since $p_{X_1}(1)p_{X_2}(1) = (0.50)(0.45) = 0.225$ is not equal to $p(1,1) = 0.15$, X_1 and X_2 are not statistically independent.

Example 3.16. Suppose that, for $p(x_1, x_2)$ of Example 3.12, we know that X_1 has assumed a value of 3. How does that knowledge affect the probabilities associated with X_2?

Solution. First, the bivariate sample space has been *reduced*. The four cells of $p(x_1, x_2)$, where $x_1 \neq 3$, are no longer possible. In Chapter 2, we saw how to deal with reduced sample spaces and conditional probabilities. Exactly the same techniques will serve to answer this question. Thus,

$$P(X_2 = x_2 \mid X_1 = 3) = \frac{P(X_2 = x_2 \cap X_1 = 3)}{P(X_1 = 3)}$$

Therefore,

$$P(X_2 = 1 \mid X_1 = 3) = \frac{P(X_2 = 1 \cap X_1 = 3)}{P(X_1 = 3)} = \frac{0.10}{0.10 + 0.15} = 0.4$$

and, similarly,

$$P(X_2 = 2 \mid X_1 = 3) = \frac{P(X_2 = 2 \cap X_1 = 3)}{P(X_1 = 3)} = 0.6$$

From this information, we may state that the *conditional probability distribution function* of X_2, given that $X_1 = 3$, is

x_2	1	2
$p(x_2 \| X_1 = 3)$	0.40	0.60

In general, if we have additional knowledge about either of the variables X_1 and X_2, we can use it to assist us in getting "updated" estimates of the probabilities associated with the other variable. Suppose that a *condition*, $C(X_1)$, is placed on X_1—such as $X_1 = 3$ or $a \leq X_1 \leq b$—then the conditional probability distribution of X_2 given $C(X_1)$ is

$$P(X_2 = x_2 \mid C(X_1)) = \frac{P(X_2 = x_2 \cap C(X_1))}{P(C(X_1))}$$

In like manner,

$$P(X_1 = x_1 \mid C(X_2)) = \frac{P(X_1 = x_1 \cap C(X_2))}{P(C(X_2))}$$

This concludes our introductory discussion of discrete probability distributions. In the next chapter, we turn our attention to continuous probability distribution functions.

EXERCISES

3.1. A crate contains six housings that have, uniquely, either 1, 2, 3, 4, 5, or 6 defects. Two housings are drawn at random from this crate without replacement. Let the random variable, X, be equal to the larger of the two numbers of defects on the two housings. What is the probability distribution of X?

3.2. Box I contains five circuit boards, of different sizes, numbered from 1 to 5. Box II contains three circuit boards, numbered 1, 2, and 3, and two circuit boards, each numbered 4. One circuit board is drawn at random from Box I, and one circuit board is drawn at random from Box II. Let the random variable, X, equal the smaller of the numbers on these two circuit boards. What is the probability distribution of X?

3.3. Bin I contains two solenoids, one numbered 1 and the other numbered 2. Bin II contains five solenoids, two numbered 0, one numbered 1, one numbered 2, and one numbered 3. One solenoid is drawn at random from Bin I and is then placed in Bin II. Two solenoids are then drawn at random without replacement from Bin II. Let the random variable, X, equal the product of the numbers on these two solenoids. What is the probability distribution of X?

3.4. Consider the experiment of rolling a pair of fair dice 4 times. After each roll, the sum of the two numbers facing upward on the dice is calculated. Let the random variable, X, equal the number of times in the 4 rolls that a sum of 3 is calculated. What is the probability distribution of X?

3.5. A manufacturing process is intended to produce electrical fuses with no more than 1% defective fuses. This process is checked every hour by trying 10 fuses selected at random from the hour's production. If 1 or more of these 10 fuses fail, the process is halted and carefully examined.

Suppose that the probability that this process will produce a defective fuse is *exactly* 0.01. (Therefore, the process is performing as intended.) What is the probability that this process will be examined? (Notice that the process is needlessly being examined because it is actually performing as intended.)

3.6. Determine the probability distribution of the number of rolls of a fair die until the first 3. What is the probability that a 3 will occur on or before the fourth toss?

3.7. The rhino charging at you will require 3 hits from your 0.22 caliber pistol before it will be sufficiently discouraged to go elsewhere. Under the circumstances, your shooting ability will yield a 0.42 probability of a hit. Assuming you will stop shooting after you have achieved 3 hits, what is the probability that 7 or fewer shots will be required?

3.8. A study of Prussian horse-cavalry soldiers noted that the probability of such a soldier being killed by the kick of a horse was 0.5% per year. In a group of 5000 such soldiers, how likely are more than 30 to be killed in one year?

3.9. In a group of 17 bottle caps, there are 7 with defective seals. How likely is it that a sample of 8 caps taken from the 17 at random will have more than 2 defective seals?

3.10. In early prototype production, the probability of a good unit is 0.65. How likely is it that a sample of 200 will provide between 110 and 135 good units?

3.11. Derive the mathematical form of the negative binomial distribution given in Table 3.2.

3.12. Suppose, in Example 3.3, that the probability of failure of a backup switch is increased to 0.5. Determine the probability distribution function and the cumulative probability distribution function of the number of switches required until one switch functions successfully. Show that your answer is a probability distribution function and determine the number of switches required to ensure a 0.95 probability of successful switching.

3.13. A certain discrete random variable has the probability distribution function, $p(x) = k/x^2$ for $x = -3, -2, -1, 1, 2, 3, 4$. Determine the value of k and the probability that X will assume a value in the range from -3 to 1, inclusive.

3.14. A shelf supports 20 ceramic resistors. Three of the resistors are faulty. If 6 resistors are sampled from the shelf *with replacement*, what is the probability that 4 or more of the sample will be faulty? Answer the same question if the resistors are sampled without replacement.

3.15. Tru-Round Piston Rings, Inc., has an average of 0.5% defective parts from its GMC line. As a part of their quality control program, they sample 40 units from each shipping box of 500 units. What is the probability that a sample will contain 2, 3, or 4 defective units?

3.16. Three Russian cosmonauts are in a stationary orbit over the Pacific Ocean, just off the coast of California. Unfortunately, they have had a fuel line failure, and all their propellant has escaped into space. They have successfully repaired the fuel line but find themselves marooned without maneuvering capability. The Russians are reluctant to attempt to resupply the needed fuel because of the proximity of the spacecraft to the continental United States, even with the apparent end of the Cold War.

(a) As a humanitarian gesture, NASA has volunteered to provide the needed fuel. The probability of any individual U.S. spacecraft achieving a successful docking with the Russian craft, even with the aid of Russian telemetry, is 0.63. How many spacecraft should be prepared?

(b) If two U.S. spacecraft must reach the stranded Russian craft rather than only one, how many spacecraft should be prepared?

3.17. The probability that a B-29 can successfully avoid the German antiaircraft fire and then make a successful bombing run on the Rhine bridge is 0.578. If 3 successful bombing runs are required to knock out the bridge, how many B-29s should be readied for tonight's raid?

3.18. Last night's raid (Exercise 3.17) resulted in 2 successful bombing runs. One more is required. Owing to damage inflicted on the German artillery positions, the probability of a B-29 making a successful bombing run has increased to 0.683. How many B-29s should be readied for tonight's raid?

3.19. A simple generalization of the binomial distribution is the *multinomial distribution function,*

$$p(x_1, x_2, x_3, \ldots, x_k) = \frac{n!}{x_1!\, x_2!\, x_3! \cdots x_k!} p_1^{x_1} p_2^{x_2} p_3^{x_3} \cdots p_k^{x_k}$$

where there are n repetitions of the experiment and k mutually exclusive outcomes that can occur in each repetition. Further, the probability of outcome k is p_k; X_k is the random variable denoting the possible number of times that outcome k will occur in the n repetitions and $\sum_{i=1}^{k} x_i = n$.

Use the multinomial distribution function to determine the probability of 3 bad bearings, 4 fair bearings, and 6 good bearings in a sample of 13 bearings from a large barrel of bearings, from which the probability of drawing a bad bearing is 0.05, a fair bearing is 0.25, and a good bearing is 0.70.

3.20. A coin collector has decided to purchase 3 coins from a set of 15 "Roman" coins. Expert evidence supports the conjecture that 20% of the coins on the market that are similar to these coins are counterfeit. What is the probability that 2 or more of the coins our collector has purchased are counterfeit?

3.21. The probability that a $5\frac{1}{4}$-inch floppy disk will last more than 40 hours of read or write time is 0.175. Find the probability that, in a set of 35 such disks, (a) more than 30 disks will last more than 40 hours; (b) between 15 and 25 disks will last more than 40 hours; and (c) less than 10 disks will last more than 40 hours.

3.22. The Auzi, a foreign luxury automobile, has developed a rare problem with its carburetor, which has caused a recall of all 75,000 Auzis manufactured during the last three years. The probability is 0.0001 that the Auzi carburetor will malfunction on engine startup, causing the engine to backfire with a subsequent fire and possible explosion under the hood.

Four thousand customers have failed to return their Auzi to the dealer to have a corrective kit put on the carburetor. What is the probability that Auzi International will have more than seven lawsuits brought against the company as a direct result of this carburetor problem?

3.23. "Take you on at pool," said Cue to Ball.
"Right you are," said Ball. "We'll play 5 games."
"I ought to beat you, Ball," said Cue.
"Of course," said Ball, "you'll have to give me odds."
"How do my chances of winning compare to yours?" asked Cue.
"Well," replied Ball, "I'll tell you. Your chances of winning either 3 or 4 of the games is exactly the same."
Assuming that each game must have a winner and that Ball's probability of winning a game is neither 0 nor 1, what is the probability that Cue will win all 5 games?

3.24. The probability of a defective tire failing in the first 45,000 miles is 35%. If 17 defective tires are sold, what is the probability that no more than 4, or more than 9, survive?

3.25. Forty Ziladium class rings have just arrived at Sparkling Jewelers, Inc. The ring supplier has just called and said that 25% of the rings were fabricated with radioactive metal. When you inform him that you have already sold 15 of the rings, he says, "Sorry, I must have the wrong number." Determine the probability that no more than 3 of the rings that were sold are radioactive.

3.26. The probability of a bearing lasting 60 hours in a selected sandy environment is 62%. If 37 bearings are simultaneously placed in such an environment, what is the probability that no more than 11 fail?

3.27. Forty lots of silicon wafers have been delivered to AAA Electronics. The vendor has just informed you that 30% of the lots are bad as a result of a metallic contamination. Fifteen lots have already been used in the production line. Determine the probability that no more than 2 of the lots that were used were bad.

3.28. In a group of 32 seals, 12 have blue eyes. How likely is it that, in a random sample of 7 of the 32 seals, fewer than 4 will have blue eyes?

3.29. Eighteen 1-megabyte RAMs are on a test table. Five have gate leakage problems and will fail in only moderately stressful environments. If 3 of these RAMs are selected at random for a particular assembly, what is the probability that (a) all 3 RAMs are bad and (b) no more than 1 RAM is bad?

3.30. Thirteen vials of vitamin B_1 are available. Three are old and have lost their potency. If 4 vials are selected at random, what are the probabilities that (a) exactly 3 vials are good and (b) no more than 2 vials are good?

3.31. Fourteen cotton balls are tossed into 3 wastebaskets. Let the random variable, X, be the number of balls that fall in the second basket. Determine the probability distribution function of X.

3.32. The following joint distribution function governs the number of cosmetic blemishes, X_1, and the number of casting imperfections, X_2, on the plastic case for the Mag-80 hard disk drive produced by Winesap Microcomputer Peripherals. Based on this information, would you say that the number of cosmetic and casting problems depend on one another? What is the probability that the total number of blemishes and imperfections will exceed 4? If a case has 2 blemishes, what is the probability that it will have no casting imperfections?

		x_1			
		0	1	2	3
	0	0.59	0.06	0.02	0.01
x_2	1	0.10	0.05	0.04	0.01
	2	0.06	0.05	0.01	0.00

3.33. The joint distribution given next governs two statistically independent random variables. What is the apparent relationship between the different rows and columns? (*Hint*: Review the material on statistically independent events.) Do you think this relationship holds for all bivariate distribution functions when the random variables are statistically independent?

		x_1		
		0	1	2
	0	0.20	0.12	0.08
x_2	1	0.10	0.06	0.04
	2	0.20	0.12	0.08

3.34. The following joint distribution was produced by UTELEM Communications, Inc., as a part of their investigation into the initial ownership of videocassette recorders (VCRs) and the frequency of rentals of movies on videotape. Here X_1 is the number of months that the recorder was owned, and X_2 is the number of movies rented in an average week.

		x_1			
		0	1	2	3
	0	0.05	0.03	0.02	0.03
	1	0.06	0.06	0.04	0.01
x_2	2	0.13	0.08	0.05	0.02
	3	0.17	0.12	0.09	0.04

(a) Obtain the marginal distributions of X_1 and X_2. (b) What is the probability that a person who has owned a VCR for 3 months will rent 3 movies in an average week? (c) Are there any comments or helpful criticism that you can make about the survey? (Should you invest in a video rental store?)

3.35. On a multiple choice exam with 6 options for each of the 15 questions, what is the probability that a student will answer correctly 10 or more of the questions just by guessing? What is the probability that he or she will answer correctly 5 or less?

3.36. Five radar stations are set up to detect incoming aircraft. If a radar station detects an aircraft 90% of the time independently of the other stations, what is the probability that a plane is detected? What is the probability it will not be detected?

3.37. A salesman makes a sale to a customer with probability of 0.3. If he must sell 3 items, what is the probability that doing so will take fewer than 6 customers? Exactly 3 customers?

3.38. Thirty percent of the applicants for a job have advanced computer training. If applicants are randomly selected for interviewing, what is the probability that the first applicant with such training is the fifth person interviewed ? What is the probability that the third such applicant is found on the tenth interview?

3.39. ASME made 2 boxes of tensile bars for ME students; each box contains 10 bars, labeled 1 to 10. One is drawn from each box. Let the random variable, X, be the smaller number of the 2 bars. What is the probability distribution function of X?

3.40. Determine the probability distribution of the number of times you pick a card from the deck and replace it until you get one of the two red kings. What is the probability that you will get a red king in the first 3 tries? What is the probability you will get a red king on the third try?

3.41. Joe's "Hit or Miss" garage has a 23% chance of fixing your car on the first try. If you bring in 12 of your Dad's Italian roadsters with overheating problems, what is the probability that 5 or more of them will be fixed on the first try?

3.42. Fifty fishes grow into full size out of 25,000 eggs laid. Find the probability that 3 or more fishes grow into adults in a sample of 1000.

3.43. A certain type of aluminum sheet is manufactured for sale to the general public. It is known from past experience that there will be X cosmetic flaws, with the following probability distribution function (pdf).

x	0	1	2	3	4
$p(x)$	0.45	0.20	0.15	0.10	0.10

Another type of aluminum sheet will have cosmetic flaws according to the following pdf.

x	0	1	2	3
$p(x)$	0.65	0.15	0.15	0.05

If 2 of the first type are produced for every 1 of the second, what is the distribution of the random variable, Y, if Y is the number of flaws in a randomly sampled sheet of aluminum?

3.44. There are two sets, I and II, as follows.

set I: $\{1, 1, 2\}$ set II: $\{1, 2, 3, 4, 5\}$

Suppose we draw a number at random from set I and place it in set II, and then we draw at random two numbers from set II and add them together. What is the distribution of X, the sum of the two numbers drawn from set II?

3.45. Suppose a student is taking a true/false test and that, owing to her extraordinary studying habits, she has a 90% chance of getting any particular question correct. On an exam of fifty questions, what is the probability that she will answer exactly 40 questions correctly? 40 or more?

3.46. If Bob has a 30% chance of having a dance partner on any given song, what is the probability that he will have danced 5 or fewer times by the end of the evening if the band plays 35 songs?

3.47. Since Bob and Teresa are late for school every day, they must park their car illegally to get to class on time. If they get a parking ticket with probability 0.05, how many consecutive days can they be late and still have an 85% chance of not getting a ticket?

3.48. Out of the last 3960 random tests of a production process, 192 of the sampled items were flawed. What is the probability that at least the next 30 items will be produced without imperfections? What is the probability that more than 80 will be produced before 2 flaws? What is the probability that more than 120 will be produced before 3 flaws?

3.49. From an ordinary deck, an engineering student draws a card, replaces it, shuffles the deck, and repeats the process. What is the probability that she draws three diamond cards on her first 3 tries and a non-diamond on her fourth try? What is the probability that she draws 6 red cards before 2 black ones?

3.50. The following random variable is one of the six described in Figure 3.2. What type of random variable is X and what is(are) its parameter value(s)?

x	0	1	2	3	4	5	6
$p(x)$	0.00506	0.05395	0.20230	0.34329	0.27893	0.10299	0.01348

3.51. For any particular batch of circuits, 12% will need more time in the heat-treating facility. If a batch contains 100 circuits, what is the probability that, from a sample of 30, 4 or fewer circuits will need more heat treatment?

3.52. Consider successive rolls of a fair die. Let rolling a 2 be considered a success and all other outcomes be failures for each roll. What is the probability of

a) getting the sequence ($FSFFFSF$)?
b) getting at least three 2s in 5 rolls?
c) getting a 2 for the first time, on either the sixth, seventh, or eighth roll?
d) getting a 3 or greater on the next roll, given that the last number rolled was a 3?

COMPUTER-BASED EXERCISES

C3.1. Evaluate and plot the following distributions.

Binomial with $n = 10$ and $p = 0.1$
Binomial with $n = 15$ and $p = 0.06667$
Binomial with $n = 20$ and $p = 0.05$
Binomial with $n = 25$ and $p = 0.04$
Binomial with $n = 30$ and $p = 0.033333$
Poisson distribution with $\lambda = 1$

At what point do you think the Poisson distribution forms a suitable approximation to the binomial distribution? What are the major differences between the Poisson and binomial distributions when the approximation is not suitable?

C3.2. Discuss the effect that a change in the value of λ over a range from 0.1 to 6 has on the shape and form of the Poisson distribution.

C3.3. Discuss the effect that changing the value of p has on the shape and form of the geometric distribution.

C3.4. The probability that there are 5 or fewer imperfect steam valves in a sample of 30 valves taken from a lot of 1000 valves is 0.77793. Determine, to four significant digits, the probability of an imperfect valve.

C3.5. Determine the number of valves (Exercise C3.4) that would have to be checked if you want to have a probability of 0.965 of obtaining 20 or more bad valves.

C3.6. The probability that there are 11 or fewer imperfect steam valves in a sample of 30 valves taken from a lot of 100 valves is 0.36329. Determine the number of imperfect valves in the lot.

CHAPTER 4

CONTINUOUS RANDOM VARIABLES

The following concepts and procedures form the body of knowledge that you should gain from this chapter.

- Understand what is meant by the terms *continuous probability density function* and *continuous cumulative distribution function*
- When given a probability density function for a random variable, X, be able to calculate the probabilities associated with various subsets of the sample space for X
- Know what μ and σ determine for a *normal distribution*
- Be familiar with the differences in the shapes of different normal distributions and with the properties that are common to all normal distributions
- Know what is meant by the terms *standard normal random variable* and *standard normal distribution*
- Be able to transform any normal random variable into a standard normal random variable
- Be able to find the area under any part of the standard normal curve
- Understand why the normal distribution is the *most important distribution* in statistical analysis
- Be able to use the normal distribution to solve problems like those found in the chapter
- Be able to appropriately apply the χ^2, t, and F distributions
- Be able to appropriately apply the uniform, beta, gamma, exponential, log-normal and Weibull distributions

- Understand what is meant by a *continuous bivariate probability distribution function*, $f(x_1, x_2)$
- When given an $f(x_1, x_2)$, be able to obtain the *cumulative* bivariate probability distribution function, $F(x_1, x_2)$, and the *marginal* probability distribution function for either X_1 or X_2
- When given an $f(x_1, x_2)$ and additional information about one of the random variables, be able to obtain the *conditional* probability distribution function for the other random variable
- When given an $f(x_1, x_2)$, be able to determine if X_1 and X_2 are statistically independent

4.1 GENERAL PROPERTIES OF CONTINUOUS RANDOM VARIABLES

In the preceding chapter, we have seen that a discrete random variable can assume only a finite or countably infinite number of distinct values. Often, however, we want to consider *continuous* random variables, which can assume an uncountably infinite number of values. The probability distribution of a continuous random variable, X, is described by a *continuous probability density function*, $f(x)$, such that $f(x) \geq 0$ for all possible x values. The range of possible values may be from $-\infty$ to $+\infty$. (By definition, $f(x) = 0$ for any value x that is not within the range of possible values.) The *cumulative probability distribution function*, $F(x)$, is defined as $F(x) = \int_{-\infty}^{x} f(u)du$, and, like its discrete counterpart, gives the probability that the random variable will achieve a value less than or equal to x.

Suppose that the graph of $f(x)$ has the shape presented in Figure 4.1. The probability of any event is equal to the area *above* the horizontal axis and *below* $f(x)$. The shaded area in the figure denotes the probability of event A, where $A = \{13 \leq X \leq 15\}$.

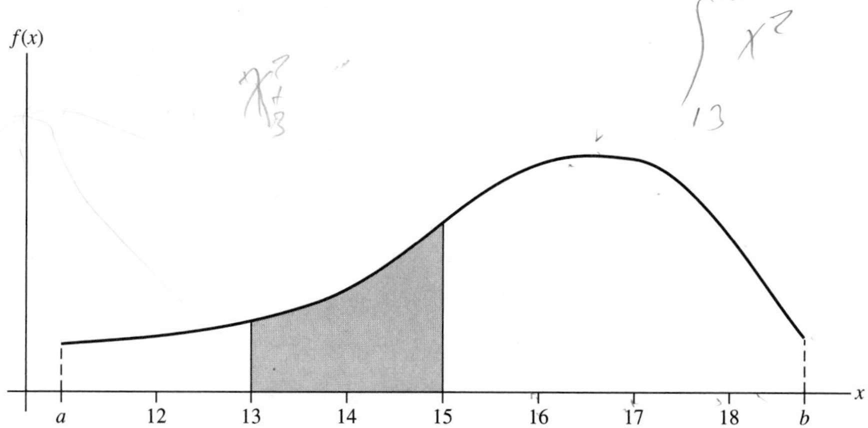

FIGURE 4-1
A continuous probability density function.

In addition to the requirement that all $f(x) \geq 0$, the total area under $f(x)$ over the range of possible values must equal 1; i.e., $\int_{\text{all } x} f(x)dx = 1$. Notice that these two requirements are the *basic properties* that must be fulfilled before a continuous mathematical function is a probability density function. They are very similar to the two basic properties associated with a *discrete* probability distribution function, as given in Equation 3.1. Indeed, the only difference is that an integral sign replaces a summation sign in the second property.

In Figure 4.1, the probability that $X > b$ or $X < a$ is equal to 0. Therefore, the probability that X will be in the interval from a to b must be equal to 1 and the corresponding area under $f(x)$ must be 1.

Continuous random variables do not have the same upper-bound restrictions on the values that a probability distribution function may take at a specific value of x; i.e., f(x) *may exceed* a value of 1 for a continuous random variable. For example, $f(x) = 10, 0 \leq x \leq 0.1$, is a valid probability density function because f(x) ≥ 0 for all x and $\int_0^{0.1} 10\, dx = 1$.

Clearly, a different interpretation is required for the probability distribution function for continuous random variables. The appropriate interpretation is that for "well-behaved" continuous probability distribution functions, $f(x)$ gives a relative measure of the *intensity*, or *density*, of the *probability mass* at x and in the "neighborhood" of x for continuous $f(x)$. In conceptual terms, this may be interpreted to mean that the higher the value of $f(x)$, the more likely the values in the neighborhood of x are to occur.

However, *before* the experiment is performed, the only meaningful probability that can be assigned to any one *particular* value, x, of a continuous random variable, X, is zero. An event associated with a continuous random variable has probability greater than zero only when a contiguous interval of length greater than zero is included in the definition of the event. This is different from discrete random variables, which have probability mass defined only at specific and unique x values.

Intuitively, because a continuous random variable is defined on a continuum, there are an *uncountably infinite* number of outcomes over any interval of greater than zero length. When we attempt to allocate, or "spread out," the total probability mass of 1 unit across the uncountably infinite set of outcomes, we effectively divide 1 by infinity, yielding a *prior probability* of zero.

The preceding discussion does not mean that a continuous random variable, X, will not achieve one of the possible values in its range when the experiment is performed. Indeed, one of the values, x, *will occur*. It is only from the viewpoint prior to the experiment that it is impossible to assign a probability to a specific value x.

Example 4.1. If $f(x) = 10$ for $0 \leq x \leq 0.1$, obtain $P(X = 0.005), P(X = 0.008)$, and $P(0.005 \leq X \leq 0.008)$.

Solution. $P(X = 0.005) = \int_{0.005}^{0.005} 10\, dx = 0, P(X = 0.008) = 0$, and $P(0.005 \leq X \leq 0.008) = \int_{0.005}^{0.008} 10\, dx = 0.03$.

4.2 THE NORMAL DISTRIBUTION

The most important and widely used continuous probability distribution function is the Gaussian, or normal, distribution, which was discovered by Abraham de Moivre around 1733. This distribution was reintroduced near the beginning of the nineteenth century by Carl Friedrich Gauss in connection with the theory of errors of physical measurements.

The probability density function of any Normal distribution is

$$f(x) = \frac{1}{\sigma\sqrt{2\pi}} e^{-(x-\mu)^2/(2\sigma^2)} \qquad -\infty < x < \infty \tag{4.1}$$

where the *mean*, μ, and the *standard deviation*, σ, are two parameters whose values completely determine $f(x)(e = 2.7183$ is the base of the Naperian, or *natural*, logarithms). The standard deviation is the square root of the *variance*, σ^2; i.e., $\sigma = \sqrt{\sigma^2}$.

Many persons know the normal distribution as the *bell-shaped curve*, as illustrated in Figure 4.2. Some of the reasons that the normal distribution is the most important continuous distribution are as stated in the following five items.

1. The normal distribution occurs naturally; i.e., there are many things in the physical world that are distributed normally. (This is a direct consequence of the *central limit theorem* and its extensions. These are discussed in detail in Chapter 7.)
2. Certain other random variables can be approximated by a normal distribution. A specific example of this is presented at the end of this section.

3404
ME318
Fluids,
Solid

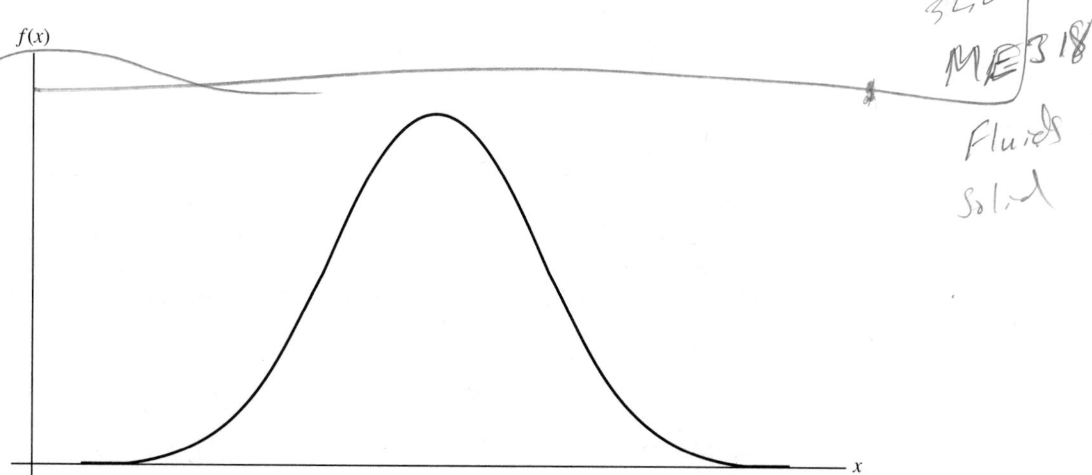

FIGURE 4-2
The bell-shaped curve.

3. Some random variables that are not even approximately normally distributed can easily be transformed into normally distributed random variables.
4. Many results and analysis techniques that are useful in statistical work are strictly correct only when the associated random variables are normally distributed. (Examples of such techniques are presented in Chapters 11 through 14.)
5. Even if the distribution of the original population is far from normal, the distribution associated with sample averages from this population tends to become normal, under a wide variety of conditions, as the size of the sample increases.

Items 3 and 4 are discussed next; discussion of item 5 is deferred to Chapter 7.

Let us examine the use of the normal distribution. As may be observed in Figure 4.3, the function is *symmetric* about the mean, μ; i.e., the half of the distribution that lies on the right of the mean is the *mirror image* of the half that lies on the left of the mean. Additionally, about two-thirds of all events occur within 1 standard deviation of the mean, about 95% occur within 2 standard deviations of the mean, and just about everything (99.74%) occurs within $\pm 3\sigma$ of the mean. Therefore, μ is the location of

1. the *middle point* of a normal distribution

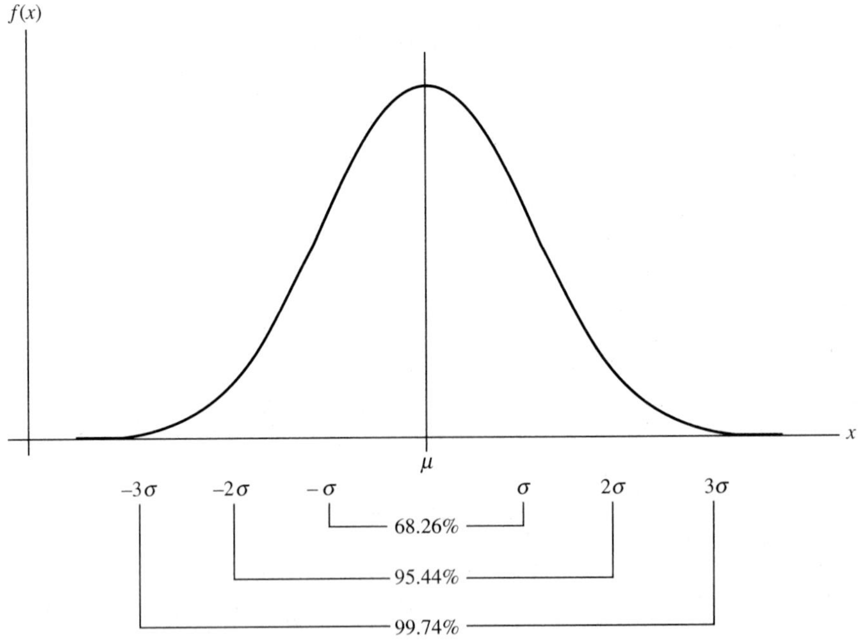

FIGURE 4-3
The standard deviation, σ, and the area under a normal distribution.

2. the *highest point* of a normal distribution
3. the *balance point* of a normal distribution

σ determines how widely *spread* the distribution will be; i.e., a larger σ implies a more disperse $f(x)$.

Example 4.2. The resistance of one type of circuit is normally distributed with a mean of 100 ohms and a standard deviation of 5 ohms. What is the probability that the resistance of one of these circuits selected at random is (a) greater than 100 ohms? (b) less than 95 ohms? (c) greater than 110 ohms?

Solution

(a) Since the mean is 100 and the distribution is symmetric about the mean, the probability of a value over 100 is 0.50.

(b) Since the standard deviation is 5, a resistance of 95 is 1 standard deviation below the mean. Therefore, 68.26% of the probability mass is between 95 and 105, and 31.74% is outside of those values, half above 105 and half below 95. Therefore, the probability of a value less than 95 is 0.1587.

(c) The same type of reasoning can be used for this part of the problem. A resistance of 110 is 2σ above the mean. Of the possible values, 95.44% are in the range 90 to 110, and 4.56% are outside that range, both above and below. Thus 2.28% are above 110.

In Example 4.2, suppose that we desire to know the probability that the resistance is above 107 ohms. Two methods that can be used. First, you can use the normal distribution evaluator program in your library of computer programs. If you choose this method, all that you would have to provide the computer would be the values of the mean and the variance and the specific value of X, which is equal to 107 for our current example. Having done so, the computer would provide the following information.

```
MEAN = 100
STANDARD DEVIATION = 5
X = 107
HEIGHT OF DENSITY FUNCTION = .0299
PR(X ≤ 107) = .9192
PR(X ≥ 107) = .0808
```

The second method is to use Table A.3, the table of the standard normal distribution, which has $\mu = 0$ and $\sigma = 1$. In order to do this, we first need to determine the *number of standard deviations* that 107 ohms is located away from the mean. One way to determine this is by calculating the z *transform*.

$$z = \frac{x - \mu}{\sigma} = \frac{107 - 100}{5} = 1.4 \tag{4.2}$$

The z transform has an interesting property. If X is $N(\mu, \sigma)$, i.e., normally distributed with mean, μ, and standard deviation, σ, then $Z = \frac{(X-\mu)}{\sigma}$ is $N(0, 1)$.

Thus Z is distributed according to the *standard normal distribution*, which has $\mu = 0$ and $\sigma = 1$.

The fact that the shape of a normal curve is completely determined by its standard deviation enables us to reduce all normal distributions to a standard normal curve by the simple change of variable contained in Z. It is also interesting that, no matter what the distribution of X, $Z = \frac{(X-\mu)}{\sigma}$ will always result in a random variable that has a mean of 0 and a standard deviation of 1. However, *only* when X is a normally distributed random variable can we conclude that Z is normally distributed.

Pay particular attention to the descriptive figure at the top of Table A.3. It presents a clear description of the contents of the table inside the upper marginal rows and leftmost marginal columns. As you can see, the numbers in the interior of the table give the probability that a standardized normal variable will achieve a value less than or equal to the corresponding value of z. You can use Table A.3 to determine directly the probability that a standardized normal random variable is less than $z = 1.4$ (or any other value of z). This is exactly the same as determining the probability that *any* normal random variable is less than 1.4 (or z) standard deviations from *its* mean. If you desire the probability that a standard normal variable will exceed a certain value of z, you need only subtract from 1.0 the number obtained from Table A.3.

In using Table A.3, you find the value of z, to the nearest *first* decimal place, in one of the two z columns. For our current example, we find the value $+1.4$ in the z column of the *positive* (right) side of the table. Having found that value, we now move to the right in the same row until we find the remaining part of the z value to the nearest *second* decimal place in the column headings at the *top* of the table. For a value of 1.4, we need move only one column to the right to the 0.00 column, where we read 0.9192. Since this is a *cumulative* distribution table, we interpret this, as described previously, to mean that $P(Z \leq 1.4) = 0.9192$. Therefore, the required probability is $P(Z \geq 1.4) = P(X \geq 107) = 1-0.9192 = 0.0808$.

Example 4.3. Suppose that, for the circuit of Example 4.2, we require the probability of a resistance value less than 98.25.

Solution. Again using the computer program or computing the z transform and using the table, we find that $P(X \leq 98.25) = 0.3632$.

Example 4.4. Suppose that we want to know the probability that a resistance will be between 96.72 and 101.17; i.e., we desire $P(96.72 \leq X \leq 101.17) = P(-0.656 \leq Z \leq 0.234)$.

Solution. $P(X \leq 96.72) = P(Z \leq -0.656) = 0.2559$, and $P(X \leq 101.17) = P(Z \leq 0.234) = 0.5925$. Therefore, as illustrated in the following plot of the $N(0,1)$ distribution, $P(96.72 \leq X \leq 101.17) = P(-0.656 \leq Z \leq 0.234) = 0.5925 - 0.2559 = 0.3366$.

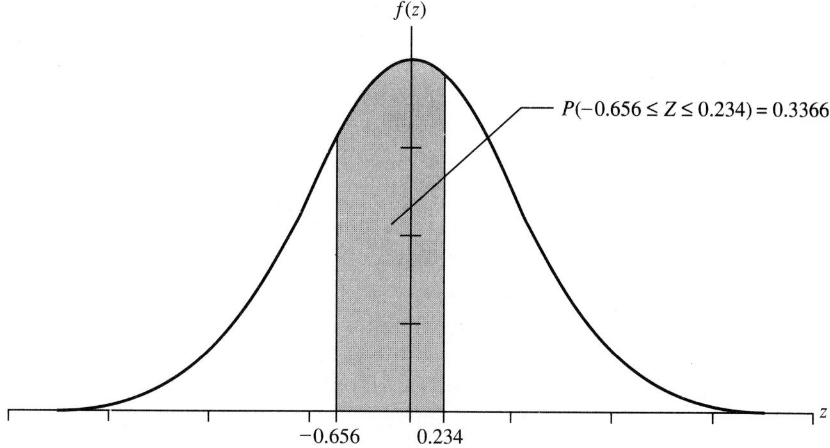

Examples 4.2, 4.3, and 4.4 have illustrated what is known as a *forward table lookup*.

Example 4.5. Suppose that we ask, What stipulated resistance value will exceed the resistance of 99.43% of our circuits?

Solution. In order to answer this question, we must first perform an *inverse table lookup*. To perform our required inverse table lookup, using Table A.3, find the value of 0.9943 in the body of the table. Having done that, read the z value in the same row in the left marginal z column, and add to that z value the number in the same column in the top marginal row. We find our probability value of 0.9943 in the row with $z = +2.5$ and in the column labeled 0.03. Therefore, we may conclude that $P(Z \leq 2.53) = 0.9943$. Remember that $P(Z \leq 2.53) = 0.9943$ means that, for any normal distribution, a value less than 2.53 standard deviations to the right of the mean will occur 99.43% of the time.

Next, we must convert this information to the equivalent ohms of resistance. Since the mean is 100 ohms and the standard deviation is 5 ohms, the value that is 2.53 standard deviations to the right of the mean is $x = 100 + 2.53(5) = 112.65$. Therefore, 99.43% of our circuits will have resistances less than or equal to 112.65 ohms.

What we have done here is to use the *inverse z transform*,

$$x = \mu + z\sigma \tag{4.3}$$

which is obtained from the z transform simply by solving for *x in terms of z*.

Of course, not all probabilities are found in Table A.3. If a probability falls between the values in the table, linear interpolation must be performed to obtain the associated value of z.

This problem may be more easily solved by using the normal distribution evaluator program. If you choose this method, the computer asks you for the mean, the standard deviation, and the *right-hand-tail probability*, i.e., the complement of the cumulative probability. In the current problem, the mean is 100, the standard deviation is 5, and the right-hand-tail probability is $0.0057 = 1 - 0.9943$, where the cumulative probability of 0.9943 is the *left-hand-tail probability*. After you have provided the information, the computer will respond with

```
MEAN = 100
STANDARD DEVIATION = 5
p = .0057
PR(X ≥ 112.65) = .0057
xp = 112.65
```

Item 2 of our list of things that make the normal distribution important is associated with approximating other distributions using the normal distribution. One of these is the *approximation* for the binomial distribution when the number of trials is large and the probability of success is *not* near 0 or 1. If we use $N(n\,p,\ \sqrt{np(1-p)}) = N(\mu, \sigma)$ to approximate a binomial distribution with parameters n and p, the approximation is reasonably good as long as $n\,p \geq 5$ when $p \leq 0.5$ or $n(1-p) \geq 5$ when $p \geq 0.5$.

If we are evaluating $P(a \leq X \leq b)$ for a binomial random variable, an even better approximation can be made by using

$$P(a \leq X \leq b) = P\left(Z \leq \frac{b + 0.5 - np}{\sqrt{np(1-p)}}\right) - P\left(Z \leq \frac{a - 0.5 - np}{\sqrt{np(1-p)}}\right) \tag{4.4}$$

The additions of $+0.5$ and -0.5 in the above formula are a "continuity correction" for the fact that we are approximating a discrete random variable with a continuous random variable. (When you are not sure that the normal approximation is valid, it is always correct—but possibly more cumbersome—to use the binomial distribution evaluator program.)

Example 4.6. As an example of using this approximation, consider a binomial distribution with $p = 0.4$ and $n = 20$, and suppose that we desire $P(4 \leq X \leq 13)$.

Solution. According to Equation 4.4,

$$P(4 \leq X \leq 13) = P\left(Z \leq \frac{13 + 0.5 - 8}{\sqrt{4.8}}\right) - P\left(Z \leq \frac{4 - 0.5 - 8}{\sqrt{4.8}}\right)$$

$$= P(Z \leq 2.510) - P(Z \leq -2.054) = 0.994 - 0.020 = 0.974$$

Evaluating this probability directly from the binomial distribution yields

$$P(4 \leq X \leq 13) = P(X \leq 13) - P(X \leq 3) = 0.994 - 0.016 = 0.978$$

Thus the error in the approximation for this example is quite small at a value of 0.004.

4.3 SOME POPULAR CONTINUOUS DISTRIBUTIONS AND THEIR RELATIONSHIPS

Figure 4.4 presents an overview of the various relationships that join the probability distribution functions discussed in this chapter. Table 4.1 gives their mathematical forms along with associated information of considerable interest to us in this and following chapters. (In Table 4.1 and at other points in the book, the notation $\exp(f(\bullet))$ is used to represent $e^{f(\bullet)}$ when it will enhance the clarity and readability of the expression; i.e., $\exp(-\beta^2) \equiv e^{-\beta^2}$.)

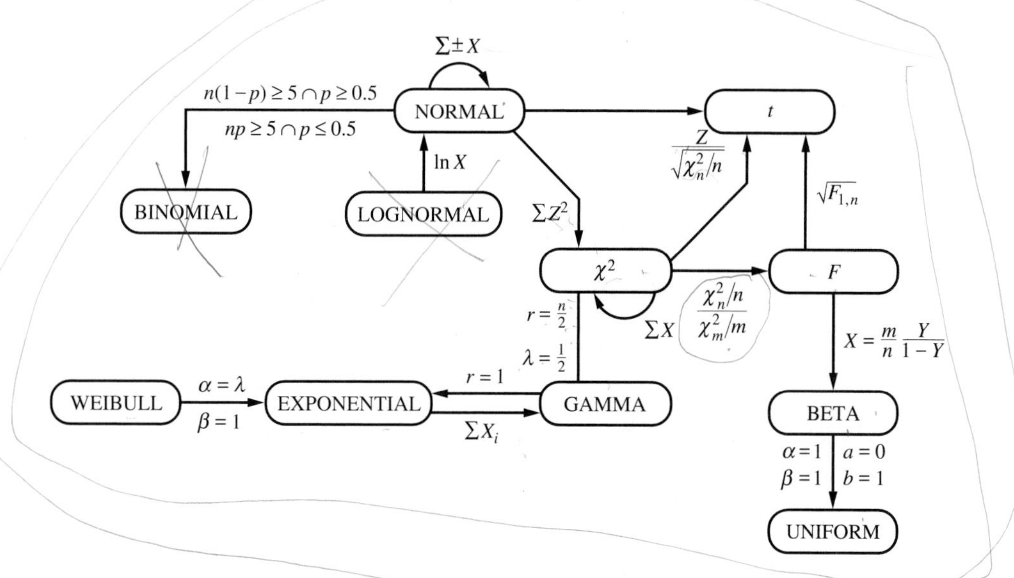

FIGURE 4-4
Relationships among the distributions of Chapter 4.

Item 4 on our list of reasons why the normal distribution is important states that many results useful in statistical work are not strictly true unless the underlying probability distribution of the characteristic being studied is a normal distribution. This statement is made primarily because several important probability distributions, *including the normal distribution itself*, are generated directly from the normal distribution. Let us first consider the various distributions that may be considered "descendants" of the normal distribution and study their relationships to the normal distribution.

As indicated in Figure 4.4, the Normal distribution generates, or replicates, itself under algebraic addition; i.e., if you take several normal random variables, X_i, with $i = 1, 2, \ldots, n$, and form weighted sums and/or differences among them, the end result will always be another normal random variable.

In general, for *any* set of multipliers and for any set of independent random variables,

$$\text{The sum } Y = \sum_{i=1}^{n} a_i X_i \quad \text{is distributed with} \quad \mu_Y = \sum_{i=1}^{n} a_i \mu_i$$

$$\text{and} \quad \sigma_Y^2 = \sum_{i=1}^{n} a_i^2 \sigma_i^2 \tag{4.5}$$

If the X_i are normally distributed, then so is Y, regardless of the size of n.

Example 4.7. Suppose X_1, X_2, and X_3 are normally distributed and statistically independent. What is the distribution of $Y = X_1 + 2X_2 - 3X_3$?

Solution. According to Equation 4.5, Y is normally distributed and has a mean equal to $\mu_1 + 2\mu_2 - 3\mu_3$ and variance equal to $\sigma_1^2 + 4\sigma_2^2 + 9\sigma_3^2$.

TABLE 4.1
Some frequently encountered continuous probability distribution functions

Probability Law	Probability Density Function		Mean	Variance
Normal	$f(x) = \frac{1}{\sigma\sqrt{2\pi}} \exp\left[-\frac{1}{2}\left(\frac{x-\mu}{\sigma}\right)^2\right]$	$-\infty < \mu < \infty$ $\sigma > 0$	μ	σ^2
χ^2	$f(x) = \frac{x^{(n/2)-1}e^{-(x/2)}}{2^{n/2}\Gamma(n/2)}$	$x > 0$ $n > 0$	n	$2n$
t	$f(x) = \frac{1}{\sqrt{\pi n}} \frac{\Gamma((n+1)/2)}{\Gamma(n/2)} \left(1 + \frac{x^2}{n}\right)^{[(n+1)/2]}$	$n > 0$	0	$\frac{n}{n-2}$ $(n>2)$
F	$f(x) = \frac{n^{n/2}m^{m/2}}{\beta(n/2, m/2)} x^{(n-2)/2}(m+nx)^{-(n+m)/2}$	$x > 0$ $m,\ n > 0$	$\frac{m}{m-2}$ $(m > 2)$	$\frac{2m^2(n+m-2)}{n(m-2)^2(m-4)}$ $(m > 4)$
Gamma	$f(x) = \frac{\lambda}{\Gamma(r)}(\lambda x)^{r-1}e^{-\lambda x}$	$x > 0$ $\lambda, r > 0$	$\frac{r}{\lambda}$	$\frac{r}{\lambda^2}$
Exponential	$f(x) = \lambda e^{-\lambda x}$	$\lambda > 0$ $x > 0$	$\frac{1}{\lambda}$	$\frac{1}{\lambda^2}$
Weibull	$f(x) = \alpha\beta x^{\beta-1}\exp(-\alpha x^{\beta})$	$x > 0$ $\alpha, \beta > 0$	$\alpha^{-1/\beta}\Gamma\left(1+\frac{1}{\beta}\right)$	$\alpha^{-\frac{2}{\beta}}\left\{\Gamma\left(1+\frac{2}{\beta}\right) - \Gamma^2\left(1+\frac{1}{\beta}\right)\right\}$
Lognormal	$f(x) = \frac{1}{\beta\sqrt{2\pi}} x^{-1}\exp\left[-(\ln x - \alpha)^2/2\beta^2\right]$	$x > 0$ $\beta > 0$	$\exp\left(\alpha + \frac{\beta^2}{2}\right)$	$\exp(2\alpha + \beta^2)\left[\exp(\beta^2) - 1\right]$
Beta	$f(x) = \frac{\Gamma(\alpha+\beta)}{\Gamma(\alpha)\Gamma(\beta)} x^{\alpha-1}(1-x)^{\beta-1}$	$0 < x < 1$ $\alpha, \beta > 0$	$\frac{\alpha}{\alpha+\beta}$	$\frac{\alpha\beta}{(\alpha+\beta)^2(\alpha+\beta+1)}$
Uniform	$f(x) = \frac{1}{b-a}$	$a \le x \le b$ $-\infty < a < b < \infty$	$\frac{a+b}{2}$	$\frac{(b-a)^2}{12}$

A *chi-square random variable with n degrees of freedom*, χ_n^2, is governed by the following probability density function.

$$f(x) = \frac{x^{n/2-1}e^{-x/2}}{2^{n/2}\Gamma(n/2)} \qquad x > 0, \quad n > 0^*$$

One way to form a chi-square random variable is by adding squared $N(0, 1)$ random variables. Thus if the X_i, $i = 1, \ldots, n$, are $N(\mu_i, \sigma_i)$, then

$$\chi_n^2 = \sum_{i=1}^{n}\left(\frac{X_i - \mu_i}{\sigma_i}\right)^2 \tag{4.6}$$

---

*The *gamma function*, which appears in the denominator of the chi-square probability density function, is defined as

$$\Gamma(u) = \int_0^{\infty} w^{u-1}e^{-w}\,dw \qquad u > 0$$

and has the following properties.

(a) $\Gamma(J + 1) = j! \qquad j \ge 0$ and integer

(b) $\Gamma\left(\frac{j}{2}\right) = \left(\frac{j}{2} - 1\right)\left(\frac{j}{2} - 2\right)\cdots\left(\frac{3}{2}\right)\left(\frac{1}{2}\right)\sqrt{\pi} \qquad j > 2$, odd, and integer

TABLE 4.1
(Continued)

β_1	β_2
0	3
$\frac{8}{n}$	$3\left(\frac{4}{n}+1\right)$
0	$\frac{6}{n-4}+3 \quad (n>4)$
$\frac{8(2n+m-2)^2(m-4)}{n(m-6)^2(n+m-2)} \quad (m>6)$	$\frac{(m-2)^3(m-4)(n+6)(n+4)(n+2)}{4(m-6)(m-8)(n+m-2)^2n^2} - \frac{8(m-4)(2n+m-2)}{(m-6)(n+m-2)} - \frac{3n(m-4)}{(n+m-2)} - \frac{n^2(m-4)}{4(n+m-2)^2}$ $(m>8)$
$\frac{4}{r}$	$\frac{6}{r}+3$
4	9
$\left\{\Gamma\left(1+\frac{3}{\beta}\right)-3\Gamma\left(1+\frac{1}{\beta}\right)\Gamma\left(1+\frac{2}{\beta}\right)+2\Gamma^3\left(1+\frac{1}{\beta}\right)\right\}^2$	$\Gamma\left(1+\frac{4}{\beta}\right)-4\Gamma\left(1+\frac{1}{\beta}\right)\Gamma\left(1+\frac{3}{\beta}\right)+6\Gamma^2\left(1+\frac{1}{\beta}\right)\Gamma\left(1+\frac{2}{\beta}\right)-3\Gamma^4\left(1+\frac{1}{\beta}\right)$
$\left[\Gamma\left(1+\frac{2}{\beta}\right)-\Gamma^2\left(1+\frac{1}{\beta}\right)\right]^3$	$\left[\Gamma\left(1+\frac{2}{\beta}\right)-\Gamma^2\left(1+\frac{1}{\beta}\right)\right]^2$
$\left[\exp\left(\beta^2\right)-1\right]\left[\exp\left(\beta^2\right)+2\right]^2$	$\left[\exp\left(\beta^2\right)-1\right]\left[\exp\left(3\beta^2\right)+3\exp\left(2\beta^2\right)+6\exp\left(\beta^2\right)+6\right]+3$
$\frac{4(\beta-\alpha)^2(\alpha+\beta+1)}{\alpha\beta(\alpha+\beta+2)^2}$	$\frac{3(2\alpha^2+\alpha^2\beta-2\alpha\beta+\alpha\beta^2+2\beta^2)(\alpha+\beta+1)}{\alpha\beta(\alpha+\beta+2)(\alpha+\beta+3)}$
0	1.8

Note that the number of terms in the sum determines the degrees of freedom. Further, there is a separate and unique chi-square probability distribution for each value of n. For smaller degrees of freedom, the chi-square distribution has the form pictured in Figure 4.5. The chi-square distribution approaches symmetry only for relatively large ($n \geq 30$) degrees of freedom. The χ^2 distribution also *replicates* itself but only under the *positive* addition of statistically independent χ^2 random variables; i.e., $\chi_6^2 = \chi_1^2 + \chi_2^2 + \chi_3^2$, if χ_1^2, χ_2^2, and χ_3^2 are statistically independent. In general,

$$\chi_N^2 = \sum_{i=1}^{n} \chi_{\nu_i}^2 \qquad \text{where} \qquad N = \sum_{i=1}^{n} \nu_i \tag{4.7}$$

for statistically independent $\chi_{\nu_i}^2$.

As we see in Chapter 9, the $N(\mu, \sigma)$ distribution is used to test conjectures about the mean or average value of a random variable, and the χ^2 distribution is used to test conjectures about the variance or dispersion of the probability distribution function of a normal random variable.

A table of the χ^2 distribution is presented in Table A.4. The left marginal column gives the value of the degrees of freedom parameter, ν, associated with each row, and the top marginal row gives the right-hand-tail probability, α, associated with each column. Referring to the descriptive figure at the top of the table, we see that the numbers in the body of the table give the values of a χ^2 random variable with ν degrees of freedom, which has α probability to its right.

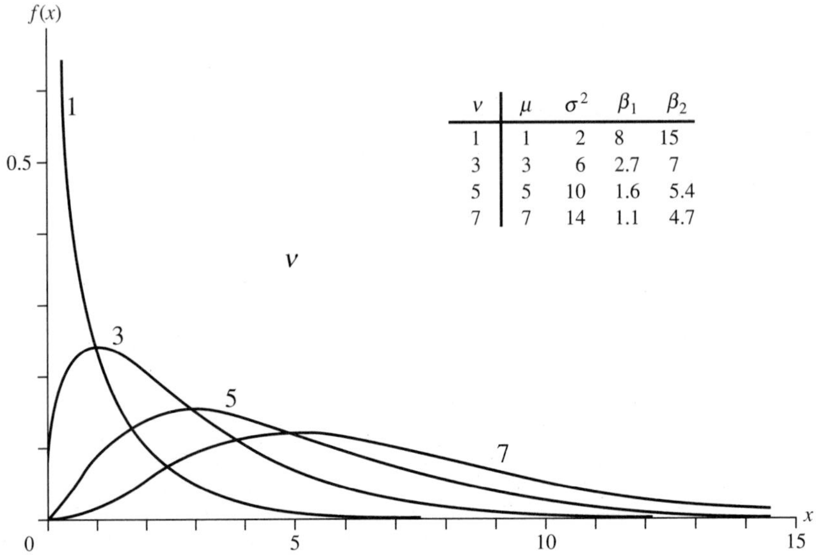

FIGURE 4-5
Representative χ^2 distribution plots with low degrees of freedom.

Example 4.8. What is the probability that a χ^2 random variable with 12 degrees of freedom will achieve a value of at least 23.337? What value of a χ^2 random variable with 17 degrees of freedom will be equaled or exceeded only 5% of the time?

Solution. From Table A.4, we may observe that a χ^2 random variable with 12 degrees of freedom will achieve a value of 23.337 *or greater* only 2.5% of the time. The inverse table lookup implied by the second question is equally straightforward. Again using Table A.4 we find that the answer to the question is 27.587.

However, Table A.4 has a serious limitation. Only eight values of the right-hand-tail probability are given in Table A.4, and they are situated in the far left- and right-hand tails of the distribution. If you desire information that requires an α value anywhere in the interval from 0.95 to 0.05, the table will not provide you with the required answers. The chi-square evaluator program in your software package does not have this limitation and performs forward and inverse table lookups for arbitrary values of ν and α.

Example 4.9. An environmentally sound insecticide for mosquitoes is dispersed in a mist from a centrally located fogger. The dispersion index for a single application is distributed according to a chi-square distribution with 3 degrees of freedom. If the effect on the dispersion index of repeated applications is strictly cumulative, what is the probability that the index will exceed 22.5 after four applications?

Solution. Since the effect on the index is cumulative, we can use Equation 4.7 to conclude that, after four applications, the dispersion index will be governed by a χ^2_{12} distribution. Using the chi-square distribution evaluator program, we find that $P(\chi^2_{12} > 22.5) = 0.0323$.

The probability density function for *Student's t distribution* is

$$f(x) = \frac{1}{\sqrt{\pi n}} \frac{\Gamma((n+1)/2)}{\Gamma(n/2)} \left(1 + \frac{x^2}{n}\right)^{-[(n+1)/2]} \qquad -\infty < x < \infty, n > 0$$

A random variable governed by a t distribution with n degrees of freedom* can be generated by forming the ratio of an $N(0, 1)$ random variable and the square root of a statistically independent χ^2 random variable divided by its degrees of freedom; i.e.,

$$T_n = \frac{Z}{\sqrt{\chi_n^2/n}} \tag{4.8}$$

Note that the degrees of freedom for the t distribution are taken from the χ^2 variable in the denominator. As before, each different value of n yields a unique and different probability distribution function.

One use of the t distribution is to test hypothetical statements about the mean of a normal random variable when the *variance of the distribution is unknown* and the *sample size is small*. Like the normal distribution, the t distribution is *symmetric* about its mean and is usually compared to the standardized normal distribution, $N(0, 1)$. At lower values of n, the t distribution is flatter, possesses a larger variance, and has fatter tails than the $N(0, 1)$ distribution. In the limit, as n becomes large, the t distribution approaches $N(0, 1)$. For most practical purposes, the $N(0, 1)$ and t distributions are approximately the same for $n \geq 30$. Figure 4.6 pictures the relationship between the t and the $N(0, 1)$ distributions for small values of n.

The t distribution table is presented in Table A.5. Once more, the marginal column gives each row's degrees of freedom, and the marginal top row gives the tabled values of right-hand-tail probabilities. In this case, five possibilities are presented, ranging from 0.1 to 0.005. Because of the symmetry of the t distribution, we may easily access five more values of the right-hand tail corresponding to the complements of the right-hand-tail probabilities given explicitly in the table; i.e., $\alpha = 0.995, 0.99, 0.975, 0.95$, and 0.9.

Example 4.10. Find the probability that a random variable governed by a t distribution with 14 degrees of freedom will exceed a value of (a) 2.145 and (b) -2.145.

Solution. (a) Looking at Table A.5 at 14 degrees of freedom, we see that the probability is 0.025. (b) By symmetry, the probability is $1 - 0.025 = 0.975$.

Between 0.1 and 0.9, α values are not accessible through Table A.5. In addition, the use of Table A.5 will frequently require you to perform linear interpolation to obtain desired results. The t distribution evaluator program will perform forward and inverse table lookups for *any* specified values of α and ν.

---

*In deference to traditional statistical symbology, the t distribution is referred to with a lowercase symbol. The random variable governed by the t distribution is denoted T_n.

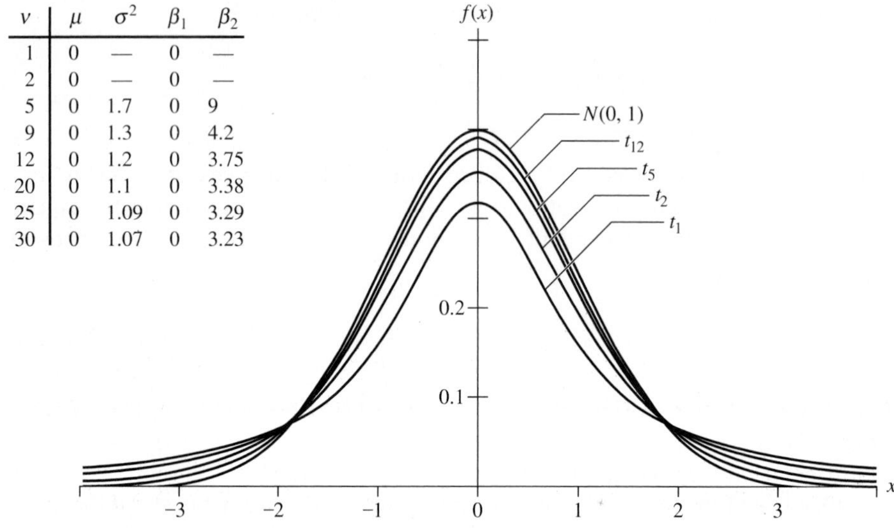

ν	μ	σ^2	β_1	β_2
1	0	—	0	—
2	0	—	0	—
5	0	1.7	0	9
9	0	1.3	0	4.2
12	0	1.2	0	3.75
20	0	1.1	0	3.38
25	0	1.09	0	3.29
30	0	1.07	0	3.23

FIGURE 4-6
The relationship between the standard normal distribution and the t distribution.

Example 4.11. A random variable X is governed by an $N(10, 1)$ distribution. What is $P(X > \sqrt{\chi_9^2} + 10)$?

Solution. One way to answer a question like this is to perform algebraic manipulations on the *argument*, i.e., $X > \sqrt{\chi_9^2} + 10$, to obtain (if possible) a mathematical form that possesses a known distribution. Once found, the appropriate evaluator may be used to obtain the required probability. Subtracting 10 from both sides of the argument and dividing by 3 yield

$$\frac{1}{3}(X - 10) > \frac{\sqrt{\chi_9^2}}{3}$$

which implies (after division by $\sigma = 1$)

$$\frac{1}{3}Z > \sqrt{\frac{\chi_9^2}{9}}$$

and yields

$$T_9 = \frac{Z}{\sqrt{\chi_9^2/9}} > 3$$

Using the t distribution evaluator program, we find that the probability that a t random variable with 9 degrees of freedom will exceed 3 is 0.0075. Therefore, $P(X > \sqrt{\chi_9^2} + 10) = 0.0075$.

The F *distribution* is governed by

$$f(x) = \frac{n^{n/2} m^{m/2}}{\beta(n/2, m/2)} x^{(n-2)/2} (m + nx)^{-(n+m)/2} \qquad x, m, n > 0$$

where

$$\beta(a, b) = \frac{\Gamma(a)\Gamma(b)}{\Gamma(a+b)}$$

One way to construct an F random variable is to take the ratio of two *independent* χ^2 random variables, both divided by their respective degrees of freedom; i.e.,

$$F_{n,m} = \frac{\chi_n^2/n}{\chi_m^2/m} \tag{4.9}$$

Note that the F distribution has *n degrees of freedom in the numerator* and *m degrees of freedom in the denominator*. Each different possible pairing of the values of n and m yields a *unique* and different probability distribution. Figure 4.7 presents plots of several selected F distributions.

One application of the F distribution is in *comparing the variances* taken from two different normal distributions. This is considered in detail in Chapter 10.

The F distribution is somewhat more difficult to use because the values of *two parameters,* n and m, must be considered. The effect of the one additional parameter, compared to the t and χ^2 distributions, is dramatic. Now a *separate table* is required for each selected value of α. Table A.6*a* gives, in the body of the table, the values of the F random variable with ν_1 and ν_2 degrees of freedom associated with a right-hand-tail probability of 0.05. Notice that ν_1 gives the

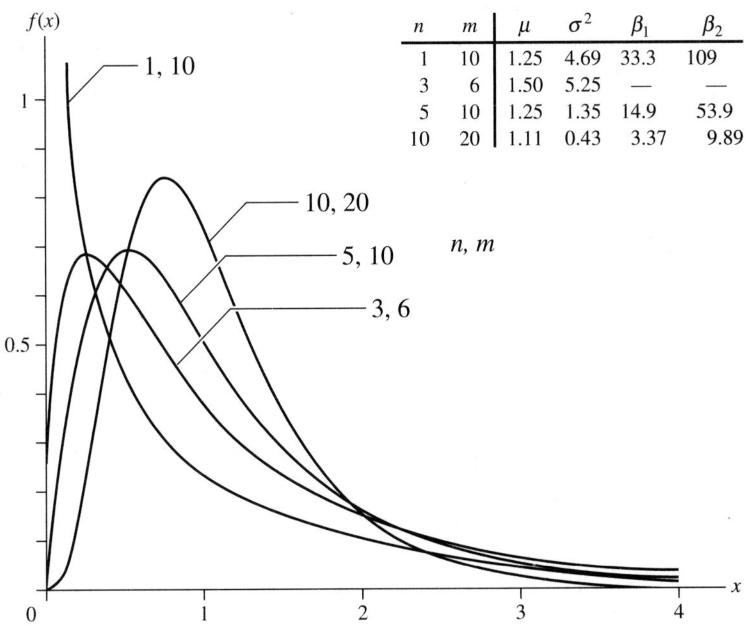

n	m	μ	σ^2	β_1	β_2
1	10	1.25	4.69	33.3	109
3	6	1.50	5.25	—	—
5	10	1.25	1.35	14.9	53.9
10	20	1.11	0.43	3.37	9.89

FIGURE 4-7
Plots of selected F distributions.

degrees of freedom for the numerator and ν_2 gives the degrees of freedom for the denominator. For example, the probability that an F random variable with 15 and 17 degrees of freedom will equal or exceed a value of 2.31 is 0.05. The value of 2.31 is obtained by reading the number that resides at the intersection of the column labeled 15 and the row labeled 17.

Table A.6*b* gives the F distribution for a right-hand-tail probability of 0.01. Once again, only selected values of the degrees of freedom parameters are given. Should you desire a value for an F variable whose degrees of freedom do not explicitly appear in the table, you will be forced to perform at least one interpolation to ascertain an approximation of your required value. If neither parameter value appears, you must perform a *double interpolation*; i.e., obtaining the F value associated with a right-hand tail of 0.05 for $\nu_1 = 27$ and $\nu_2 = 27$ degrees of freedom would require such a procedure. Further, Table A.6 has only two right-hand-tail values, 0.05 and 0.01. The F distribution evaluator program in your software package allows arbitrary values for n, m, and the right-hand-tail value.

Example 4.12. The strengths, X_S, of a particular type of tensile member in a bridge truss are distributed according to a chi-square distribution with 9 degrees of freedom, χ_9^2. Once the bridge is built, the stresses, X_s, that a member experiences will be distributed according to a chi-square distribution with 4 degrees of freedom, χ_4^2. How likely is such a member to fail owing to any single stress?

Solution. What we desire to find in this problem is the probability that the stress will exceed the strength. One way to express this mathematically is to use the ratio of stress to strength and ask, What is $P((X_s/X_S) > 1) = P((\chi_4^2/\chi_9^2) > 1)$? Dividing each chi-square variable by its respective degrees of freedom yields

$$P\left(\frac{\chi_4^2}{\chi_9^2} > 1\right) = P\left(\frac{\chi_4^2/4}{\chi_9^2/9} > \frac{9}{4}\right) = P\left(F_{4,9} > \frac{9}{4}\right) = 0.1436$$

The probability that a member will fail owing to a single stress is 0.1436. The tensile members currently selected appear to be too weak to bear the projected stresses.

The *Gamma distribution*,

$$f(x) = \frac{\lambda}{\Gamma(r)}(\lambda x)^{r-1}e^{-\lambda x} \qquad x, \lambda, r > 0$$

in its most general form, is not directly generated from the normal distribution. However, if we set $\lambda = \frac{1}{2}$ and $r = \frac{n}{2}$, the gamma distribution becomes identically the χ^2 distribution; i.e., the χ^2 distribution is a special case of the gamma distribution. The gamma distribution with integer values of the parameter, r, may be viewed as the result of adding r *independent, identically distributed* exponential random variables. That is, if $X_i = 1, 2, \ldots, n$ are distributed as independent exponential random variables, all with the same parameter value of λ, then

$$Y = \sum_{i=1}^{n} X_i \tag{4.10}$$

is distributed as a gamma random variable with parameters λ and $r = n$. In addition, the *exponential distribution* is a special case of the gamma distribution with $r = 1$.

The gamma and the exponential distributions have many applications in the area of *reliability theory*. The exponential distribution is often called the *distribution of decay*, since it may be used to characterize the proportion of a population still surviving, after a certain period of time, under the condition of a constant failure rate, λ. The "average time between failures" can be shown to be the reciprocal of the failure rate.

The exponential distribution,

$$f(x) = \lambda e^{-\lambda x} \qquad x, \lambda > 0$$

has the property of being memoryless. This property means that, if the time to failure, T, is an exponential random variable, then the probability of T being less than t minutes—given that it has already lasted exactly τ minutes—is equal to the probability of T being less than $t - \tau$ minutes when the experiment has just begun. Mathematically, this property is stated as $P(T \leq t \mid T \geq \tau) = P(T \leq t - \tau \mid T \geq 0)$. Figure 4.8 shows the gamma distribution with $\lambda = 1$ for several values of r. (The exponential distribution is also pictured, where $r = 1$.)

Example 4.13. Two components form a standby redundant system. Component A is placed in service first. When it fails, Component B is immediately brought on-line by the process control computer. The time-to-failure density functions for the components are both gamma distributions with $\lambda = 2$ failures per month, where $r_A = 3$ and $r_B = 5$. What is the time-to-failure distribution of the system? What is the probability that the system will last for at least 10 months?

Solution. By Equation 4.10, the distribution of system time to failure is gamma with $\lambda = 2$ and $r = 8$ (the sum of eight independent, identically distributed exponential random variables). To answer the second part of the problem, we need to find the area under the curve of the system time-to-failure distribution from 10 to $+\infty$. Substituting $\lambda = 2$ and $r = 8$ into the equation for the gamma distribution given in Table 4.1 and forming the integral, we use the gamma distribution evaluator program to obtain

$$\int_{10}^{\infty} \frac{2}{\Gamma(8)}(2x)^7 e^{-2x}\, dx = 0.0008$$

The probability that the system will last for 10 months or more is 0.0008.

The *Weibull distribution*,

$$f(x) = \alpha\beta x^{\beta-1} e^{-\alpha x^{\beta}} \qquad x, \alpha, \beta > 0$$

is also very popular in the area of reliability theory. It is a "robust" distribution that can represent all three types of component failure—the early burnout, the chance failure, and the wearout modes. By the addition of a single parameter, it

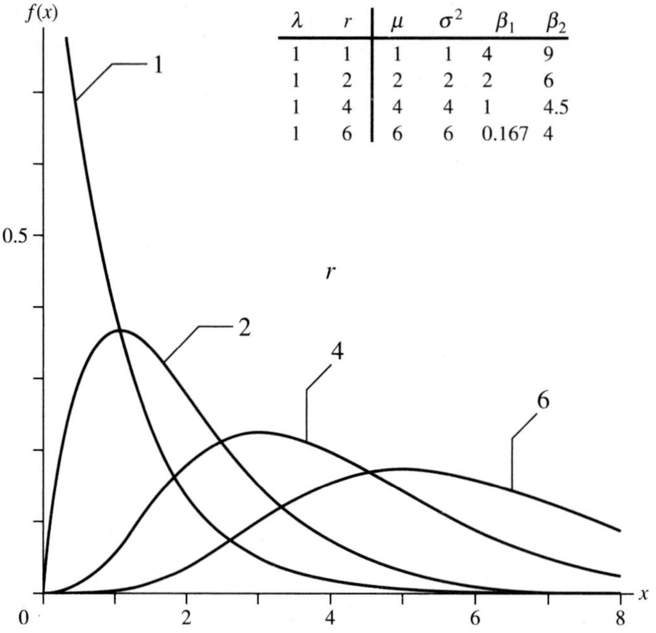

FIGURE 4-8
The gamma distribution for small r.

is also possible to have the random variable bounded below by a nonzero value. As we have seen in Figure 4.4, the exponential distribution is a special case of the Weibull distribution when $\alpha = \lambda$ and $\beta = 1$.

Example 4.14. The index of waste product dilution in a secondary treatment process follows a Weibull distribution with $\alpha = 1$ and $\beta = 3$. What is the probability that the index will be less than 0.5 in secondary treatment?

Solution. To answer this question, we use the Weibull distribution evaluator program to compute the area under the curve from zero to 0.5; i.e.,

$$\int_0^{0.5} 3x^2 \exp(-x^3) = 0.1175$$

Thus the probability is 0.1175 that the index will be less than 0.5.

The *lognormal distribution* is

$$f(x) = \frac{1}{\beta\sqrt{2\pi}} x^{-1} e^{-[(\ln x - \alpha)^2/2\beta^2} \qquad x, \beta > 0$$

A lognormal random variable is simply a random variable whose natural logarithm, ln (log base e), is distributed as a normal random variable; i.e.,

$$Y = \ln X \tag{4.11}$$

is distributed normally if X is distributed lognormally. The parameters, α and β, are the mean and standard deviation of $Y = \ln X$, respectively. Typical lognormal distributions are seen in Figure 4.9.

Example 4.15. The bacteria index level in the flow to a primary sewage treatment plant follows a lognormal distribution with $\alpha = 2$ and $\beta = 1$. What is the probability that, in a random measurement, the index level will exceed 12.5?

Solution. The easiest way to answer this question is to use the lognormal distribution evaluator program to integrate the lognormal density function (with $\alpha = 2$ and $\beta = 1$) between the limits 0 to 12.5 and to subtract the result from 1.0. However, we may also use the normal distribution evaluator program and take advantage of the monotonic relationship between X and $\ln X$; i.e.,

$$P(X \geq 12.5) = P(\ln X \geq \ln 12.5) = P(Y \geq 2.5257)$$

Since $Y = \ln X$ is a normal random variable with $\mu = 2$ and $\sigma = 1$, we may use the normal distribution evaluator program to determine that

$$P(Y \geq 2.5257) = P(X \geq 12.5) = 0.2995$$

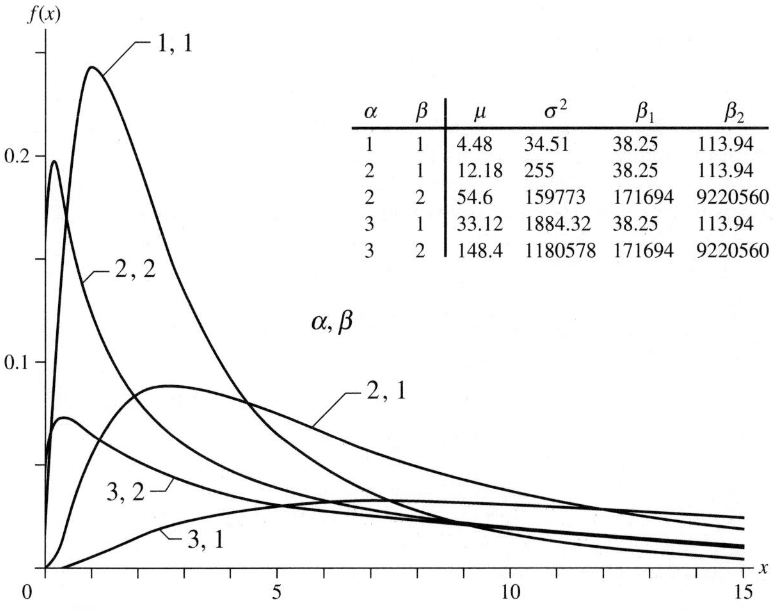

α	β	μ	σ^2	β_1	β_2
1	1	4.48	34.51	38.25	113.94
2	1	12.18	255	38.25	113.94
2	2	54.6	159773	171694	9220560
3	1	33.12	1884.32	38.25	113.94
3	2	148.4	1180578	171694	9220560

FIGURE 4-9
The lognormal distribution for selected values of α and β.

The *uniform distribution*

$$f(x) = \frac{1}{b-a} \qquad a \leq x \leq b$$

is often called the *distribution of no knowledge* because it implies that all that is known about a random variable is that it will take a value between two specified limits. However, it is *equally likely* to assume any value between those limits.

The *beta distribution* is defined by

$$f(x) = \frac{\Gamma(\alpha+\beta)}{\Gamma(\alpha)\Gamma(\beta)} x^{\alpha-1}(1-x)^{\beta-1} \qquad 0 < x < 1, \quad \alpha, \beta > 0$$

If $\alpha = 1$ and $\beta = 1$, the beta distribution becomes equivalent to the uniform distribution with $a = 0$ and $b = 1$, with the single difference that the limiting values of 0 and 1 are not included in the defined range of X. The beta distribution can also be used to generate the *generalized beta distribution*, which can lie between any two selected points. That is, if X is a beta random variable, then $Y = a + (b - a)X$ is a generalized beta variable lying between the values of $a < Y < b$. The beta distribution is very popular because it is *bounded* by two finite values; it is very rich in its ability to assume different shapes according to the value of the parameters α and β. If Y is a beta random variable with $\alpha = n/2$ and $\beta = m/2$, the transformation $X = (m/n)[(Y/1 - Y)]$ yields X as an F random variable with n and m degrees of freedom.

Example 4.16. The proportion of dissolved salts in well water in Smalltown, Texas, is known to follow a beta distribution with $\alpha = 1$ and $\beta = 10$. What is the probability that the water pumped from the well will have between 1% and 30% dissolved salts?

Solution. Substituting $\alpha = 1$ and $\beta = 10$ into the beta density function, we obtain $f(x) = 10(1 - x)^9$. Integrating $f(x)$ over the limits from 0.01 to 0.3 yields the desired answer of 0.876. This is easily verified using the beta distribution evaluator program.

4.4 CONTINUOUS BIVARIATE PROBABILITY DENSITY FUNCTIONS

As in Chapter 3, the last section of the chapter is reserved for the discussion of bivariate probability distribution functions.

Example 4.17. The *joint* probability density function of X_1 and X_2, $f(x_1, x_2) = 1$, with $0 \leq x_1 \leq 1$ and $0 \leq x_2 \leq 1$ (pictured in Figure 4.10), is a bivariate generalization of the uniform distribution given in Table 4.1. In order to be a valid continuous bivariate probability density function, $f(x_1, x_2)$ must be defined such that $f(x_1, x_2) \geq 0$ for all defined (x_1, x_2), and $\iint_{\text{all } x_1, x_2} f(x_1, x_2)\, dx_1 dx_2 = 1$. These properties are satisfied by the $f(x_1, x_2)$ of Figure 4.10.

As before, the prior probability associated with any *point* (x_1, x_2), is zero. Only contiguous *areas* in the $x_1 - x_2$ plane have nonzero probabilities. The prob-

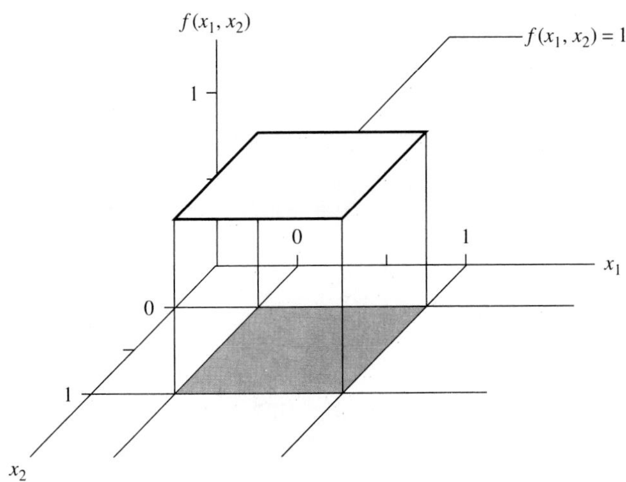

FIGURE 4-10
$f(x_1, x_2)$ for Example 4.17.

ability associated with such an area is equal to the *volume* of probability mass that resides over that contiguous area and is bounded above by the surface defined by the bivariate probability density function.

Example 4.18. For the $f(x_1, x_2)$ of Example 4.17, what is $P(0.25 \leq X_1 \leq 0.75 \cap 0.3 \leq X_2 \leq 0.5)$?

Solution. An analytic solution to this question is provided by performing a double integration of $f(x_1, x_2)$ over the area defined by the preceding question; i.e.,

$$\int_{0.3}^{0.5}\int_{0.25}^{0.75} f(x_1, x_2)\,dx_1 dx_2 = \int_{0.3}^{0.5}\int_{0.25}^{0.75} (1)\,dx_1 dx_2 = 0.1$$

The physical interpretation associated with this double integral is pictured in Figure 4.11.

The *cumulative bivariate probability density function,* $F(x_1, x_2)$, is defined as

$$F(x_1, x_2) = P(X_1 \leq x_1 \cap X_2 \leq x_2) \tag{4.12}$$

which is read the *probability that* X_1 *is less than or equal to* x_1 and X_2 *is less than or equal to* x_2.

Example 4.19. Find $F(0.6, 0.35)$ for the $f(x_1, x_2)$ of Example 4.17.

Solution. Once more, a double integral is required.

$$\int_{-\infty}^{0.35}\int_{-\infty}^{0.6} f(x_1 x_2)\,dx_1\,dx_2 = \int_{0}^{0.35}\int_{0}^{0.6} (1)\,dx_1\,dx_2 = 0.21.$$

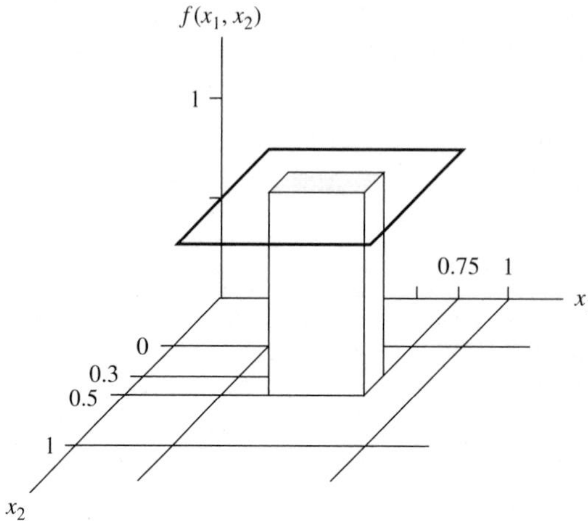

FIGURE 4-11
The probability volume for Example 4.18.

The *marginal probability distribution function* of X_1, $f_{X_1}(x_1)$, is a univariate probability distribution function derived from $f(x_1, x_2)$ by accumulating the probabilities over all values of x_2. Mathematically,

$$f_{X_1}(x_1) = \int_{\text{all } x_2} f(x_1, x_2)\, dx_2$$

Similarly,

$$f_{X_2}(x_2) = \int_{\text{all } x_1} f(x_1, x_2)\, dx_1 \tag{4.13}$$

Example 4.20. Obtain the marginal probability distributions of X_1 and X_2 for $f(x_1, x_2) = \exp[-(x_1 + x_2)]$, with $x_1, x_2 > 0$.

Solution

$$f_{X_1}(x_1) = \int_0^\infty f(x_1, x_2)\, dx_2 = \int_0^\infty \exp[-(x_1 + x_2)]\, dx_2 = \exp(-x_1)$$

In like manner, $f_{X_2}(x_2) = \exp(-x_2)$. Both of these marginal density functions are valid univariate probability density functions.

The random variables, X_1 and X_2, are *statistically independent* if and only if

$$f(x_1, x_2) = f_{X_1}(x_1) f_{X_2}(x_2) \tag{4.14}$$

This means that the product of the marginal probability density functions must equal the joint bivariate probability density function. If this *not* true, we must conclude that X_1 and X_2 are not statistically independent.

Example 4.21. Determine whether X_1 and X_2 for the joint density of Example 4.20 are statistically independent.

Solution. Since $f_{X_1}(x_1)f_{X_2}(x_2) = \exp(-x_1)\exp(-x_2) = \exp[-(x_1 + x_2)] = f(x_1, x_2)$, X_1 and X_2 are statistically independent.

Example 4.22. Suppose that $f(x_1, x_2) = x_1 + x_2, 0 \leq x_1 \leq 1$ and $0 \leq x_2 \leq 1$, and suppose that we know that X_1 will assume a value between 0.5 and 0.75. How does that knowledge affect the probabilities associated with X_2? That is, what is the conditional probability density function of X_2 given that $0.5 \leq X_1 \leq 0.75$?

Solution. As may be seen in Figure 4.12, the bivariate sample space has been *reduced* and we need to consider only those (x_1, x_2) that are shaded in Figure 4.12 (where $0.5 \leq X_1 \leq 0.75$). Just as with discrete bivariate random variables, we may use the techniques of conditional probability from Chapter 2. Thus,

$$\begin{aligned} f(x_2 \mid 0.5 \leq X_1 \leq 0.75) &= \frac{P(x_2 \cap 0.5 \leq X_1 \leq 0.75)}{P(0.5 \leq X_1 \leq 0.75)} \\ &= \frac{\int_{0.5}^{0.75}(x_1 + x_2)\,dx_1}{\int_0^1 \int_{0.05}^{0.75}(x_1 + x_2)\,dx_1\,dx_2} \\ &= \frac{5}{9} + \frac{8}{9}x_2 \end{aligned}$$

Example 4.23. Suppose that we again have the joint density of Example 4.22, but we now know that $X_1 = 0.5$. What is the conditional probability density of X_2 *given* that $X_1 = 0.5$?

Solution

$$f_{X_1}(x_1) = \int_0^1 (x_1 + x_2)\,dx_2 = x_1x_2 + \left.\frac{x_2^2}{2}\right|_0^1 = x_1 + 0.5$$

Hence,

$$f(x_2 \mid X_1 = 0.5) = \frac{f(X_1 = 0.5, x_2)}{f(X_1 = 0.5)} = \frac{0.5 + x_2}{0.5 + 0.5} = 0.5 + x_2$$

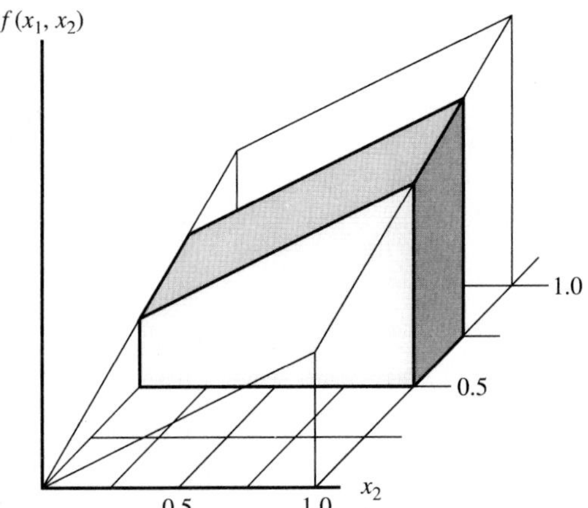

FIGURE 4-12
The joing density function, $f(x_1, x_2) = x_1 + x_2$.

When one of the variables is *known with certainty*, the conditional density function of the other random variable is obtained by taking the ratio of the joint density function and the marginal density function of the given variable and substituting the known value throughout the ratio.

Obviously, there is much more in the area of bivariate and multivariate distributions that could be included in the last sections of Chapters 3 and 4. However, much of that material is beyond the scope of this book, and the material that is presented is more than sufficient for the uses of bivariate random variables that you will encounter later in the text. Likewise, in regard to *univariate* random variables, there are many other named probability distributions that could have been presented. However, those that have been presented are the most important for the purposes of this book. If you understand them and their uses, the other distributions that you might encounter should give you little or no difficulty.

It is strongly suggested that you pay particular attention to the exercises at the end of this chapter. It is only in the performing of those exercises that you will truly begin to grasp the versatility and power of application that is contained in the distributions of Table 4.1 and Figure 4.4.

EXERCISES

4.1. Suppose that a continuous random variable X has the following probability density function.

$$f(x) = \begin{cases} 2x & 0 \le x \le 1 \\ 0 & \text{otherwise} \end{cases}$$

(a) Find $P(X \le 0.5)$.
(b) Find $P(0.25 \le X \le 0.75)$.
(c) Find $P(0.25 \le X \le 4)$.

4.2. Suppose that a continuous random variable X has the following probability density function.

$$f(x) = \begin{cases} x - 6 & 6 \le x < 7 \\ 8 - x & 7 \le x \le 8 \\ 0 & \text{otherwise} \end{cases}$$

(a) Find $P(6.5 \le X \le 7.5)$.
(b) Find $P(X \le 6.5)$.
(c) Find $P(X \ge 7)$.

4.3. Suppose that a continuous random variable X has the following probability density function.

$$f(x) = \begin{cases} Ke^{-0.25x} & x > 0 \\ 0 & \text{elsewhere} \end{cases}$$

(a) Find the value of K for which $f(x)$ is a valid probability density function.
(b) Find $P(2 \le X \le 8)$.
(c) Find $P(X \ge 7)$.
(d) Find $P(X \le 1)$.

(e) Find the number c such that $P(X \geq c) = 0.05$.
(f) Find the number d such that $P(X \leq d) = 0.025$.

4.4. Let Y be normally distributed with $\mu = 3$ and $\sigma = 2$. Suppose that $Y = -12$ corresponds to $Z = a$, where Z is the standard normal random variable. What is the value of a?

4.5. Suppose that Z is the standard normal random variable.
(a) Find $P(Z \geq 2.63)$.
(b) Find $P(Z \leq -1.72)$.
(c) Find $P(Z \leq 1.28)$.
(d) Find $P(-1.54 \leq Z \leq 2.66)$.
(e) Find the number c such that $P(Z \geq c) = 0.0096$.
(f) Find the number d such that $P(Z \leq d) = 0.9082$.
(g) Find the values a and b such that $P(Z < a) + P(Z \geq b) = 0.01$ and $P(Z \leq a) = P(Z \geq b)$.

4.6. Suppose that the random variable X is normally distributed with $\mu = 14$ and $\sigma = 4$.
(a) Find $P(8 \leq X \leq 18)$.
(b) Find $P(-1 \leq X < 10)$.
(c) Find the number c such that $P(X \geq c) = 0.05$.
(d) Find the numbers a and b such that $P(X \geq b) + P(X \leq a) = 0.05$ and $P(X \geq b) = P(X \leq a)$.

4.7. If the height of the adult population of the United States is a normal random variable with $\mu = 67$ inches and $\sigma = 3$ inches, what percentage of U.S. adults are shorter than 75 inches?

4.8. If the diameters of ball bearings produced by a certain machine are normally distributed with $\mu = 0.6105$ inches and $\sigma = 0.0026$ inches, what percentage of the ball bearings produced by this machine have diameters greater than 0.6160 inches?

4.9. The length of a drive shaft is made up of three parts, A, B, and C. The lengths (in inches) of A, B, and C are statistically independent and normally distributed with the following means and variances.

	μ	σ^2
A	3	0.0025
B	5	0.0049
C	4	0.0036

What is the probability that the total length, $Y = A + B + C$, will exceed 12.2 inches?

4.10. In early prototype production, the probability of a good unit is 0.65. How likely is it that a sample of 200 will provide between 110 and 135 good units?

4.11. The Acme Air Drop Company drops supplies to arctic oil exploration teams. Drop locations are marked by big orange X's on the ice. Based on extensive experience, it is known that drops vary from the target in the east-west direction in accordance with a normal distribution with $\mu = 0$ and $\sigma = 3$ yards. Further, variations from the target in the north-south direction are also governed by a normal distribution with $\mu = 0$ and $\sigma = 3$ yards. If the deviations in the two directions are statistically independent, what probability distribution governs the distance by which any drop will miss the target?

4.12. Suppose that the load, in tons, to which a machine member will be subjected is distributed according to a χ^2 distribution with 3 degrees of freedom. Further, the strength, in tons, of the machine member is distributed according to a χ^2 distribution with 2 degrees of freedom. What specific distribution should be used to characterize the ratio of load to strength? How would you determine the probability that a machine member selected at random will survive?

4.13. The service time, T_i, for a person in a grocery line is distributed according to an exponential distribution with parameter $\lambda = 5$ customers per hour. What distribution governs the time you must wait to begin to be served if 3 customers are in front of you? What is the probability that you must wait more than $\frac{1}{2}$ hour?

4.14. The probability that a car will enter the flow of traffic is related to the gap between 2 adjacent cars in the stream of traffic nearest the entering car. Suppose that the probability of gap acceptance (entering the traffic) is governed by a lognormal distribution on T, the number of seconds of gap, where the mean of $\ln T$ is 1, and the variance of $\ln T$ is 0.3^2. What length of gap will be accepted 90% of the time?

4.15. The proportion of people that are infected by a certain influenza strain in any previously uninfected population is governed by a beta distribution with $\alpha = 4$ and $\beta = 2$. What is the probability that more than 40% of any new population will be infected?

4.16. The time-to-failure density function, $f(t)$, of a particular type of ball bearing is a Weibull distribution with $\alpha = 1$ and $\beta = 4$. At what number of units of time do we expect 80% of all such bearings to fail?

4.17. Show that the memoryless property of the exponential distribution, $P(T \leq t \mid T \geq \tau) = P(T \leq t - \tau)$, is true.

4.18. The average time between buses at the corner of Elm and 42nd is approximately governed by a normal distribution with a mean of 25 minutes and a standard deviation of 12 minutes. Find the probability that two adjacent bus arrivals will be separated by more than 22 minutes and less than 45 minutes.

4.19. The breaking strength of a type of nylon thread is distributed according to a normal distribution with $\mu = 10.06$ ounces and $\sigma = 1.54$ ounces. What strength value will exceed the strength of 98% of such threads?

4.20. As pictured, a 1-inch needle is pinned exactly in its middle, precisely in the middle of two parallel lines.

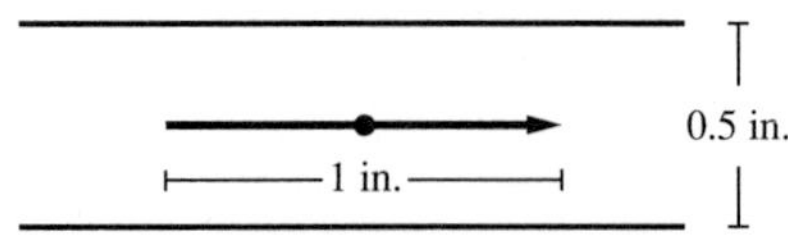

The needle may spin freely on its pin. When the needle is spun, it stops in a random position. What is the probability distribution governing the position in which the needle stops? What is the probability that the needle will stop in a position that crosses the parallel lines?

4.21. A specific hybrid mathematical form, which is discrete over part of its range and continuous over the rest of its range, has the following expression.

x	1	2	3	4
$p(x)$	$\frac{1}{6}$	$\frac{1}{5}$	$\frac{1}{10}$	$\frac{1}{3}$

and $f(x) = x^2 \qquad 5 \leq x \leq k$

Is it possible for this mathematical form to be a valid probability distribution function? Why or why not? If it is possible, state the specific conditions where this would be true.

4.22. Suppose that the life, X, of a certain component is governed by the following probability distribution function.

$$f(x) = \begin{cases} 0 & x < 100 \\ \dfrac{k}{x^2} & x \geq 100 \end{cases}$$

(a) Find the necessary value of k.

(b) What is the probability that a component of this type will last at least 500 hours?

4.23. The Cydroog Company manufactures laser cylinders for the USAF. It is known that the diameter of the cylinders is normally distributed with a mean of 0.6 inch and a variance of 0.000016 inch-squared. What is the probability that one of these cylinders, selected at random, will have a diameter between 0.595 and 0.605 inch?

4.24. The proportion of people that are affected by a certain high-frequency signal is governed by a beta distribution with $\alpha = 5$ and $\beta = 2$. What is the probability that less than 500 of a randomly selected group of 1600 people will be affected?

4.25. A drive shaft will suffer fatigue failure, on the average, after 40,000 hours of use. If it is known that the probability of failure before 36,000 hours is 0.4 and that the distribution governing time to failure is a normal distribution, what is the standard deviation of the time-to-failure distribution?

4.26. Suppose that the probability that an experimental aircraft will succeed in its first flight is 0.62. What is the probability that, in 250 such aircraft, at least 85 and no more than 162 will succeed in their first flights?

4.27. A brass bushing encircles a steel shaft at a support point in an electrical water pump. The nominal outside diameter of the shaft is 0.500 inch; and we know, from long experience, that the standard deviation is 0.003 inch. The nominal inside diameter of the bushing is 0.508 inch with a standard deviation of 0.002 inch. What is the probability that shaft and bushing, selected at random, will not fit?

4.28. Let X be the random variable defined as the number of units of time until an Acme 100-watt bulb burns out. The probability density function governing X is

$$f(x) = \frac{b}{x^3} \qquad 1000 \leq x \leq 2000$$

What is the probability that such a bulb will last no more than 1800 time units?

4.29. Let $X_i, i = 1, 2, \ldots, 5$, and $Y_j, j = 1, 2, \ldots, 11$, be statistically independent random samples from normal distributions, both with zero means and with variances of $\frac{1}{4}$ and $\frac{1}{3}$, respectively. Find

$$P\left\{6\sum_{i=1}^{5} X_i^2 > 4\sum_{j=1}^{11} Y_j^2\right\}$$

4.30. Suppose that the following bivariate density function governs the relative amounts of two different chemicals found in the solution taken from an *in situ* mining process.

$$f(x_1, x_2) = k(x_1 + x_2^2) \qquad 0 \leq x_1 \leq 1, \qquad 0 \leq x_2 \leq 1$$

(a) Evaluate k.

(b) Find the marginal density functions governing X_1 and X_2.

(c) Are X_1 and X_2 statistically independent?
(d) Find the conditional probability density function of X_1 given that $0.4 \le X_2 \le 0.6$.

4.31. The joint density of two random variables is $f(x_1, x_2) = k(2x_1 + x_2)$, with $0 \le x_1 \le 2$ and $2 \le x_2 \le 3$.
(a) Evaluate k.
(b) Find the marginal density functions governing X_1 and X_2.
(c) Are X_1 and X_2 statistically independent?
(d) Find the conditional probability density function of X_2 given that $0.5 \le X_1 \le 1.5$.

4.32. A bivariate density function is given as $f(x_1, x_2) = cx_1x_2$, with $0 \le x_1 \le 4$ and $1 \le x_2 \le 5$.
(a) Evaluate c.
(b) Find the marginal density functions governing X_1 and X_2.
(c) Are X_1 and X_2 statistically independent?
(d) Find the conditional probability density function of X_2 given that $0.5 \le X_1 \le 1.5$.

4.33. The impact point of arrows about the center of the target for a certain professional archer is governed by the following joint density function.

$$f(x_1, x_2) = \frac{0.5}{\pi} \exp[-0.5(x_1^2 + x_2^2)] \qquad 0 < x_1, \quad 0 < x_2$$

where X_1 and X_2 are the horizontal and vertical deviations from the center of the target (in feet).
(a) Are X_1 and X_2 statistically independent?
(b) Find the conditional probability density function of X_2 given that $0.5 \le X_1 \le 1.5$.

4.34. The joint density of two random variables is $f(x_1, x_2) = 8x_1x_2$, with $0 \le x_1 \le 1$ and $0 \le x_2 \le x_1$.
(a) Find the marginal density functions governing X_1 and X_2.
(b) Are X_1 and X_2 statistically independent?
(c) Find the conditional probability density function of X_2 given that $X_1 = 0.5$.

4.35. A specific hybrid mathematical form, which is discrete over part of its range and continuous over the rest of its range, has the following expression.

x	1	2	3	4
$p(x)$	0.1667	0.2	0.16	0.14

$$f(x) = x^2 \qquad 5 \le x \le k$$

(a) Is it possible for this mathematical form to be a valid probability distribution function? Why or why not?
(b) If it is possible, state the specific conditions under which this would be true.

4.36. If $f(x) = \frac{1}{150}(2x + 5)$, with $0 \le x \le 10$,
(a) Find $P(X \ge 8)$.
(b) Find $P(2 \le X \le 8)$.
(c) Find d such that $P(4 \le X \le d) = 0.25$.

4.37. The grade point averages of Big State U students are distributed normally with a mean of 2.4 and variance of 0.64.
(a) What fraction of BSU students have a GPA higher than 3.5?
(b) What fraction of them have a GPA lower than 2.5?

4.38. If X is a normal random variable with mean 3 and standard deviation 1 and Y is another normal random variable with mean 4 and standard deviation 2, then what is the probability that an observation of X will be less than an observation of Y?

4.39. If A is a χ^2 random variable with 3 degrees of freedom, if B is another χ^2 random variable with 4 degrees of freedom, and if A and B are independent; find the probability that $X = A + B$ will exceed the value of 14.

4.40. The results of a reliability test are distributed as a χ^2 random variable with 4 degrees of freedom, and the results of a second test are distributed as a χ^2 random variable with 6 degrees of freedom. What is the probability that the value obtained from the second test will be at least three times greater than that obtained in the first test?

4.41. The resistance of a circuit is normally distributed with a mean of 200 ohms and a standard deviation of 10 ohms. What is the probability that the resistance of one of these circuits selected at random is (a) greater than 200 ohms, (b) less than 190 ohms, (c) greater than 220 ohms?

4.42. If X is distributed normally with a mean of 6 and variance of 4, what is the probability that X will yield a result greater than 5.8 on two consecutive trials? (Assume that the trials are statistically independent.)

4.43. A continuous random variable X has probability density function shown in the following, where f is an unknown constant.

$$f(x) = \begin{cases} x - 10 & 10 \leq x < 11 \\ 12 - f & 11 \leq x < 12 \\ 0 & \text{otherwise} \end{cases}$$

(a) Find f. (b) Find $P(8 \leq X \leq 11.5)$. (c) Find $P(X \geq 11.8)$.

4.44. Three cartons are loaded with three different types of loads. The distribution of strength for the cartons is chi-square with 3 degrees of freedom. The three loads are distributed like chi-square random variables with 4, 5, and 6 degrees of freedom, respectively. If we load each of the three cartons with a different load, then what is the probability that no carton will break?

4.45. Suppose that a continuous function has the following form.

$$f(x) = 2x + q \qquad 0 \leq x \leq 3$$

Is there a value of q that will make $f(x)$ a probability density function? Explain.

4.46. Suppose that a new IQ scale is developed around a normal distribution with a mean value of 20 and a standard deviation of 8.

a) Find $P(\text{IQ} \geq 35)$.

b) What boundaries would exclude the top and bottom 5% of the population?

4.47. A soft drink machine can be regulated so that it discharges an average of m ounces per cup. If the fill is distributed normally with a variance of 0.09, find m such that the 8-ounce cups will overflow only 1% of the time.

COMPUTER-BASED EXERCISES

C4.1. Plot and superimpose the following normal distributions: $N(0, 1)$, $N(0, 2)$, $N(2, 1)$, and $N(2, 2)$. What general properties can you observe from these plots?

C4.2. Plot $f(x) = 0.5x^2e^{-x}$, with $0 < x$. Is this a probability distribution function? What is the highest point on the distribution? What other interesting observations may be made about $f(x)$?

C4.3. Superimpose the plot of an $N(15, \sqrt{6})$ distribution onto the plot of a binomial distribution with $n = 25$ and $p = 0.6$. What observations may be made about the plots? Use this plot to formulate an intuitive graphical justification for the "continuity correction" that has been presented at the end of Section 4.3.

C4.4. Investigate the effect on the shape of the F distribution when (a) n is varied and m is held constant and (b) m is varied and n is held constant. Summarize your answers to parts (a) and (b) in a single descriptive statement. Is there a general statement that can be made about simultaneous changes in both n and m? If so, give that statement.

C4.5. Investigate the effect of changing the value of λ on the shape of the gamma distribution. Summarize your findings in a single sentence.

C4.6. Investigate the effect of the parameters α and β on the shape of the beta distribution. Summarize your findings.

C4.7. Investigate the effect of the parameters α and β on the shape of the Weibull distribution. Summarize your findings.

C4.8. Verify your answer to Exercise 4.13 by plotting the gamma distribution with $r = 3$ and $\lambda = 5$. Shade the area under the distribution that corresponds to your answer to Exercise 4.13.

CHAPTER 5

THE MEAN, VARIANCE, EXPECTED VALUE OPERATOR, AND OTHER FUNCTIONS OF RANDOM VARIABLES

When you finish your study of this chapter, you should understand the ideas and be able to perform the operations described in the following list.

- When given a probability distribution function, be able to compute the *mean, variance, standard deviation, median, mode, skewness*, and *kurtosis* of the random variable, X
- Understand the definition and the use of *Tchebycheff's inequality*
- When given a function, $g(X)$, and the probability distribution function of a random variable, X, be able to calculate $E(g(X))$
- Understand the properties of the *expected value operator, E,* and be able to apply those properties to the solution of expected value problems
- Know what is meant by the terms *mode* and *median*
- Have a clear idea of what the mean, variance, skewness, and kurtosis of a random variable represent
- When given a joint bivariate probability distribution, be able to compute both the covariance and correlation between the two random variables

- Thoroughly understand the fact that *any function of one or more random variables* is also a random variable
- When given the probability distribution function of a discrete random variable, X, be able to find the probability distribution of any function of X, $g(X)$
- Understand the concept of a discrete convolution and be able to perform a convolution of two discrete probability distributions for any defined convolution operator
- Be able to work problems like those found in this chapter and in the exercises

5.1 INTRODUCTION

Nothing can substitute for a complete mathematical statement of the probability distribution function of a random variable. Unfortunately, as we see in the next chapter, the probability distribution of a physical quantity of interest is *rarely* known. If the probability distribution function is not known, we must take data from the physical system and extract as much information as possible from that data.

Data-based methods have been developed that allow us to sketch the probability distribution in a rough way. Other data-based methods allow us to compute *sample estimates* of quantities that characterize where most of the distribution lies, how disperse the distribution is, whether the distribution is lopsided, and so on. These data-based methods are *not useful unless* we first understand their fundamental meanings as viewed in the *context of general theoretical probability distributions*. This fact is the major reason that Chapters 3 and 4 have been included in this book and that Chapters 3, 4, and 5 are placed before the other chapters in the text.

5.2 MEASURES OF CENTRALITY: THE MEAN, MEDIAN, AND MODE

Let us consider three theoretical quantities that are measures of the *location* or *centrality* of a probability distribution function, the *mean, median,* and *mode.*

The mean is defined mathematically as

$$\begin{aligned} \mu &= \sum_{\text{all } x} x\, p(x) \qquad \text{for discrete distributions} \\ \mu &= \int_{\text{all } x} x f(x) dx \qquad \text{for continuous distributions} \end{aligned} \tag{5.1}$$

Conceptually, the mean is nothing more than a *weighted sum* of all possible values of X. This observation leads to an interesting physical interpretation.

Many engineers will recognize the formulas of Equation 5.1 to be identical, in mathematical content, to the *first moment about the origin* of a set of discrete points of mass or of a continuum of distributed mass (for the discrete and continuous distributions, respectively). Carrying this analogy a little farther, the *probability mass,* $p(x)$, for discrete distributions, is located x units away from the origin. Thus $p(x)$ contributes $xp(x)$ units of *torque* about the origin, where x is the *moment arm* of the "mass," $p(x)$. This means that a small probability mass can have just as much effect on the value of the mean as a large mass. Of course, this will be true only if the smaller mass is more distant from the mean than the larger mass; i.e., if $P(X = w) = 0.1$ and $P(X = v) = 0.2$, then w must be twice as far from the mean as v if the two masses are to have the same effect on the value of the mean. If w and v are on opposite sides of the mean and w is twice as far from the mean as v, the masses at w and v will exactly cancel one another's effect.

Example 5.1. The mean of the binomial distribution of Figure 5.1 may be directly computed in two ways—by applying Equation 5.1,

$$\mu = (0)(0.2373) + (1)(0.3955) + (2)(0.2637) + (3)(0.0879) + (4)(0.0146) + (5)(0.0010) = 1.25$$

or by applying the formula for the mean of a binomial distribution as given in Table 3.2; i.e.,

$$\mu = np = (5)(0.25) = 1.25.$$

As illustrated in Figure 5.1, μ may also be obtained by plotting any of the distributions.

Viewing $p(x)$ as a system of weights and moment arms, as in Figure 5.2, we see that an equivalent amount of torque is provided by putting all the probability mass (a total of 1 unit of mass) at a distance of μ from the origin. The system of masses would perfectly balance if a razor edge support were placed at μ and the pin at the origin were removed. Thus the mean is also the *center of gravity* of the probability masses.

$$p(x) = {}_5C_x\, 0.25^x 0.75^{5-x}$$

x	$p(x)$
0	0.2373
1	0.3955
2	0.2637
3	0.0879
4	0.0146
5	0.0010

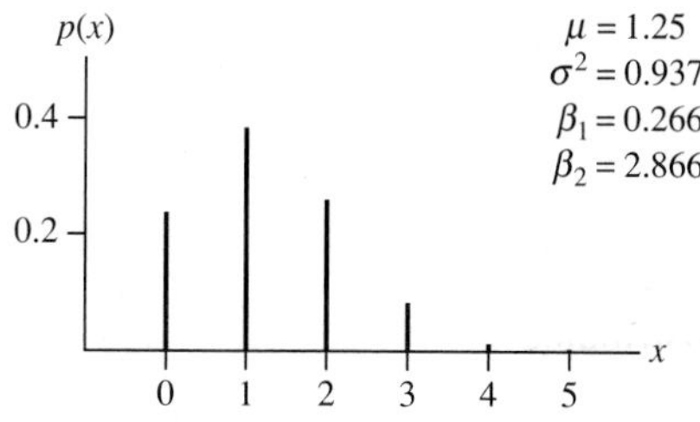

FIGURE 5-1
The binomial distribution with $n = 5$ and $p = 0.25$.

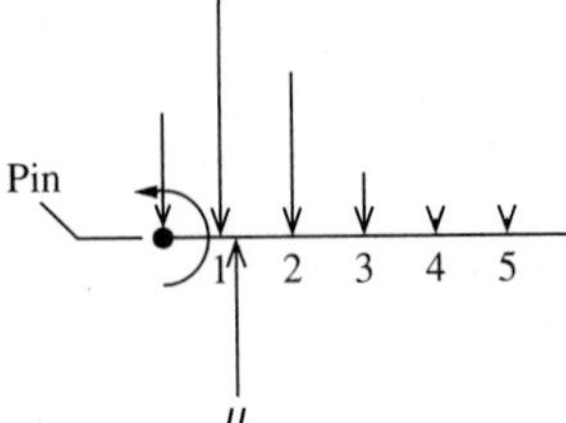

FIGURE 5-2
The binomial distribution viewed as a system of weights.

It is equally important to recognize that the mean is also the *long-term* average value of the distribution. In other words, suppose that we take repeated random observations of the characteristic governed by $p(x)$ and compute the average value of the observations. As the number of observations becomes large, that average value would approach the mean, μ. We consider this idea in detail in Chapter 6.

Example 5.2. Suppose that we are given the Weibull distribution function, $f(x) = 0.5x\exp(-0.25x^2)$, with $x > 0$ ($\alpha = 0.25$, $\beta = 2$), as pictured in Figure 5.3. The mean of $f(x)$ is

$$\mu = \int_0^\infty x f(x) dx = \int_0^\infty 0.5x^2 \exp(-0.25x^2) dx$$

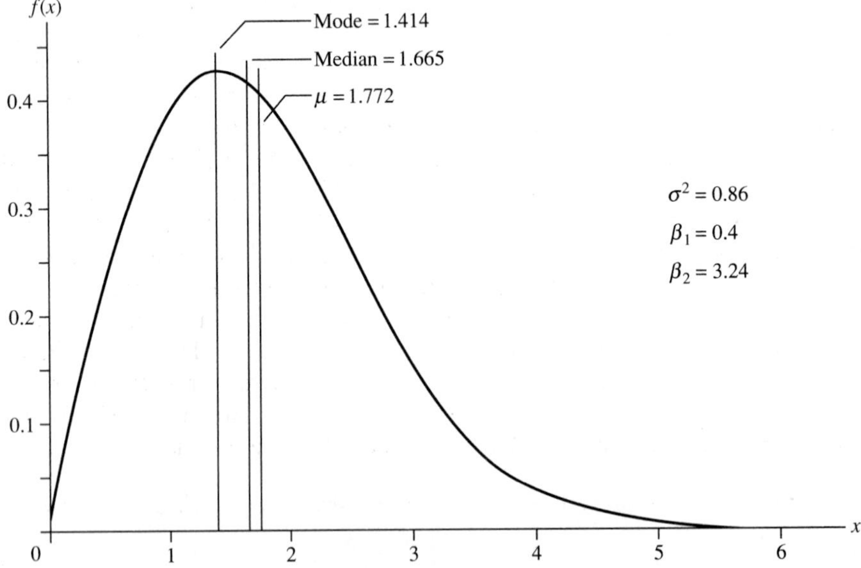

FIGURE 5-3
The Weibull distribution with $\alpha = 0.25$ and $\beta = 2$.

Let $y = 0.25x^2$, which implies that $dy = 0.5x\,dx$. Substituting this result yields

$$\mu = 2\int_0^\infty y^{0.5}e^{-y}dy$$

Recognizing this integral as a special case of the well-known gamma function,

$$\int_0^\infty w^{1/r}e^{-w}dw = \Gamma(1 + (1/r))$$

we find that $\mu = 2\Gamma(1.5) = 2(0.5\sqrt{\pi}) = \sqrt{\pi} = 1.772$. This is verified much more easily with the formula for the mean of a Weibull distribution in Table 4.1; i.e.,

$$\mu = \alpha^{-1/\beta}\Gamma(1 + (1/\beta)) = 0.25^{-0.5}\Gamma(1.5) = 2\Gamma(1.5) = 1.772$$

If you desire additional pictorial examples, refer to the numerous figures of Chapters 3 and 4 for plots of various distributions with not only the values of their means, μ, but also with the values of other important measures, which are discussed in detail in the following sections of this chapter.

Sometimes the mean is not as good a measure of the centrality of a distribution as we might like. Suppose that the mean per capita annual income in Oilfield County, Texas, is \$94,310. This might be surprising or even alarming to some individuals. However, this does not necessarily imply that most of the individuals in that county make such elevated salaries. The mean reflects the fact that a few individuals make tremendous incomes and those few pull up the overall average. For example, if nine people make a total of \$90,000 and one person makes \$1,000,000, the average income for those ten people is \$109,000. Clearly, some more descriptive measure is needed for this kind of situation.

One answer is the *median,* the point that exactly divides the distribution into two equal probability masses. If someone tells you the median income in Oilfield County, Texas, is \$15,000, this would certainly seem to be a more reasonable measure of location, or centrality, for those incomes. The median is also the *50th percentile* or the *second quartile* of a distribution. A *percentile* is the point that equals or exceeds a specific percent of the probability mass of the distribution. For example, if 22 is the 79th percentile of $f(x)$, $F(22)=0.79$. The *first quartile* is the same as the 25th percentile, the *3rd quartile* is the 75th percentile, and the *interquartile* range is the difference between the third and first quartile.

A mass balance interpretation, like that for the mean, can be made for the median. In the case of the median, every element of probability mass is given the *same moment arm*. For simplicity, this common moment arm length may be assigned a value of 1 unit; i.e., the moment arm has a value of -1 if x is less than the median and a value of $+1$ if x is greater than the median.

Example 5.3. The median for the Weibull distribution of Figure 5.3 is that value x_m such that

$$\int_0^{x_m} f(x)dx = \int_0^{x_m} 0.5x\exp(-0.25x^2)dx = 0.5 \quad \text{or} \quad 1 - \exp(-0.25x_m^2) = 0.5$$

which yields $x_m = 1.665$; i.e., F(1.665) =0.5. This may be verified by using the inverse table lookup option of the Weibull distribution evaluator program using a tail probability of 0.5.

In this case, the median is less than the mean, reflecting the asymmetry of this Weibull distribution with the long, thin tail to the right.

The computation of the median for a discrete distribution is not necessarily as clean-cut from a conceptual view. Consider the binomial distribution of Figure 5.1. The cumulative distribution gives $P(X \leq 1) = 0.6328$ and $P(X \leq 0) = 0.2373$. There is no distinct point at which the probability mass of the distribution is divided into two equal parts. Happily, this presents no real difficulty in data analysis, as we see in Chapter 6.

A *mode* of a probability distribution function, $f(x)$, is present at *any local maximum* of $f(x)$, i.e., at any x_0 such that all values of $f(x)$ in the *near neighborhood* of x_0 are *no more* than $f(x_0)$. Therefore, a probability distribution may have one, two, or more modes. The probability distributions presented in Figures 5.1 and 5.3 are unimodal; i.e., they have *one* mode. In Figure 5.1, the mode is at $x = 1$.

Example 5.4. While a plot is useful in determining the approximate location of the mode, the exact value of the mode of the Weibull distribution in Figure 5.3 is somewhat less obvious. However, from differential calculus, we know that a local maximum of *f*(*x*) must occur at the point where the slope, or first derivative, becomes zero. The mode of the Weibull distribution of Figure 5.3 is found as follows:

$$\frac{\partial f(x)}{\partial x} = \frac{\partial}{\partial x} 0.5x\exp(-0.25x^2) = 0.5x\exp(-0.25x^2)(-0.5x) + 0.5\exp(-0.25x^2)$$

$$= 0.5\exp(-0.25x^2)(1 - 0.5x^2) = 0$$

Thus the slope is zero at $x = +\infty$ and at $x = \sqrt{2} = 1.414$. Obviously, the mode is at $x = 1.414$.

5.3 MEASURES OF VARIABILITY: THE RANGE AND THE VARIANCE

Let us consider two *measures of variability,* (also called *dispersion*)—the *range* and the *variance*. The range is the difference between the smallest possible value of the random variable and the largest possible value. Thus the binomial distribution of Figure 5.1 has a range of 5 - 0 = 5. In like manner, the range of the Weibull distribution of Figure 5.3 is $+\infty - 0 = \infty$. From these two examples, we see that the range is *rarely helpful* in describing variability. Virtually all of the binomial distribution of Figure 5.1 is between 0 and 3, and 99.8% of the Weibull distribution of Figure 5.3 is between 0 and 5.

The *variance* is a much more useful measure of variability. We have already encountered the variance in our discussion of the normal distribution in Chapter 4. However, the variance is a *general characteristic* of all probability distribution functions and is defined mathematically as

$$\sigma^2 = \sum_{\text{all } x} (x - \mu)^2 p(x) \qquad \text{for discrete distributions}$$

$$\sigma^2 = \int_{\text{all } x} (x - \mu)^2 f(x)dx \qquad \text{for continuous distributions} \tag{5.2}$$

Notice that σ^2 is actually a weighted sum of the squared differences, or deviations, of all possible values of X from the mean, μ. For this reason , σ^2 will *never* have a value less than zero (and σ^2 will achieve a value of zero when there is *only one* possible value for X.) A mass balance interpretation similar to that for the mean or the median can also be made for the variance. In this case, however, the moment arm for each element of probability mass is the *square* of the distance from the mean. This implies that a probability mass located 10 units away on either side of the mean has a moment arm of 100 equivalent *positive* units of distance. Therefore, when working with the variance, a probability mass of 0.001 located 10 units from the mean will have the same effect as a probability mass of 0.1 located 1 unit from the mean.

Often, the *standard deviation,* σ, is used in preference to σ^2. The reason for this preference is that σ is expressed in the *same units of measure* as X and μ. For example, if X is measured in inches, σ^2 is expressed in *inches-squared.* Although the dimensionality of σ^2 is difficult to grasp at any intuitive level, σ is expressed in inches and has direct conceptual meaning. The variance is a relative quantity: a larger variance means that the random variable is more spread with reference to its mean, and a smaller variance means the opposite.

We saw in Chapter 4 that a knowledge of the values of μ and σ for a normal distribution gives us complete knowledge of the probability distribution function. A natural question to ask is, How much information is provided by knowing the values of μ and σ when X is not normally distributed?

Of course, the answer to this question depends on the form of $f(x)$. However, even if *nothing* is known about $f(x)$—other than the fact that X is a random variable—*Tchebycheff's inequality* states that

$$P(\mu - k\sigma \leq X \leq \mu + k\sigma) > 1 - \frac{1}{k^2} \tag{5.3}$$

for any $f(x)$. Therefore, $P(\mu - 2\sigma \leq X \leq \mu + 2\sigma) > 0.75$ for any $f(x)$.

In lieu of any knowledge about $f(x)$, Equation 5.3 is a very powerful general result. However, Equation 5.3 will often produce results very inferior to those that can be obtained when we have information about $f(x)$. For example, for a normal distribution, $P(\mu - 2\sigma \leq X \leq \mu + 2\sigma) = 0.9544$, and, for an exponential distribution, $P(\mu - 2\sigma \leq X \leq \mu + 2\sigma) = 0.9502$. Both of these results are more powerful and more precise than the result from Equation 5.3.

Example 5.5. The variance of the Binomial distribution of Figure 5.1 is

$$\begin{aligned}\sigma^2 &= (0 - 1.25)^2(0.2373) + (1 - 1.25)^2(0.3955) + \ldots + (5 - 1.25)^2(0.001)\\ &= .9375\\ \sigma &= 0.9682\end{aligned}$$

Example 5.6. The variance of the Weibull distribution of Figure 5.3 is

$$\sigma^2 = \int_0^\infty (x - \sqrt{\pi})^2 0.5x \exp(-0.25x^2)dx$$

$$= \int_0^\infty (x^2 - 2x\sqrt{\pi} + \pi)0.5x \exp(-0.25x^2)dx$$

$$= \int_0^\infty 0.5x^3 \exp(-0.25x^2)dx - \int_0^\infty x^2\sqrt{\pi}\exp(-0.25x^2)dx$$

$$+ \int_0^\infty 0.5x\pi \exp(-0.25x^2)dx$$

Consider the first of the three integrals. Letting $u = 0.25x^2$ implies that $du = 0.5x$, which yields $4\int_0^\infty ue^{-u}du = 4\Gamma(2) = 4$. The second integral is very similar to that evaluated for the mean, yielding

$$\int_0^\infty x^2\sqrt{\pi}\exp(-0.25x^2)dx = 4\sqrt{\pi}\Gamma(1.5) = 2\pi$$

The third integral is simply π times the integral of the probability distribution function over its entire range. Since the integral of the probability distribution function over its entire range must have a value equal to 1, $\sigma^2 = 4 - 2\pi + \pi = 0.8584$.

This agrees with the general formula given in Table 4.1—recalling that $\alpha = 0.25$ and $\beta = 2$,

$$\sigma^2 = \alpha^{(-2/\beta)}\{\Gamma(1 + (2/\beta)) - \Gamma^2(1 + (1/\beta))\} = 0.25^{-1}\{\Gamma(1 + 1) - \Gamma^2(1 + 0.5)\}$$

$$= 4\{1 - (0.5\sqrt{\pi})^2\} = 4 - \pi = 0.8584$$

As before, using the software to plot the distribution of any random variable yields the variance as a by-product.

5.4 THE EXPECTED VALUE OPERATOR

The mean and variance of a probability distribution function can be related to a more general context, that of the *expected value operator*. The *expected value of any function*, $g(X)$, of a random variable, X, is defined as

$$E\{g(X)\} = \sum_{\text{all } x} g(x)p(x) \qquad \text{for discrete distributions}$$

$$= \int_{\text{all } x} g(x)f(x)dx \qquad \text{for continuous distributions} \tag{5.4}$$

Note that, if $g(X) = X$, then $E\{g(X)\} = \mu$ and that, if $g(X) = (X - \mu)^2$, then $E\{g(X)\} = \sigma^2$. Therefore, $E(X) = \mu$, and $E\{(X - \mu)^2\} = \sigma^2$.

The expected value operator has the following fundamental properties (where c is a constant).

1. $E\{c\} = c$
2. $E\{g(X) + c\} = E\{g(X)\} + c$
3. $E\{cg(X)\} = cE\{g(X)\}$
4. $E\{g_1(X) + g_2(X)\} = E\{g_1(X)\} + E\{g_2(X)\}$

$\mu = E(X) = \int_{\text{all } x} xf(x)dx = \int_{\text{all } x} (x-0)^1 f(x)dx$ is the first moment about the origin ($x = 0$). It is natural, therefore, to call $\sigma^2 = E\{(X-\mu)^2\} = \int_{\text{all } x}(x-\mu)^2 f(x)dx$, the *second moment about the mean.* In this context, the *kth moment about an arbitrary point, a,* is defined as

$$E\{(X-a)^k\} = \int_{\text{all } x} (x-a)^k f(x)dx$$

For our purposes, however, we will limit our attention to moments about the origin, μ'_k, and moments about the mean, μ_k, where

$$\begin{aligned} \mu'_k = E\{X^k\} &= \sum x^k p(x) && \text{for discrete distributions} \\ &= \int x^k f(x)dx && \text{for continuous distributions} \end{aligned} \qquad (5.5)$$

$$\begin{aligned} \mu_k = E\{(X-\mu)^k\} &= \int (x-\mu)^k f(x)dx && \text{for continuous distributions} \\ &= \sum (x-\mu)^k p(x) && \text{for discrete distributions} \end{aligned} \qquad (5.6)$$

The μ_k and the μ'_k are also known as the *central and noncentral moments,* respectively.

In Exercise 5.1a, you are asked to show that $\sigma^2 = E\{(X-\mu)^2\} = E(X^2) - \mu^2 = \mu'_2 - \mu^2$. This and other relations between the central and noncentral moments are useful computational aids.

5.5 TWO ADDITIONAL MEASURES OF A PROBABILITY DISTRIBUTION FUNCTION: SKEWNESS AND KURTOSIS

We have already discussed $\mu'_1 = \mu$ and $\mu_2 = \sigma^2$. Two other quantities are of particular interest, μ_3 and μ_4. A natural question to ask at this point is, Why do we need further measures in addition to the ones already discussed? The answer is best given by an example. In Figure 5.4, we see two probability distributions that are markedly different and yet possess the same mean and variance. Thus if we have *only* the mean and variance of these two distributions, we are unable to differentiate between them. One way to differentiate is to compare the values of their respective third moments about the mean. The quantity μ_3 is very useful in characterizing the skewness—i.e., the *asymmetry,* or lopsidedness—of a probability distribution.

Example 5.7. Let us compute μ_3 for the binomial distribution of Figure 5.1:

$$\begin{aligned} \mu_3 &= (0-1.25)^3(0.2373) + (1-1.25)^3(0.3955) + \cdots + (5-1.25)^3(0.001) \\ &= 0.46875 \end{aligned}$$

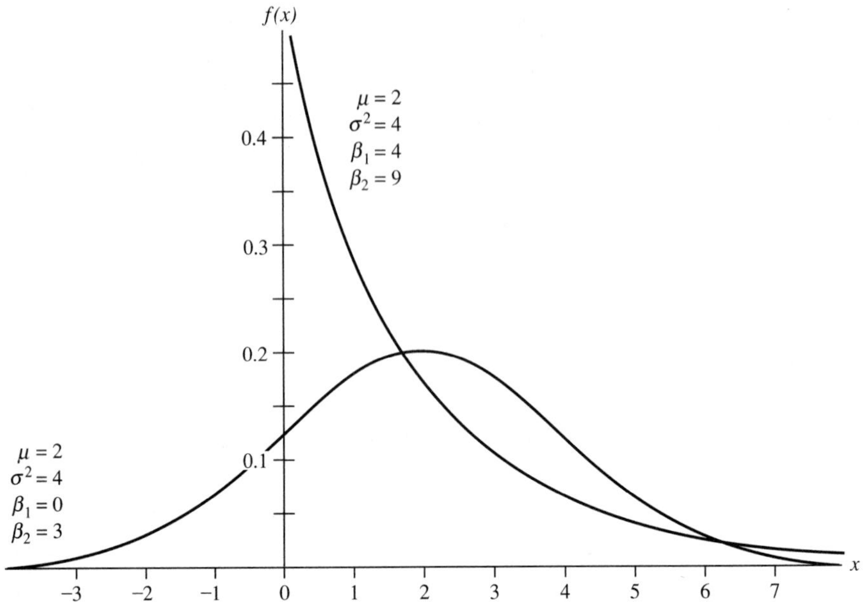

FIGURE 5-4
Exponential and normal distributions with the same μ and σ^2.

The value of μ_3 for the Weibull distribution of Figure 5.3 is obtained through techniques very similar to those used for the evaluation of σ^2. However, the evaluation is cumbersome and nothing new would be shown.

Example 5.8. Instead, let us consider μ_3 for the exponential distribution of Figure 5.4, where $\lambda = 0.5$. Thus $f(x) = 0.5e^{-0.5x}$ for $x > 0$, and

$$\mu_3 = \int_0^\infty (x-\mu)^3 0.5e^{-0.5x}\,dx = \int_0^\infty (x^3 - 3\mu x^2 + 3\mu^2 x - \mu^3)0.5e^{-0.5x}\,dx$$

$$= 0.5\left(\int x^3 e^{-0.5x}\,dx - 3\mu\int x^2 e^{-0.5x}\,dx + 3\mu^2\int x e^{-0.5x}\,dx - \mu^3\int e^{-0.5x}\,dx\right)$$

Recalling that

$$\int_0^\infty x^n e^{-ax}\,dx = \frac{\Gamma(n+1)}{a^{n+1}} = \frac{n!}{a^{n+1}} \qquad \text{and} \qquad \mu = \frac{1}{\lambda} = 2$$

the above integral expression is equivalent to

$$\mu_3 = 0.5\left(\frac{3!}{0.5^4} - \frac{3\mu(2!)}{0.5^3} + \frac{3\mu^2}{0.5^2} - \frac{\mu^3}{0.5}\right) = 0.5(96 - 96 + 48 - 16) = 16$$

Observe that μ_3 is always equal to *zero* for *any symmetric distribution*. If the distribution is concentrated in its leftmost part with a long, thin tail pointing to the right, μ_3 is positive . If μ_3 is positive, the distribution is *positively skewed,* or skewed to the right. Conversely, if the distribution is concentrated in its rightmost part with a long, thin tail pointing to the left, μ_3 is negative. If μ_3 is negative, the distribution is *negatively skewed,* or skewed to the left. Examples of positively skewed distributions are given in Figures 4.5, 4.8, 5.1, and 5.3; examples of negatively skewed distributions are presented in Figure 5.6. The mirror image of any positively skewed distribution is negatively skewed, and the mirror image of any negatively skewed distribution is positively skewed.

Dimensionless measures of skewness are also used and are defined in the following two equivalent ways.

$$\beta_1 = \frac{\mu_3^2}{(\sigma^2)^3} = \frac{\mu_3^2}{\sigma^6} \quad \text{or} \quad \gamma_1 = \frac{\mu_3}{\sigma^3} = \sqrt{\beta_1} \tag{5.7}$$

For the purposes of this book, we will use the more popular of the two measures, β_1.

Example 5.9. For the binomial distribution of Figure 5.1,

$$\beta_1 = \frac{(0.46875)^2}{(0.9375)^3} = 0.26667$$

which yields the same result as the formula in Table 3.2

$$\beta_1 = \frac{0.75}{5(0.25)} + \frac{0.25}{5(0.75)} - \frac{2}{5} = 0.26667$$

Example 5.10. For the exponential distribution of Figure 5.4,

$$\beta_1 = \frac{\mu_3^2}{\sigma^6} = \frac{16^2}{4^3} = 4$$

which we know is correct, because, by Table 4.1, *any* exponential distribution has $\beta_1 = 4$.

Like μ and σ^2, the value of β_1 is an automatic part of the information provided by the plotting software for any probability distribution.

One minor problem with β_1 is that the sign of μ_3 is not preserved; i.e., as defined, β_1 is always positive. In the figures in this book, whenever μ_3 associated with any β_1 is negative, the value of β_1 is immediately followed with the notation $(-)$. Figures 5.5 and 5.6 are provided to foster an intuitive appreciation of the shapes of distributions that possess various values of β_1.

β_1 provides a method of differentiating between distributions such as those illustrated in Figure 5.4. However, the probability distributions presented in Figure 5.7 possess not only the same mean and variance but also the same value of β_1. One way to differentiate between distributions such as those of Figure 5.7 is to compute a measure of *kurtosis*, or *peakedness*, that uses the value of the fourth moment about the mean, μ_4.

Figure 4.6 has presented several t distributions plotted on the same graph with the standardized normal distribution. It may be shown that any normal distribution has $\mu_4 = 3\sigma^4$, and thus the standard normal distribution has $\mu_4 = 3$. The t distribution has

$$\mu_4 = \left(\frac{n}{n-2}\right)^2 \left(\frac{6}{n-4} + 3\right) \text{ for } n > 4$$

Dist	n	p	μ	σ^2	β_1	β_2
1	30	0.03	0.9	0.87	1.01	3.95
2	30	0.05	1.5	1.42	0.57	3.50
3	30	0.075	2.25	2.08	0.35	3.28
4	30	0.10	3.0	2.70	0.24	3.17
5	30	0.15	4.5	3.83	0.13	3.06
6	30	0.20	6.0	4.80	0.08	3.01
7	30	0.30	9.0	6.30	0.03	2.96
8	30	0.50	15	7.50	0.00	2.93

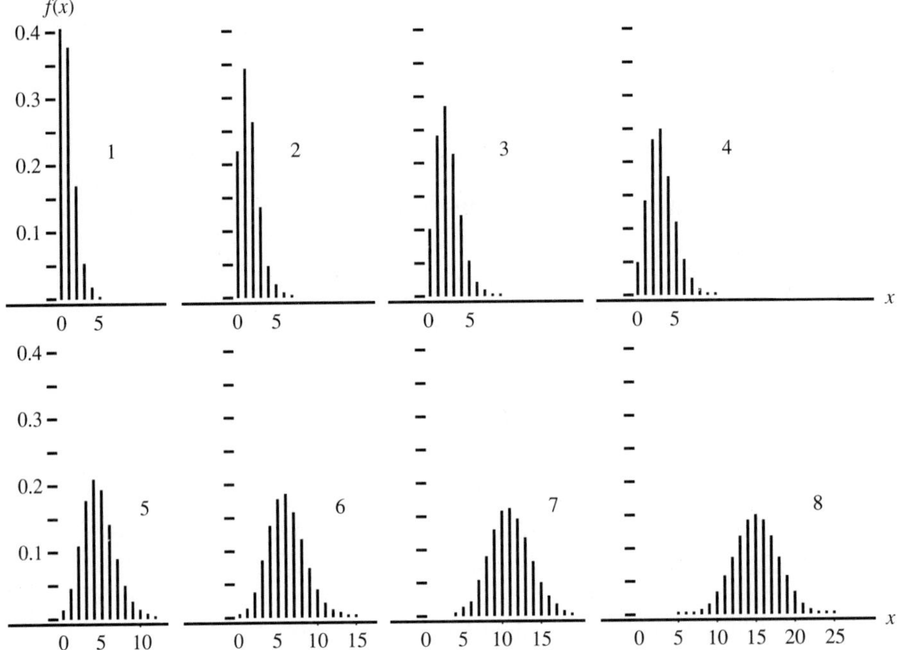

FIGURE 5-5
The binomial distribution with $n = 30$ and selected values of p.

which approaches, from *above*, the normal distribution's value of $\mu_4 = 3$ as n approaches ∞.

Two dimensionless measures of kurtosis are also used and are equivalently defined as

$$\gamma_2 = \frac{\mu_4}{\sigma^4} - 3 \qquad \text{and} \qquad \beta_2 = \frac{\mu_4}{\sigma^4} = \gamma_2 + 3 \tag{5.8}$$

For any normal random variable, $\beta_2 = 3$ and $\gamma_2 = 0$. We will use the more popular measure, β_2. This measure is used to characterize the kurtosis of a distribution, i.e., the relation of the "fatness of the tails" of the distribution to the central portion of the distribution. The β_2 measure was originally designed to measure departures of symmetric unimodal distributions from the normal distribution.

Dist	α	β	μ	σ^2	β_1	β_2
a	5	4	0.56	0.02	0.02(−)	2.52
b	5	3	0.63	0.03	0.10(−)	2.59
c	5	2	0.71	0.03	0.36(−)	2.88
d	5	1	0.83	0.02	1.40(−)	4.2
e	5	0.5	0.91	0.01	3.74(−)	7.25

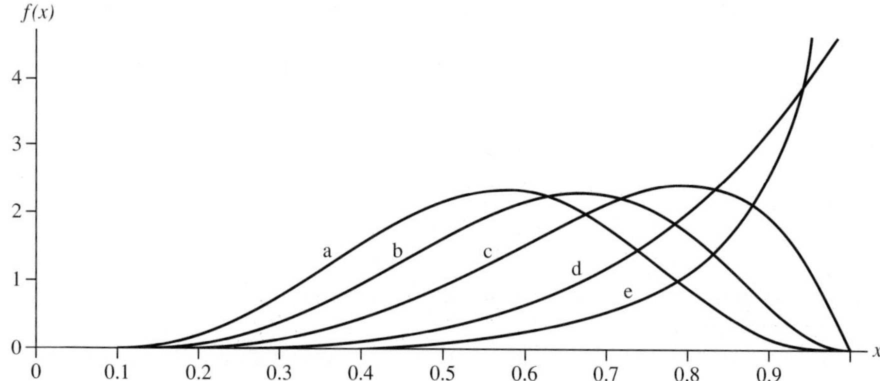

FIGURE 5-6
The beta distribution with selected parameter values.

For this reason, β_2 is more easily interpreted in the context of symmetric distributions. Figure 5.8 illustrates the shape differences between symmetric distributions with values of β_2 equal to, greater than, or less than 3. Notice that distribution A in Figure 5.8 is more "peaked" than distributions B and C. This shape causes β_2 to be larger for distribution A than for the other two distributions

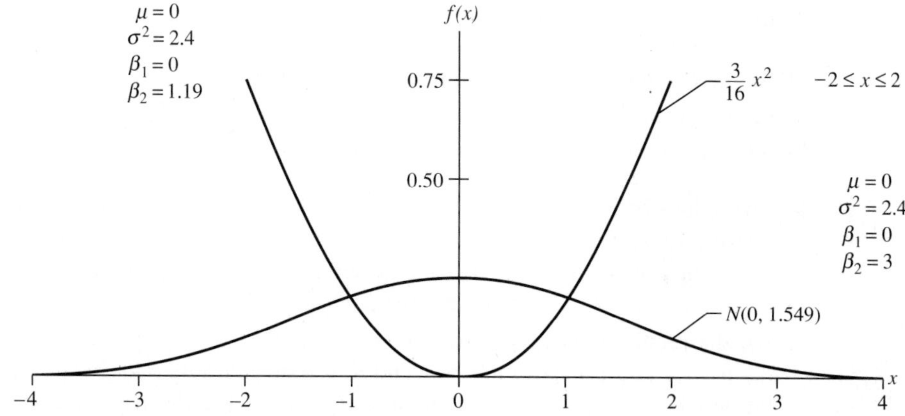

FIGURE 5-7
Two distributions with the same mean, variance, and β_1.

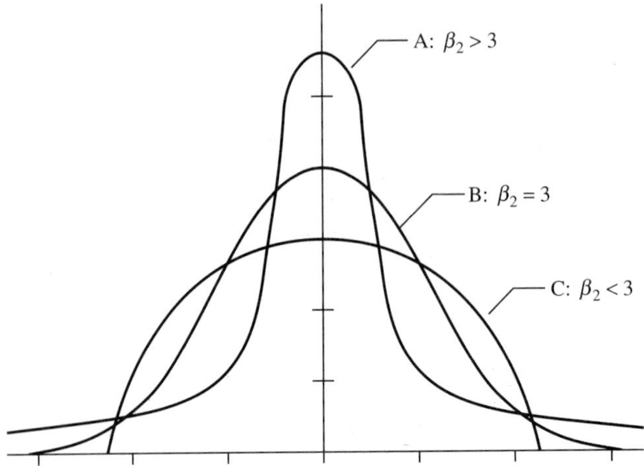

FIGURE 5-8
Symmetric unimodal distributions differing in kurtosis.

because the greater peakedness causes the distribution to have fatter tails farther away from the mean.

Like μ, σ^2, and β_1, the value of β_2 is an automatic part of the information provided by the plotting software for any probability distribution.

The following are reasons why β_1 and β_2 are used in preference to the simple third and fourth moments about the mean.

1. β_1 and β_2 are dimensionless; i.e., the ratios that define β_1 and β_2 contain identical powers of the unit of measure for the random variable, X, in both the numerator and denominator. Hence, β_1 and β_2 are *invariant* to the selection of the units of measure of X.
2. β_1 and β_2 are scaled in reference to the variance of the distribution. This implicitly filters out any confusion that might be caused by differing levels of dispersion in the distribution.

These two items allow interesting general facts and relations to be brought to light. As examples of such facts and relations, consider the values found in the β_1 and β_2 columns in Tables 3.2 and 4.1:

1. The values of β_1 and β_2 are *constants* for the uniform, normal, and exponential distributions. This means that, as the parameters of these distributions change, the values of σ, μ_3, and μ_4 change in such a way that the relations stated in Equations 5.7 and 5.8 remain constant.
2. All the relations indicated in Figures 3.2 and 4.4 are supported by the formulas for β_1 and β_2 in Tables 3.2 and 4.1. For example, (a) β_1 and β_2 for the binomial distribution become identical to those for the Bernoulli distribution when $n = 1$, (b) β_1 and β_2 for the χ^2 distribution become identical to those of the gamma distribution when $r = n/2$ and $\lambda = 0.5$, and (c) β_1 and β_2 for the

exponential distribution become identical to those of the Weibull distribution when $\beta = 1$ and $\alpha = \lambda$.

3. Several of the distributions in Tables 3.2 and 4.1 have values of β_1 and β_2 that approach those of the normal distribution (0 and 3, respectively) when a selected parameter value grows large. Examples of this phenomenon are seen (a) when n is increased in the binomial, χ^2, and t distributions and (b) when r is increased in the negative binomial and gamma distributions. Notice that, with the exception of the t distribution, these five distributions are all formed by adding other random variables together. (This result may be attributed directly to the central limit theorem, which we discuss in detail in Chapter 7.)

In closing this section of the chapter, we note that distributions that have different values of β_1 and β_2 will usually (not always) present different shapes when plotted. However, the converse is not true. Indeed, any normal, exponential, or uniform distribution always has constant values of β_1 and β_2 characteristic of its distribution form. As we have seen, these three distributions can present widely different shapes. This fact has been illustrated further by Figure 4.9, where Lognormal distributions of widely varying shapes are seen to have the same values of β_1 and β_2. (Note that β_1 and β_2 depend only on the parameter β for the Lognormal distribution. Therefore, the values of β_1 and β_2 will be identical for Lognormal distributions with the same value of β regardless of the value of α).

5.6 THE COVARIANCE AND CORRELATION OF BIVARIATE DISTRIBUTION FUNCTIONS

The mean, variance, and higher-order measures of univariate probability distributions have direct extensions to bivariate and multivariate joint probability distributions. The description of these extensions is most easily presented by using the expected value function. For example, the expected value of any bivariate function, $g(X_1, X_2)$ is defined as

$$E\{g(X_1, X_2)\} = \sum\sum_{\text{all } x_1, x_2} g(x_1, x_2)p(x_1, x_2) \qquad \text{for discrete distributions}$$

$$= \int\int_{\text{all } x_1, x_2} g(x_1, x_2)f(x_1, x_2)\, dx_1\, dx_2 \qquad \text{for continuous distributions} \tag{5.9}$$

With Equation 5.9, we can compute all of the univariate functions of the marginal distributions associated with either X_1 or X_2 by appropriately defining the forms of $g(X_1, X_2)$. For example, setting $g(X_1, X_2) = X_1$ would cause the operations of Equation 5.9 to yield $E(X_1) = \mu_1$, the mean of the marginal distribution of X_1.

One form of $g(X_1, X_2)$ that is of particular interest is $g(X_1, X_2) = \{(X_1 - \mu_1)(X_2 - \mu_2)\}$. The expression

$$\text{Cov}(X_1, X_2) = E\{(X_1 - \mu_1)(X_2 - \mu_2)\} = E\{X_1X_2\} - \mu_1\mu_2 \tag{5.10}$$

is the *covariance* between X_1 and X_2. The covariance is a measure of the relationship that exists between X_1 and X_2. As X_1 and X_2 assume their values, if X_1 tends to be large when X_2 is large or X_1 tends to be small when X_2 is small, the covariance between X_1 and X_2 is positive. If the converse is true, the covariance between X_1 and X_2 is negative.

If X_1 and X_2 are statistically independent random variables, then $E\{X_1X_2\} = E\{X_1\}E\{X_2\}$. This implies, by Equation 5.10, that the *covariance of statistically independent random variables is zero*. (However, the converse is *not* true; i.e. a covariance of zero does *not* imply that two random variables are statistically independent. This fact is illustrated in Exercise 5.11k.)

Because the units of measure for the covariance suffer from the same problems as the variance, it is often convenient to have a dimensionless form of the covariance. One such form is the correlation coefficient between X_1 and X_2, defined as

$$\rho = \text{Corr}(X_1, X_2) = \frac{\text{Cov}(X_1, X_2)}{\sqrt{V(X_1)V(X_2)}} \tag{5.11}$$

where $V(X_1)$ and $V(X_2)$ are the variances of X_1 and X_2, respectively. The correlation coefficient is bounded between -1 and $+1$ and takes the same sign as the covariance. We make extensive use of the correlation coefficient in Chapters 11 and 12, where we discuss fitting equations to data. Let us now consider two examples of the calculation of the covariance and correlation coefficient.

Example 5.11. Suppose that we have the joint distribution of Example 3.10,

		x_1		
		1	2	3
x_2	1	0.15	0.20	0.10
	2	0.35	0.05	0.15

and that we want to compute the covariance between X_1 and X_2, $E\{X_1X_2\} - \mu_1\mu_2$.

The marginal probability distributions are

x_1	1	2	3
$p_{X_1}(x_2)$	0.50	0.25	0.25

and

x_2	1	2
$p_{X_2}(x_2)$	0.45	0.55

and their means and variances are $\mu_1 = 1.75$, $\mu_2 = 1.55$, $V(X_1) = \sigma_1^2 = 0.69$, and $V(X_2) = \sigma_2^2 = 0.25$. The computation of $E\{X_1X_2\}$ is straightforward,

$$E\{X_1X_2\} = \sum\sum x_1x_2p(x_1, x_2) = (1)(1)(0.15) + (1)(2)(0.35) + (2)(1)(0.20)$$

$$+(2)(2)(0.05) + (3)(1)(0.10) + (3)(2)(0.15) = 2.65$$

Therefore, $\text{Cov}(X_1, X_2) = E\{X_1X_2\} - \mu_1\mu_2 = 2.65 - (1.75)(1.55) = -0.0625$. The correlation coefficient is obtained in one additional step as

$$\text{Corr}(X_1, X_2) = \frac{-0.0625}{\sqrt{(0.69)(0.25)}} = -0.1505$$

Example 5.12. Suppose that we have the joint density of Example 4.14,

$$f(x_1, x_2) = x_1 + x_2 \qquad 0 \le x_1 \le 1, \quad 0 \le x_2 \le 1$$

and that we want to compute the covariance between X_1 and X_2, $E\{X_1X_2\} - \mu_1\mu_2$.

Since the marginal probability distributions are $f_{X_1}(x_1) = x_1 + 0.5$ and $f_{X_2}(x_2) = x_2 + 0.5$, $\mu_1 = \mu_2 = 0.583$ and $\sigma_1^2 = \sigma_2^2 = 0.076$.

$$E\{X_1X_2\} = \int\int x_1x_2f(x_1, x_2)\,dx_1\,dx_2$$

$$= \int\int x_1x_2(x_1 + x_2)dx_1dx_2 = 1/3 = 0.333$$

which leads directly to

$$\text{Cov}(X_1, X_2) = E\{X_1X_2\} - \mu_1\mu_2 = 0.333 - (0.583)(0.583) = -0.00689$$

On substitution into Equation 5.11, we find that the correlation coefficient is

$$\text{Corr}(X_1, X_2) = 0.00689/0.076 = -0.091$$

The means and variances of the marginal probability distribution functions in Examples 5.11 and 5.12 have been obtained by using the plotting programs in your software package for user–provided distributions.

5.7 FUNCTIONS OF ONE OR MORE RANDOM VARIABLES

In practical analysis, there is often a need to obtain or to approximate the probability distribution of a function of one or more random variables. If $Y = g(X)$ is a function of a random variable, X, i.e., $g(X)$ is an expression like $2X$, $X^2 + 4X$, or $\sin(X)$, then $Y = g(X)$ is *also* a random variable and its probability distribution function can be derived from the probability distribution function of X. Similarly, if $Y = h(V, W)$ is a function of the random variables V and W, then Y is a random variable and its probability distribution can be derived from the joint probability distribution of V and W. Carrying this argument to its logical conclusion, we may state that *any function of one or more random variables is a random variable itself.*

Deriving the probability distribution function of our new random variable from the distribution(s) of the original variable(s) is a straightforward process when X is a discrete random variable. When X is a continuous random variable, the process usually requires *much* more sophisticated mathematical methods. For this reason, we limit our attention to functions of *discrete* random variables. The first kind of function we consider is a simple *algebraic transformation* of *one* random variable.

Example 5.13. Consider the experiment of tossing a coin 2 times and recording the outcome of the tosses. The sample space for this experiment could be described as HH, HT, TH, and TT. Let the random variable, X, be the number of tails occurring in the 2 tosses. The probability distribution for X is

x	0	1	2
$p(x)$	0.25	0.5	0.25

Suppose that $Y = g(X) = X^2$. We say that there is a *one-to-one mapping* between X and $Y = g(X)$; i.e., each value of X is associated with only one value of Y, and each value of Y is associated with only one value of X. More specifically,

$$\begin{aligned} X = 0 \rightarrow X^2 = 0 & \qquad & P(X^2 = 0) = P(X = 0) = 0,25 \\ X = 1 \rightarrow X^2 = 1 & \qquad \text{which implies that} \qquad & P(X^2 = 1) = P(X = 1) = 0.50 \\ X = 2 \rightarrow X^2 = 4 & \qquad & P(X^2 = 4) = P(X = 2) = 0.25 \end{aligned}$$

Thus, the probability distribution of $Y = X^2$ is

y	0	1	4
$p(y)$	0.25	0.5	0.25

Example 5.14. Consider another random variable, X, with the following probability distribution function.

x	−2	−1	0	1	2	3
$p(x)$	0.0625	0.1875	0.25	0.125	0.25	0.125

Again, let $Y = X^2$. Although X^2 can assume only the values 0, 1, 4, or 9, more than one value of X will yield a specific value of Y. Therefore, a one-to-one mapping is *not* present. Specifically,

$$\begin{aligned} X = 3 \rightarrow X^2 = 9 & \qquad & P(X^2 = 9) &= P(X = 3) = 0.125 \\ X = 2, -2 \rightarrow X^2 = 4 & \qquad & P(X^2 = 4) &= P(X = 2 \text{ or } X = -2) \\ & \qquad \text{yields} & &= 0.25 + 0.0625 = 0.3125 \\ X = 1, -1 \rightarrow X^2 = 1 & \qquad & P(X^2 = 1) &= P(X = 1 \text{ or } X = -1) \\ X = 0 \rightarrow X^2 = 0 & \qquad & &= 0.125 + 0.1875 = 0.3125 \\ & & P(X^2 = 0) &= P(X = 0) = 0,25 \end{aligned}$$

Thus the probability distribution of $Y = X^2$ is

y	0	1	4	9
$p(y)$	0.25	0.3125	0.3125	0.125

Since Y is a random variable, we may compute its expected value as follows.

$$E\{Y\} = \sum y_i p(y_i) = (0)(0.25) + (1)(0.3125) + (4)(0.3125) + (9)(0.125) = 2.6875$$

As we have seen, $E\{g(X)\} = E\{X^2\}$ may also be determined as follows.

$$\begin{aligned} E\{X^2\} = \sum x_i^2 p(x_i) &= (4)(0.0625) + (1)(0.1875) +)(0)(0.25) + (1)(0.125) \\ &+ (4)(0.25) + (9)(0.125) = 2.6875 \end{aligned}$$

Notice that these two methods of computing $E(Y) = E(X^2)$ are *equivalent*. Either method correctly gives the value for the mean of the random variable, X^2. These two approaches yield identical results for $E\{g(X)\}$ for any random variable $g(X)$. Next, let us consider the *convolution* of two *independent* discrete random variables.

Example 5.15. Suppose that a device consists of components 1 and 2. The device will function successfully if and only if *both* components work; i.e., if either component fails, the device fails. If component 1 succeeds, the random variable X_1 is equal to 1, and if it fails, X_1 is 0. In like manner, if component 2 succeeds, the random variable X_2 is equal to 1, and if it fails, X_2 is 0. Suppose that the following Bernoulli distributions govern X_1 and X_2, respectively.

	Value	
Component	**0**	**1**
x_1	0.7	0.3
x_2	0.4	0.6

Let $Y = X_1 + X_2$. If X_1 and X_2 are statistically independent, the four possible value combinations of x_1 and x_2 that can occur, along with their respective probabilities are the following.

(x_1, x_2)	**Probability**	$y = x_1 + x_2$
0,0	(0.7) (0.4) = 0.28	0
0,1	(0.7) (0.6) = 0.42	1
1,0	(0.3) (0.4) = 0.12	1
1,1	(0.3) (0.6) = 0.18	2

The probability distribution of y is

y	0	1	2
$p(y)$	0.28	0.54	0.18

Thus we see that the probability that the device will function successfully is $p(2) = 0.18$.

To obtain the probability distribution of Y, we have considered every possible value of X_1 in conjunction with every possible value of X_2. In so doing, we have performed a *summation convolution* of the probability distributions of X_1 and X_2. (Recall that we have also performed a summation convolution in Example 3.1, where we find the probability distribution of the sum of two fair dice.) In our current example, the convolution has yielded two combinations that result in the same value of y. Before the probability distribution of Y could finally have been determined, the probabilities associated with the *different ways* to obtain $Y = 1$ had to be added together.

We are *not limited* to performing *summation* convolutions. Indeed, a great number of mathematical operations can be used in performing a convolution. Thus we may write $Y = X_1 \otimes X_2$, where $\otimes$ is a symbol for the *general convolution operator*. When a summation convolution is performed, $\otimes = +$. A nonexhaustive list of convolution operators is given in Table 5.1.

TABLE 5.1
Some convolution operators

Operation	**Operator**
Summation	+
Difference	−
Quotient	/
Product	*
Exponentiation	^

Example 5.16. It is also possible to convolve a random variable with itself. Suppose that we want to know the probability distribution function of $W = X_3 * X_3$, where the probability distribution function of X_3 is

x_3	-3	1	3
$p(x_3)$	0.8	0.1	0.1

This would require the performance of a *product convolution* of all possible values of X_3 with one another, as follows.

Combination (x_3, x_3)	**Probability**	$w = x_3 \times x_3$
$-3, -3$	$(0.8)(0.8) = 0.64$	9
$-3, 1$	$(0.8)(0.1) = 0.08$	-3
$-3, 3$	$(0.8)(0.1) = 0.08$	-9
$1, -3$	$(0.1)(0.8) = 0.08$	-3
$1, 1$	$(0.1)(0.1) = 0.01$	1
$1, 3$	$(0.1)(0.1) = 0.01$	3
$3, -3$	$(0.1)(0.8) = 0.08$	-9
$3, 1$	$(0.1)(0.1) = 0.01$	3
$3, 3$	$(0.1)(0.1) = 0.01$	9

Adding together the probabilities of the outcomes that have yielded the same value w, we find that the probability distribution of W is

w	-9	-3	1	3	9
$p(w)$	0.16	0.16	0.01	0.02	0.65

We have just performed the product convolution of X_3 with itself; i.e., $W = X_3 * X_3$. At first glance, this procedure might appear to be the same as the algebraic transformation $U = X_3^2$. After all, from a strict algebraic viewpoint, $(x_3)(x_3)$ *is* equal to x_3^2. However, in our current context, the product convolution $W = X_3 * X_3$ is *not* the same as the transformation $U = X_3^2$.

Example 5.17. In order to obtain the probability distribution of $U = X_3^2$ for the X_3 of Example 5.16, we take each possible specific value X_3, square it, and sum the probabilities of any common resulting values. For this example, this procedure yields

u	1	9
$p(u)$	0.1	0.9

In this case, the 3 x_3 values, when squared, yield only 2 u values, and the resulting probability distribution is *markedly* different from the result of the product convolution $W = X_3 * X_3$.

The conceptual understanding of *convolutions* and *algebraic transformations* is not difficult. However, the actual implementation of such procedures can quickly become cumbersome. For this reason, the software package contains programs for performing both transformations and convolutions. With these programs, it is very easy to perform sequential operations, using convolutions and transformations, to find the probability distribution function of discrete random variables such as $Y = X_1 * X_2 + X_3^3$. You are given several opportunities in the exercises to use the programs for transformation and convolution. After completion of those exercises, you should have a very good understanding of how to form various *algebraic combinations* of discrete random variables.

5.8 A COMMENT ABOUT THE "ROAD AHEAD"

The end of this chapter marks, in a sense, the end of the part of this book that is designed to give you a sound theoretical base in probabilistic phenomena. This is not to say that what has gone before is purely theoretical, with no practical application. Almost nothing could be farther from the truth. However, the chapters that follow have more emphasis on practical applications. For the most part, the remaining chapters contain material whose mastery will enable you to

- correctly obtain data from physical processes of interest
- analyze that data fully and appropriately
- derive information from that analysis that will enhance your ability to arrive at correct conclusions about the process under study and to make correct decisions about that process

EXERCISES

At this point in the book, the differentiation between exercises and computer-based exercises is discontinued because it is no longer as meaningful as in the earlier chapters. Almost all the exercises will have parts that can be performed with the aid of the software package.

5.1. Using the information contained in Equations 5.5 and 5.6, show that
(a) $\sigma^2 = \mu_2 = \mu_2' - \mu^2$
(b) $\mu_3 = \mu_3' - 3\mu\mu_2' + 2\mu^3$
(c) $\mu_4 = \mu_4' - 4\mu\mu_3' + 6\mu^2\mu_2' - 3\mu^4$

5.2. Suppose that the discrete random variable, X, has the following probability distribution function.

x	0.5	5	7	11
$p(x)$	0.333	0.167	0.375	0.125

(a) What is the mean of X? the variance of X? the median of X? the mode of X?
(b) What is the value of β_1? of β_2?
(c) What is the expected value of the function $h(X)$, where $h(X) = X^2 + X + 1$?
(d) What is the probability distribution of $Y = h(X) = X^2 + X + 1$?
(e) What is the mean of Y? the variance of Y? the median of Y? the mode of Y?
(f) For Y, what is the value of β_1? of β_2?

5.3. Let the probability density function of X be defined as

$$f(x) = \begin{cases} 0; x < 4 \\ kx; 4 \le x \le 8 \\ 0; x > 8 \end{cases}$$

(a) What value must k assume?
(b) Given that k has the value in part (a), what is the mean of X? the variance of X? the median of X? the mode of X? the standard deviation of X?
(c) What is the value of β_1? of β_2?
(d) Suppose that $g(X) = 2X$. Calculate $E\{g(X)\}$ [k = the value found in part (a)].
(e) Using a discrete approximation of $f(x) = kx$, with at least 20 points, construct an approximation to the distribution of $Y = 2X$. [Using continuous transformation methods, the distribution of $Y = 2X$ for $f(x) = kx$, may be shown to be $f(y) = 0.25ky$ for $8 \le y \le 16$.]
What are the similarities and differences that you can observe between your discrete approximation and the theoretical continuous result?

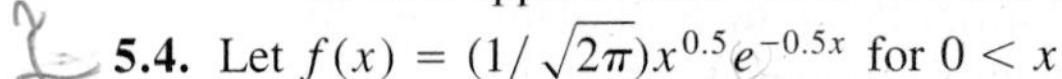

5.4. Let $f(x) = (1/\sqrt{2\pi})x^{0.5}e^{-0.5x}$ for $0 < x$.
(a) What is the mean of X? the variance of X? the median of X? the mode of X? the standard deviation of X?
(b) What is the value of β_1? of β_2?
(c) Suppose that $g(X) = X^{-0.5}$. Calculate $E\{g(X)\}$.

5.5. Show that μ, σ^2, β_1 and β_2 are identical for each case.
(a) For the beta and uniform distributions when $a = 0$, $b = 1$, $\alpha = 1$ and $\beta = 1$.
(b) For the Weibull and gamma distributions when $\alpha = \lambda$, $\beta = 1$, and $r = 1$.

5.6. Show that in the limit as n grows large and p grows small but $\lambda = np$ is *held constant*, μ, σ^2, β_1 and β_2 become identical for the binomial and Poisson distributions.

5.7. Perform summation convolutions on the following distributions to achieve the probability distribution for the random variable that results when X is added to itself twice (i.e., $X + X$), 4 times, 8 times, and 16 times. Plot each of the distributions, and compute their values of μ, σ^2, β_1, and β_2. Are any general results observable? If so, state them.

(a)

x	1	2	3	4	5
$p(x)$	0.2	0.2	0.2	0.2	0.2

(b)

x	1	2	3	4	5	6	7	8
$p(x)$	0.5	0.25	0.125	0.0625	0.03125	0.015625	0.0078125	0.0078125

(c)

x	1	2	3	4	5	6	7	8
$p(x)$	0.0004	0.0096	0.02	0.03	0.06	0.13	0.25	0.5

5.8. Obtain and plot the distribution that governs the random variable $Y = X_1 * X_2 + X_3^3$ if the distributions governing X_1, X_2, and X_3 are as given in the following.

The distribution of X_1 is a binomial distribution with $n = 8$ and $p = 0.35$.
The distribution of X_2 is as follows.

x_2	1	2	3	4	5
$p(x_2)$	0.01456	0.165013	0.435035	0.326276	0.059040

The distribution of X_3 is as follows.

x_3	1	2	3	4	5	6	7	8
$p(x_3)$	0.0004	0.0096	0.02	0.03	0.06	0.13	0.25	0.5

5.9. Form a discrete approximation to an $N(0, 1)$ distribution using at least 30 points, and transform the distribution to $Y = X^2$. How close is your result to a χ_1^2 random variable? Next, take your result and perform a summation convolution of that result with itself to obtain an approximation of a χ_2^2 random variable. How do your results agree with the theoretical distribution?

5.10. Show that the first moment about the mean is *always* zero.

5.11. Compute the covariance and correlation coefficient for each of the following joint distributions.

(a) Exercise 3.32
(b) Exercise 3.33
(c) Exercise 3.34
(d) Example 4.9
(e) Example 4.12
(f) Exercise 4.30
(g) Exercise 4.31
(h) Exercise 4.32
(i) Exercise 4.33
(j) Exercise 4.34
(k)

v \ u	−4	−2	2	4
−2	0.00	0.25	0.00	0.00
−1	0.00	0.00	0.00	0.25
1	0.25	0.00	0.00	0.00
2	0.00	0.00	0.25	0.00

5.12. Verify that, for an exponential distribution, $P\{\mu - 2\sigma \le X \le \mu + 2\sigma\} = 0.9502$.

5.13. Suppose that you have a probability distribution function but all you know about it is that $\mu = 4.5$ and $\sigma = 2.25$. Use Tchebycheff's inequality to find the symmetric range about μ that is certain to contain 99% of the distribution's values.

5.14. Evaluate—in general—μ, σ^2, β_1, and β_2 for the Bernoulli distribution function; i.e., show that the functions of p and q given in Table 3.2 are correct.

5.15. Evaluate, in general,

(a) μ and σ^2 for the binomial distribution function
(b) μ and σ^2 for the Poisson distribution function
(c) μ for the geometric distribution function
(d) μ for the hypergeometric distribution function

Hints: $\sum_{x=0}^{n} {}_nC_x a^x b^{n-x} = (a + b)^n$ and $\sum_{x=0}^{\infty} \frac{m^x}{x!} = e^m$

5.16. Evaluate, in general, μ, σ^2, β_1, and β_2 for the exponential density function.

5.17. Evaluate, in general, μ and σ^2 for the
(a) uniform density function
(b) beta density function
(c) Weibull density function
(d) normal density function
(e) gamma density function

5.18. Evaluate μ and σ^2 for each of the following distribution functions.
(a) Exercise 4.1
(b) Exercise 4.2
(c) Exercise 4.3
(d) Exercise 4.20
(e) Exercise 4.21
(f) Exercise 4.22
(g) Exercise 4.28
(h) Exercise 3.1
(i) Exercise 3.2
(j) Exercise 3.3
(k) Exercise 3.13
(l) X_1 in Exercise 3.32
(m) X_2 in Exercise 3.33
(n) X_2 in Exercise 4.30
(o) X_1 in Exercise 4.33
(p) Exercise C4.2

5.19. Compute the variances for the t distribution functions with 1 degree of freedom and with 2 degrees of freedom.

5.20. Evaluate, in general, the mode for the
(a) uniform density function
(b) gamma density function
(c) χ^2 density function

5.21. The Pareto distribution function has the general form,

$$f(x) = \psi\eta^{\psi}x^{-(\psi+1)} \qquad x \geq \eta \text{ and } \psi, \eta > 0$$

(a) Show that this is a valid probability distribution for all ψ and η satisfying the stated conditions.
(b) Plot $f(x)$ for selected values of the parameters ψ and η, and then describe, in general, the effects that ψ and η have on the shape of a Pareto distribution.
(c) Compute, in general, μ and σ^2 for $f(x)$.
(d) If $\psi \leq 1$, what can be said about μ?
(e) If $\psi = 2$, what can be said about σ^2?

5.22. The Cauchy distribution is defined by

$$f(x) = \frac{1}{\sigma\pi}\frac{1}{1 + \left(\frac{x-\mu}{\sigma}\right)^2} \qquad -\infty < x < \infty$$

(a) Determine whether this is a valid probability distribution function.
(b) Does this distribution possess a defined mean or variance?
(c) How do the parameters μ and σ affect the shape of the distribution?

5.23. Given the following probability distribution,

$$f(x) = \frac{2}{\zeta}\left(1 - \frac{x}{\zeta}\right) \qquad 0 < x < \zeta$$

(a) Show that $f(x)$ forms a valid probability distribution.
(b) Plot $f(x)$ for selected values of ζ, and then describe, in general, the effect that ζ has on the shape of the distribution.
(c) Compute, in general, μ and σ^2 for $f(x)$.
(d) Compute, in general, the median and the mode of $f(x)$.

5.24. Find L such that $p(x) = L/x!$, where $x = 0, 1, 2, \ldots$, is a valid discrete probability distribution function. Compute, in general, μ and σ^2 for $p(x)$.

5.25. In Example 5.6, it is stated that $\int_0^\infty x^2\sqrt{\pi}\exp(-0.25x^2)dx = 2\pi$. Show that this is true.

5.26. Develop and state the probability distribution function of the *three* random variables—specifically, the sum of the uppermost faces of k fair dice in a single roll of the dice, where k is set equal to 3, 4, and 5.

5.27. Show by an example that the binomial distribution with parameters n and p is equivalent to the distribution of the sum of n Bernoulli random variables with parameter p. (Use $p = 0.4$ and $n = 5$.)

5.28. Show by an example that the negative binomial distribution with parameters r and p is equivalent to the distribution of the sum of r geometric random variables with parameter p. (Use $p = 0.4$ and $r = 3$.)

CHAPTER 6

CLASSIFICATION AND DESCRIPTION OF SAMPLE DATA

The following concepts and procedures are essential to a good understanding of the contents of this chapter.

- Be able to construct and interpret a *frequency table, a cumulative frequency table, a percentage table, a cumulative percentage table, a histogram,* and *a frequency polygon* for any given data set
- Be able to obtain $\overline{x}$, s^2, and s for a given data set
- Understand the ideas behind the concept of *degrees of freedom*
- Be able to obtain and interpret *sample estimates* of the mode, median, range, skewness, and kurtosis based on a given data set

6.1 INTRODUCTION

Few people can make any sense out of the 100 numbers presented in Table 6.1 just by looking at them.

Often, *thousands* of numbers—not just a hundred—must be analyzed in practical problems. For this reason, we must first operate on the numbers in a *logical and planned* manner so that we achieve a *condensation,* or *summary,* that preserves the important properties of the data set. The part of statistics that deals with this activity and with the collection of the data is called *descriptive statistics*. In the efforts associated with descriptive statistics, we make *no* attempt to draw conclusions from the data. Drawing conclusions is the concern of the other branch of statistics, *inferential statistics*, which we will consider in detail in the later chapters.

TABLE 6.1
One hundred data values

−1.54	1.71	−0.51	0.59	−1.08	−1.18	−0.80	0.46	0.82	0.80
0.69	−1.41	−0.87	−0.27	0.80	−0.81	1.47	0.11	1.60	−0.42
0.98	0.09	−0.48	−0.48	0.25	−0.62	−0.81	−0.36	1.45	−0.13
0.78	1.33	0.03	−1.32	−0.09	0.17	0.47	−0.45	0.34	−0.72
−0.47	0.93	0.89	0.35	0.22	−1.92	0.20	−0.34	−1.05	2.18
−0.60	−0.91	−0.23	−1.50	0.47	0.41	−0.77	0.70	1.10	−1.31
0.79	−0.39	−1.16	−0.50	−0.60	−2.15	−0.73	−1.24	1.35	0.32
−0.21	0.29	0.07	−1.00	0.28	1.56	0.43	1.41	0.04	0.05
0.88	0.12	−1.13	1.09	−1.04	−0.92	−0.69	0.30	0.08	0.30
1.89	0.05	−1.08	−1.50	0.84	1.92	−0.68	−0.11	0.71	1.78

6.2 THE FREQUENCY TABLE AND ITS OUTGROWTHS

One way to achieve a summary of a data set is by forming a *frequency table,* or *frequency distribution.* A frequency table is a collection of *classes* defined by *class limits.* The classes cover the entire range of the data values and do not overlap—and all classes are of the same size. The common size of all classes is the *class interval.* The midpoint of any class is the *class mark.* The *class frequency* is the number of data points that fall into a particular class. Note that the sum of all class frequencies must equal the size of the sample.

Example 6.1. One possible set of classes for the data in Table 6.1—each class with its associated class frequency—is given in Table 6.2. In the data analysis program in your software package, the program asks you to provide the lower class limit of the lowest class (−2.50 for Table 6.2) and the common class interval (0.50 for Table 6.2). The first class includes data values, x, such that $-2.50 \leq x < -2.00$ (note the *strict inequality* defining the upper limit on the class), the second class includes values such that $-2.00 \leq x < -1.50$, and so on. This definition of classes removes any ambiguity about the class membership of any particular data point; the 10 classes cover all values of the data from the minimum of −2.15 to the maximum of 2.18, they do not overlap, and they all have the same class interval of 0.50 unit.

TABLE 6.2
A frequency table

Class	Class Frequency
−2.50 to < −2.00	1
−2.00 to < −1.50	2
−1.50 to < −1.00	13
−1.00 to < −0.50	16
−0.50 to < 0.00	15
0.00 to < 0.50	25
0.50 to < 1.00	14
1.00 to < 1.50	7
1.50 to < 2.00	6
2.00 to < 2.50	1

TABLE 6.3
A cumulative frequency table

Class	Frequency
< -2.50	0
< -2.00	1
< -1.50	3
< -1.00	16
< -0.50	32
< -0.00	47
< 0.50	72
< 1.00	86
< 1.50	93
< 2.00	99
< 2.50	100

A *cumulative frequency table (distribution)* tabulates the number of data points that are members of the current or any lower class. It is constructed by summing the class frequencies from the lowest class up to and including the class of current interest.

Example 6.2. The cumulative frequency table for the data of Table 6.1 is given in Table 6.3. Notice that a value of zero is reported for an *implied* class, which is one class interval lower than our lowest nonempty class. This implied class is useful in the graphical techniques described later.

Prior to the availability of statistical software packages, frequency tables were constructed by hand, a cumbersome and time-consuming process. For this reason, other less sophisticated and less complicated methods were developed and used as *preliminary* analysis techniques. Two of these techniques are the *dot diagram* and the *stem-and-leaf* table. The dot diagram, in which each value in the data set becomes its own class, is most useful when many repeated values are present in the data set.

Example 6.3. The following data are the lengths (in inches) of 30 cotter pins taken from a single day's production.

3.39	3.41	3.38	3.39	3.42	3.40	3.41	3.39	3.44	3.43
3.38	3.40	3.37	3.41	3.40	3.41	3.38	3.42	3.43	3.39
3.40	3.41	3.46	3.39	3.37	3.42	3.41	3.40	3.38	3.41

Here is a dot diagram for this data set:

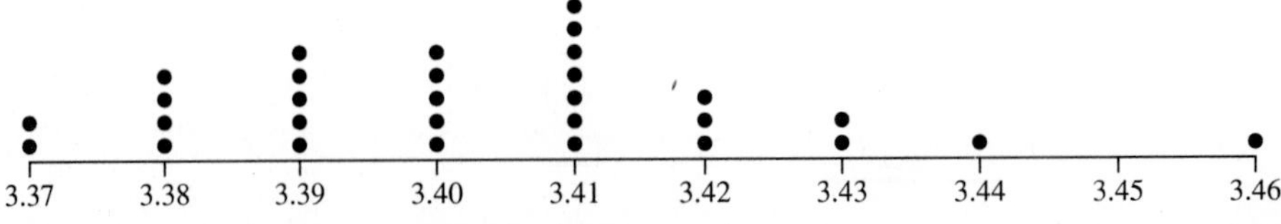

A stem-and-leaf table splits the values of the data set into two parts. Usually the "stem" consists of the first one or two digits of the value, and the "leaves" are the remaining digits.

Example 6.4. Let us select the first two digits as the stem of a stem-and-leaf table and apply the technique to the following data concerned with the internal diameter of cast-iron pipe (in inches).

2.165	2.829	2.348	2.443	2.642	2.287	2.065	2.274	2.248	2.211
2.509	2.537	2.123	2.049	2.154	2.374	2.546	2.225	2.039	2.728
2.203	2.344	2.163	2.168	2.202	2.643	2.622	2.965	2.539	2.700
2.003	2.424	2.112	2.379	2.020	2.460	2.290	2.039	2.911	2.426
2.925	2.984	2.073	2.192	2.119	2.171	2.455	2.473	2.242	2.154
2.699	2.407	2.193	2.297	2.030	2.381	2.119	2.222	2.896	2.113
2.180	2.511	2.860	2.979	2.065	2.664	2.738	2.088	2.254	2.295
2.194	2.615	2.040	2.094	2.378	2.873	2.289	2.023	2.814	2.649
2.959	2.271	2.046	2.275	2.864					

The stem-and-leaf plot of this data is given next.

2.0 ‖ 49, 39, 73, 30, 40, 23, 46, 94, 88, 65, 03, 20, 39, 65
2.1 ‖ 65, 23, 54, 63, 68, 12, 92, 19, 71, 54, 19, 13, 80, 93, 94
2.2 ‖ 87, 74, 11, 25, 03, 02, 42, 97, 22, 54, 95, 89, 71, 75, 90, 48
2.3 ‖ 48, 74, 44, 79, 81, 78
2.4 ‖ 43, 24, 60, 26, 55, 73, 07
2.5 ‖ 09, 37, 46, 39, 03, 11
2.6 ‖ 42, 43, 22, 99, 64, 15, 49
2.7 ‖ 28, 00, 38
2.8 ‖ 29, 96, 60, 73, 14, 64
2.9 ‖ 65, 11, 25, 84, 79, 59

Examples 6.3 and 6.4 show that a dot diagram or a stem-and-leaf table can provide more insight into the data. However, you will usually obtain superior information through the frequency table and its by-products. The data analysis program in your software package performs all of the clerical tasks associated not only with a frequency table but also with all of the remaining tables and graphical techniques directly related to the frequency table.

Percentage tables and *cumulative percentage tables* are useful extensions of the frequency and cumulative frequency tables. These extensions are formed by dividing each class frequency or cumulative class frequency by the total number of data points.

Example 6.5. Table 6.4 presents the output from the data analysis package for the 79 values in the data set given below. The fourth and fifth columns are called the percentage and cumulative percentage tables for the selected set of classes.

43	52	53	54	45	54	59	54	48	40	56	53
59	42	43	49	51	51	44	57	51	45	50	42
51	56	57	50	50	51	55	50	53	48	46	47
51	51	57	52	48	56	50	47	53	57	55	50
50	49	52	48	51	60	48	53	62	51	49	61
42	55	51	60	51	50	57	55	49	53	56	49
46	50	56	47	50	47	54					

TABLE 6.4
Output from the data analysis program for Example 6.5

Class	Frequency	Cumulative Frequency	Percentage	Cumulative Percentage
40 to <44	6	6	7.59	7.59
44 to <48	9	15	11.39	18.99
48 to <52	31	46	39.24	58.23
52 to <56	17	63	21.52	79.75
56 to <60	12	75	15.19	94.94
60 to <64	4	79	5.06	100.00

The following are two primary reasons for the percentage and cumulative percentage tables.

1. They are useful for comparing two data sets that have unequal sample sizes.
2. When plotted in the manner described in Section 6.3, they form a *graphical approximation* of the underlying probability and cumulative probability distribution functions. This approximation is better for large samples and is more accurate for values in the center of the distribution than in the tails of the distribution.

6.3 GRAPHICAL PRESENTATIONS OF THE FREQUENCY TABLE

The tabular methods just described can be clarified and amplified by graphical means. One graphical method is the *frequency polygon,* which is a graph of the class frequencies against the class marks.

Example 6.6. The frequency polygon for the frequency table of Table 6.2 is given in Figure 6.1. Each plotted point is connected to its adjacent points by a straight-line segment. (This forms a *polygon,* hence the name.) In addition, two implicit classes, with frequencies of zero, are included at either end of the polygon. This provides a visually appealing closure to the frequency polygon, and, in addition, it makes the polygon more closely resemble the shape of the underlying probability distribution function. A *cumulative frequency polygon,* or *ogive*, may be formed by plotting the class marks against the cumulative class frequencies (with the implicit lower class interval as described in Example 6.2).

Percentage polygons and *cumulative percentage* polygons are similar to frequency polygons and cumulative frequency polygons. The only difference is that the vertical axis has been *scaled* to range between 0 and 1 by dividing all class frequencies by the total number of observations.

Although a *pie chart* is not a direct outgrowth of the frequency polygon, the information contained in a percentage polygon can also be presented in such a chart. A pie chart is a disk for which the wedge sizes are allocated in direct proportion to the percentages represented.

Example 6.7. A pie chart based on the percentages of Table 6.2 is given in Figure 6.2.

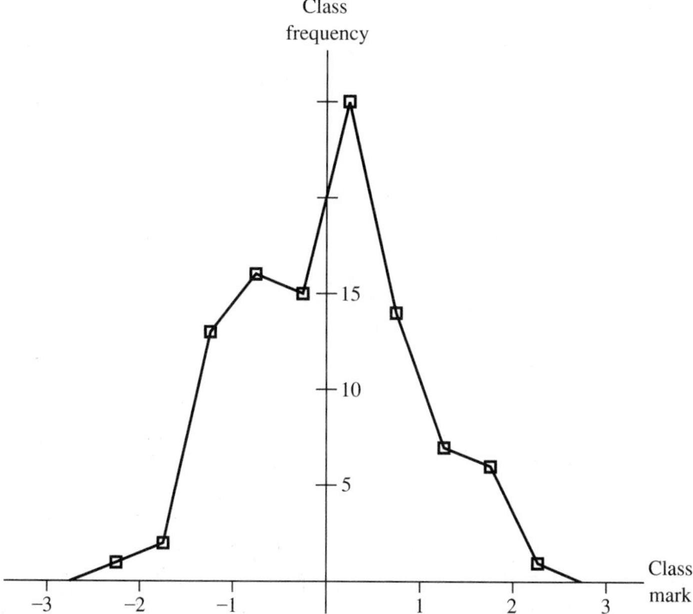

FIGURE 6-1
A frequency polygon for Table 6.1.

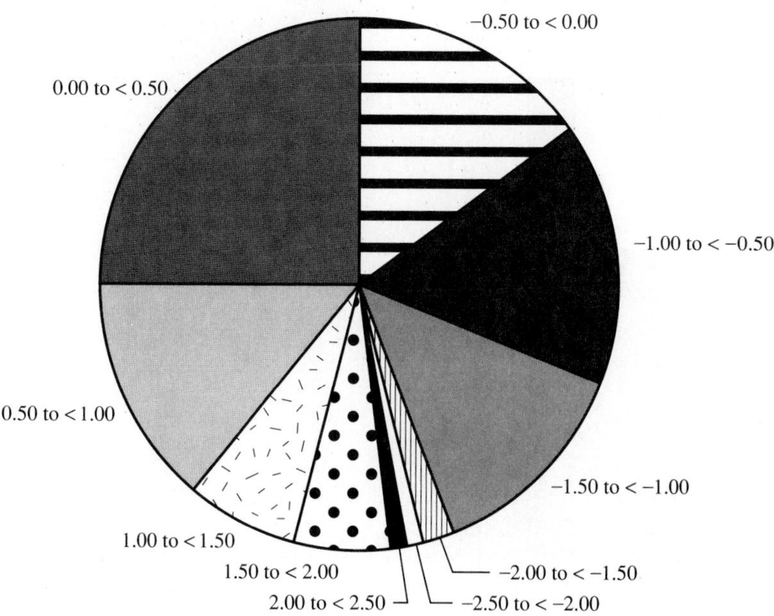

FIGURE 6-2
A pie chart for Table 6.2.

A *histogram,* or *bar graph,* is another way to illustrate the contents of a frequency polygon. It has a rectangle, for each class, that spans the class interval at the height of the class frequency.

Example 6.8. A histogram for Table 6.1 is shown in Figure 6.3.

It is a rare occasion when the first set of selected class intervals provides an acceptable frequency table and frequency polygon or histogram. By varying the lowest class limits and the common value of the class interval, an analyst can usually arrive at an acceptable frequency table. Of course, this process would be prohibitively time-consuming without a data analysis program like the one accompanying your text.

Example 6.9. An example of a better set of classes is presented in Figure 6.4, which was constructed by setting the lowest class limit to −2.75 while retaining the old class interval of 0.50 (or, equivalently, shifting the previous classes 0.25 units to the left). The frequency polygon of Figure 6.4 is much "smoother" than the frequency polygon of Figure 6.1. The superiority of the polygon in Figure 6.4 is further verified by the knowledge that the data of Table 6.1 was drawn from a standardized normal, N(0, 1), distribution. In general, if you have too many class intervals, the frequency polygon will be too "jagged," or "sawtoothed." If there are too few class intervals, the frequency polygon will be too smooth and will be unable to capture any detail about the underlying distribution governing the data.

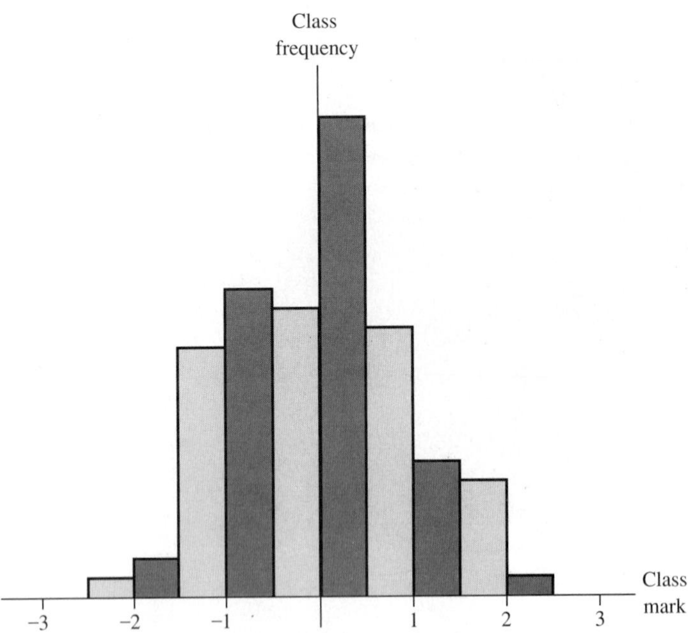

FIGURE 6-3
A histogram of Table 6.1.

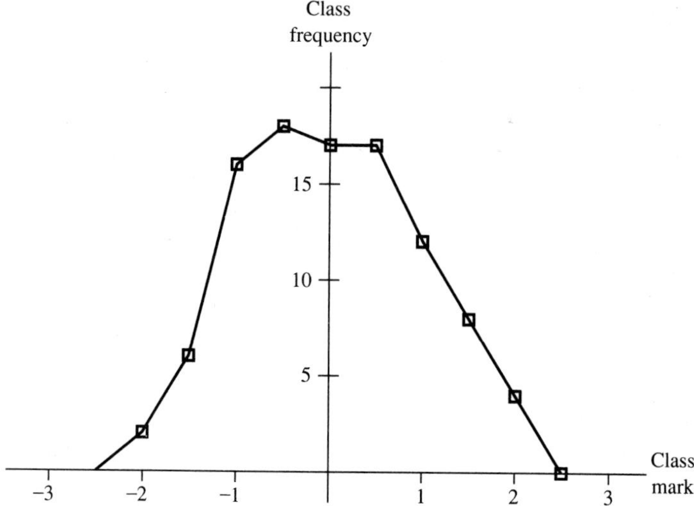

FIGURE 6-4
An improved frequency polygon for Table 6.1.

Each of the univariate methods presented may be generalized to bivariate data, and the frequency table may be generalized to multivariate data. Of course, the bookkeeping procedures and the selection of classes become more complicated as the dimensionality of the data increases. A fictitious example of a bivariate histogram is presented in Figure 6.5.

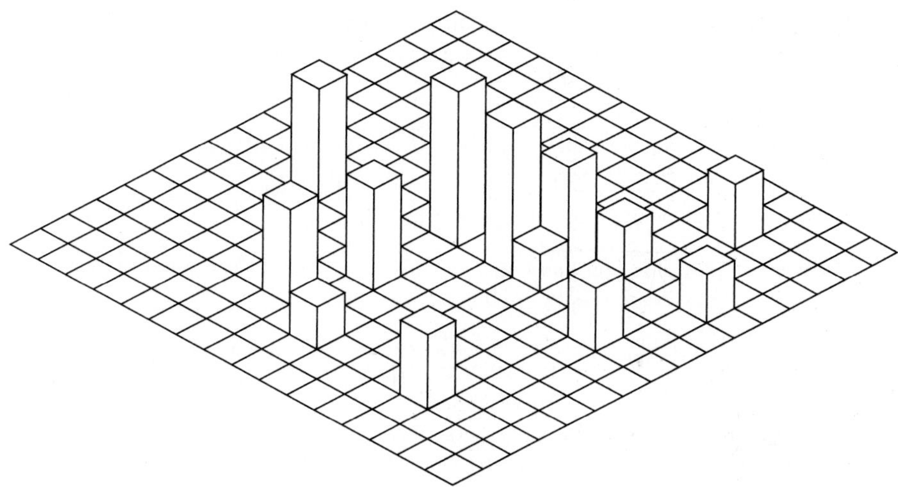

FIGURE 6-5
A two-dimensional histogram.

6.4 SAMPLE ESTIMATES

Although frequency table methods constitute a major step in summarizing data in an understandable form, it is generally useful to obtain even more *succinct descriptors* of the data set. As we have seen in Chapter 5, such theoretical quantities as the mean, median, mode, variance, range, skewness, and kurtosis are beneficial in summarizing general characteristics of a theoretical probability distribution function. It is a direct consequence of this fact that *sample estimates* of these same quantities give us similar insight into the underlying physical process and probability distribution from which the data was drawn.

Suppose that we have a sample data set of *size n*, with observed values $x_1, x_2, \ldots, x_n$ $(x_i, i = 1, 2, \ldots, n)$. The sample mean (or arithmetic average) for a data set is given by

$$\overline{x} = \frac{x_1 + x_2 + \cdots + x_n}{n} = \frac{\sum_{i=1}^{n} x_i}{n} \tag{6.1}$$

$\overline{x}$ (read "x-bar") is also a sample estimate of the first moment about the origin. If we assign a weight of $p_i = \frac{1}{n}$ to each x_i, then $\overline{x} = \sum_{i=1}^{n} p_i x_i$. This expression is a weighted sum of x_i and is very similar to Equation 5.1, where the theoretical mean, μ, for a probability distribution function is defined. Like the theoretical mean, the sample mean for a data set is a measure of centrality because it indicates where the center of most of the data is likely to be.

The sample variance for a data set is given by

$$s^2 = \frac{\sum_{i=1}^{n}(x_i - \overline{x})^2}{n - 1} \tag{6.2}$$

where s^2 is our best estimate of σ^2. The sample standard deviation for a data set is $s = \sqrt{s^2}$. It might seem that the divisor in Equation 6.2 should be n rather than $n - 1$. (It *was* n for the sample mean). The reason that the divisor is $n - 1$ is that we have *lost a degree of freedom*, i.e., an independent piece of information, by using the data to compute $\overline{x}$.

Example 6.10. Suppose that you have 10 data points but you know only the following nine values.

$$2, 4, 6, 8, 10, 10, 4, 6, 8$$

Further, suppose that you know that for this data, $\overline{x} = 6$. The missing data value *must* be 2. We can see this by solving the equation

$$\frac{\sum_{i=1}^{10} x_i}{10} = 6 = \frac{2 + 4 + 6 + 8 + 10 + 4 + 6 + 8 + x_{10}}{10} \rightarrow x_{10} = 2$$

If we do *not know* the value of $\overline{x}$, x_{10} could possibly assume *any value*. With the knowledge of $\overline{x}$, x_{10} can assume *only one value*. In the same way, if we know prior to an experiment that we are going to take 10 data points, each of those 10 points could conceivably achieve any one of an infinite set of possible values. Thus it is said that 10 degrees of freedom remain for that data set. If, however, the data are taken and $\overline{x}$ computed, only 9 degrees of freedom remain—because the knowledge of $\overline{x}$ and 9 values of the x_i *strictly define* what the value of the remaining x_i *must be*. We return to the idea of degrees of freedom in Chapter 8.

A pattern can be developed for estimates of *all* of the moments about the mean. The general form of the estimate of the kth moment about the mean for a data set is

$$u_k = \frac{\sum_{j=1}^{n}(x_j - \overline{x})^k}{n - 1} \tag{6.3}$$

If $k = 2$, we have the sample variance. If $k = 3$ or $k = 4$, we have the sample estimates associated with skewness and kurtosis, respectively. It follows directly that

$$b_1 = \frac{u_3^2}{s^6} = g_1^2 \quad \text{and} \quad b_2 = \frac{u_4}{s^4} = g_2 + 3 \tag{6.4}$$

Caution is advised in the *stringent* interpretation of b_1 and b_2 when the sample size is small; b_1 and b_2 are more greatly influenced than s^2 by values far from $\overline{x}$, and they require fairly large sample sizes before their values approach the true values associated with the distribution underlying the data. Exactly how large a sample size must be varies from case to case, and the general study of that question is beyond the scope of this book.

The sample range for a data set is the difference between the maximum and minimum values of the data set. The sample range for the data set of Table 6.1 is $R = 2.18 - (-2.15) = 4.33$. The sample range has a distinctly *unfortunate property*: The larger the size of the data set, the larger the range is likely to be. The reason is that the larger the data set, the more likely it will contain the rare extremely high or extremely low values. Therefore, the range is not a dependable relative measure of variability because its value is very closely tied to the size of the data set.

The sample variance, s^2, does not tend to increase with the sample size because of the divisor of $n - 1$ in Equation 6.2. In fact, as we see in Chapters 7 and 8, the larger the sample size, the better s^2 estimates the true population variance, σ^2.

The sample median for a data set, $\tilde{x}$, is the midpoint of the data set after it has been ordered by ascending value, from the smallest to the largest data point.

Example 6.11. The following 9 data values have a sample median equal to 10.

$$3\ 3\ 6\ 8\ \underline{10}\ 13\ 14\ 14\ 19$$

If there is an even number of data points, the sample median for the data set is the average of the two middle points. For example, suppose that only the first 8 data points are present:

$$3\ 3\ 6\ \underline{8\ 10}\ 13\ 14\ 14$$

The sample median would have a value of $(8 + 10)/2 = 9$.

The sample mode for a data set is defined in two ways. It is often defined to be equal to the most frequently occurring data value. However, when a sufficient

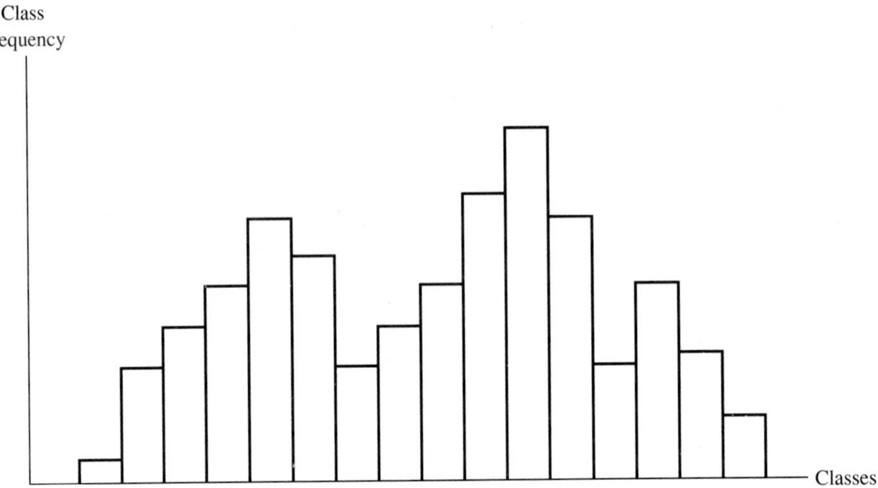

FIGURE 6.6
A histogram exhibiting two secondary modes.

number of data points are present, most practitioners define the mode in the *context of a frequency table*. In that context, the sample mode is the class mark of the class having the *highest class frequency*. The latter definition is the best one to use in most applications. It is possible that two or more modes may exist. If we use the first definition for the mode, the data set of nine values used in Example 6.11 would have 2 modes, one at 3 and the other at 14.

Using the second definition, consider the histogram in Figure 6.6. Many practitioners would say that this histogram exhibits three modes. Although there is only one maximum, there are three relative maxima. The relative maxima of lesser values are also called *secondary modes* of the data set.

In concluding the discussion on descriptive measures, a popular data-based sample estimate, with no real theoretical base, is the *sample coefficient of variation*, $cv = (s/\overline{x})100\%$. As might be inferred, cv gives a rough measure of relative variation by expressing the ratio of s as a percentage of $\overline{x}$.

Example 6.12. The following table presents the results from the sample estimate portion of the data analysis program in your software package when it is applied to the data of Table 6.1.

```
SAMPLE SIZE = 100
MINIMUM = -2.15,
MAXIMUM = 2.18
RANGE = 4.33
MEDIAN = .05
ARITHMETIC MEAN = .00359
VARIANCE = .89194
    (STANDARD DEVIATION = .94443)
COEFFICIENT OF VARIATION = 26307%
SKEWNESS -- B(1) = .02102(+)
KURTOSIS -- B(2) = 2.39519
```

Since we know that the data was sampled from an $N(0, 1)$ distribution, we may compare the *true values* with the values of the associated sample estimates.

	True Value	Estimated Value
Mean	0	0.00359
Variance	1	0.89194
Mode	0	−0.5 (from frequency polygon of Figure 6.4)
Median	0	0.05
Skewness	0	0.021(+)
Kurtosis	3	2.395

With the exception of the measure of kurtosis, the estimates appear to be "reasonably good." The discrepancy in the kurtosis measure serves to emphasize the caution needed in the presence of small sample sizes.

The next three chapters introduce methods that will enable you to logically quantify the "goodness" or accuracy of your sample estimates of the mean and variance, $\overline{x}$ and s.

6.5 SOME SUGGESTIONS FOR FURTHER STUDY

The material presented in the first four sections of Chapter 6 will be more than adequate for most engineering applications for which a classification and a description of data are required. However, a great many other techniques for exploratory data analysis, both graphical and statistical, have not been discussed in this chapter. There are three primary reasons for their exclusion: (1) Many of these other techniques have only historical interest and are inferior to the methods presented; (2) many of these methods are very specialized to the arena of analysis for which they have been designed, and they do not always work well in a general context; and (3) the remainder of these methods require concepts and sophistication beyond the scope of this book.

If you are interested in further study, I suggest the excellent text *The Visual Display of Quantitative Information,* authored by Edward R. Tufte [1983]. Tufte comprehensively discusses both the history of graphical methods of data analysis and the theory of data graphics, including quantitative methods of assessing the superiority of one method of graphical display over another. The works of Koopmans [1981] and Vellman and Hoaglin [1981] are also good starting points for further study into this subject area.

EXERCISES

6.1. The following table gives the selling prices of 62 houses in Mostun, Texas, in the spring of 1981.

280	275	225	195	178	156	154	149	144	139
134	129	129	125	124	122	119	114	110	104
96	95	94	94	93	92	89	88	87	85
79	79	77	76	74	74	74	73	71	71
70	69	69	67	66	65	62	62	61	60
59	57	56	55	54	53	52	51	51	49
48	39								

(a) Construct a frequency table, and draw a frequency polygon, using a class interval of 30 and starting the first class at 20.
(b) Draw a histogram of the results from part (a).
(c) Draw a cumulative percentage polygon from (a).
(d) Compute the sample range, median, mode, $\overline{x}$, s^2, b_1, and b_2 for the data set.
(e) Are there better class intervals, which provide a better insight into the data set? If so, illustrate and explain why they are better.

6.2. For each of the following three data sets,
(a) Construct your most descriptive frequency table, and draw a frequency polygon of that frequency table. Compute the sample range, median, mode, $\overline{x}$, s^2, b_1, and b_2 for the data set.
(b) Is the distribution that governs this data set one of the named distributions that we have studied in this book?

Data Set A: 100 Values

0.469	0.363	0.351	0.196	0.551	0.171	0.971	0.325	0.603	0.525
0.478	0.834	0.184	0.775	0.143	0.082	0.334	0.417	0.618	0.958
0.229	0.566	0.806	0.446	0.561	0.665	0.095	0.075	0.913	0.365
0.135	0.478	0.989	0.958	0.995	0.466	0.308	0.889	0.794	0.168
0.175	0.185	0.281	0.286	0.336	0.714	0.494	0.972	0.739	0.848
0.815	0.362	0.802	0.271	0.053	0.487	0.525	0.752	0.114	0.455
0.606	0.881	0.585	0.969	0.163	0.364	0.464	0.853	0.350	0.381
0.286	0.884	0.622	0.725	0.572	0.708	0.095	0.211	0.238	0.850
0.946	0.717	0.468	0.562	0.147	0.392	0.239	0.664	0.261	0.789
0.946	0.849	0.574	0.375	0.891	0.228	0.842	0.218	0.187	0.880

Data Set B: 100 Values

1.185	0.531	0.146	0.146	0.172	1.707	0.799	0.736	3.217	0.001
0.150	0.198	0.316	0.226	0.068	0.032	0.366	0.120	0.644	0.929
0.549	0.074	0.876	0.584	0.135	0.592	0.589	0.567	0.038	0.247
1.869	0.020	0.280	0.126	1.736	0.496	1.287	0.581	0.321	0.343
0.249	0.101	1.270	0.335	1.220	0.601	1.383	0.675	0.032	0.182
0.533	1.286	0.063	0.152	0.703	0.810	0.936	1.085	0.054	0.185
0.260	0.493	0.715	1.018	0.057	0.410	0.999	0.197	0.074	0.399
1.298	0.910	0.432	0.895	0.011	0.964	1.632	1.130	0.115	0.870
0.212	0.263	0.562	0.559	0.683	0.686	0.325	0.189	0.924	0.011
0.390	0.814	1.036	0.276	0.423	0.424	0.567	0.867	2.212	0.641

Data Set C: 100 Values

2.592	6.177	3.052	2.636	5.462	5.465	4.297	7.547	0.942	1.642
3.539	0.666	7.046	3.621	11.254	11.864	1.035	3.945	17.172	5.796
15.362	6.423	3.378	5.501	4.426	7.088	0.963	3.240	8.575	5.528
3.329	4.138	12.648	8.237	0.358	1.505	4.233	7.534	7.951	2.736
1.923	2.097	2.845	6.055	4.721	2.382	2.829	1.480	6.909	10.433
4.483	4.518	5.685	3.512	4.768	3.106	3.698	6.494	6.420	5.857
5.795	5.641	7.553	2.561	3.072	9.484	6.492	2.804	1.357	1.827
3.637	3.954	3.183	2.022	1.626	8.858	4.407	8.845	12.200	3.165
7.171	8.013	3.169	1.778	18.109	1.638	4.171	3.250	8.247	2.376
11.560	7.280	12.780	4.522	6.342	3.363	1.948	5.563	0.778	4.818

6.3. For each of the given distributions, use its associated distribution simulator program in your computer package to generate samples of sizes 50, 100, 200, and 400. For each of the four samples, formulate your best frequency polygon, and compute the sample range, sample median, sample mode, $\overline{x}$, s^2, b_1, and b_2 for the data set. Using your knowledge of the theoretical distributions, discuss how well the contents of each data set reflects the generator distribution and how the sample size affects the agreement between the data set and the generator distribution.

(a) The standardized normal distribution.
(b) The gamma distribution with $r = 2$ and $\lambda = 2$.
(c) The beta distribution with $\alpha = 5$ and $\beta = 1$.

Are there any general statements that you can make based on this analysis?

6.4. The following data has been taken from the output of the production lines at SPACER Washers, Ltd. Unfortunately, the identity of which line produced which unit was not preserved. What, if anything, can you tell about the lines based on this data set?

125 Data Values

10.06726	10.35188	9.892497	9.639807	10.29235	9.417468	9.964650	9.452679
7.257992	9.588139	8.709900	7.187894	8.947398	7.640567	7.616752	8.260061
8.143322	7.184432	7.828551	8.299376	7.743091	8.379128	7.784816	7.971300
10.98962	9.909413	9.857012	9.369839	9.974536	9.803609	10.64654	10.59562
6.669064	9.424267	7.902421	8.930561	8.825352	8.482026	7.610155	6.887304
11.07191	10.77518	10.05939	9.721624	9.803664	9.699136	10.21799	10.21728
9.167899	6.735599	7.324095	8.457249	8.604888	8.319174	8.451744	9.355298
10.59945	10.37250	9.820396	10.15401	10.10916	9.991127	9.790497	10.76527
7.070581	8.439549	9.442536	7.256726	6.453430	7.272583	6.719518	7.797897
10.65326	10.70469	9.857050	10.47840	9.967582	9.776708	9.554341	10.21515
7.698979	7.956040	8.663272	6.680142	7.725085	8.680919	7.748480	7.559114
9.601135	10.05255	9.935870	10.86006	9.191732	7.456713	8.628185	7.365546
6.104943	8.795546	9.975698	6.940441	8.012416	6.710139	7.860485	8.879116
9.940590	7.078428	7.485160	9.644786	8.540267	8.819004	8.001250	8.625236
7.526873	6.904415	6.603111	8.261885	8.524504	7.206530	7.444095	9.975379
10.60354	10.20488	9.476544	9.569002	10.17083			

6.5. The following data are the recorded times that identical lengths of steel plates lasted before failing in a vibration fatigue test. Construct your most descriptive frequency table of this data set, and draw a frequency polygon of that frequency table. Compute the sample range, sample median, sample mode, $\overline{x}$, s^2, b_1, and b_2 for the data set. What does this analysis indicate to you? What would be the value that you would select as the 75th percentile of the time-to-failure distribution of the steel plating samples? the 95th percentile?

68 Data Values

130.6627	116.0904	120.5248	121.1449	126.4982	122.0014	103.7083
119.0590	109.6592	119.1960	128.5335	103.9567	119.1048	116.9843
118.5632	105.0407	122.5049	100.7252	109.2863	119.6508	127.2017
119.7483	119.7249	113.4724	124.4559	103.6276	121.5718	90.95787
88.90195	117.4161	110.2306	107.5966	126.3976	106.8430	112.8014
102.6377	97.71074	112.9669	116.7805	122.1396	107.4140	105.3853
117.6061	115.6994	106.4181	109.3490	126.6105	128.4759	118.1854
110.2727	99.16590	109.9687	110.3935	109.5466	102.7687	113.3863
130.1150	116.4510	124.4811	120.1628	128.8074	116.5194	112.0747
111.8857	113.9223	119.0236	116.5696	101.6396		

6.6. One hundred circuit boards were to be placed in a laboratory life test under stressful conditions of humidity and heat. Unfortunately, one package of 6 circuit boards was lost during shipment. The remaining 94 circuit boards yielded the following continuous-service times prior to failure in an accelerated life test. Construct your most descriptive frequency table of this data set, and draw a frequency polygon of that frequency table. Compute the sample range, sample median, sample mode, $\overline{x}$, s_2, b_1, and b_2 for the data set. What does this analysis indicate to you? What would be the value that you would select as the 80th percentile of the time-to-failure distribution of the circuit boards? the 95th percentile?

94 Data Values

230.5800	2699.703	1348.345	181.3645	2376.629	2850.462	131.2434
766.1620	5850.501	1761.624	1373.664	838.8182	2768.530	1033.769
149.7692	965.5808	1884.231	628.0235	1425.901	284.2001	279.2453
210.0941	943.0424	484.5742	725.7719	78.13717	903.6336	296.9648
185.8454	49.04858	985.5012	63.52300	1636.325	1355.031	673.7493
482.9469	1932.677	355.3122	595.8227	458.9726	69.24088	2451.392
929.5786	1040.447	1337.819	141.5606	10.14482	4.293229	1199.899
2269.268	737.4523	1626.065	51.83372	423.2133	536.7179	168.7216
1037.787	439.9714	1357.867	1848.761	392.7366	1870.487	582.4540
4647.731	2298.168	1696.782	215.5969	670.6698	2189.788	164.5893
37.94855	178.6182	6672.794	178.2145	1281.151	1203.948	453.8529
2197.737	1732.920	375.2959	752.3557	149.8070	908.4772	185.7449
194.1585	347.2265	896.3394	823.2396	972.6251	1880.074	8.634668
1739.809	42.14361	1238.563				

6.7. Fifteen microcomputer input devices ("mice") are sampled from each of 56 crates on the shipping dock. Each mouse is inspected for the presence of one or more serious cosmetic defects. The following data set gives the number of mice that were found to have defects in each of the 56 samples. Construct your most descriptive frequency table of this data set, and draw a frequency polygon of that frequency table. Compute the sample range, sample median, sample mode, $\overline{x}$, s^2, b_1, and b_2 for the data set. What does this analysis indicate to you? What is your best guess at the actual proportion of mice that have one or more serious cosmetic defects?

56 Data Values

0	0	0	2	0	0	0	0	0	0	1	0	0	1	0	0	0	0	1	0	0	2
2	1	0	1	2	1	0	0	1	1	0	1	0	1	0	2	1	0	0	0	1	0
0	1	2	0	0	2	1	1	0	0	0	0										

6.8. The automatic teller machine next to the Philosophy and Mathematics Building at Big State University is being studied by Capitol City Savings to determine whether its level of use warrants its continued presence there. The following data give the recorded number of customers (in $\frac{1}{2}$-hour periods) that used the teller machine between the hours of 10:00 A.M. and 4:00 P.M. for the last nine days. (On the first day, only a 2:00 P.M.-to-4:00 P.M. period was observed.) Construct your most descriptive frequency table of this data set, and draw a frequency polygon of that frequency table. Compute the sample range, sample median, sample mode, $\overline{x}$, s^2, b_1, and b_2 for the data set.

If bank officials have stated that the teller machine must have, on the average, at least 10 customers per hour to justify its continued presence, what would you recommend to them?

100 Data Values

4	5	12	5	3	10	7	7	7	6	2	5	7	5	4	5	5	6	5	1	7	3
1	10	4	1	2	7	5	7	5	2	3	8	4	11	5	9	10	4	6	4	7	5
2	8	2	4	3	3	4	4	6	5	5	6	8	5	3	5	3	5	6	5	2	5
5	5	7	3	4	5	8	4	6	4	3	6	3	6	3	3	8	6	10	4	6	5
6	3	6	8	5	6	6	5	3	4	5	7										

6.9. Mansat Receiver Station 10 is suspected of sporadic malfunction. In comparison with a more expensive laboratory test receiver, the following data, giving the number of successful Mansat signal receptions before a failure, was collected.

80 Data Values

14	11	1	0	22	1	8	17	10	2	1	0	7	5	1	2	24	2	14	8	14	5
1	2	4	2	3	2	3	13	8	6	13	7	12	2	11	9	3	1	0	2	4	4
2	14	0	1	13	7	0	3	5	2	9	6	0	12	0	36	9	1	5	8	8	6
2	4	8	2	3	10	4	2	14	1	7	6	0	2								

Because of the prevailing transmission conditions, a long-term 90% success rate on any individual signal is considered acceptable. If you were in charge, would you have the Mansat Receiver Station 10 replaced?

6.10. Seventy-five boxes of 20 PROTEK automobile fuses rated at 20 amps are tested in the ΓM Research Laboratory. The following data gives the number of fuses in a sample of 8 fuses from each box that blew before the rated amperage was reached. If an acceptable level of such failures is 7%, should the ΓM labs pass the PROTEK fuses?

75 Data Values

2	0	0	1	1	0	1	0	2	1	0	1	1	1	0	1	0	1	0	1	0	2
0	1	2	1	0	0	1	1	0	1	1	2	0	0	0	0	0	0	2	1	0	1
1	1	1	1	1	1	0	1	2	0	0	0	1	1	0	1	2	0	0	2	0	1
0	1	1	2	1	0	2	1	1													

6.11. The following data gives the clearances between randomly selected pairs of keys and keyways in a riding mower torque converter (where negative values indicate that the key was too big for the keyway). Based on this data set, formulate your best estimate of the proportion of randomly selected pairs of keys and keyways that will not fit.

50 Data Values

.00717	.00721	−.00022	.00742	.00517	.00116
.00473	.00510	.00752	.00368	.00418	.00283
.00303	.00415	.00197	.00688	.00255	.00391
.00590	.00170	.00690	.00430	.00193	.00550
.00300	.00466	.00379	.00674	.00361	.00634
.00338	.00174	.00560	.00071	−.00160	.00677
.00386	.00705	.00308	.00229	.00517	−.00133
.00496	.00453	.00244	.00118	.00376	.00391
−.00170	.00559				

6.12. As an adjunct to the study discussed in Exercise 6.11, 72 sets of 10 randomly selected pairs of keys and keyways yielded the following numbers of pairs that would not mate. Are the two data sets consistent? Why or why not?

72 Data Values

2	1	1	1	0	0	0	1	0	1	2	1	0	1	0	0	2	1	0	0	0	2	2	0
1	1	0	0	0	2	1	0	0	1	0	0	0	0	1	2	0	0	0	1	1	1	0	0
0	1	0	0	0	1	0	1	1	0	1	1	0	2	0	1	0	2	0	1	0	0	0	3

6.13. The following data gives the number of engineers recruited annually by 45 medium-sized engineering companies over a period of one year.

59 73 70 79 79 45 71 67 69 61 59 75 71 67 69 61 59 75 71 67 69 61
59 75 86 68 75 79 54 57 61 71 72 64 76 76 44 69 74 66 67 60 74 66
76 77 77 68 79 46 85 77 51 59 90 87 67

(a) Compute the sample range, median, mode, sample mean, sample variance, skewness, and kurtosis for the data set.
(b) Construct a frequency table and the frequency polygon using a class interval of 11 units.
(c) Draw a histogram of the results from part (b).
(d) What conclusions can you draw regarding the type of distribution that this data set represents?

6.14. Your automobile company wishes to launch a new vehicle. The distance (in thousands of miles) traveled by each of 75 vehicles, before developing any kind of problem, is given in the following.

32 42 36 31 38 28 38 35 37 33 40 40 27 35 38 37 32 41 40 40 29 36 33 36
29 25 34 29 35 31 45 42 38 43 35 32 37 38 42 38 39 30 47 33 48 34 37 34
28 40 41 38 33 40 37 40 38 40 46 37 36 39 44 35 34 28 32 34 31 36 39 41
39 46 40

(a) Compute the sample range, median, mode, mean, variance, and coefficient of variation for the data set.
(b) Construct your best frequency table and polygon, and give your reasons for selecting it.
(c) If a rival automobile firm already has a vehicle that travels an average of 45 thousand miles before developing any kind of problem, do you think that your company should go ahead with the launching of the vehicle? Why?

6.15. The scores obtained by 55 students in a particular class are given in the following.

42 48 44 34 45 77 81 80 80 83 79 77 76 75 81 75 80 32 39 48 77 33
57 27 46 49 76 74 77 85 79 42 32 46 49 56 45 39 53 32 50 44 53 47
38 46 48 42 33 39 39 40 33 72 83 82 82 75 76 81 85

(a) construct your best frequency table and the frequency polygon.
(b) What conclusions can you draw from this polygon?
(c) If you are asked to assign grades to these students, what intervals would you select for grades A, B, C, D, and F?

6.16. A light bulb manufacturing firm has supplied 55 lots of bulbs over a period of one year to a particular customer. The quality assurance department has the following record of the number of defective bulbs in a sample size of 10 drawn from each of the 55 lots.

0 0 2 0 1 0 0 1 0 1 0 1 0 0 1 2 2 1 0 1 2 1
1 0 1 1 1 1 1 1 1 1 1 0 0 0 2 1 0 1 0 1 2 1
0 1 2 0 1 0 0 0 1 2 2

If the quality assurance department decides that the average level of success should be 95% for all future consignments, is the next lot supplied by the bulb manufacturer likely to be acceptable? What if it is decided that the average level of success should be 90%?

6.17. Two manufacturers of a particular class of industrial valves are currently supplying them to a chemical company. Sixty-five samples of 13 each are selected from supplies made by each manufacturer in order to investigate conformance to required quality norms. It is found that the number of defective valves in these samples of 13 valves for the two manufacturers are

0 1 1 3 0 1 0 0 1 2 1 0 1 1 1 0 0 3 1 2 2 4
2 1 1 1 0 2 0 0 0 1 1 1 2 2 1 0 1 1 1 2 0 1
1 2 1 2 2 2 0 1 0 1 1 0 2 2 1 1 2 1 1 0 0

and

1 2 2 1 2 3 1 0 2 4 0 0 2 3 4 4 1 1 2 3 2 5
4 1 6 2 2 0 4 2 2 2 2 1 1 2 1 3 3 3 0 1 1 0
1 3 2 2 3 2 2 2 2 2 1 6 4 0 4 1 2 2 1 1 2

Which manufacturer would you say supplies better-quality valves? Why?

6.18. One hundred pieces of yarn were tested for breakage as a result of overtwisting during fabric weaving. The number of cycles of strain to breakage was recorded for each sample.

15	20	21	38	38	40	40	42	52	56	61	61
65	74	76	90	93	98	106	121	121	124	125	131
135	135	136	143	146	88	149	151	157	166	169	175
176	180	180	180	182	183	184	186	188	86	81	192
194	195	196	198	199	202	204	211	220	224	228	229
236	238	244	245	246	249	250	251	262	264	264	265
276	279	282	285	285	291	292	315	321	325	337	339
341	351	353	364	394	395	399	400	401	422	497	565
571	597	653	828								

(a) Construct your best frequency polygon for this data set.
(b) What type of distribution do you think governs the data set? Why?
(c) If an ability to withstand at least 100 cycles is required, what proportion of the yarn samples would be acceptable?

6.19. The following data set refers to the amount of time that 65 springs lasted before failure as a result of stresses in a test carried out in a research laboratory.

396.91	13.49	181.08	203.75	137.62	371.97
6.66	29.93	60.37	161.75	150.77	10.29
12.90	61.99	3.07	289.06	0.49	46.02
15.94	217.93	53.76	137.58	115.45	72.88
6.53	134.40	12.10	50.60	6.45	62.36
73.31	57.38	69.34	106.27	165.93	174.43
36.06	310.22	905.36	215.74	28.26	49.77
155.21	112.23	319.83	215.56	2.42	7.06
16.85	26.24	164.35	31.44	77.11	69.53
6.37	328.36	15.23	291.48	6.84	57.71
24.78	41.64	339.97	66.08	177.02	

(a) Construct your most descriptive frequency table and the frequency polygon.
(b) Compute the sample range, median, mode, skewness, and kurtosis for the data set.
(c) What value will you select as the 85th percentile of the time-to-failure distribution of the springs?

6.20. In the first round of a shooting contest, the number of bull's-eyes scored by 70 contestants before their first miss is given in the following.

9	10	5	6	2	5	17	0	13	3	1	9	27	14	29	6	12	2	2	1	5	11
6	5	3	12	7	8	13	8	1	3	23	1	3	7	35	1	1	7	7	6	2	10
2	1	9	5	1	0	5	1	13	3	3	6	18	8	4	1	1	3	2	1	0	9
2	7	3	2																		

(a) If only 13 of them can advance to the second round in the contest, what should be the minimum score of the contestant in the first round?
(b) If in order to enter the second round a contestant must have a score of at least 9, what proportion of the contestants will advance to the second round?
(c) What can you say about the kind of distribution that this data set represents?

CHAPTER

7

SAMPLING DISTRIBUTIONS: RANDOM SAMPLING, THE SAMPLE MEAN AND SAMPLE VARIANCE, AND THE CENTRAL LIMIT THEOREM

The following concepts and procedures form the major content of this chapter.

- Be able to define the term *random sample* for (a) sampling without replacement from a finite population and (b) sampling from an infinite population
- Be able to describe how to implement random sampling in practical situations
- Given the probability distribution function $p(x)$ of a discrete random variable, X, be able to find the probability of obtaining a particular random sample *realization* $(x_1, x_2, \ldots, x_n)$ from the associated population
- Thoroughly understand the fact that $\overline{X}$, the *sample mean*, is a random variable
- Given $p(x)$, be able to find the probability distribution function of $\overline{X}$, i.e., be able to obtain $p_{\overline{X}}(\overline{x})$
- Understand the relationship between the population standard deviation, σ, the standard deviation of $\overline{X}$, $\sigma_{\overline{X}}$, and the sample size, n
- Memorize the *central limit theorem* and be able to use it in solving problems

- Understand the generalization of the central limit theorem that removes the restriction of *identically distributed* random variables
- Understand when the *t distribution* should be used when working with $\overline{X}$
- Thoroughly understand the fact that S^2, *the sample variance*, is a random variable
- Understand why $[(n-1)S^2/\sigma^2]$ is distributed according to a χ^2 random variable with $n - 1$ degrees of freedom
- Understand why S_1^2/S_2^2 for two normally distributed populations with the same variance is distributed as an F random variable

7.1 RANDOM SAMPLING FROM FINITE AND INFINITE POPULATIONS

In the early days of statistics, statisticians were primarily concerned with populations of people or animals. However, when modern statisticians use the term *population*, they mean any set of things under study, and, as we have seen, this can embrace a large area of interest. Suppose that we are investigating the average daily number of rejected units produced at our factory in the last 12 months. This population is finite and includes production from all lines in the factory. As another example, suppose that we are interested in some characteristic that is associated with the ROM chips manufactured by a particular company. From a pragmatic conceptual view, this population is infinite, since it is possible that the company will continue making ROM chips into the distant and unlimited future. Similarly, the number of times that a fair die may be rolled forms an infinite population for practical considerations.

In statistical analysis, we make a *fundamental assumption:*

> The values that members of *any* population can assume are governed by an *underlying* probability distribution function.

For example, the distribution of the strengths of $\frac{1}{4}$-inch carriage bolts is likely to be some normal distribution, $N(\mu, \sigma)$. If we count the number of defective units in a sample of five units and we collect statistics on 100 such samples, we are sampling observations that are governed by a binomial distribution. Often, statisticians use a verbal shorthand and say that they are sampling "from a binomial distribution." If we repeatedly sample from a production process and count the units sampled until the first defective unit is found, we are sampling from a geometric distribution.

Engineers frequently need to study physical systems in which the distribution function of a population characteristic is not known. In such situations, we must rely on sample data to assist us in describing the properties of that distribution function. The techniques developed in Chapter 6 are especially useful in performing that description.

Inferences about a population based on sampled data will be no better than the data itself. The sampled data must be *representative* of the population about which we wish to infer. A famous example of incorrect data sampling occurred in a presidential election during the 1930s. The Republican candidate's pollsters assured their candidate that he would win by a comfortable margin. Unfortunately, the Republican's statisticians obtained the names of voters in their sample from telephone books. At that time, telephones were a luxury and were far more likely to be possessed by Republican supporters than by Democratic supporters. The Democratic candidate won because a great majority of the voters who did not have phones voted for him.

Before a sample is taken, the values that will be observed may be considered as an ordered set of random variables, $X_1, X_2, X_3, \ldots, X_{n-1}, X_n$, which can assume any of the possible values defined by the probability distribution function governing the population. After the sample is taken, the observed values of $X_1, X_2, X_3, \ldots, X_{n-1}, X_n$ (also called the *observed data*, or the *realization of the sample*), will be denoted $x_1, x_2, x_3, \ldots, x_{n-1}, x_n$. To ensure that the realization is *representative* of the population, it must be based on a *random* sample.

> For an *infinite population*, $X_1, X_2, X_3, \ldots, X_{n-1}, X_n$ constitute a *random sample* from the population if (a) the X_i's are statistically independent and (b) each of the X_i's is governed by the probability distribution function governing the population.

X_i that satisfy properties (a) and (b) in this definition are said to be *independent and identically distributed* (IID) random variables. If the sampled population is finite and sampling is performed with replacement—i.e., after an item is selected and observed, it is returned to the population—then properties (a) and (b) are satisfied.

If the sampled population is finite and sampling is performed without replacement, then property (a) is not satisfied. * For this reason, a different definition for a random sample is required for sampling without replacement from a finite population. As we have seen in Chapter 3 in our discussions on the hypergeometric distribution, when sampling is performed without replacement the probability that any item becomes a member of the sample, after the first item is selected, depends on what was observed in the previous trials. We have also seen that the total number of unique samples of size n that can be drawn from a population of N unique things is ${}_NC_n$.

> For a *finite population* when sampling is performed *without replacement*, $X_1, X_2, X_3, \ldots, X_{n-1}, X_n$ constitute a *random sample* from the population if the sampling is conducted in such a way that all ${}_NC_n$ possible samples have an equal probability of being selected.

There are various methods of random sampling from a finite population when sampling is performed without replacement. One of the best ways is to use a *random number table*. A portion of such a table (generated by the uniform distribution simulator program in your software package) is given in Table 7.1. A

---

*If N is large relative to n, properties (a) and (b) can be *approximately* satisfied when sampling without replacement from a finite population.

TABLE 7.1
An excerpt from a table of random numbers

16877	04419	75939	95567	76431	75407	55762	41403	59269	38793	44722	76792	54146	13258
10017	59657	41389	41858	19566	83203	25112	58415	34693	24872	41984	75478	30434	68066
67782	83840	39074	11320	20061	54138	03669	69839	15450	44299	04691	45855	58802	86620
09682	99583	01346	45697	26120	87905	94501	43849	56457	77606	37816	85013	62796	47911
36554	44275	76938	87829	77741	72567	94176	30586	22398	15413	99549	98825	95166	17253
56877	73549	51618	14552	68883	53175	13723	90442	99543	86379	89355	17895	75800	02103
47327	17425	21591	01873	13358	06921	65441	68481	72461	40677	98687	73916	00755	07952
62215	43386	61520	76541	79393	59959	56900	50176	51870	09301	31056	71787	94366	71147
15066	99305	78263	47625	30216	26783	08182	21246	03892	01224	51510	82592	18323	80343
19581	64977	84887	87365	45306	84322	77292	14318	57050	99454	53994	43204	91183	37182
32017	61646	09583	78794	81939	28078	17275	54003	39554	17143	70561	12497	94291	85775
53784	08284	57387	75811	32880	20046	43688	82321	59269	71696	76143	14808	69132	08420
50907	39432	29459	98341	59445	62483	90288	81597	07391	41514	92464	92251	09700	50767
13802	90811	95760	68282	04579	96269	34114	18143	25767	44410	05675	41460	47522	85299
33932	89196	81633	12352	77242	67362	56036	42927	45131	58162	63462	36022	61545	07622
57296	41128	04836	50691	14907	61966	96725	16243	35655	18677	24674	02584	73016	74864
70063	01379	18743	85973	91541	92304	88433	75208	39577	72718	44900	21280	53727	87501
09188	20069	58948	15491	58479	59348	23332	69536	19198	64938	03456	38187	10858	16440
65574	10941	65416	73041	03510	94507	30214	33850	06520	45618	69737	84566	38083	80978
88968	37084	07888	04595	80303	47051	89414	51183	31468	15296	02093	15653	42019	63498
58671	06142	17417	51070	09650	30848	22272	78427	63619	99919	50943	91511	64132	06647
88640	34360	47410	53359	98505	07936	54445	97750	95489	51985	57750	14107	93029	27938
05854	25803	85412	58406	44907	74444	38854	86282	42426	19186	34229	14796	07326	86339
34234	52029	63257	23187	59707	43850	84551	41359	92229	61773	74328	31954	18490	29945
99352	93787	72851	35810	02654	53647	30630	05157	95711	61128	21788	11494	43978	15385
56923	52437	13874	88569	48652	50648	06329	90895	78656	17075	49328	83438	28720	42843
17448	61850	51771	29912	00578	06220	05554	41672	64227	43156	79537	01514	18388	67789
16798	78262	53130	83134	79229	79732	15847	24623	30695	47714	24278	52454	38352	23816
38107	97002	32727	71794	54758	30236	87285	45081	14714	33750	41636	07860	41593	50889
96457	32608	27205	89443	58694	09215	42031	78385	87580	41685	18514	84028	59061	34066
18746	53722	13278	48466	66554	73387	74719	80044	54169	01943	90287	65254	48436	02884
18896	34802	23694	47707	93575	64493	38949	28424	86894	25711	70414	33707	72851	56068
39361	86988	23360	54951	16946	65661	24348	57719	59767	29687	51695	99250	89402	20762
17457	18481	14113	62462	02798	54977	48349	59481	53814	75971	56677	82129	49188	69477
03704	36872	83214	59337	01695	60666	97410	47018	10718	15750	34737	39857	11211	56327
21538	86497	33210	60337	27976	70661	08250	56299	53469	28622	04268	50981	60659	41346
57178	67619	98310	70348	11317	71623	55510	88036	92199	93626	13575	25883	61883	85940
31048	68772	79573	51414	71255	39404	83702	26676	90690	75642	29406	49113	97189	39595
44704	73876	25033	58490	26938	03599	16773	02583	91629	00844	79644	68587	65934	16955
16787	20905	38374	09991	55357	12478	38821	45583	80337	31517	93412	12391	74153	56508
43050	70680	43594	26484	40571	82928	22211	73516	98335	58985	07372	66017	68254	96804
11684	97749	17465	88090	45649	74712	91113	43693	42806	77037	22759	16638	14720	41748
98771	70289	98855	24602	53183	49047	29403	33836	53411	33086	43347	83497	58432	44976
82680	53774	06799	23867	95505	41895	62268	97967	36650	97861	02030	49697	99370	11553
47425	72285	03232	90424	49400	47001	45068	55056	74171	76499	01305	80389	54545	88031
82574	74215	92724	71176	91208	04210	86275	45784	93449	06638	26601	63192	29607	21881
39967	62964	20922	18918	28526	24550	04542	29104	16731	85725	27727	36834	02492	69240
99019	90482	91606	63533	31958	02084	55570	48592	81943	42173	66624	15234	18562	39342
13979	95989	61212	88129	45926	66162	97908	57304	71438	70758	34080	51625	70205	48804
18107	23929	72521	07737	69123	03661	67603	71255	39404	83702	26676	90690	64347	50232

random number table is characterized by the fact that, *prior* to the formation of the table, every location in the table has an equally likely chance of having any of the digits 0, 1, 2, 3, 4, 5, 6, 7, 8, or 9.

Example 7.1. Suppose that we want to randomly sample $n = 10$ stress gauges from the 200 available. One method of performing random sampling is to uniquely assign one of the numbers 0,1,2, . . . ,199 to each of the gauges. Next, go to the random number table, and, starting at *any position* in the table, proceed in *any direction*, reading 3-digit numbers. Referring to Table 7.1, suppose you elect to start in the first (leftmost) position of the 34th row from the top; you move horizontally to the right across that row (the first seven 5-digit columns) and continue on succeeding rows, moving downward.

The first 3-digit number is 174. Thus, gauge 174 will contribute the first measurement to the data set, x_1. The next number, 571, is ignored because it is not in the set of numbers that range from 0 to 199. Continuing in this fashion, the following gauges are selected to provide the desired observed data, the x_i, $i = 1, \ldots, 10$.

$$174, 114, 113, 145, 21, 60, 66, 108, 131, 103$$

(The numbers just obtained are italicized and underlined in Table 7.1.) If you should happen to obtain a repeat of a particular number, simply ignore it and continue. If you have a computer-generated set of random numbers, use the numbers in the order that they were generated. (You are given an opportunity to do this in the exercises at the end of the chapter.)

Since it is not possible to number the members of an infinite population, other approaches must be used. The use of mechanical and other artificial devices can sometimes be employed with success. If at all possible, you should *avoid human judgment*. The presence of judgment in your sampling procedure virtually ensures that both conscious and unconscious biases will affect your sample. Happily, you will rarely encounter a situation in practical analysis where sampling from an infinite population is required or even possible. For example, even though the population of cylindrical bearings made by a particular manufacturer may be conceptually infinite, an analyst simply cannot wait until the end of time to allow the infinite population to come into existence. Rather, she or he is forced to work with a finite subset of that infinite population so that answers can be provided in a timely fashion. When you take a random sample from a physical process, be careful that you do not inadvertently introduce bias into your sample even when you use a random number table.

Example 7.2. Suppose that you want to sample from a week's production of an electromagnetic coil used in the ringer of telephone sets. These coils are extremely delicate and must be wound by hand. One approach would be to select, using a random number table, random points in time during the week. At any of the selected times, the coil currently in production would become a member of the sample.

Unfortunately, this technique would favor the inclusion of coils that take longer to wind. Longer winding times imply a higher defect rate because they usually occur when a problem is encountered in the winding process. Thus, your sample would be biased toward having a larger proportion of defective coils in the sample than the proportion in the population.

Once you have obtained the x_i, $i = 1, \ldots, n$, you usually make *inferences* about the sampled, or *parent*, population. *Qualitative inference* can be made from such things as the percentage polygon or the cumulative percentage polygon. *Quantitative statistical inference* can be made with parameter estimates such as $\overline{x}$ and s. Statistical inference is discussed in Chapter 9.

7.2 THE DISTRIBUTION OF THE SAMPLE MEAN, $\overline{X}$

Chapter 6 introduces $\overline{x}$, the sample mean for a data set, which is computed using the observed values of x_i, $i = 1, \ldots, n$ that are available after the random sampling process is completed. In this section, we discuss $\overline{X} = \frac{1}{n}\sum_{i=1}^{n} X_i$, the *sample mean*, which is the random variable associated with $\overline{x}$; i.e., $\overline{x}$ is the realization of the random variable $\overline{X}$. Let us consider a finite population of purple widgets.

Example 7.3(a). The five widgets pictured in Figure 7.1 are the *only* purple widgets in existence, and each is labeled with its worth. The probability distribution function of X, the worth of a purple widget, is

x	1	3	5	7	9
$p(x)$	0.2	0.2	0.2	0.2	0.2

with $\mu = 5$ and $\sigma = \sqrt{8}$ (calculated using Equations 5.1 and 5.2).

Suppose that we sample 2 widgets, X_1 and X_2, *without replacing* the first widget drawn and that we compute the sample mean, $\overline{X} = (X_1 + X_2)/2$. What is the probability distribution of $\overline{X}$, $p_{\overline{X}}(\overline{x})$?

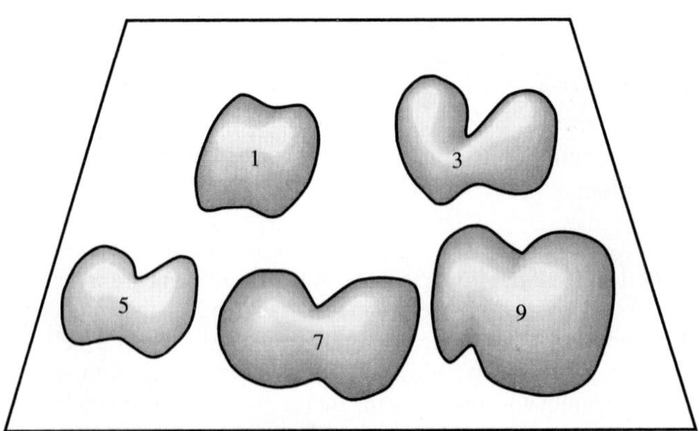

FIGURE 7-1
The only purple widgets in existence.

Solution. There are ${}_5P_2 = 20$ possible samples of size 2, and they yield

	x_1	x_2	$\overline{x}$		x_1	x_2	$\overline{x}$		x_1	x_2	$\overline{x}$		x_1	x_2	$\overline{x}$
1	1	3	2	**6**	3	5	4	**11**	5	7	6	**16**	7	9	8
2	1	5	3	**7**	3	7	5	**12**	5	9	7	**17**	9	1	5
3	1	7	4	**8**	3	9	6	**13**	7	1	4	**18**	9	3	6
4	1	9	5	**9**	5	1	3	**14**	7	3	5	**19**	9	5	7
5	3	1	2	**10**	5	3	4	**15**	7	5	6	**20**	9	7	8

With random sampling, each of these 20 possible samples of size $n = 2$ is equally probable. This implies that the probability distribution of $\overline{X}$ is

$\overline{x}$	2	3	4	5	6	7	8
$p_{\overline{x}}(\overline{x})$	0.1	0.1	0.2	0.2	0.2	0.1	0.1

Using the methods of Chapter 5, we obtain

$$\mu_{\overline{X}} = (0.1)(2) + (0.1)(3) + (0.2)(4) + (0.2)(5) + (0.2)(6) + (0.1)(7) + (0.1)(8) = 5$$

$$\sigma^2_{\overline{X}} = 0.1(2-5)^2 + 0.1(3-5)^2 + 0.2(4-5)^2 + 0.2(5-5)^2$$
$$+ 0.2(6-5)^2 + 0.1(7-5)^2 + 0.1(8-5)^2 = 3 \qquad (\sigma_{\overline{X}} = \sqrt{3})$$

Thus $\mu_{\overline{X}} = \mu = 5$ and $\sigma_{\overline{X}} = \sqrt{3} < \sigma = \sqrt{8}$. In addition, Figure 7.2 shows that the range of $p_{\overline{X}}(\overline{x})$ is smaller than that of $p(x)$, and it shows that $p_{\overline{X}}(\overline{x})$ is more concentrated about the mean than $p(x)$.

Our finding that $\sigma/\sigma_{\overline{X}} = \sqrt{8}/\sqrt{3}$ can be extended to a general result: If $\mu_X = \mu$ and $\sigma^2_X = \sigma^2$, the mean and standard deviation of the distribution of $\overline{X}$ are, respectively,

$$\mu_{\overline{X}} = \mu \qquad \text{and} \qquad \sigma_{\overline{X}} = \frac{\sigma}{\sqrt{n}}\sqrt{\frac{N-n}{N-1}} \tag{7.1}$$

In Example 7.3(a) (as we verified using basic analysis methods),

$$\sigma_{\overline{X}} = \frac{\sqrt{8}\,\sqrt{5-2}}{\sqrt{2}\,\sqrt{5-1}} = \frac{\sqrt{8}\,\sqrt{3}}{\sqrt{2}\,\sqrt{4}} = \sqrt{3}$$

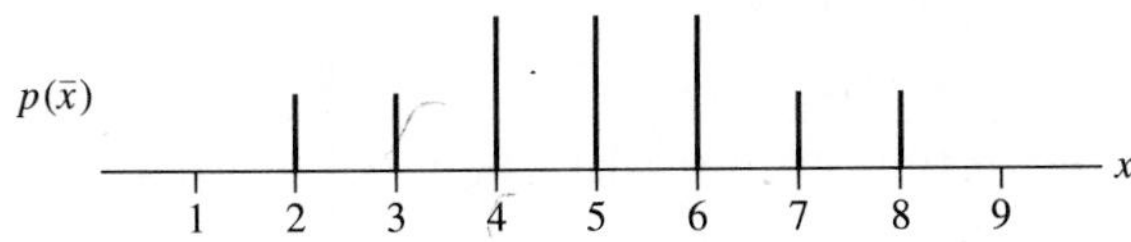

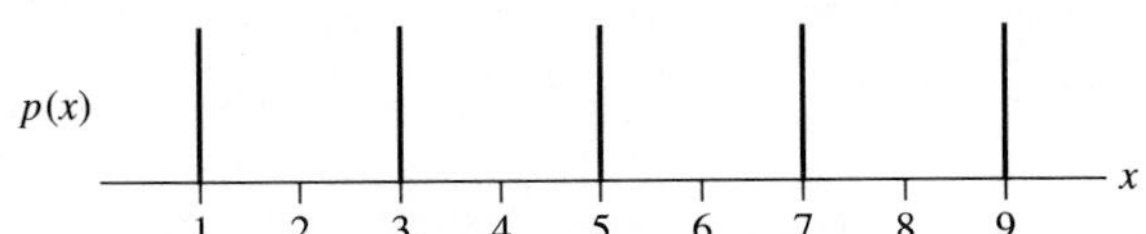

FIGURE 7-2
The probability distributions of $p(\overline{x})$ and $p(x)$.

The term $\sqrt{(N-n)/(N-1)}$ is the *finite population correction factor*. It accounts for the fact that we might sample a *large proportion* of a finite population. In one extreme case, we could sample each and every member of the population; i.e., $n = N$, and the correction factor would equal 0. This yields the expected result of $\sigma_{\overline{X}} = 0$. If we sample all members of the population, nothing about the population is unknown, and, therefore, the standard deviation of $\overline{X}$ should equal 0.

At the other extreme, we might sample only an insignificant proportion of the total population. This is *always* the case for infinite populations (or when we sample *with replacement*). If the population is infinite, the correction factor is equal to 1. This may be shown by evaluating the limit of the correction factor as $N \to \infty$; i.e.,

$$\lim_{N\to\infty} \frac{N-n}{N-1} = 1$$

Example 7.3(b). Suppose that the population of purple widgets is not limited to the five members pictured in Figure 7.1. Let us suppose that there is an infinite number of widgets and that the values are still governed by

x	1	3	5	7	9
$p(x)$	0.2	0.2	0.2	0.2	0.2

What is the probability distribution of $\overline{X}$ for $n = 2$?

Solution. The only difference between the solution to this question and the one posed in Example 7.3(a) is that the realizations, x_1 and x_2, may achieve the same value; i.e., with an infinite number of widgets with value 1, even if the first widget sampled is a 1, the probability of getting a 1 on the second widget is still 0.2 [*not* zero as in Example 7.3(a)]. Since x_1 can equal x_2, in addition to the 20 outcomes given in Example 7.3(a) the following five additional possibilities must be considered.

	x_1	x_2	$\overline{x}$
21	1	1	1
22	3	3	3
23	5	5	5
24	7	7	7
25	9	9	9

All 25 outcomes are equally likely and yield

$\overline{x}$	1	2	3	4	5	6	7	8	9
$p_{\overline{X}}(\overline{x})$	0.04	0.08	0.12	0.16	0.20	0.16	0.12	0.08	0.04

Once again, using the methods of Chapter 5, we obtain

$$\mu_{\overline{X}} = 5 \quad \text{and} \quad \sigma_{\overline{X}}^2 = 4 \quad (\sigma_{\overline{X}} = \sqrt{4} = 2)$$

Thus $\mu_{\overline{X}} = \mu = 5$ and $\sigma_{\overline{X}} = \sigma/\sqrt{n} = \sqrt{8}/\sqrt{2} = \sqrt{4} = 2$ as dictated by Equation 7.1 for infinite N.

Notice that if we sample with replacement, this identical result is obtained even if there are *only* five widgets, as described in Example 7.3(a).

As a matter of convenience, the correction factor is usually omitted if $N > 20n$, since the correction factor will then be near enough to 1.0 to make no appreciable difference.

In Example 7.3(a), the value of the correction factor is $\sqrt{3/4} = 0.866$. The effect of the correction factor is *not linearly proportional*. For example, for $N = 10$ and for the same sample size of $n = 2$, the correction factor would be $\sqrt{(10-2)/9} = \sqrt{8/9} \doteq 0.943$. Intuitively we expect this new value to be larger, reflecting the fact that there is a great deal more uncertainty in 8 unsampled elements than in 3 unsampled elements. On the other hand, if we sample $n = 200$ from $N = 500$, the correction factor is $\sqrt{(500-200)/(500-1)} = 0.775$. We have sampled the same proportion as in Example 7.3(a), where $n = 2$ and $N = 5$. In this case, however, the impact of sampling 200 members of the population outweighs the effect of the larger population and causes a decrease in the correction factor and a corresponding decrease in $\sigma_{\overline{X}}$.

When the population is infinite, $\sigma_{\overline{X}} = \sigma/\sqrt{n}$. Even in this case, however, if the sample size is made large enough, $\sigma_{\overline{X}}$ can be driven to as small a number as desired. Unfortunately, the *inverse square relationship* between the sample size and the reduction of error can be quite prohibitive.

Example 7.4. Suppose that $\sigma = 9$ and $n = 16$. What new sample size would reduce $\sigma_{\overline{X}}$ from its current value of $\frac{9}{4}$ to a value of $\frac{9}{8}$?

Since $\sigma_{\overline{X}} = \sigma/\sqrt{n}$, we have one equation to solve, $9/8 = 9/\sqrt{n}$, which yields $n = 64$. To reduce $\sigma_{\overline{X}}$ by one–half, we must quadruple the sample size. Halving $\sigma_{\overline{X}}$ again would require $n = 256$.

In Example 3.1, the distribution of the sum of a single roll of two dice is developed. Dividing the random variable of that distribution by 2 yields $\overline{X} = \Sigma X/2$. Figure 7.3 presents the distribution of $\overline{X}$ for n independent rolls of a fair die, with $n = 1, 2, 3, 5$, and 10. The practical range of distribution decreases as n grows larger, and the shape of the distribution looks more and more like a normal distribution.

This graphically observed fact is formalized in *the central limit theorem:*

> If X is distributed with mean μ and standard deviation σ, then $\overline{X}$ obtained from a random sample of size n will have a distribution that approaches
>
> $$N(\mu,\ \sigma/\sqrt{n}) \qquad \text{as } n \to \infty$$

Memorize these for Exam 2 + Final

The central limit theorem is a *very important result*. It means that, no matter what distribution governs the sampled population and as long as it has a finite variance, the distribution of $\overline{X}$ will become closer and closer to a normal distribution as the sample size grows. An equivalent expression of the central limit theorem deals with the sum of the X_i:

> If X is distributed with mean μ and standard deviation σ, then $\sum X_i$ obtained from a random sample of size n will have a distribution that approaches
>
> $$N(n\mu, \sigma\sqrt{n}) \qquad \text{as } n \to \infty$$

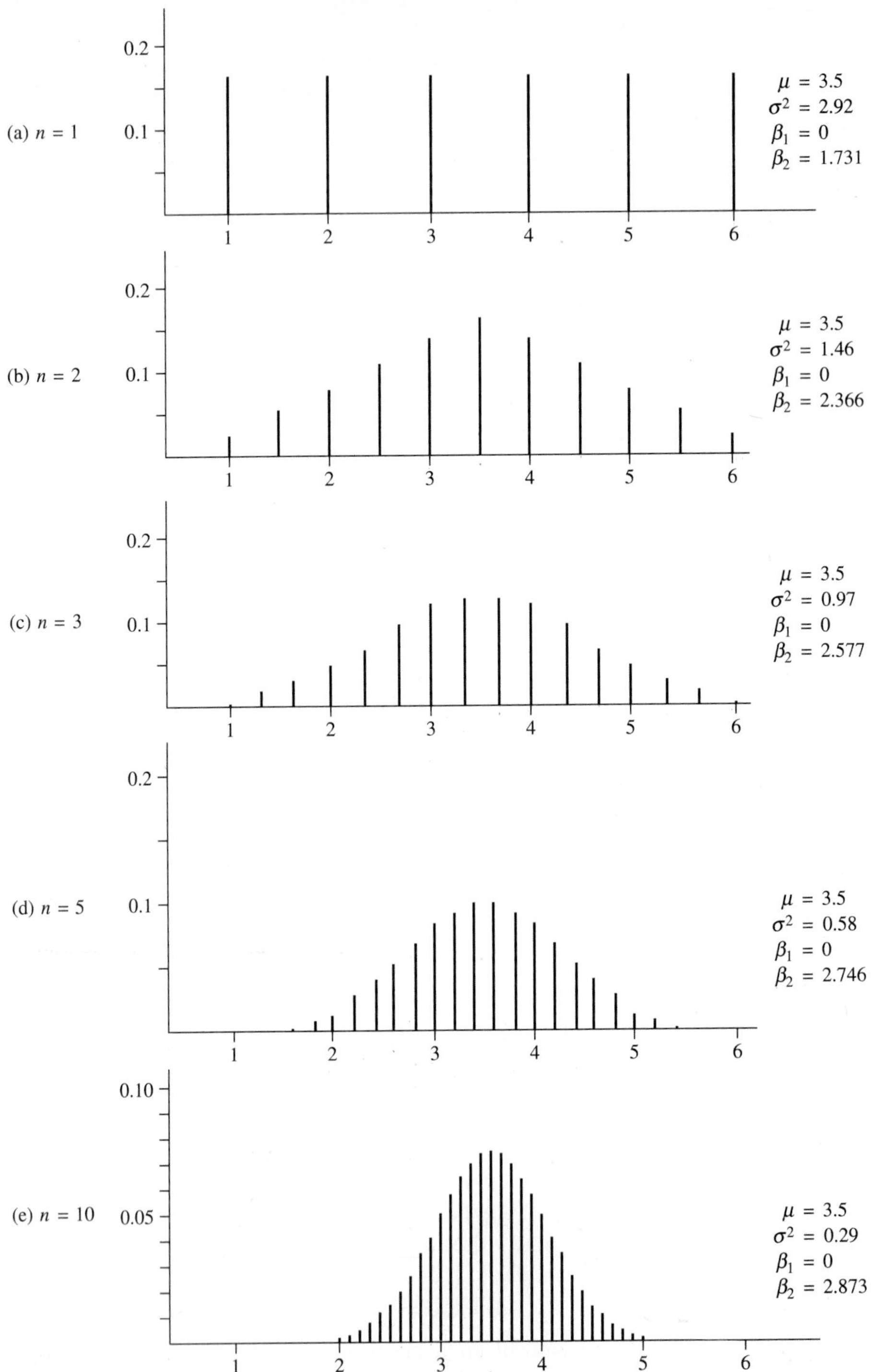

FIGURE 7-3
The function $p(\overline{x})$ for selected numbers of rolls of a fair die.

This second form follows directly from the operational rules for the distribution of a sum of *independent* random variables; i.e.,

$$\text{If each } X_i \text{ is a random variable with mean } \mu_i \text{ and variance } \sigma_i^2, \text{ then } \sum X_i \text{ has mean } \sum \mu_i \text{ and variance } \sum \sigma_i^2 \tag{7.2}$$

(We saw an empirical illustration of this second form of the central limit theorem in Exercise 5.7.) A logical question that arises from the foregoing discussion is, What sample size is required before I can assume that $\overline{X}$ is close enough to a normal random variable for practical purposes? Unfortunately, that answer varies from use to use and from distribution to distribution. However, for all but the most pathological distributions, a sample size of 30 should suffice. Further, symmetric distributions require a smaller sample size than skewed distributions. Figure 7.4 indicates that $n = 4$ is very nearly sufficient for the continuous uniform and triangular (a beta distribution with $\alpha = 1$ and $\beta = 2$) distributions. In both of these examples, using the simulation programs in the software package to perform random sampling, 2000 sets of observed data of size $n = 4$ have been obtained and $\overline{x}$ has been computed for each set. Next, a frequency table was set up, and finally the percentage polygon points were plotted against the associated theoretical normal distribution. As you can see, both distributions have remarkable agreement for such a small sample as $n = 4$. ($\hat{\mu}$ and $\hat{\sigma}^2$ are estimates of the mean and variance associated with the percentage polygon.)

As the limiting case, we recall that the distribution of $\overline{X}$ is exactly normal—no matter how small the value of n (> 0)—as long as X is distributed normally. Notice that the central limit theorem requires all members of the sample to

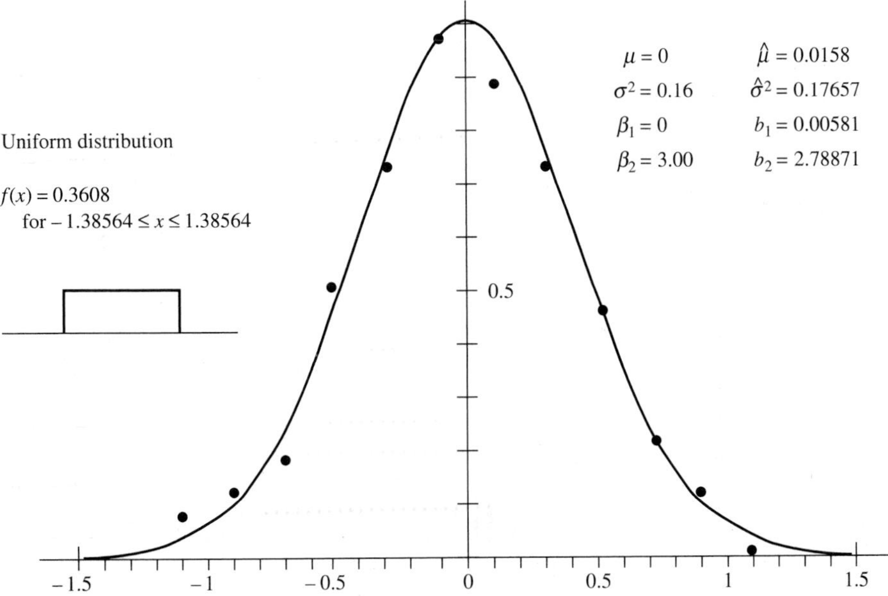

FIGURE 7-4
Empirical distributions of $\overline{x}$ for $n = 4$.

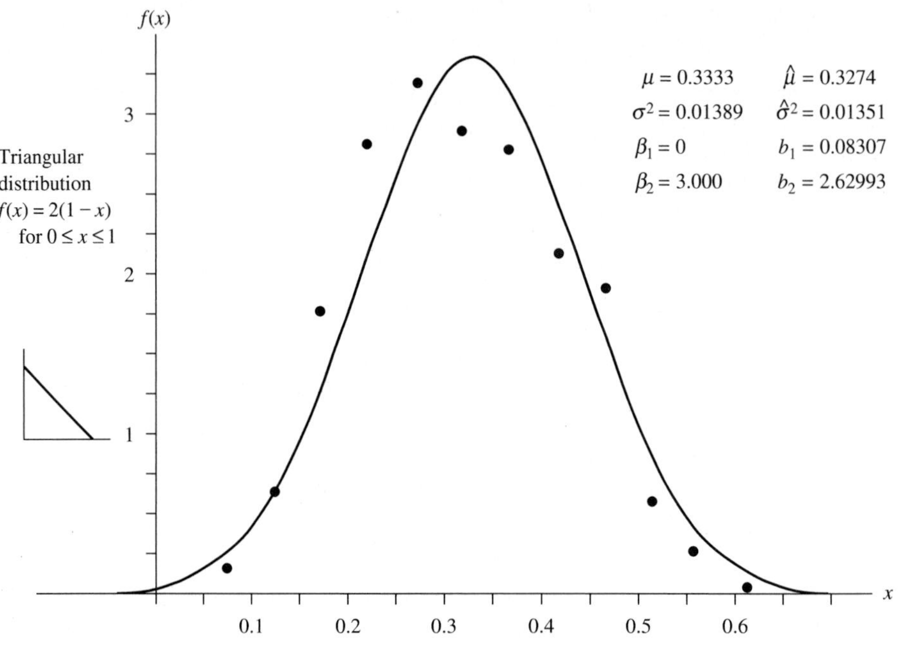

FIGURE 7-4
(Continued)

be independent and identically distributed (IID). Fortunately, in many cases, the identically distributed requirement is not necessary. If no contributor, X_i, is "large" relative to the sum of the X_i, then $\overline{X}$ will be distributed according to a normal distribution with

$$\text{a mean of } \textstyle\sum_{i=1}^{n} \mu_i/n \quad \text{and a standard deviation of } \sqrt{\textstyle\sum_{i=1}^{n} \sigma_i^2}/n \quad \text{as } n \to \infty \tag{7.3}$$

This *generalization* of the central limit theorem now allows us to characterize $\overline{X}$ as approximately normally distributed even if the X_i come from different distributions. We see this phenomenon every day in natural physical processes. The fact that the heights of individuals are normally distributed is accounted for in this generalization. A person's height is actually the sum of the lengths, breadths, or depths of many parts of the human anatomy. Certainly the contributors to a person's height do not all possess the same distribution.

Example 7.5. The daily amount of a corrosive fluid by-product from a manufacturing process is governed by a continuous uniform probability distribution with a mean of 25 gallons and a standard deviation of 2 gallons. The holding tank for the collection of the by-product has a capacity of 875 gallons. If we never want the holding tank to overflow, what is a practical maximum number of days between the occasions when the tank is emptied for disposal?

Solution. We can apply the second form of the central limit theorem to answer this question. The distribution governing the total amount of fluid in the tank, X, is a function of the number of days, n, since the last time the tank was emptied.

Using Equation 7.2, X will have a mean of $25n$ and a standard deviation of $2\sqrt{n}$. Using the normal distribution evaluator program, the following results have been obtained.

n	μ	σ	$P(x > 875)$
33	825	11.49	0.0000
34	850	11.66	0.0160
35	875	11.83	0.5000

If you are willing to tolerate a slightly greater than 1% chance of overflow, an emptying cycle of 34 days is indicated. Otherwise, the tank should be emptied every 33 days.

Up to this point, we have assumed that we knew the value of σ or the values of the σ_i. We have also assumed that either the sampled distribution or distributions were normal or the sample size was large enough to invoke the central limit theorem. What if we do *not* know the value of σ and desire to know the distribution of $\overline{X}$?

If the sample size exceeds 30, then we can be *assured* that $(\overline{X} - \mu)/(S/\sqrt{n})$ is distributed approximately as

$$N(0, 1) \quad \text{as } n \to \infty \tag{7.4}$$

where S is defined as in Section 7.3. This distribution statement does not assume that X is a normal random variable but does assume that the sample is large.

If we do *not* know σ, if the sample size is *small*, and if X *is* normally distributed, then it may be shown that

$$\frac{\overline{X} - \mu}{S/\sqrt{n}} \tag{7.5}$$

is governed by a t distribution with $\nu = n - 1$ degrees of freedom. Since the t distribution closely approaches the $N(0, 1)$ distribution when ν equals 30, $\nu = 30$ is a good cutoff point for the use of the t distribution in this context.

Referring to a single t distribution is not really correct. It would be correct to refer to a *family* of t distributions. For every value of ν, the degrees-of-freedom parameter, there exists a completely unique probability distribution function. The property joining members of this family is that they all share the same mathematical formula, as given in Table 4.1, and the difference is that each value of ν causes a different $f(t)$ to result.

7.3 THE DISTRIBUTION OF THE SAMPLE VARIANCE, S^2

We have learned in Chapter 6 that σ^2 is estimated by

$$s^2 = \frac{\sum(x_i - \overline{x})^2}{n - 1}$$

where s^2 is the realization of the random variable.

$$S^2 = \frac{\sum(X_i - \overline{X})^2}{n - 1}$$

Suppose that we operate on the definition of S^2 by multiplying both sides by $n - 1$ and dividing both sides by σ^2. The result is

$$\frac{(n-1)S^2}{\sigma^2} = \frac{\sum(X_i - \overline{X})^2}{\sigma^2} = \sum \frac{(X_i - \overline{X})^2}{\sigma^2}$$

If we could replace $\overline{X}$ with μ in the last expression, we would have $W = \sum \frac{(X_i - \mu)^2}{\sigma^2}$, the sum of n squared standardized random variables. If the X_i are $N(\mu, \sigma)$, we have the sum of n squared standardized normal random variables. From Chapter 4, we know that W is distributed as a χ^2 random variable with n degrees of freedom, χ^2_n.

As we have seen in Example 6.10, the fact that we cannot replace $\overline{X}$ with μ simply causes the loss of a single degree of freedom. Therefore,

$$\frac{(n-1)S^2}{\sigma^2} = \sum_{i=1}^{n} \frac{(X_i - \overline{X})^2}{\sigma^2} \quad \text{is distributed as} \quad \chi^2_{n-1} \tag{7.6}$$

Example 7.6. As an example of the use of this relationship, let us obtain the probability that a χ^2_{23} random variable will lie between the values of 10.196 and 35.172 ($\chi^2_{23,0.99} = 10.196$ and $\chi^2_{23,0.05} = 35.172$). As pictured in Figure 7.5, $P(10.196 \leq \chi^2_{23} \leq 35.172) = 0.94$. If S^2 is computed on the basis of a sample of $n = 24$, $23S^2/\sigma^2$ is distributed as χ^2_{23} and $P(10.196 \leq 23S^2/\sigma^2 \leq 35.172) = 0.94$. This statement will be very useful when we discuss interval estimates associated with σ^2 in Chapter 8.

If we consider S_1^2 and S_2^2 to be obtained from different statistically independent random samples of sizes n_1 and n_2, we have (where $\sim$ is read "is distributed as")

$$\frac{(n_1-1)S_1^2}{\sigma_1^2} \sim \chi^2_{n_1-1} \quad \text{and} \quad \frac{(n_2-1)S_2^2}{\sigma_2^2} \sim \chi^2_{n_2-1}$$

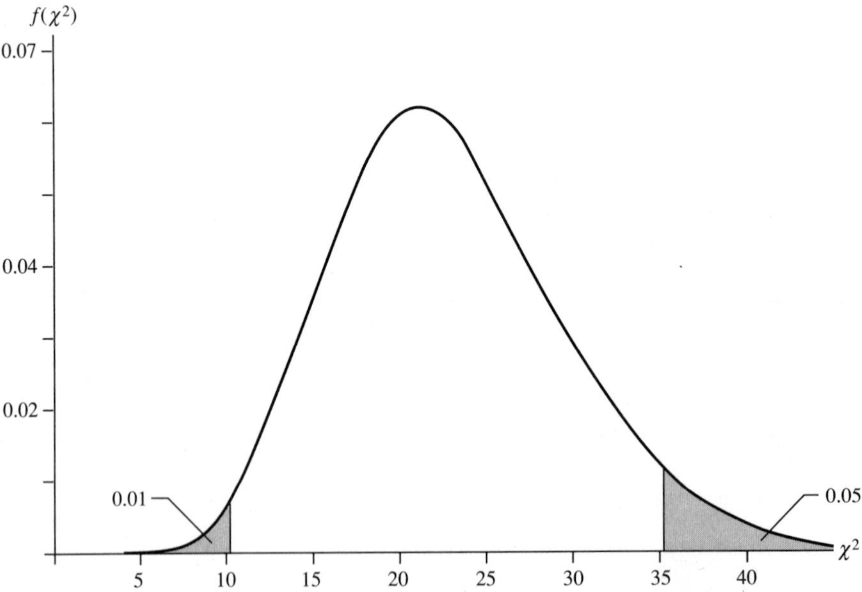

FIGURE 7-5
The χ^2_{23} distribution.

Placing these two quantities in ratio, we obtain

$$\frac{(n_1 - 1)S_1^2}{\sigma_1^2} \bigg/ \frac{(n_2 - 1)S_2^2}{\sigma_2^2} \sim \frac{\chi^2_{n_1-1}}{\chi^2_{n_2-1}}$$

All we need do to form an F *random variable* is divide the numerator by $n_1 - 1$ (the degrees of freedom for the numerator χ^2 variable) and divide the denominator by $n_2 - 1$ (again the associated degrees of freedom).

$$\frac{S_1^2}{\sigma_1^2} \bigg/ \frac{S_2^2}{\sigma_2^2} \sim \frac{\chi^2_{n_1-1}}{n_1 - 1} \bigg/ \frac{\chi^2_{n_2-1}}{n_2 - 1} \sim F_{n_1-1,n_2-1} \tag{7.7}$$

If we further assume that $\sigma_1^2 = \sigma_2^2$, then

$$\frac{S_1^2}{S_2^2} \sim F_{n_1-1,n_2-1} \tag{7.8}$$

We make extensive use of the relationships given in Equations 7.7 and 7.8 in Chapters 10 through 14.

EXERCISES

7.1. If a sample of size $n = 100$ from a population of size $N = 40,000$ implies that $\sigma_{\overline{X}} = 4$, what sample size would be required to reduce $\sigma_{\overline{X}}$ to a value of 1? What would be the required sample size if N is *not* 40,000, but rather 5000?

7.2. A sample of size $n = 36$ is to be taken from a beta distribution with parameters $\alpha = 2$ and $\beta = 3$. How likely is $\overline{X}$ to exceed a value of 0.45?

7.3. A sample with $n = 16$ has $\overline{x} = 32.2$ and $s = 2.5$. Is it reasonable to say that this sample came from a population with $\mu = 30.75$?

7.4. The variance of a process is required to be less than or equal to 12. Suppose that a decision rule deems the process in violation of this requirement if a value of $s^2 = 22.56$ or greater is obtained from a sample of size 10. How likely is σ^2 of that process to be 12 or less when $s^2 = 21.96$?

7.5. From Table 4.1, the t distribution with 1 degree of freedom is

$$f(x) = \frac{1}{\sqrt{\pi}} \frac{\Gamma(1)}{\Gamma(\frac{1}{2})} \frac{1}{1 + x^2} = \frac{1}{\pi(1 + x^2)}$$

Show that $t_{1,0.01} = 31.82$; i.e., $F(31.82) = 0.99$.

The t distribution with 2 degrees of freedom is

$$f(x) = \frac{1}{\sqrt{2\pi}} \frac{\Gamma(\frac{3}{2})}{\Gamma(1)} (1 + \frac{x^2}{2})^{-3/2} = \frac{1}{\sqrt{2\pi}} \frac{\sqrt{\pi}}{2} (1 + \frac{x^2}{2})^{-3/2} = \frac{1}{2\sqrt{2}} (1 + \frac{x^2}{2})^{-3/2}$$

Verify that $t_{2,0.05} = 2.92$; i.e., $F(2.92) = 0.95$.

7.6. Derive the distribution of $\overline{X}$ obtained from a random sample of size $n = 3$ for the roll of a fair die under the restriction that no number appears more than once.

7.7. The χ^2 distribution with n degrees of freedom is

$$f(x) = \frac{1}{2^{0.5n}\Gamma(0.5n)} x^{0.5(n-2)} e^{-0.5x}$$

Setting $n = 2$ yields $f(x) = 0.5e^{-0.5x}$. Verify that $X^2_{2,0.05} = 5.991$; i.e., $F(5.991) = 0.95$.

7.8. A sample of size $n = 13$ is to be taken from a normal population with a variance of 16. Find $P(S^2 < 4.39 \cup S^2 > 6.83)$.

7.9. The F distribution with 4 numerator degrees of freedom and 4 denominator degrees of freedom is

$$f(x) = \frac{\Gamma(4)}{\Gamma(2)\Gamma(2)}(4^2)(4^2)\frac{x}{(4 + 4x)^4} = 6\frac{x}{(1 + x)^4}$$

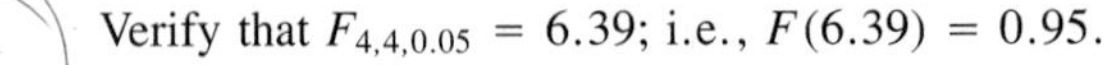

Verify that $F_{4,4,0.05} = 6.39$; i.e., $F(6.39) = 0.95$.

7.10. Two samples of size $n_1 = 13$ and $n_2 = 7$ are to be taken from a normal population with the same variance. What is $P(S_1^2/S_2^2 > 5.62)$?

7.11. As an empirical study of the central limit theorem, use the program in your software library that simulates the distributions of Tables 3.2 and 4.1 to randomly sample from the exponential distribution with $\lambda = 5$, the beta distribution with $a = 5$ and $b = 0.5$, and a uniform distribution bounded between 0 and 1.
(a) Plot each distribution.
(b) Generate a sample of size $n = 500$ from each distribution. Save the generated data set for later use with the data analysis program to ensure that the deviates are suitably representative of the "parent distribution." Form subsamples of size 5 from these 500 values, and compute $\overline{x}$ from each of the 100 subsamples. Save the 100 values of $\overline{x}$ for later analysis by the data analysis program.
(c) Are the 500 values representative of the sampled distributions? Why or why not?
(d) Is the effect of the central limit theorem apparent in the analysis of the 100 sample means (with sample size = 5)? Why or why not?
(e) Is the effect stronger in any of the three sampled distributions? Why or why not?

7.12. Show that Equation 7.5 is correct.

7.13. Using the uniform distribution simulator program to generate your random numbers (do *not* use Table 7.1), repeat the random sampling procedure of Example 7.1 and take a sample of 15 of the 200 gauges.

7.14. In Example 7.2, an incorrect method of sampling ringer coils is presented. Design a correct method to randomly sample those coils.

7.15. The following discrete probability distribution governs the number of failures in 100 tests of a particular type of guidance device under very harsh environmental conditions:

x	1	2	3	4	5	6	7
$p(x)$	0.1	0.2	0.1	0.3	0.05	0.1	0.15

Determine the probability distribution of $\overline{X}$ for 3, 5, and 10 such sets of 100 tests.

7.16. The Wearmauser Company is a dominant member of the lumber and wood products industry. Two items of interest to company executives are the average length and diameter of logs processed at their two largest lumber mills. Discuss how you would approach obtaining this information for the executives.

7.17. You are the newest analyst for Gallumping Pollsters, Inc. You have been assigned the task of conducting a poll of the voters in Boggy Creek County for the Whig party candidate. There are 3357 registered voters in the county. Discuss how you would conduct such a poll.

7.18. A sample of size $n = 45$ is to be drawn from a gamma distribution with $\lambda = 2$ and $r = 6$. What is $P(\overline{X} \geq 6)$?

7.19. The weights of cartons of conical bearings used in the assembly of gravity-driven conveyor systems is governed by a distribution with $\mu = 25$ pounds and $\sigma = 2.5$ pounds. The motorized cart used to transport the bearings from the loading dock to the assembly point is rated at a maximum of 890 pounds. If efficiency in delivering the cartons is important, how many cartons should be loaded on the cart for each trip?

7.20. The standard deviation of the weight of bags of Pure Grain Company barley is 0.75 pound. A sample of 19 bags of barley is to be obtained. What is $P(S > 1)$?

7.21. Two samples of size $n_1 = 29$ and $n_2 = 12$ are to be taken from a normal population with the same variance. What is $P(S_1^2/S_2^2 < 1)$?

7.22. A sample of size 36 is to be drawn from a Poisson distribution with parameter $\lambda = 4$. What is $P(\overline{X} > 4.3)$?

7.23. Two samples of sizes 16 and 11 are to be taken from a normal population with the same variance. Find the number c such that $P(S_1^2/S_2^2 \leq c) = 0.95$.

7.24. A bottling machine is set so that it discharges a mean of 10 ounces per bottle. The amount of fill dispensed by the machine is normally distributed with $\sigma = 1$ ounce. A sample of 16 filled bottles is taken from the output of the machine on a given day. What is $P(9.6 < \overline{X} < 10.4)$?

7.25. A sample of size 9 is to be taken from the standard normal distribution; find a number c such that $P(\sum_{i=1}^{9} Z_i^2 \leq c) = 0.9$.

7.26. The service times for customers at a supermarket checkout are independent random variables with a mean of 2 minutes and a variance of 1 minute. What is the probability that 50 customers can be serviced in less than 2 hours?

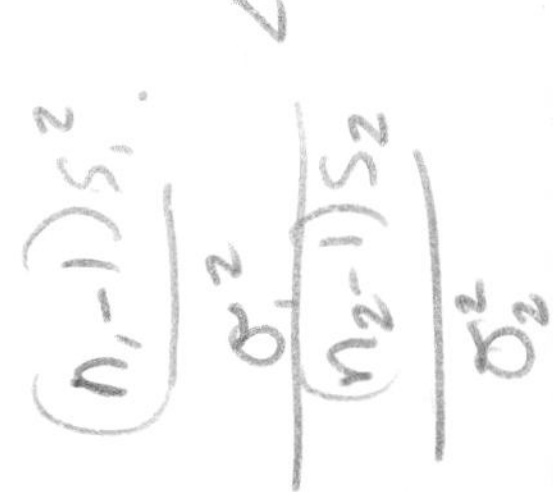

7.27. Suppose that SAT test scores from all high school seniors in the "Crabgrass State" have a mean of 1000 and a variance of 1600. If a specific high school class of $n = 36$ students has an average score of $\overline{x} = 985$, is there any evidence to suggest that this high school class' score is inferior to the state average?

7.28. Thirty heat lamps are connected in a greenhouse so that when one lamp fails, another takes over immediately. The lamps operate independently and each has a mean life of 100 hours and a standard deviation of 10 hours. What is the probability that a lamp will still be burning if the greenhouse is not checked for 3100 hours after the lamp system is turned on?

7.29. A large corporation has an average salary of \$5.00 per hour with a standard deviation of \$0.50. A sample of 36 workers is taken from the Lufkin plant. These workers have an average salary of \$4.80 per hour. Do you think workers at the Lufkin plant are paid, on the average, less than workers throughout the corporation's facilities?

7.30. Two samples of size $n_1 = 16$ and $n_2 = 21$ are to be drawn from normal populations with variances 36 and 25, respectively. What is $P(S_1^2/S_2^2 < 5)$?

7.31. Let A be governed by an F distribution with degrees of freedom d_1 and d_2. Prove that $1/A$ has an F distribution with degrees of freedom d_2 and d_1.

7.32. The variance of a particular product is known to be 25. If a sample of the product of size $n = 16$ is to be taken, what is $P(S > 4)$?

CHAPTER 8

POINT AND INTERVAL ESTIMATORS AND THE ESTIMATION OF THE MEAN AND THE VARIANCE

The following concepts and procedures form the main content of this chapter.

- Be able to give definitions of a *point estimate*, a *point estimator*, an *interval estimate*, and an *interval estimator*.
- Be able to give definitions of an *unbiased estimator* and a *most efficient unbiased estimator*.
- Understand why $\overline{x}$ is used to estimate the *true mean*, μ, of a probability distribution.
- Be able to discuss the important properties associated with an interval estimate. Know why an interval estimate gives more information than a point estimate.
- Understand the similarities and differences between the three general types of confidence intervals: the equal-tail, symmetric, and minimal width confidence intervals.
- Know what is meant by a $1 - \alpha$ *confidence interval*.
- Understand the relationships between the width of a confidence interval, the value of α, and the sample size.

- Understand Figure 8.1: It tells when to apply the t distribution and when to apply the standardized normal distribution in the setting of confidence intervals about a mean.
- Know how to obtain the *appropriate sample size* for a confidence interval about a mean when σ is known and you are given the required precision and the level of confidence.
- Know how to set confidence intervals about the true but unknown variance and standard deviation of a normal probability distribution.
- Be able to set a confidence interval about the probability of success, p, for a binomial distribution when the sample size is large.

8.1 POINT ESTIMATORS

The *Oxford American Dictionary* defines an *estimate* as "a judgement of a thing's approximate value." Most people estimate things during every day of their lives. These estimates are often based on incomplete information. A teacher may estimate that there are 22 "A" students in the class; a student could estimate that she has a probability of 0.9 of earning an A in that same class. If this is at the start of a new semester, both estimates are based on a lack of information. Regardless of how these values are determined, they are *point estimates* of the true but unknown values of the number of A students and the probability that the student will make an A.

In our mathematical approach to estimation problems, we require the capability of randomly sampling from the physical process being studied. Given that a random sample will be available, we may construct a *point estimator*, which is a function of the random variables contained in the sample. Let us consider some specific examples of such estimators.

Example 8.1. Let X be the number of successes in $n = 20$ Bernoulli trials where the probability of success, p, is unknown. $\hat{P} = X/n$ is a point estimator* for p (where, as discussed in Chapter 3, X is the sum of the 20 Bernoulli random variables that compose the random sample). Suppose that 13 successes occur in the 20 trials; i.e., $x = 13$. Our point estimate of p is $\hat{p} = x/n = 13/30 = 0.65$.

Example 8.2. Given a random sample of size n from a normal distribution whose mean, μ, and standard deviation, σ, are unknown, $\overline{X} = (1/n)\sum_{j=1}^{n} X_j$ is a point estimator for the unknown mean, μ, and $S^2 = [1/(n-1)]\sum_{j=1}^{n}(X_j - \overline{X})^2$ is a point estimator for the unknown variance, σ^2.

Since any point estimator is a function of one or more random variables, it is also a random variable. Once a random sample has been realized and the sample values have been observed, point estimators generate a single point value (realization)

*A point estimator, like $\hat{P}$, which may be computed using information only from the random sample, is often also called a *statistic*.

as an estimate of the *parameter* of interest. Thus $\overline{x}$, s^2, and $\hat{p}$ are called *point estimates*.

A point estimator can possess a variety of properties. If possible, we would prefer a point estimator that is both *accurate and efficient*, i.e., an estimator that generates estimates that hit close to the quantity being estimated most of the time and does not require an unreasonably large sample size to do so. A *point estimator* is accurate if its expected value is equal to the parameter being estimated. If an estimator possesses this property, it is an *unbiased estimator*, and its realization is called an *unbiased point estimate*. Let us consider three examples of unbiased estimators.

Example 8.3. If X is distributed according to a binomial distribution, then $\hat{P} = X/n$ is an unbiased estimator of the probability of success, p; i.e., $E(\hat{P}) = \frac{1}{n}E(X) = \frac{1}{n}(np) = p$. (As given in Table 3.2, $E(X) = \mu = np$ is the expected value or mean of the binomial distribution.)

Example 8.4. If the elements of the random sample, the X_i, $i = 1, \ldots, n$, are taken from any probability distribution, then

$$E(\overline{X}) = \frac{1}{n}\sum_{i=1}^{n} E(X_i) = \frac{1}{n}n\mu = \mu$$

Example 8.5. Since $(n-1)S^2/\sigma^2$ is a χ^2_{n-1} random variable, $E[(n-1)S^2/\sigma^2] = E(\chi^2_{n-1}) = n-1$ (from Table 4.1). It follows that $[(n-1)/\sigma^2]E(S^2) = n-1$, which implies that $E(S^2) = \sigma^2$.

From the discussion of expected values in Chapter 5, we may conclude that the *long-term average* associated with any unbiased estimator will equal the parameter that it estimates. Stating this another way, we are assured that, as the sample size grows larger, an unbiased estimate will approach closer and closer to the true value of the parameter being estimated.

That is a very desirable property. However, knowing that an estimator is unbiased is often not sufficient when we are searching for the best estimator to use. Suppose that you are presented with two or more unbiased estimators of the same parameter. The *most efficient* (best) unbiased estimator from that set is the one whose distribution has the *smallest variance*. Suppose that the point estimators $\hat{\Theta}_1$ and $\hat{\Theta}_2$ are both unbiased estimators of the parameter θ. If the variance of $\hat{\Theta}_1$ is less than the variance of $\hat{\Theta}_2$, then $\hat{\Theta}_1$ is *more efficient* than $\hat{\Theta}_2$. In other words, if $\theta = E(\hat{\Theta}_1) = E(\hat{\Theta}_2)$ and $\text{var}(\hat{\Theta}_1) < \text{var}(\hat{\Theta}_2)$, then we should use $\hat{\Theta}_1$.

Example 8.6. As an example of a more efficient estimator, consider $\overline{X}$, and $\tilde{X}$, the point estimator of the median, x_m, computed from a normally distributed random sample of size n. Since the underlying distribution is symmetric, both of these estimators are unbiased estimators of the true but unknown mean, μ. We know from past work that $\text{var}\{\overline{X}\} = \sigma^2/n$. It can be shown that $\text{var}(\tilde{X}) = 1.57\sigma^2/n$.

Since $\overline{X}$ is a more efficient estimator of μ than $\tilde{X}$, $\overline{X}$ should be used in preference to $\tilde{X}$. In terms of its *lesser* efficiency, $\tilde{X}$ will require a sample 1.57 times larger than the sample for $\overline{X}$ in order to achieve the same variance as $\overline{X}$. In other words, unless the sample size for $\tilde{X}$ is at least 1.57 times larger than the sample size for $\overline{X}$, $\overline{x}$ is, in probability, expected to be closer to μ than $\tilde{x}$.

8.2 INTERVAL ESTIMATORS

Unfortunately, even the best of all possible point estimators has serious shortcomings. It gives only a single point value, which can often be very misleading in the framework of decision making.

Example 8.7. Suppose that you have an investment opportunity that would pay an expected return of \$2 for every \$1 invested. Further, this return is tied to a period of 1 year, and the minimum investment is \$3000. How eager are you to invest your hard-earned cash?

Solution. Consider the possible ways that this expected return might be true. There are *two obvious extremes* that would yield the *same stated expected return*. First, the probability that you would receive the 2-to-1 return could have a value of 1—i.e., only one outcome is possible; you are certain to receive the 2-to-1 return. Second, suppose that two outcomes are possible. One outcome is that you receive a 100-to-1 return on your investment, and the other is that you get *none* of your money back. If the probability of getting no money back is 0.98, then the expected return is *also* 2 to 1; i.e., (100)(0.02) + (0)(0.98) = 2.

This kind of situation occurs frequently in practical situations, and a point estimate is *not sufficient*. The ultimate in decision-making information would be a complete characterization of the probability distribution governing the item of interest. Not surprisingly such a characterization is rarely available. Recognizing this, statisticians have developed an intermediate approach, the *interval estimate*, that provides a great amount of additional information at a comparatively small cost. As an example of an interval estimate, if you know that investments similar to the one described in Example 8.7 have a return ranging between 2.2-to-1 and 1.6-to-1 in 95% of 5000 documented cases, you might be more willing to invest your money.

Like the point estimate, an interval estimate is any pair of values, regardless of the selection method, that are claimed to contain the true value of the quantity being estimated. Once more, our mathematical approach to interval estimation requires a more defendable foundation, presented in the form of a *confidence interval*. A confidence interval for a parameter θ with *confidence coefficient* $1-\alpha$ is specified by the pair of values θ_L and θ_U. The values of θ_L and θ_U give the *lower* and *upper confidence limits* on θ such that

$$(\theta_L, \theta_U) \text{ is the } 1-\alpha \text{ confidence interval on } \theta \tag{8.1}$$

The *confidence interval width*, $(\theta_U - \theta_L)$, and the confidence coefficient, or *level of confidence*, $1-\alpha$, jointly reflect the intrinsic *reliability*, or *precision*, of the estimation technique being used.

8.3 INTERVAL ESTIMATES FOR POPULATION MEANS

Let us consider how to construct a confidence interval for a true but unknown population mean, μ. The several different cases that must be considered in setting confidence intervals on means are schematically detailed in Figure 8.1. As may be observed, selection of the correct distribution is determined by whether the value of σ is known, by the sample size, and by whether the sampled population is normally distributed. We discuss the easiest case first, i.e., when σ is *known* and the population is *normally distributed*.

When σ is known and the population is normal, it does not matter whether the sample size is large or small. Regardless of the sample size, the fact that $\overline{X}$ is governed by an $N(\mu, \sigma/\sqrt{n})$ distribution implies that $Z = (\overline{X} - \mu)/(\sigma/\sqrt{n})$ is distributed according to an $N(0, 1)$ distribution.

The fact that $P\{Z \leq z_{\alpha/2}\} = 1 - (\alpha/2)$ implies that $P\{-z_{\alpha/2} < Z < z_{\alpha/2}\} = 1 - \alpha$. On substitution, we have

$$P\left\{-z_{\alpha/2} < \frac{\overline{X} - \mu}{\sigma/\sqrt{n}} < z_{\alpha/2}\right\} = 1 - \alpha$$

which leads directly to

$$P\left\{-z_{\alpha/2}(\sigma/\sqrt{n}) < (\overline{X} - \mu) < z_{\alpha/2}(\sigma/\sqrt{n})\right\} = 1 - \alpha \tag{8.2}$$

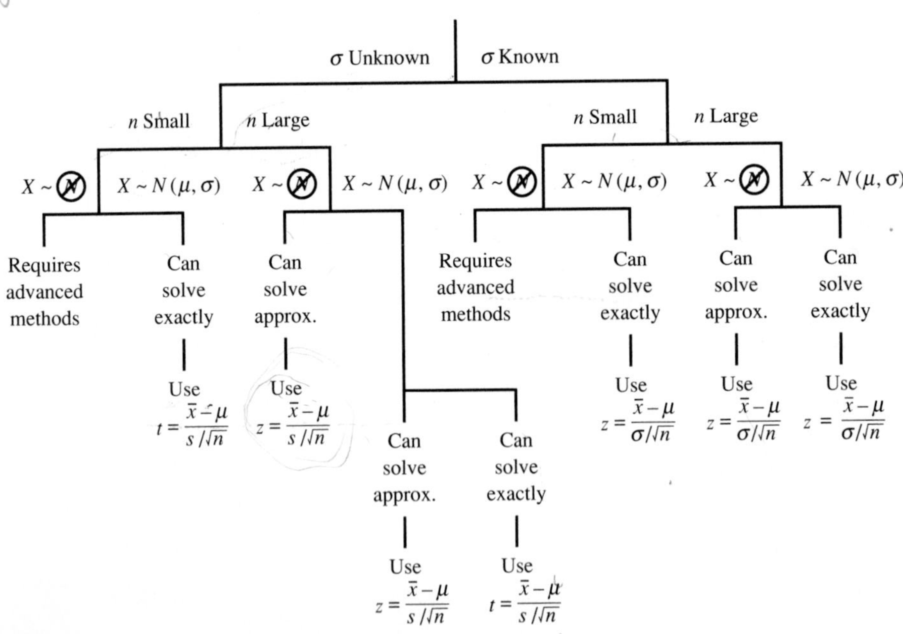

FIGURE 8-1
The eight situations associated with estimating means of normally distributed random variables.

Solving both inequalities for μ yields

$$P\left(\overline{X} - z_{\alpha/2}\left(\sigma/\sqrt{n}\right) < \mu < \overline{X} + z_{\alpha/2}\left(\sigma/2\right)\right) = 1 - \alpha \tag{8.3}$$

This random interval has a $1 - \alpha$ probability of containing the true but unknown value of μ. Notice that the random interval, $(\overline{X} - z_{\alpha/2}(\sigma/\sqrt{n}), \overline{X} + z_{\alpha/2}(\sigma/\sqrt{n}))$, is bounded by random variables. If, after the random sample has been realized, we substitute $\overline{x}$ for $\overline{X}$ in those random bounds, we obtain the following deterministic interval:

$$\left(\overline{x} - z_{\alpha/2}, (\sigma/\sqrt{n}), \overline{x} + z_{\alpha/2}(\sigma/\sqrt{n})\right) \tag{8.4}$$

which is the $1 - \alpha$ confidence interval for μ. It can also be called the $100(1 - \alpha)\%$ *confidence interval*.

Equation 8.4 is superior to any other confidence interval that can be constructed. Here are three important observations that can be made about this confidence interval.

1. The higher the level of confidence, $1 - \alpha$, the wider the interval will be. This follows directly from the fact that, as $1 - \alpha$ increases, α decreases and $z_{\alpha/2}$ becomes larger.
2. As n increases, the confidence interval width decreases. However, this decrease is hampered in its effect by the inverse square relationship discussed in Chapter 7; i.e., $\sigma_{\overline{X}} = \sigma/\sqrt{n}$.
3. The smaller the population variance, σ^2, the smaller the confidence interval will be. Often the most economical approach to decreasing the interval width is to take direct action on the physical process being analyzed by removing identifiable sources of variability.

It may be tempting to interpret Equation 8.4 as declaring that "the probability that μ will be contained within the upper and lower confidence limits from a particular set of observed data is $1 - \alpha$." Unfortunately, such a reading and interpretation would be *wrong*. For example, suppose that some celestial source of information told us that the mean of a population was 25. However, we did not gain this knowledge until after we had set a confidence interval of (25.2, 27.2). What is the probability that the confidence interval that we constructed contains the mean? Of course, the answer is 0. In order to interpret the confidence interval correctly, we must return to the classical *frequency interpretation* of probability.

Suppose that we construct a large number of confidence intervals, each from its own data set, using Equation 8.4. In each interval constructed, the observed $\overline{x}$ will differ and no two intervals will be the same. What the confidence interval statement is actually telling us is that the expected proportion of those intervals that will contain the true mean is $1 - \alpha$. For example, if 1000 intervals are constructed and α is set equal to 0.05, we would expect about 950 of the intervals to contain the mean. From this interpretation, we see that considering a single confidence interval in probabilistic terms is meaningless. A single confidence interval will either contain the true mean or it will not!

It is important to understand that this rationale fully embraces the possibility of the so-called "bad sample," i.e., the low-probability sample that is made up primarily of very low (or very high) values, all from the same low-probability tail of the population's distribution. If such an unlikely sample is obtained by our random sampling procedure, we cannot expect the true nature of the population to be reflected. Therefore, we would not expect the interval based on this anomalous sample to contain the true mean. We expect to get such bad samples, in the long run, α proportion of the time.

An alternate way of generating Equation 8.4 presents a different aspect in regard to confidence intervals and is useful in developing and understanding confidence intervals for other parameters. Suppose that we solve both inequalities in Equation 8.2 for $\overline{X}$, yielding

$$P\left\{\mu - z_{\alpha/2}(\sigma/\sqrt{n}) < \overline{X} < \mu + z_{\alpha/2}(\sigma/\sqrt{n})\right\} = 1 - \alpha$$

As illustrated in Figure 8.2, the left- and right-hand bounds in the above expression define the range of values that $\overline{X}$ will assume with probability of $1-\alpha$. These bounds are useful only when we know the value of μ—a rare situation. However, the expression's factual content can be made very useful if we simply change our point of view. Most of the time we do not know the value of μ, but we will know $\overline{x}$ after the random sample has been observed. For our current discussion, we also know the sample size n and the true value of σ.

- For a normal population with standard deviation, σ, what values of the true mean, μ, would be likely to yield any particular value of $\overline{x}$?
- Indeed, with a probability of 0.95, what is the range of possible values of μ that would yield the value of $\overline{x}$ that *was obtained?*

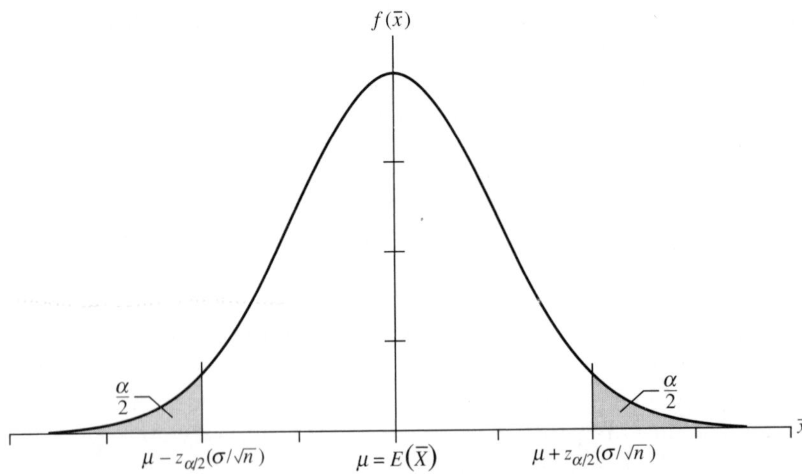

FIGURE 8-2
The distribution of $\overline{X}$.

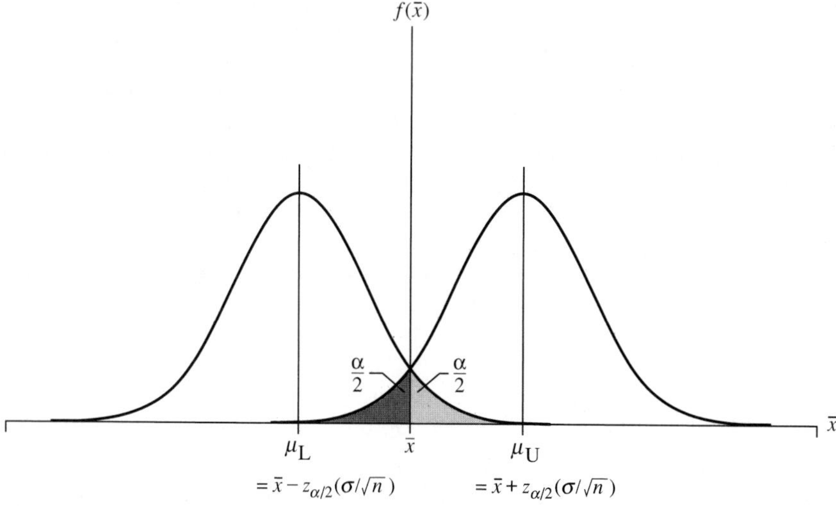

FIGURE 8-3
The upper and lower confidence limits on μ.

Figure 8.3, which pictures two normal distributions, $N(\mu_L, \sigma/\sqrt{n})$ and $N(\mu_U, \sigma/\sqrt{n})$, helps to answer the latter of these two questions. Suppose that μ_L was the value of the unknown mean, μ. Let us select the value of μ_L so that the generation of a value of the sample mean as *large or larger* than $\bar{x}$ will have a probability of $\alpha/2$. As presented in Figure 8.3, such a selection of μ_L implies that $\bar{x} = \mu_L + [z_{\alpha/2}(\sigma/\sqrt{n})]$, which leads to $\mu_L = \bar{x} - [z_{\alpha/2}(\sigma/\sqrt{n})]$. Similarly, μ_U is the value of the unknown mean, μ, that would cause a value of the sample mean as *small or smaller* than $\bar{x}$ to be generated with a probability of $\alpha/2$. The fact that $\bar{x} = \mu_U - [z_{\alpha/2}(\sigma/\sqrt{n})]$ implies that $\mu_U = \bar{x} + [z_{\alpha/2}(\sigma/\sqrt{n})]$.

Thus the values of the true but unknown mean, μ, that bound the observed $\bar{x}$ form the interval (μ_L, μ_U), which is identical to Equation 8.4:

$$\left(\bar{x} - z_{\alpha/2}\frac{\sigma}{\sqrt{n}}, \bar{x} + z_{\alpha/2}\frac{\sigma}{\sqrt{n}}\right) \tag{8.4}$$

Readers who are interested in more detail on this alternate, general way of viewing confidence intervals are referred to Mood and Graybill [1963].

Let us consider the following example of setting a confidence interval about a mean under our current assumptions—that σ is known and the population is normally distributed.

Example 8.8. Nine lengths of drilling pipe are stressed to failure in a testing laboratory. The following readings were obtained: 4500, 5000, 6500, 5200, 4800, 4900, 5125, 6200, and 5375. It is *known*, from long experience in working with similar pipe, that $\sigma = 300$. Construct a 95% confidence interval for the mean failure strength.

Solution. Use the confidence interval program in your program library. After accessing the program, you first state whether the true standard deviation is known (YES) and whether the sampled population is governed by a normal distribution (YES). Next, you are asked to input the sample size (9), the value of the true standard deviation (300), the value of $\overline{x}$ (5288.89), and the value of α (0.05).

Because the sample size of 9 is small, we must assume that the sampled population is governed by a normal distribution. Otherwise, by Figure 8.1, we must know the sampled population distribution and we must use more sophisticated methods than those provided by our software package. Since the value of the true standard deviation is *known*, the program uses the normal distribution (as opposed to the t distribution, which would be correct if σ is not known). The program then obtains $z_{\alpha/2} = z_{0.025} = 1.96$ and uses Equation 8.4 to obtain the following 95% confidence interval for the mean.

$$\left(5288.89 - (1.96)\left(\frac{300}{3}\right), 5288.89 + (1.96)\left(\frac{300}{3}\right)\right) = (5092.9, 5484.9)$$

Suppose that, instead of knowing the value of the population standard deviation, we are forced to use the sample estimate of σ^2, s^2. This would be the case if the pipe being tested is a new kind of pipe and little experience had been gained in its use. Even with no prior knowledge of σ^2, we are still sampling (by assumption) from a normal population. Since the sample is small, we use the random variable $T_{n-1} = (\overline{X} - \mu)/(S/\sqrt{n})$, which is governed by a t distribution with $n-1$ degrees of freedom. Using a similar approach to that for Equation 8.4, the $1-\alpha$ confidence interval for the mean for small n from a normal population when σ is unknown may be shown to be

$$\left(\overline{x} - t_{n-1,\alpha/2}\frac{s}{\sqrt{n}}, \ \overline{x} + t_{n-1,\alpha/2}\frac{s}{\sqrt{n}}\right) \tag{8.5}$$

where, as pictured in Table A.5, $t_{n-1,\alpha/2}$ is the value for the t distribution with $n-1$ degrees of freedom such that the area under the density function greater than $t_{n-1,\alpha/2}$ is $\alpha/2$.

There is an old saying that "you don't get something for nothing." This confidence interval is an illustration of that idea. We have already seen that the t distribution is more variable than the standardized normal distribution. For this reason, $t_{n-1,\alpha/2}$ is greater than $z_{\alpha/2}$ for finite values of n. Therefore, if s is approximately the same magnitude as σ, we expect that the confidence interval of Equation 8.5 will be wider than the confidence interval of Equation 8.4. Intuitively, this makes sense because of the greater amount of uncertainty present when σ is unknown and must be estimated from the data set.

Example 8.9. Assuming that the true value of σ is not known, construct a 95% confidence interval for the mean failure strength of the drilling pipe of Example 8.8.

Solution. Once more, use the confidence interval program. Your responses are the same as for Example 8.8, with two exceptions: You answer NO when asked whether the true standard deviation is known, and you input 655.28 (obtained by the data analysis program for the data of Example 8.8) when asked for the value of the *sample* standard deviation, s. The program obtains $t_{n-1,\alpha/2} = t_{8,0.025} = 2.306$ and uses Equation 8.5 to obtain the following 95% confidence interval for the mean.

$$\left(5288.89 - (2.306)\left(\frac{655.28}{3}\right), 5288.89 + (2.306)\left(\frac{655.28}{3}\right)\right) = (4785.2, 5792.6)$$

If n is large, the central limit theorem may be invoked, and the distribution of the sampled population becomes unimportant. For most practical applications, $n \geq 30$ can be considered large, and Equation 8.4 can be used. If n is large and the value of σ is unknown, simply substitute the value of s everywhere that σ appears in Equation 8.4.

If n is small and we have reason to believe that the population is *not* distributed normally, the setting of a confidence interval requires advanced methods. An example of such a method is provided by Example 8.13 in Section 8.7.

8.4 DETERMINING AN ADEQUATE SAMPLE SIZE FOR INTERVAL ESTIMATION OF A MEAN

The width of the confidence interval given in Equation 8.4 is $2E = 2z_{\alpha/2}(\sigma/\sqrt{n})$. Therefore, the precision of the interval estimate in Equation 8.4 is determined by

$$E = z_{\alpha/2}\frac{\sigma}{\sqrt{n}} \tag{8.6}$$

If a stipulated precision E is required, simple algebraic operations yield the appropriate sample size as

$$n = \left(\frac{z_{\alpha/2}\sigma}{E}\right)^2 \tag{8.7}$$

If the computed value of n is not an integer, the sample size should be rounded *up* to the next higher integer. This is the conservative approach and will make the interval somewhat smaller than required.

Example 8.10. Suppose that we needed to set a confidence interval on the strength of the drilling pipe discussed in Example 8.8 and that the interval needs to be no greater than 150 units. The sample size would need to be computed on the basis of $2E = 150$, or $E = 75$, units. Substituting the appropriate values into Equation 8.7 yields $n = ((1.96)(300)/75)^2 = 61.47$. Thus the sample size would be set at $n = 62$.

If the value of σ is not known, iterative procedures can be used to obtain appropriate sample sizes for confidence intervals for means; readers interested in this topic are referred to Berry and Lindgren [1990].

8.5 CONFIDENCE INTERVALS FOR THE VARIANCE

We have already stated that S^2 is an unbiased estimator of σ^2. It may also be shown that S^2 is the most efficient estimator of σ^2. Of course, point estimators of σ^2 suffer from the same deficiencies that point estimators of μ possess.

Earlier in this text it has been shown that, if the original population is distributed normally, $\left[(n-1)S^2\right]/\sigma^2$ is governed by a χ^2_{n-1} distribution. This relationship allows us to write

$$P\left(\chi^2_{n-1,1-\alpha/2} \le \frac{(n-1)S^2}{\sigma^2} \le \chi^2_{n-1,\alpha/2}\right) = 1-\alpha \tag{8.8a}$$

which, after some manipulation, yields

$$P\left(\frac{(n-1)S^2}{\chi^2_{n-1,\alpha/2}} \le \sigma^2 \le \frac{(n-1)S^2}{\chi^2_{n-1,1-\alpha/2}}\right) = 1-\alpha \tag{8.8b}$$

Substituting s^2 for S^2 yields the following equal-tail $1-\alpha$ *confidence interval on* σ^2.

$$\left(\frac{(n-1)s^2}{\chi^2_{n-1,\alpha/2}}, \frac{(n-1)s^2}{\chi^2_{n-1,1-\alpha/2}}\right) \tag{8.9}$$

It should be noted that the confidence interval given in Equation 8.9 is *neither symmetric nor the shortest possible interval* for this level of confidence. Equation 8.9 is based on the fact that there are *equal-tail areas* stipulated in Equation 8.8. Because the χ^2 distribution is skewed to the right, the lower confidence limit is necessarily closer to the sample estimator than the upper confidence limit. The equal-tail confidence interval of Equation 8.9 is usually selected for its ease of computation.

Example 8.11. Use the data of Example 8.8 to set a 95% confidence interval on σ^2.

Solution. Use the confidence interval program, and select the option for confidence intervals for variances. After accessing the program, you are asked to input the sample standard deviation, s (655.28), the value of α (0.05), and the sample size (9). The necessary values of the χ^2 distribution are obtained by the program—($\chi^2_{n-1,1-\alpha/2} = \chi^2_{8,0.975} = 2.1797$ and $\chi^2_{n-1,\alpha/2} = \chi^2_{8,0.025} = 17.5346$)—and, using Equation 8.9, the program obtains the following 95% confidence interval on σ^2.

(195,906, 1,575,945)

If a confidence interval is desired for σ, take the square root of the limits of Equation 8.9. For Example 8.11, that procedure, as performed by the program, yields the following 95% confidence interval on σ: (442.6, 1255.4).

8.6 ESTIMATING PROPORTIONS

Suppose that we are interested in estimating the proportion of successes, p, for a binomial population. We already know that the best point estimator for p is $\hat{P} = X/n$, where X is the number of successes in a sample of size n. Additionally, for large n and p not near 0 or 1,

$$Z = \frac{\hat{P} - p}{\sqrt{\hat{P}(1 - \hat{P})/n}}$$

is approximately distributed as $N(0, 1)$. Thus, just as in Equation 8.4, we may state the following $1 - \alpha$ confidence interval for p.

$$\left(\hat{p} - z_{\alpha/2}\sqrt{\frac{\hat{p}(1 - \hat{p})}{n}},\ \hat{p} + z_{\alpha/2}\sqrt{\frac{\hat{p}(1 - \hat{p})}{n}}\right) \tag{8.10}$$

Example 8.12. Suppose that in the random sampling of 500 people, 280 say that they support an upcoming bond issue. Set a 90% confidence interval on the proportion of persons supporting the bond issue.

Solution. Use the confidence interval program, and select the option for confidence intervals for proportions. After accessing the program, you are asked to input $\hat{p}$ (0.56), the value of α (0.1), and the sample size (500). The program obtains $z_{\alpha/2} = z_{0.05} = 1.645$ and uses Equation 8.10 to yield (0.523, 0.597), the 90% confidence interval required.

The theory for means, as discussed in Section 8.5, can easily be extended to apply to proportions when one is interested in determining an adequate sample size for setting a confidence interval on a proportion. It can be shown that the maximum sample size that might be required for a precision of E is

$$n = 0.25\left(\frac{z_{\alpha/2}}{E}\right)^2 \tag{8.11}$$

8.7 THE THREE GENERAL TYPES OF CONFIDENCE INTERVALS

There are three general types of confidence intervals: the confidence interval with equal tails, the symmetric confidence interval, and the confidence interval of minimal width. Let $\hat{\Theta}$ be the estimator of the parameter θ. Like the random intervals presented in Equations 8.3 and 8.8, we may construct a random interval about the general parameter θ:

$$P(L(\hat{\Theta}, 1 - p_1) < \theta < U(\hat{\Theta}, p_2)) = 1 - \alpha \tag{8.12}$$

where $L(\hat{\Theta}, 1 - p_1)$ and $U(\hat{\Theta}, p_2)$ are functions of the estimator $\hat{\Theta}$ and form the lower and upper random bounds on θ. Further, we consistently enforce the restriction that $p_1 + p_2 = \alpha$.

The associated confidence interval is achieved by replacing $\hat{\Theta}$ with its point estimate, $\hat{\theta}$, everywhere in Equation 8.12 to yield

$$(L(\hat{\theta}, 1 - p_1), U(\hat{\theta}, p_2)) = (\theta_L, \theta_U)$$

If $p_1 = p_2 = \alpha/2$, we have a confidence interval with *equal tails* like those presented in Equations 8.4, 8.5, 8.9, and 8.10. If $(\hat{\theta} - \theta_L) = (\theta_U - \hat{\theta})$, we have a *symmetric* confidence interval, i.e., a confidence interval where the distance between the point estimate of $\theta, \hat{\theta}$, and either confidence limit is the same. A confidence interval of *minimal width* is achieved by obtaining the values of p_1 and p_2 that cause $(\theta_U - \theta_L)$ to take on the smallest possible value.

If the distribution associated with $\hat{\Theta}$ is symmetric, all three intervals are identical and form the confidence interval of minimal width (as illustrated in Equations 8.4, 8.5, and 8.10). For estimators with asymmetric distributions, neither the symmetric nor the equal-tail confidence interval form the confidence interval of minimal width.

Example 8.13. Suppose that we have obtained a single data point from an exponential distribution and its value is 2.1. Place (a) a 90% equal-tail confidence interval about λ, (b) a 90% minimal-width confidence interval about λ, and (c) a 90% symmetric confidence interval about λ.

Solution

(a) Using the alternate approach for confidence intervals introduced in Section 8.4, we ask, What true value of λ would cause our estimator (of the mean)—the random variable X—to achieve values of 2.1 or smaller with a probability of 0.05? This will yield the 90% equal-tail lower confidence limit on λ.

In setting confidence intervals about the mean and standard deviation, we know the distributions of the estimators $\overline{X}$ and S^2. Likewise, in this example the distribution of X is exponential with the parameter λ, whose value is unknown. Setting the cumulative distribution $F(x) = 1 - e^{-\lambda x} = \alpha/2$ and solving yields

$$\lambda_L = \frac{-\log_e(1 - (\alpha/2))}{x}$$

Substituting 2.1 for x and 0.1 for α yields $\lambda_L = 0.0244$. Similarly, setting the cumulative distribution $F(x) = 1 - e^{-\lambda x} = 1 - \alpha/2$ and solving yields

$$\lambda_U = \frac{-\log_e(\alpha/2)}{x}$$

which, on substitution, yields $\lambda_U = 1.4265$. Thus, (0.0244, 1.4265) is the 90% equal-tail confidence interval for λ based on our single data value of 2.1.

(b) In this case,

$$\lambda_L = \frac{-\log_e(1 - p_1)}{x} \quad \text{and} \quad \lambda_U = \frac{-\log_e(p_2)}{x}$$

The minimal-width confidence interval requires that we find the limits such that

$$(\lambda_U - \lambda_L) \quad \text{is minimized and} \quad p_1 + p_2 = \alpha \tag{8.13}$$

Trial and error may be used to minimize $(\lambda_U - \lambda_L)$, and doing so yields $p_1 = 0$, $p_2 = 0.1$, $\lambda_L = 0$, and $\lambda_U = 1.0965$. This can be verified by plotting

$$\lambda_u - \lambda_L = -\frac{\log_e(\alpha - p_1)}{x} + \frac{\log_e(1 - p_1)}{x}$$

for values of p_1 from 0 to around 0.05. (Attempts to plot for higher values of p_1 may invoke numerical difficulties.)

(c) A symmetric confidence interval, by definition, requires a point estimate, $\hat{\lambda}$. Since λ is the reciprocal of the mean of the exponential distribution, $\hat{\lambda} = 1/\overline{x}$.

For our sample of size $n = 1$, $\hat{\lambda} = 1/x = 1/2.1 = 0.4762$. We need to find the limits such that

$$(\hat{\lambda} - \lambda_L) = (\lambda_U - \hat{\lambda}) \quad \text{and} \quad p_1 + p_2 = \alpha \tag{8.14}$$

All that remains to be done is to determine the values of $p_1 + p_2$ that satisfy Equation 8.14. This may also be approached by trial and error. However, when we try the extreme value of $p_1 = 0$ ($\lambda_L = 0$), λ_U once more assumes a value of 1.0965. Since λ_L cannot assume a negative value, we must conclude that a symmetric confidence interval about $\hat{\lambda} = 0.4762$ does not exist. This only serves to emphasize that, for asymmetric distributions, the three major types of confidence intervals are not the same.

Exercise 8.8 will give you the opportunity to apply these ideas to confidence intervals about a single variance.

In this chapter, we have begun to make inferences about the properties and characteristics of the probability distribution that governs our sampled population. In the next chapter, we continue this kind of investigation by developing methods to *make decisions* about the sampled population.

EXERCISES

8.1. In a production process yielding lead sulfate as a by-product, the following amounts (in pounds) were obtained by random sampling 10 days of production: 22, 27, 26, 20, 29, 25, 31, 26, 23, and 26. Set a 90% confidence interval on the mean daily lead sulfate production. Set a 90% confidence interval on the standard deviation of the daily lead sulfate production.

8.2. Uranium ore is processed, yielding products ^{235}U and ^{238}U. Fourteen out of 40 random lots had an average of over 40% of ^{235}U. Set a 95% confidence interval on the proportion of lots that have more than 40% ^{235}U.

8.3. Government regulations require that all interval estimates on the contents of standard containers of powdered milk from a government-licensed vendor be set at the 90% confidence level and that the interval be no more than 0.8 pound in width. Determine the sample size required if past records indicate that the standard deviation of the contents of such containers is $\sigma = 0.25$ pound.

8.4. Amalgamated Metals produces ingots of nickel at the Dry Gulch plant. A random sample of 25 ingots yielded $\overline{x} = 5.06$ pounds. The last five years' production records show that the variance of the weight of nickel ingots is $\sigma^2 = 0.0036$ pounds^2. Set an 80% confidence interval on the mean weight of nickel ingots from this company.

8.5. Seventeen bushings yield $s = 0.0004$ inch in the measurement of their internal diameters. Set a 90% confidence interval on the true standard deviation of the bushings' internal diameters.

8.6. Twenty-two aluminum tension members are loaded to twice their nominal strengths. Eight break into two pieces, and 4 "neck down" but do not break. If either of these cases are classed as failures, set a 90% confidence interval on the proportion of such tension members that are capable of supporting twice their nominal strength rating.

8.7. Derive Equation 8.11.

8.8. In Example 8.11, the drilling pipe data were used to illustrate how to set a 95% equal-tail confidence interval about σ^2.

(a) Find the confidence limits for a 80% *symmetrical* confidence interval for σ^2 for the drilling pipe data.

(b) Find the *shortest* 95% confidence interval for σ^2 for the drilling pipe data.

8.9. We know from Chapter 4 that the normal approximation to the binomial distribution is valid only when n is large and p is not near 0 or 1. Suppose that $n = 20$ trials have been performed and 2 successes have occurred—implying that p is in the neighborhood of $\hat{p} = x/n = 0.1$. A binomial distribution with $p = 0.1$ and $n = 20$ may *not* be accurately approximated by a normal distribution. Construct a 95% equal-tail confidence interval on the true value of p for the sampled distribution that yielded the 2 successes in 20 trials.

8.10. A random sample of size $n = 2$, taken from an exponential distribution, yields two observed values that sum to a value of 3. Place a 95% equal-tail confidence interval on the true value of λ governing the sampled distribution.

8.11. $\overline{x}$ and s for the grade-point averages of a sample of 36 engineering seniors are calculated to be 2.6 and 0.3, respectively. Find the 95% confidence interval for the mean of the entire senior class.

8.12. The wavelengths of the light from ruby lasers manufactured by Schlage Electronics is known to be normally distributed with a certain mean. Our use requires this mean value, but we also require that the standard deviation of such wavelengths not be too large. As a first step, 26 lasers were tested and the tests yielded $s = 6.3$ angstroms. Using this data, place a 98% confidence interval on the true value of σ.

8.13. The following data are the amounts of enamel coating, in ounces, that were used to cover six metal plates of the same size: 7.1, 9.7, 6.6, 10.8, 7.5, and 8.9. Find a point estimate of σ and a confidence interval for σ with a 97.5% confidence level.

8.14. Suppose that the mean strength of a certain kind of drilling rod is *known* to be $\mu = 6000$ pounds per square inch (psi). The following data (given in psi) were obtained by testing 12 rods to failure: 5100, 4700, 5050, 4800, 5200, 5300, 4900, 4950, 4950, 4850, 5175, and 4775. Place an 85% confidence interval on σ using this data set.

8.15. A sample of 1000 barrels of a chemical gives $s^2 = 144$ milligrams^2 of impurities. Place a 92.5% confidence interval on the true standard deviation associated with the amount of impurities in such barrels.

8.16. Pull strength tests on 10 soldered leads for a semiconductor device yield the following results (in pounds of force) that were required to rupture the bonds: 15.8, 12.7, 13.2, 16.9, 10.6, 18.8, 11.1, 14.3, 17.0, and 17.5. Place a 99% confidence interval on the mean strength of the bonds. Place a 90% confidence interval on the variance of the bonds.

8.17. The thickness of nickel plating on Acme armatures is known to follow a normal distribution. The thickness of the plating on 10 such armatures was measured, yielding the following data: 4.1, 3.9, 4.7, 4.4, 4.0, 3.8, 4.4, 4.2, 4.4, and 5.0. Place a 90% confidence interval on the mean thickness of the plating on the armatures. Place a 95% confidence interval on the variance of the plating thickness.

8.18. In Exercise 6.5, 68 times to failure for metal plates in a fatigue test are given. Place a 95% confidence interval on the mean time to failure. Place a 90% confidence interval on the standard deviation of the time to failure.

8.19. In Exercise 6.6, 94 times to failure for circuit boards are given. Place a 98% confidence interval on the mean time to failure. Place a 96% confidence interval on the standard deviation of the time to failure.

8.20. In Exercise 6.7, 56 crates of input devices are examined, and the numbers of devices with one or more serious cosmetic defects in each crate are given. Place a 95% confidence interval on the average number of defective devices in a crate. Place a 90% confidence interval on the standard deviation.

8.21. Exercise 6.8 gives the numbers of customers observed during 100 half-hour periods. Place a 95% confidence interval on the average number of customers during a 30-minute period. Does this new information affect your answer to Exercise 6.8?

8.22. The ductility of the forming wax used in the "lost wax" casting procedure at the Smart Karved Class Rings company is known to be normally distributed with a certain average ductility. Twenty-two canisters of wax were tested, and the tests yielded $s = 5.3$ units. Using this information, place a 94% *one-tailed* upper confidence limit on the true value of σ.

8.23. A sample of 100 freeze plugs used on Don Jeer medium-sized tractor engines yields $\overline{x} = 0.100$ inch and $s = 0.014$ inch. Place a 99.5% confidence interval about the true mean.

8.24. A sample with $n = 16$ has $\overline{x} = 32.2$ and $s^2 = 6.25$. Is it reasonable to say that this sample came from a population with a true variance of 10? Why or why not?

8.25. Assuming that the X_i are taken from a normally distributed population, show that $\sum_{i=1}^{n} (X_i - \overline{X})^2/n$ is not an unbiased estimator of the true variance, σ^2.

8.26. Assuming that the X_i are taken from a normally distributed population, show that $\sum_{i=1}^{n} (X_i - \mu)^2/n$ does provide an unbiased estimate of the true variance, σ^2.

8.27. Suppose that you have n observations taken from a gamma distribution. Discuss how you would approach the problem of determining the values of r and λ. (*Hint:* Look at Table 4.1!)

8.28. A production process yields CO_2 as a by-product and is monitored by the government. Regulations state that we should report, with a 90% level of confidence, the average daily production of CO_2. Over a five-day period, samples were taken and recorded; they were 15, 19, 21, 16, and 17 cubic units per day. What should our report to the government say?

8.29. A man plays golf 18 times over a six-month period. His scores on the notorious par-3 sixth hole are 5, 4, 3, 5, 6, 7, 3, 4, 4, 6, 4, 3, 5, 5, 4, 5, 4, 7. What is the minimum-width 90% confidence interval on his average score on the sixth hole?

8.30. If 123 of 456 balloons fail when filled with water, set a 95% confidence interval on the true proportion of balloons that do not fail.

8.31. A sample of rifle shells of the same weight, caliber, and powder load had muzzle velocities of 2802, 2835, 2793, 2786, 2841, 2798, 2800, 2817, 2789, and 2809 ft/sec. Construct a 95% confidence interval about the mean muzzle velocity for this type of shell.

8.32. Concrete beams are tested by a highway inspector to ensure that they meet certain specifications. To pass inspection, the beams must have a flexure strength of at least 600 psi. For a highway construction project, beams were randomly tested and 10 sample measures yielded $\overline{x} = 810$ psi and $s = 140$ psi. Construct a 97% confidence interval on the mean strength of a concrete beam used on the project.

8.33. A group of engineering students were trying to determine the average number of hours a student spends studying per night. After some careful planning, a sample of size 100 was obtained. This sample yielded $\overline{x} = 5.1032$ and $s = 6.5783$. Determine a confidence interval on the average number of hours per night that an engineering student spends with the books.

8.34. A nuclear power plant used up plutonium rods in the following amounts: 26.3, 32.2, 19.7, 21.5, 27.1 and 17.2 tons per month. If a report provided by the supervisor stated that the mean number of rods used per month has an equal-tail confidence interval of (20.04, 27.96), what level of confidence is associated with the interval?

8.35. A sample of 50 physics test grades of a total of 200 yielded $\overline{x} = 65$ and $s = 10$. What is a 95% confidence interval on the mean test score? With what degree of confidence can we say that the mean score is 65 ± 1?

8.36. A worker at a plant did a statistical analysis of the quality of her products and took 10 random readings. She found that $\overline{x} = 13.78$ and that a 90% confidence interval about the mean was $13.063 < \mu < 14.497$. However, after she completed the analysis she threw away the data. Now, her boss wants to know what the readings were. If she remembers that the first 8 were 12.2, 15, 13, 13.5, 12.8, 12.0, 15.2, and 15.2, can she find out what the last two were and give the information to her boss? If so, what are the other two measurements?

CHAPTER

9

HYPOTHESIS TESTS ABOUT A SINGLE MEAN, A SINGLE PROPORTION, OR A SINGLE VARIANCE

The following concepts and procedures form the major points of this chapter.

- Understand how *hypothesis tests* are constructed and why they may be used to make decisions about a population having a specified mean, variance, or probability of success
- Understand why a *null hypothesis* has three possible *alternate hypotheses* and know when each of the alternate hypotheses is appropriate
- Be able to give definitions for *Type I* and *Type II errors* (producer's and consumer's risks) and be able to give specific examples of each kind of error
- For a hypothesis about a single mean, know how to calculate the probability of a β error when a *specific* value for the alternate hypothesis is given and the values of σ, α, and n are known
- For tests of a single mean, proportion, or variance, know how to compute the value of the *decision criterion* (or criteria) for the three different alternate hypotheses and know how to use those values to determine whether the null hypothesis should be *rejected*
- For tests of a single mean, understand how the values of n, C, α, and β are related to one another

- Know what an *operating characteristic (o.c.) curve* is and understand its meaning. Be able to use the appropriate computer program to construct an o.c. curve and determine a statistically valid sample size for tests about a single mean, variance, or proportion
- For tests of a single mean, understand the relationship between the β error and the *difference* between μ_0 and $\hat{\mu}_A$
- Know what is meant by the *distribution of the null hypothesis* and the *distribution of the alternate hypothesis*. Be able to state the type of distribution that is associated with either of these hypotheses for tests about a single mean, variance, or proportion
- Know the assumptions associated with hypothesis tests about a *single variance* and when to use the large or small sample test
- Know the assumptions associated with large sample hypothesis tests about one proportion
- Be able to perform all *five* of the steps associated with a test of hypothesis for one mean, proportion, or variance
- Know the definitions of *practical* significance and *statistical* significance and know when they can be different

9.1 INTRODUCTION

The last chapter discusses the techniques used in estimating the values of means, variances, and proportions. This is very useful knowledge when we desire to know only the approximate values of selected parameters. Often, however, we are more interested in whether a parameter meets a certain *standard*. In many cases, once we can state that the standard is satisfied, the parameter's value is of no further interest.

Example 9.1. Tri-State Brick, a masonry contractor, sells a special cinder block designed for use in heavy loading conditions. The contractor *claims* that these special blocks support 8000 pounds per square inch (psi). We, as the potential buyer of the blocks, will use that stated strength in our design calculations for any structure that is built using the blocks. From our viewpoint, it is *not important* if the blocks are too strong (although that would be perfectly acceptable). However, we are *very concerned* that the blocks have an average strength approximately equal to the rated strength of 8000 psi. Indeed, if they fall far enough below their rated strength, the stability of our buildings will be threatened.

Our company has done business with Tri-State for many years, and bricks purchased from them have rarely done anything but exceed our expectations. For this reason, we are inclined to give them the *benefit of the doubt* and reject a shipment of blocks from Tri-State if and only if it has been *conclusively* shown that the blocks do not meet the standard.

The sound, statistically based approach to this *decision-making problem* is to randomly sample the blocks and base our decision on the results gained from the observed data. (Of course, this does not remove the chance of obtaining a bad sample and making an incorrect decision.) Because of the trust we have in Tri-State, we will assume that the blocks meet the standard of 8000 psi until *proven* otherwise. In statistical terms, the *null hypothesis* is that the average strength of the blocks is 8000 psi, or, in notational shorthand, $H_0 : \mu = \mu_0 = 8000$ psi.

Because we are not concerned with the possibility that the blocks might be too strong, the natural *alternate hypothesis* is that *the blocks possess an average strength less than 8000 psi*. Notationally, this is expressed as $H_A : \mu = \mu_A < 8000$ psi. Thus we have two competing hypotheses:

$$H_0 : \mu = 8000 \text{ psi}$$

$$H_A : \mu < 8000 \text{ psi}$$

where H_0 will be "accepted" unless it is "proven" to be wrong. (Note that this does not constitute a proof that H_0 is true.) If a proof that H_0 is false is found, we will *reject the null hypothesis* in favor of the alternate hypothesis. (In the strictest sense, we can never *prove* that H_0 is false. However, we can show beyond a reasonable statistical certainty that H_0 is false. This is similar to a judicial system in which a defendant is innocent until shown to be guilty beyond reasonable doubt.)

There are two ways that errors can be made in this situation: We can say that the blocks do not meet the standard when, in fact, they do, or we can say that the blocks meet the standard when they do not. Rejecting H_0 when it is true is often called the *producer's risk*. This name is very appropriate in Example 9.1. The producer is Tri-State Brick, and Tri-State's risk is that their bricks meet the standard and will still be rejected. Statisticians also refer to the producer's risk as a *Type I error,* or an α *error.*

Accepting H_0 when it is false is called the *consumer's risk,* which is synonymously known as the *Type II error,* or the β *error.* In Example 9.1 we are the consumer, and we risk accepting a lot of bricks that do not meet our required standard. Figure 9.1 provides additional clarification about α and β errors.

Since we know that making a Type I or a Type II error is *always possible,* the next thing to consider is the evaluation of the probabilities associated with those errors. For convenience, let us set $\alpha = P(\alpha \text{ error}) = P(\text{rejecting } H_0 \text{ when } H_0 \text{ is true})$ and $\beta = P(\beta \text{ error}) = P(\text{failing to reject } H_0 \text{ when } H_0 \text{ is false})$.

		Decision: H_0	Decision: H_A
Truth	H_0	μ is not less than 8000 H_0: $\mu = 8000$ is not rejected	μ is not less than 8000 H_0: $\mu = 8000$ is rejected
		No error is made	α error is made
	H_A	μ is less than 8000 H_0: $\mu = 8000$ is not rejected	μ is less than 8000 H_0: $\mu = 8000$ is rejected
		β error is made	No error is made

FIGURE 9-1
Possible errors in hypothesis testing.

Suppose that the design engineering group has determined that, although the 8000-psi strength is preferable, significant structural degradation will not occur unless the average strength is 7900 psi or less. For this reason, it is important that any shipment of blocks with average strength less than 7900 psi has a *high* probability of being rejected, i.e., not used in the company's construction projects. Let us consider how this information can be used in testing hypotheses about a single mean.

9.2 HYPOTHESIS TESTS ABOUT A SINGLE MEAN

Suppose that a random sample of 25 blocks is to be taken from a shipment of Tri-State blocks; the block strengths are distributed normally with a standard deviation of $\sigma = 250$ psi. If the null hypothesis of $\mu = 8000$ is *true*, the distribution of $\overline{X}$ is normal with $\mu = 8000$ psi and $\sigma_{\overline{X}} = \sigma/\sqrt{n} = 250/\sqrt{25} = 50$ psi. Therefore, the *distribution of the null hypothesis* is $N(8000,50)$.

The alternate hypothesis is $H_A : \mu < 8000$. In view of the design engineering group's statement, an average strength slightly less than 8000 psi—say, 7995 psi—is not seriously low. For this reason, consider a *specific* alternate hypothesis that *is* seriously low; i.e., let $\hat{H}_A : \mu = 7900$. If $\hat{H}_A$ is true (instead of H_0), the distribution of $\overline{X}$ would be normal with $\mu = 7900$ psi and $\sigma_{\overline{X}} = 50$ psi; i.e., the *distribution of the alternate hypothesis* is $N(7900, 50)$. The distributions of both the null hypothesis and the alternate hypothesis are pictured in Figure 9.2. Notice

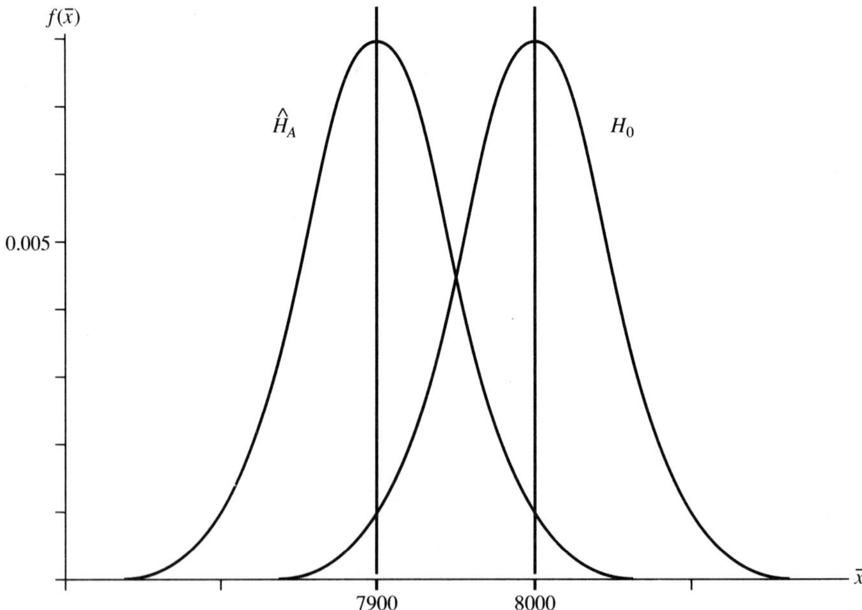

FIGURE 9-2
The distributions of the null and alternate hypotheses.

that the two distributions are identical in shape and differ only in the values of their respective means.

We want to achieve the following two conflicting goals: (1) to accomplish a high rejection rate of shipments of blocks that have average strengths of 7900 psi or less and (2) to reject only a small percentage of shipments that on average meet or exceed the rated strength of 8000 psi.

Example 9.2. What value of $\overline{x}$ from the observed data of 25 blocks should we use as the *decision criterion, C?* In other words, what values of $\overline{x}$ should cause rejection of H_0?

Let us consider some of the possible values of C. Suppose that we decide that $C = 7900$. This would mean that a sample with $\overline{x}$ less than or equal to 7900 psi would cause rejection of the sampled shipment. The effect of this criterion is shown in Figure 9.3. The value of α, as given in the figure, achieves the aim of making rejection of "good" shipments very unlikely (a probability of $\alpha = 0.0228$). Unfortunately, this achievement exacts a price by inflating the β value to equal 50%! Exactly one-half of the shipments with a mean of 7900 psi would not be rejected. This is clearly an unacceptable result. If $C = 8000$ is selected, the situation given in Figure 9.4 would arise. Once more, one of the errors is made acceptably small at the cost of greatly inflating the other one.

Since neither $C = 7900$ nor $C = 8000$ yields acceptable results, one possible decision criterion that might improve things would be to take the *average* of the mean values associated with H_0 and $\hat{H}_A$; i.e., $C = 7950$. The result of that selection is illustrated in Figure 9.5. As expected, the criterion exactly midway

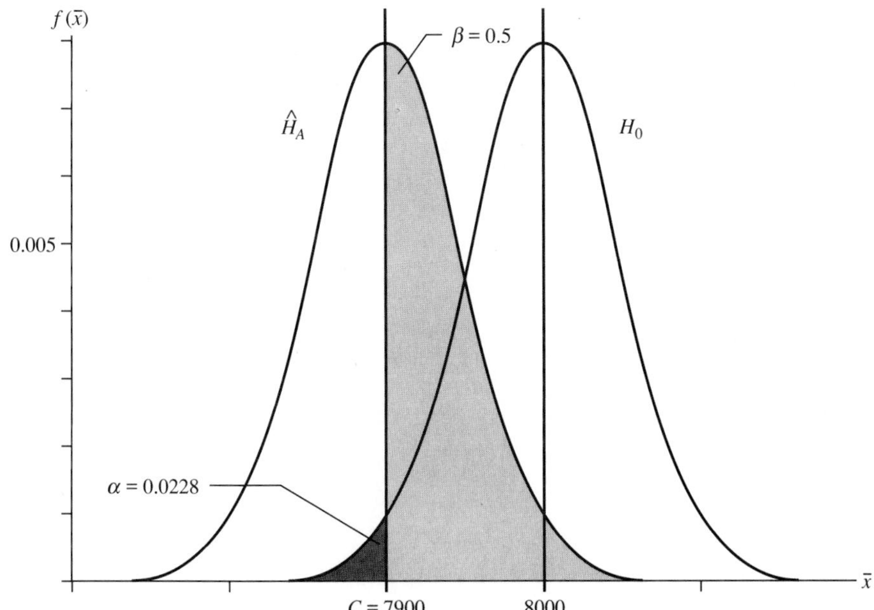

FIGURE 9-3
The effect of setting the decision criterion at $\overline{x} = 7900$.

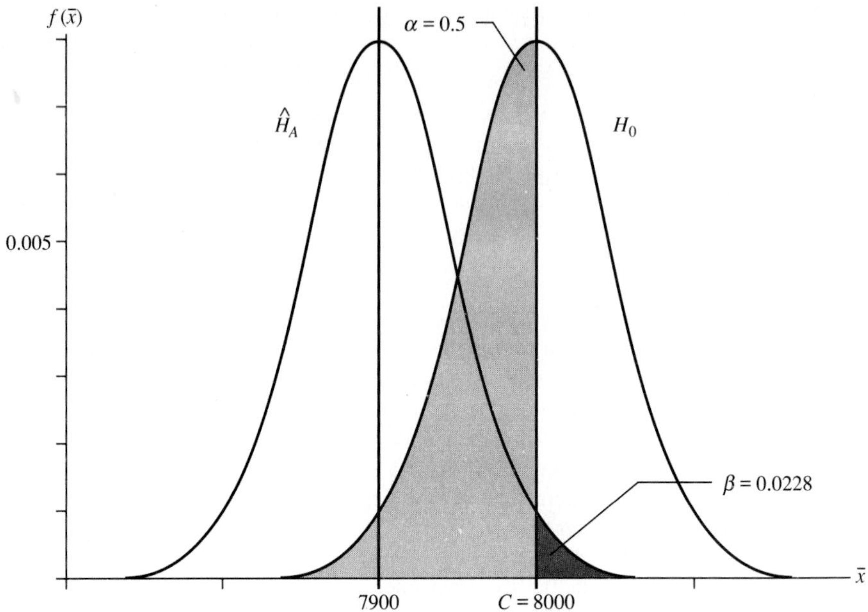

FIGURE 9-4
The effect of setting the decision criterion at $\bar{x} = 8000$.

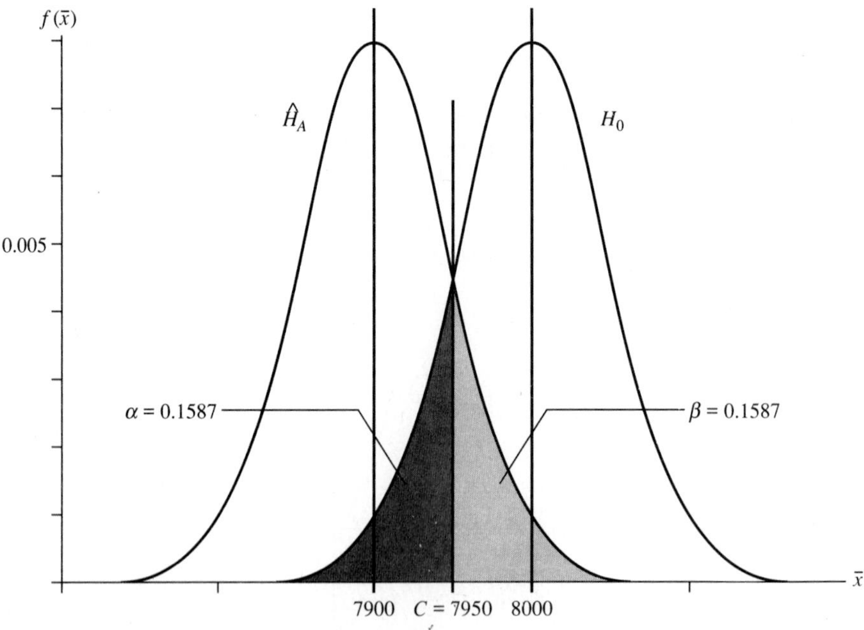

FIGURE 9-5
The effect of setting the decision criterion at $\bar{x} = 7950$.

between the means of the distributions of the null and alternate hypotheses yields $\alpha = \beta$. However, the common value of 0.1587 may be *unacceptably high*. It appears that no value of C will allow satisfaction of both conflicting goals.

In order to determine what direction to take from this point, let us consider the mathematical base that has allowed the computation of the α and β errors in Figures 9.3 through 9.5. To calculate the value of α in those figures, we used the following relationships:

$$\alpha = P(\text{reject } H_0 \text{ when } H_0 \text{ is true}) = P(\overline{X} \leq C \mid H_0 \text{ true})$$

$$= P(\overline{X} < \mu_0 - z_\alpha \sigma_{\overline{X}}) = P\left(\overline{X} < \mu_0 - z_\alpha \frac{\sigma}{\sqrt{n}}\right)$$

Therefore,

$$C = \mu_0 - z_\alpha \frac{\sigma}{\sqrt{n}} \tag{9.1}$$

In like manner, $\beta = P(\text{accept } H_0 \text{ when } \hat{H}_A \text{ is true}) = P(\overline{X} > C \mid \hat{H}_A \text{ true})$, which yields

$$C = \hat{\mu}_A + z_\beta \frac{\sigma}{\sqrt{n}} \tag{9.2}$$

Equations 9.1 and 9.2 are two equations in the four unknowns C, n, α, and β. Given the values of *any two* of these unknowns, the values of the other two are *explicitly* obtainable. By studying Equations 9.1 and 9.2 and Figures 9.2 through 9.5, the following properties may be observed.

1. If n is held constant, the values of α and β are *inversely* related; i.e., the larger the value of α (β), the smaller the value of β (α).
2. If n is held constant and C is moved toward (away from) μ_0, the value of α will increase (decrease) and the value of β will decrease (increase).
3. If C is held constant and n is increased, the values of both α and β will decrease.
4. If C and n are held constant and σ is decreased, both α and β will decrease.

We saw illustrations of properties 1 and 2 in the previous discussion. Properties 3 and 4 are direct results of decreasing the value of $\sigma_{\overline{X}}$. Since $\sigma_{\overline{X}}$ is inversely proportional to the square root of the sample size, it is often economically advantageous to attempt to reduce σ directly, when possible, by removing sources of variability from the real-world process under study.

Figure 9.6 illustrates, for Example 9.2, the effect of quadrupling the sample size to a new value of 100, where $\sigma_{\overline{X}}$ is reduced from 50 to its new value of 25. This increase in n causes α and β to achieve a new common value of 0.0228, an acceptable error level in most practical applications.

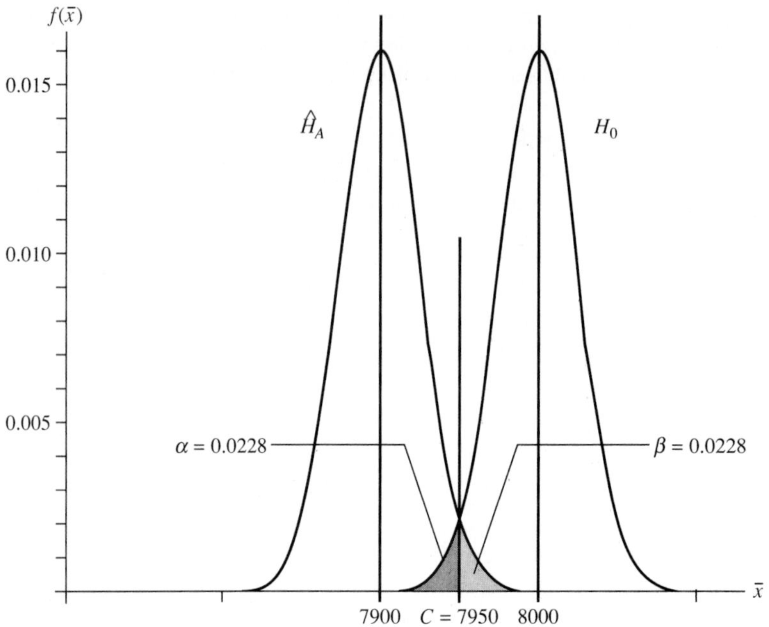

FIGURE 9-6
The effect of increasing n to a new value of 100.

Example 9.3. As an illustration of the use of Equations 9.1 and 9.2, suppose that $\alpha = 0.05$ and $\beta = 0.075$. What values of C and n would achieve such values for the Tri-State Brick example?

Solution. As a first step we require the values of $z_\alpha = z_{0.05} = 1.645$ and $z_\beta = z_{0.075} = 1.44$. Next, we use Equations 9.1 and 9.2 to solve for n in terms of α and β. This operation yields

$$n = \frac{\sigma^2(z_\alpha + z_\beta)^2}{(\mu_0 - \hat{\mu}_A)^2} \tag{9.3}$$

Substituting the appropriate values into Equation 9.3 yields $n = [(250)^2 \times (1.645 + 1.44)^2]/(8000 - 7900)^2 = 59.48$.

Rounding up, we find that the appropriate sample size is 60 blocks. To obtain the value of C, substitute the *unrounded value* of n into either Equation 9.1 or 9.2, and obtain 7946.678. This procedure will produce α and β errors that are slightly less than the stipulated values because the value of n is slightly greater than the solution of the equations indicates.

Often the exact value of an important specific alternate hypothesis, such as the value of 7900 psi in Examples 9.1 and 9.2, will not be known. To counter this lack of knowledge, the preceding computations can be performed for a *spectrum* of specific alternate possible values of the mean. If this is done while holding n and α (and therefore C) constant, the set of β values that results from the selected set of $\hat{\mu}_A$ values forms the necessary information to plot the *operating characteristic curve* (o.c. curve) for the hypothesis test. The o.c. curve is nothing more than a plot of the β error values against the associated values of $\hat{\mu}_A$. An equivalent way to present the information given in an o.c. curve is to plot a *power curve,* which plots the values of $1 - \beta$ against the associated values of $\hat{\mu}_A$. Thus, the power curve plots the probability of rejecting H_0 when $\mu = \hat{\mu}_A$. Solving for z_β in Equation 9.2 gives the relationship between z_β and the selected value of $\hat{\mu}_A$ as

$$z_\beta = \frac{\sqrt{n}(C - \hat{\mu}_A)}{\sigma} \tag{9.4}$$

Example 9.4. Let us use the o.c. curve program for tests of one mean (as described in Chapter XI of the *User's Manual*) to construct an o.c. curve for the Tri-State Brick example. First, the program uses $n = 100$, $\sigma = 250$, and $\alpha = 0.05$ in Equation 9.1 to obtain $C = 7958.87$. It then uses Equation 9.4 to obtain $z_\beta = [10(7958.87 - \hat{\mu}_A)]/250$. Table 9.1 presents the results from the program's application of Equation 9.4 to a selected set of values for $\hat{\mu}_A$ with $n = 100$ and $C = 7958.87$. Figure 9.7 illustrates the same results in a graphical form.

The following observations may be made with the help of Figure 9.7.

1. First, consider the case where $\hat{\mu}_A < \mu_0$. The β error measures the hypothesis test's ability to discriminate between two possible values of the mean, $\hat{\mu}_A$ and μ_0. The closer that μ_A gets to μ_0, the harder it is to tell them apart. This is reflected by the β error's growing larger as $\hat{\mu}_A$ approaches μ_0.
2. If $\hat{\mu}_A = \mu_0$, β will be equal to $1 - \alpha$. This is reasonable, because at that point $\hat{H}_A = H_0$, and the probability that H_0 will be accepted is $1 - \alpha$.

TABLE 9.1
Values of β for Example 9.2 with $n = 100$ and $\alpha = 0.05$

$\hat{\mu}_A$	z_β	β	$\hat{\mu}_A$	z_β	β
7850	4.355	0.0000	7970	−0.445	0.6718
7875	3.355	0.0004	7980	−0.845	0.8009
7900	2.355	0.0093	7990	−1.245	0.8935
7910	1.955	0.0253	8000	−1.645	0.9500
7920	1.555	0.0600	8010	−2.045	0.9800
7930	1.155	0.1240	8020	−2.445	0.9930
7940	0.755	0.2251	8030	−2.845	0.9978
7950	0.355	0.3633	8040	−3.245	0.9994
7960	−0.045	0.5180			

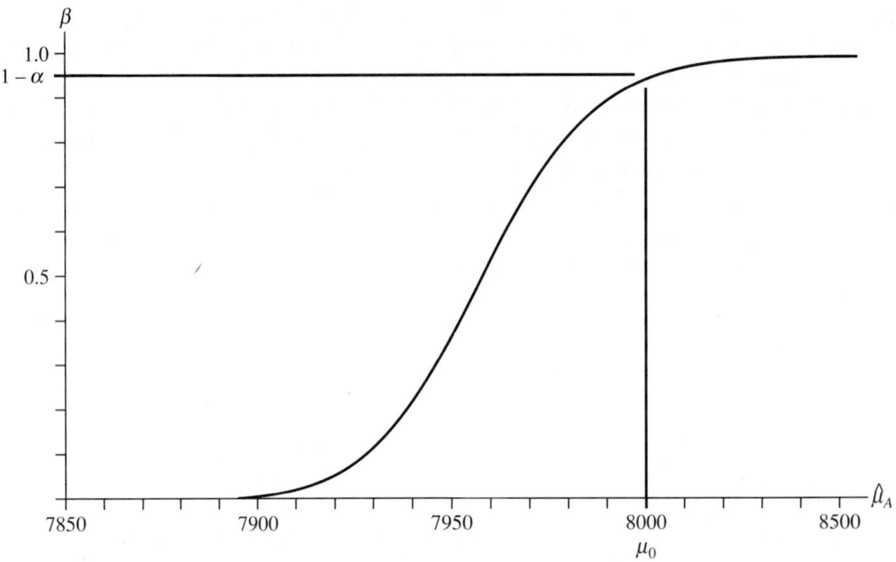

FIGURE 9-7
The o.c. curve for Example 9.2.

3. As $\hat{\mu}_A$ becomes large relative to μ_0, β approaches ever closer to its *maximum* value of 1. This implies that we are becoming ever more certain to commit the error of accepting $H_0 : \mu = \mu_0$ when H_0 is false ($\mu > \mu_0$). This is quite acceptable in our present circumstance in which we desire to detect only whether $\mu < \mu_0$ ($H_A : \mu < \mu_0$).

Example 9.5. Suppose that 100 bricks have been randomly selected. After each brick was loaded to failure, the average breaking strength was found to be $\overline{x} = 7973$ psi. Since $\overline{x}$ is not less than $C = 7958.87$, there is not sufficient evidence to reject the null hypothesis of $H_0 : \mu = \mu_0 = 8000$ psi.

A natural question to ask at this point is, Why was α set to a value of 0.05? After all, other values of α might indicate that H_0 should be rejected. Common values used in practice are $\alpha = 0.01, 0.05$, and 0.10. In some cases, the maximum available sample size will dictate a certain value of α if β is to have an acceptable value. *However,* except for special cases, there is no single answer to the question of what value should be assigned to α. Indeed, different individuals (even statisticians) often select differing values of α for the same hypothesis test.

One way around this problem is to determine the threshold value, or *p-value,* of α at which the value of $\overline{x}$ that *has been obtained* will be identical to *C;* i.e., the value of α that will just barely cause H_0 to be rejected. Knowledge of this p-value of α will be useful both when H_0 is rejected and when H_0 is accepted.

For our current example, this p-value of α is easily obtained by setting $C = \overline{x} = 7973$ in Equation 9.1 and solving for α. This process yields $z_\alpha =$

$(\mu_0 - C)/(\sigma/\sqrt{n}) = (8000 - 7973)/(250/\sqrt{100}) = 1.08$, which implies a p-value for α of 0.1401. Therefore, $\bar{x} = 7973$ will cause rejection of the null hypothesis if α has been set at a value of 0.1401 or greater. Such a large p-value for α lends strength to the conclusion not to reject the null hypothesis. If the null hypothesis is rejected, the p-value of α will likewise give added insight into how strong a conviction should be held in regard to the rejection of H_0. If the threshold value of α is *small,* then the decision maker can feel much more comfortable with the rejection of H_0. For example, suppose that you had set $\alpha = 0.05$, performed your test, and rejected H_0. Then you found that the p-value of α was 0.005. Therefore, even if you had set a much more stringent value of α prior to the experiment, you would still have rejected H_0.

In addition to determining whether H_0 should be rejected for the user-stipulated value of α, each of the hypothesis test programs in your software package automatically provides the p-value of α.

Up to this point we have discussed only alternate hypotheses of the less-than form, $H_A : \mu < \mu_0$. There are two other possibilities, $H_A : \mu > \mu_0$ and $H_A : \mu \neq \mu_0$. In the greater-than alternate hypothesis, $H_A : \mu > \mu_0$, we have a situation that is very similar to that of the less-than alternate hypothesis. The values of C and n are obtained in exactly the same manner, and the o.c. curve is the *mirror image* of the kind of o.c. curve that we developed in Example 9.4. This kind of alternate hypothesis is appropriate when we are concerned about the mean of the population under study being too large but do not care if it is too small.

Example 9.6. Suppose that an acid bath is used to clean impurities from the surfaces of metal bars used in our laboratory experiments. A chemical company provides batches of acid to our lab in large lots. The company has been instructed that the solution must not be too strong or the bars will suffer damage to their plated surfaces. The acid should have a nominal acidity index of 8.5, and it is important that the average index not exceed 8.65. The standard deviation of this kind of acid solution is known to be 0.2 unit. The producer's risk, α, is set to 0.05, and we want to reject solution lots with an average index greater than or equal to 8.65, 95 percent of the time ($\beta = 0.05$). What is the value of the decision criterion, and what number of batches should be sampled?

Solution. The null hypothesis is $H_0 : \mu = \mu_0 = 8.5$, versus the alternate hypothesis, $H_A : \mu > \mu_0$ with $\hat{H}_A : \mu = \hat{\mu}_A = 8.65$ at a rejection level of 95%. For the greater-than alternate hypothesis, Equations 9.1 and 9.2 are symmetrically changed to yield

$$C = \mu_0 + z_A \frac{\sigma}{\sqrt{n}} \qquad \text{and} \qquad C = \hat{\mu}_A - z_\beta \frac{\sigma}{\sqrt{n}}$$

respectively. Solving these two equations for n yields, once more, Equation 9.3. Substituting the required values into Equation 9.3 yields $n = [(0.2)^2(1.645 + 1.645)^2]/(8.5 - 8.65)^2 = 19.24$. Thus the required sample size is 20. The value of $C = 8.575$ is obtained by substituting $n = 19.24$ into either of the equations for C just shown.

To actually perform the test of hypothesis for the situation described in this example, we would examine 20 randomly selected batches from the lot under consideration, compute their average index, $\overline{x}$, and compare it against $C = 8.575$; $H_0 : \mu = 8.5$ would be rejected if $\overline{x}$ is greater than C.

Just as the less-than and greater-than alternatives are called *one-tailed tests,* the not-equal alternate hypothesis is called a *two-tailed test*. In a two-tailed test, the null hypothesis may be rejected if $\overline{x}$ is either too large *or* too small.

Example 9.7. Suppose that you are manufacturing light posts for use on major city streets. The bolts in the holding bracket at the base of the post are designed so that they will hold the post securely upright under ordinary stress conditions. However, to minimize damage to a vehicle should it run into a post, the bolts must break cleanly apart at specific points under large impact stress. Thus, the bolts must be neither too weak nor too strong. The desired average load-carrying capacity is 500 pounds, and it is known both that the standard deviation is 80 pounds and that the strength of the bolts is distributed normally. If $\alpha = 0.1$ and the sample size is $n = 36$, how do we decide whether to reject $H_0 : \mu = 500$?

Solution. Consider the distribution of the null hypothesis presented in Figure 9.8. Notice that the Type I error, α, is divided into two equal parts and is placed in the upper and lower tails of the distribution of H_0; i.e., $P(\overline{X} < C_1) = P(\overline{X} > C_2) = \alpha/2$ for $\overline{X} \sim N(500, 80/\sqrt{36})$. Therefore, we have *two* decision criteria, C_1 and C_2. If $\overline{x}$ is either less than C_1 or greater than C_2, H_0 is rejected in favor of H_A. The values of C_1 and C_2 are found by

$$C_1 = \mu_0 - z_{\alpha/2}\frac{\sigma}{\sqrt{n}} \tag{9.5}$$

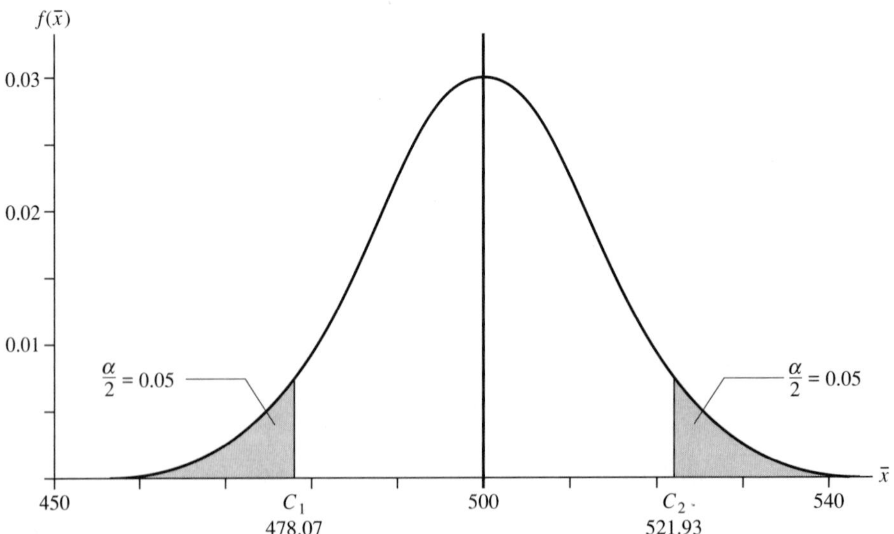

FIGURE 9-8
The distribution of H_0 for Example 9.7.

and

$$C_2 = \mu_0 + z_{\alpha/2}\frac{\sigma}{\sqrt{n}} \tag{9.6}$$

For this specific example, Equations 9.5 and 9.6 yield $C_1 = 500 - 1.645(80/6) = 478.1$ and $C_2 = 500 + 1.645(80/6) = 521.9$.

The computation of the β error for a specific alternate hypothesis and, therefore, the construction of the o.c. curve for the two-tailed test are quite similar to the one-tailed test computations. For example, if $\hat{\mu}_A = 510$, the value of β is equal to the area of the shaded portion of Figure 9.9. Mathematically, this area is given by

$$P(C_1 < \overline{X} < C_2 \mid \hat{H}_A \text{ is true}) = P(\mu_0 - z_{\alpha/2}\frac{\sigma}{\sqrt{n}} < \overline{X} < \mu_0 + z_{\alpha/2}\frac{\sigma}{\sqrt{n}}, \text{ given } E(\overline{X}) = \hat{\mu}_A = 510) \tag{9.7}$$

In Figure 9.9 $\beta = 0.8063$ is obtained by computing the area under the distribution of the alternate hypothesis, $N(510, 80/6)$, from 478.1 to 521.9.

Notice that there is *only one difference* in the computation of the β error for the two-tailed test: The area under the distribution of the alternate hypothesis corresponding to the β error is *bounded between two values*, C_1 and C_2. The one-tailed test bounds the β error area with C on *only one side* while allowing the other side to be unbounded.

If we can compute one β error, we can select a set of possible alternate hypothesis values and form an o.c. curve for a two-tailed test for a mean. Figure 9.10 shows the o.c. curve that has been obtained using the o.c. curve program cited for Example 9.7. Notice that this case yields a symmetric curve with its maximum at μ_0.

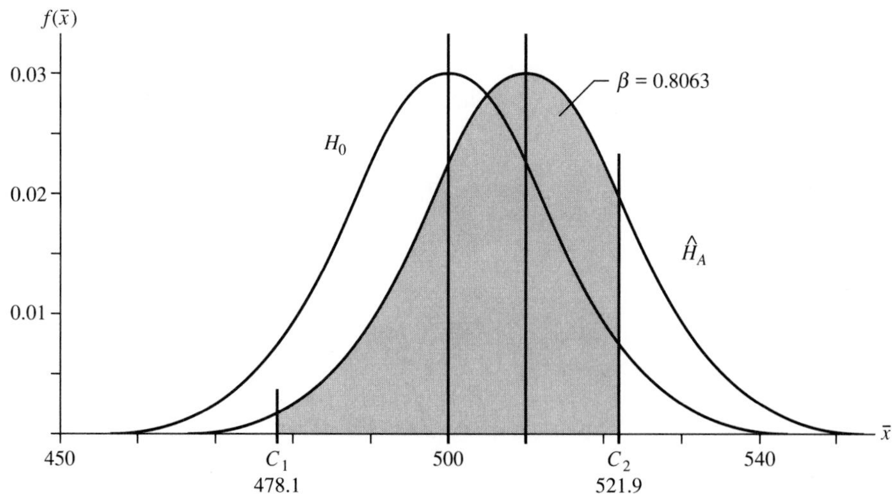

FIGURE 9-9
The β error for $\hat{\mu}_A = 510$.

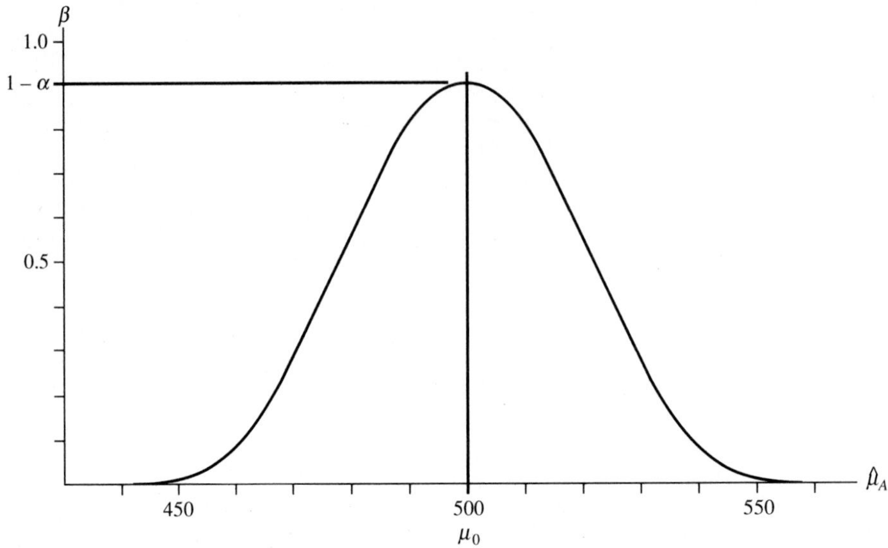

FIGURE 9-10
The two-tailed o.c. curve for Example 9.6.

The following five steps give the *operational rules* for performing a test of hypothesis about a single mean for a normal population when σ is known.

1. Determine the *appropriate alternate hypothesis* to be used.
2. Select the values of α and n that will yield an acceptable value of β for an important alternate value of the population mean.
3. Use a random sampling procedure to obtain n items from the population under study.
4. Compute the criterion value C or values C_1 and C_2.
5. Compute $\overline{x}$ for the observed data, and compare it against the criterion value(s) obtained in step 4. Determine whether to reject the null hypothesis by means of Table 9.2.

Steps 2, 4, and 5 of this procedure are most easily accomplished by using the programs that perform hypothesis tests about a single mean and calculate their associated o.c. curves.

TABLE 9.2
Decision table for $H_0 : \mu = \mu_0$ (large sample or σ known and normal population)

H_A	Reject H_0 if		
$\mu < \mu_0$	$\overline{x} < C$		
$\mu > \mu_0$	$\overline{x} > C$		
$\mu \neq \mu_0$	$\overline{x} < C_1$	or	$\overline{x} > C_2$

The procedure just described can also be used for tests of a single mean under assumptions different from those given above. If n is large, it is no longer necessary to assume that the population is normal, and prior knowledge of the value of σ is not required. When n is large, we may invoke the *central limit theorem* (this allows us to assume that $\overline{X}$ is approximately normal), and we may assume that s, the large sample estimate of σ, is sufficiently accurate for our purposes. For large n, the only required modification on our earlier procedure is to compute the sample estimate, s, and to use it wherever σ was used before.

As we have seen, there are occasions when the sample size is not large and we do not have prior knowledge of the value of σ. (You might review Figure 8.1 to refresh your memory about this.) In such cases, we *must* assume that the population under study is normally distributed. If this assumption cannot be made, special techniques, which are beyond the scope of this book, must be employed. If the population is normal, the t distribution may be used in the same way that the standardized normal distribution has been used. Indeed, all that we need to do is to use $t_{n-1,\alpha}$ wherever z_α was used before and to substitute the sample estimate s wherever σ was used.

Example 9.8. A motor oil is claimed to lessen gasoline usage in passenger automobiles. Suppose that 13 randomly selected automobiles are provided 2 gallons of gasoline in each of their tanks. The average distance traveled by each of the cars before running out of fuel was $\bar{x} = 68$ miles. The oil company claims that the average distance traveled for such a test would be 75 miles. If $s = 15$ miles, what may we conclude about the oil company's claim?

Solution. Using the five-step operational rules given earlier, we first identify the alternate hypothesis. Since we are concerned about the mileage being too small, the competing hypotheses are

$$H_0 : \mu = 75 \qquad H_A : \mu < 75$$

Since the sample size has already been determined, let us set $\alpha = 0.05$ and use the software package to perform the hypothesis test. Since the alternate hypothesis is the less-than type, C will be less than μ_0. Using the ideas that led to Equation 9.1, C is computed as

$$C = \mu_0 - t_{n-1,\alpha}\frac{s}{\sqrt{n}} = 75 - 1.782\frac{15}{\sqrt{13}} = 67.59 \tag{9.8}$$

All that remains is to use Table 9.2 to reach our conclusion about the claim. In this case, $\bar{x} = 68$ is not less than C and the null hypothesis may not be rejected. However, the p-value of α is 0.059, which means that an increase in the stipulated value of α of only 0.009 would have led to the rejection of H_0. This information should be considered in any decision made on the basis of this experiment.

Computation of the β error and the o.c. curve for the hypothesis test about a single mean when the t distribution is used is conceptually identical to the case in which the standardized normal distribution is used. Unfortunately, this similarity does not extend to the mathematical sophistication required to compute the β error. When the t distribution is used, the distribution of the H_0 is directly related to the

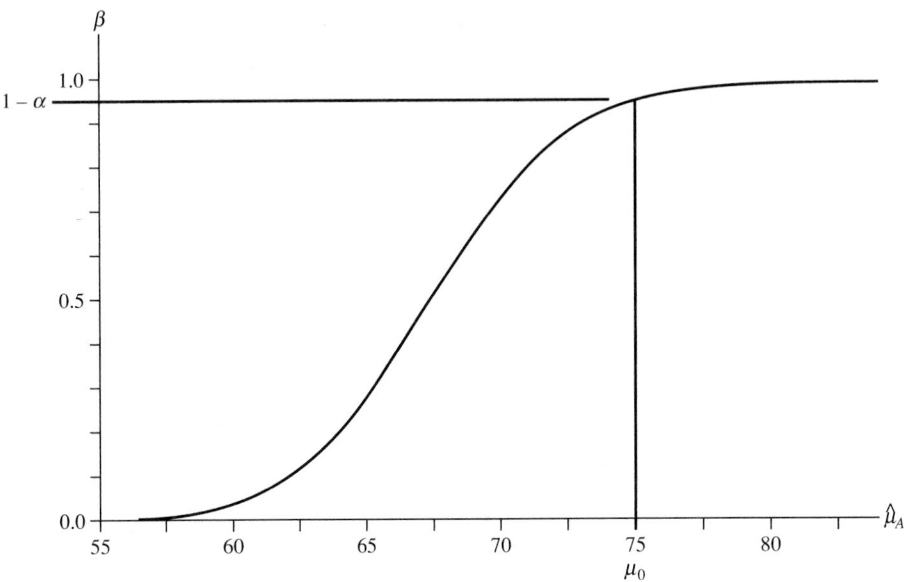

FIGURE 9-11
The o.c. curve for Example 9.8.

t distribution. However, the distribution of the alternate hypothesis is related to the *noncentral t distribution,* a complicated distribution that requires mathematical techniques beyond the scope of this book. For this reason, the specific approach used to develop the β error for this hypothesis test is not discussed. However, the same computer program that we have used to generate the o.c. curves of Figures 9.7 and 9.10 may also be used to obtain o.c. curves for tests of a single mean when σ is unknown and the sample size is small. Figure 9.11 shows the o.c. curve for Example 9.8.

Now, let us briefly consider how to perform hypothesis tests about a single proportion.

9.3 HYPOTHESIS TESTS ABOUT A SINGLE PROPORTION

It is often necessary to make inferences from data about proportions. Politicians sample the electorate to estimate the proportion of the people who will vote for them. Public health experts sample the population to estimate the proportion with smallpox vaccinations. Quality control engineers sample production runs to estimate the proportion of defective items produced.

As we have learned from Chapter 4, if the sample size is *large* and p is not near 0 or 1 ($np \geq 5$ when $p \leq 0.5$ or $n(1 - p) \geq 5$ when $p \geq 0.5$), the normal distribution forms an excellent approximation of the binomial distribution. Therefore, under the same conditions, the normal distribution can be used for a test of a hypothesis about a proportion. The expression $\hat{P} = X/n$ is the estimator of the proportion of successes, p—where n is the sample size and X is the number of successes in the sample. For large n and p not too near 0 or 1, $\hat{P}$ is approximately normal with $\mu_{\hat{P}} = p$ and $\sigma^2_{\hat{P}} = p(1-p)/n$, and *testing* a single proportion is *no different* from testing a single mean. All we need to do is use Table 9.2, substituting the appropriate parameter values (of $\mu_{\hat{P}}$ and $\sigma^2_{\hat{P}}$) to reach the required conclusions.

Example 9.9. The president of your firm has claimed that 90% of your torsion springs will last beyond the accepted maximum standard of performance. An independent testing company found that 168 springs in a sample of 200 did exceed the standard. What can be said about the performance claim? (Use $\alpha = 0.05$.)

Solution. In this particular case, $H_0: p = p_0 = 0.9$, and, because we do *not* want to place the burden of proof on the "boss," $H_A: p < p_0 = 0.9$. The following is the calculation of C.

$$C = p_0 - z_\alpha \sigma_{\hat{P}} = 0.9 - 1.645\sqrt{\frac{0.9(1-0.9)}{200}} = 0.865$$

Since $\hat{p} = 168/200 = 0.840$, the null hypothesis that 90% of the springs will last longer than the standard must be rejected. This conclusion appears even more reasonable when one considers the fact that the p-value of α for this hypothesis is 0.002; i.e., even if α is set equal to 0.002, the null hypothesis would still be rejected.

Unlike the hypothesis test, the construction of the o.c. curve for tests about a single proportion is *not the same* as constructing an o.c. curve for the test about a single mean. When we developed the hypothesis test for a single mean, one of the primary assumptions was that the variances of the distributions of the null and alternate hypotheses were *identical*. For the test of a single proportion, the variance of the distribution of the alternate hypothesis is not the same as the variance of the distribution of the null hypothesis.

When H_0 is *true*, the distribution of $\hat{P} = X/n$ is $N(p_0, \sqrt{p_0(1-p_0)/n})$. However, when H_0 is *false* and a specific alternate hypothesis, $\hat{H}_A: p = \hat{p}_A$ is true, the distribution of $\hat{p}$ is $N\left(\hat{p}_A, \sqrt{\hat{p}_A(1-\hat{p}_A)/n}\right)$. Figure 9.12 illustrates this situation for Example 9.9, where $\hat{p}_A$ is set at 0.84.

The β error may be found by noting that

$$C = \hat{p}_A + z_\beta \sqrt{\hat{p}_A(1-\hat{p}_A)/n} \tag{9.9}$$

Substituting the known values for C and $\hat{p}_A$ allows us to solve for z_β and thus for β. Note also that

$$C = p_0 - z_\alpha \sqrt{p_0(1-p_0)/n} \tag{9.10}$$

Using Equations 9.9 and 9.10 yields the sample size required to achieve the specified α and β errors,

$$n = \left(\frac{z_\alpha \sqrt{p_0(1-p_0)} + z_\beta \sqrt{\hat{p}_A(1-\hat{p}_A)}}{\hat{p}_A - p_0}\right)^2 \tag{9.11}$$

Equation 9.11 is valid for either form of the one-tailed test. To find n for a two-tailed test, substitute $z_{\alpha/2}$ for z_α in Equation 9.11.

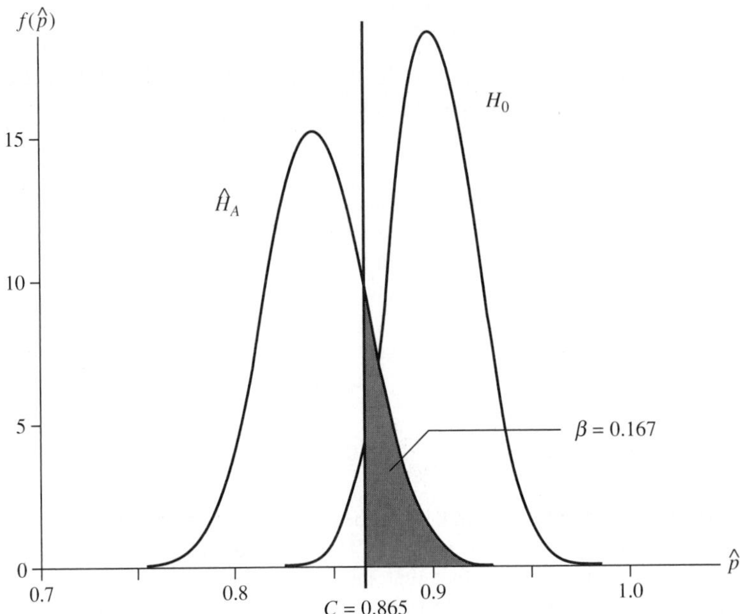

FIGURE 9-12
H_0 and $\hat{H}_A$ and the β error for Example 9.9.

In this case (not in general), the variance of the distribution of the alternate hypothesis is larger than the variance of the distribution of the null hypothesis. For $\hat{p}_A = 0.84, \beta = 0.167$. This is much greater than the incorrect value of 0.119 that would have been obtained if one had incorrectly assumed that the variance of the alternate distribution was the same as that of the null distribution. The additional complication caused by different variances presents no conceptual difficulty once it is understood. The additional computational effort is rendered unimportant by the use of the o.c. curve program for tests of a single proportion. The o.c. curve for this example is presented in Figure 9.13.

Now let us address how to test conjectures about a single variance.

9.4 HYPOTHESIS TESTS ABOUT A SINGLE VARIANCE

As Chapters 7 and 8 have shown, the sample variance, S^2, is associated with the χ^2 distribution; i.e., $(n-1)S^2/\sigma^2$ is distributed as χ^2_{n-1}. In a test of hypothesis about a single variance, the null and alternate hypotheses are

$$H_0 : \sigma^2 = \sigma_0^2 \qquad H_A : \sigma^2 \ (<, >, \neq) \ \sigma_0^2$$

The decision table for choosing between H_0 and H_A is given in Table 9.3.

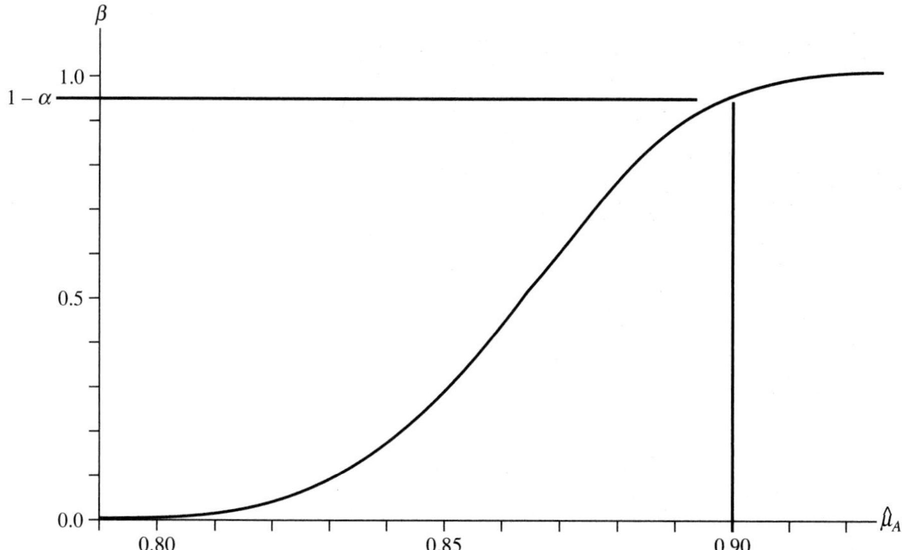

FIGURE 9-13
The o.c. curve for Example 9.9.

Example 9.10. A radio frequency generator must be capable of holding close to the bandwidth that has been set. Our supplier claims that the standard deviation of the frequency output at prescribed time points is 5 standard units. Data has been taken randomly at 17 times for a particular generator, and s^2 has been found equal to 33. What may we say about the variability of the selected unit?

Solution. The null hypothesis, $H_0 : \sigma^2 = \sigma_0^2 = 25$, is accompanied by the alternate hypothesis, $H_A : \sigma^2 > \sigma_0^2$. Since $(n-1)S^2/\sigma_0^2$ is distributed as a χ^2_{n-1} random variable if H_0 is true, the value of C at the 0.95 level of confidence is

$$C = \frac{\sigma_0^2 \chi^2_{n-1,\alpha}}{n-1} = \frac{25\chi^2_{16,0.05}}{16} = \frac{25(26.296)}{16} = 41.09$$

According to Table 9.3, $s^2 = 33$ does not constitute sufficient reason to say that the sampled frequency generator is more variable than claimed. The null hypothesis

TABLE 9.3
Decision table for $H_0 : \sigma^2 = \sigma_0^2$

H_A	Reject H_0 if
$\sigma^2 < \sigma_0^2$	$s^2 < C = \sigma_0^2 \dfrac{\chi^2_{n-1,1-\alpha}}{n-1}$
$\sigma^2 > \sigma_0^2$	$s^2 > C = \sigma_0^2 \dfrac{\chi^2_{n-1,\alpha}}{n-1}$
$\sigma^2 \neq \sigma_0^2$	$s^2 < C_1 = \sigma_0^2 \dfrac{\chi^2_{n-1,1-\alpha/2}}{n-1}$ or
	$s^2 > C_2 = \sigma_0^2 \dfrac{\chi^2_{n-1,\alpha/2}}{n-1}$

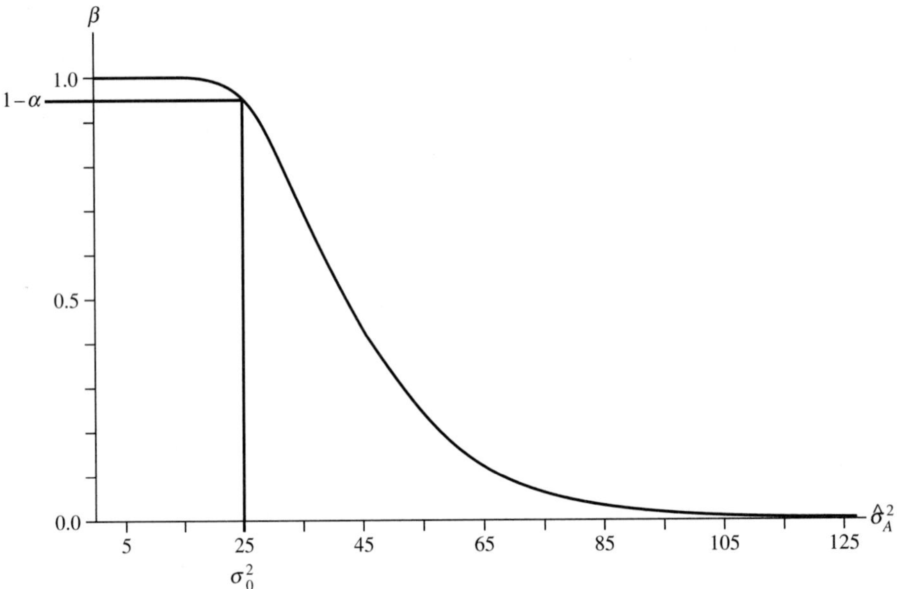

FIGURE 9-14
The o.c. curve for Example 9.10.

is *not* rejected. Indeed, it would take setting the value of α to 0.17 before the hypothesis could be rejected.

This analysis is easily performed using the program for hypothesis tests for one variance. The mathematical details of constructing an o.c. curve for the testing of a single variance are beyond the scope of this book. However, once more, this lack of detail does not prevent you from using the program in your software library for constructing such o.c. curves. Figure 9.14 presents the o.c. curve for Example 9.10.

The use of the decision rules given previously *requires* that the sampled population be governed, at least approximately, by a *normal distribution*. If this assumption cannot be supported and the sample size is small, the analyst must resort to techniques that are beyond the scope of this book. If the sample is *large*, it may be shown that S is approximately normally distributed with a mean of σ and a variance of $\sigma^2/2n$. This allows us to use the techniques of Section 9.2 to perform any required tests about the variance of the population.

Example 9.11. Thirty-five brass fasteners are tested to failure, and their failure stresses are recorded.Computing s yields a value of 3.5 pounds. Is there sufficient evidence to conclude that the manufacturer's claim of a true standard deviation of 3 pounds is false?

Solution. In this case, $H_0 : \sigma = 3$ and $H_A : \sigma > 3$. Since the sample is large, S is distributed approximately as

$$N\left(\sigma_0, \sqrt{\frac{\sigma_0^2}{2n}}\right) = N\left(3, \sqrt{\frac{9}{70}}\right) \qquad \text{if } H_0 \text{ is true}$$

Let $\alpha = 0.05$. The decision criterion is

$$C = \sigma_0 + z_\alpha \sqrt{\frac{\sigma_0^2}{2n}} = 3 + 1.645\sqrt{\frac{9}{70}} = 3.59$$

Since $s = 3.5$ is not greater than $C = 3.59$, there is insufficient evidence for the rejection of the null hypothesis. Of course, you could reach this same conclusion using the computer program for hypothesis tests about a single variance.

9.5 SOME COMMENTS ON THE DIFFERENCE BETWEEN STATISTICAL SIGNIFICANCE AND PRACTICAL SIGNIFICANCE

Given a large enough sample size, you will *almost always reject the null hypothesis*. The reason that this statement is true is intuitive and simple. The more data you possess about the sampled population, the smaller a difference from the null hypothesis you will be able to detect.

For simplicity (and without loss of generality), consider the hypothesis test about a single mean when the value of the sampled population's standard deviation is known. Further, suppose that the alternate hypothesis is $H_A : \mu < \mu_0$. In this case, $C = \mu_0 - z_\alpha \sigma_{\overline{X}} = \mu_0 - z_\alpha(\sigma/\sqrt{n})$. Notice that as n becomes larger, C becomes closer and closer to μ_0. Thus, as n increases, our hypothesis test will be able to detect much smaller differences between the true mean of the population, μ, and the hypothesized value, μ_0.

Assume that the desired, or *nominal,* specification about some mean is that it be equal precisely to 10 and that the known value of σ is 1. In addition, any value of the mean in the range from 9.90 to 10 is acceptable. Thus a value of 9.95 is not considered a serious departure from the nominal value of 10. Suppose that your boss comes into your office and tells you that a sample of 10,000 items has yielded $\overline{x} = 9.95$. What may you conclude from this? First, you would probably compute C (with $\alpha = 0.05$) to be $C = 10 - 1.645(1/\sqrt{10{,}000}) = 9.98355$. According to Table 9.2, you should reject the null hypothesis! Unfortunately, from the practical considerations given earlier, you know that rejecting the null hypothesis when $\overline{x} = 9.95$ is *not correct*.

The position in which you find yourself is similar to that of the botanist who would prefer to use a magnifying glass to study the color variations in a flower petal but who possesses only an expensive high-powered microscope! This situation is one of the primary reasons for the emphasis in this chapter on the proper design of a hypothesis test through consideration of the α and β errors and the o.c. curve. By stipulating the α and β errors and a difference $\hat{\mu}_A - \mu_0$ that is *practically significant,* we will not find ourselves in the ironical position of rejecting a null hypothesis because a *statistically significant difference* has been detected—when, in fact, *no practically significant difference* is present.

Constructing a hypothesis test that possesses too much discrimination power because of too large of a sample size is just as bad an error as constructing a test that has too little power, because of too small a sample size. In fact, from an economic viewpoint, taking a sample that is too large is a more serious error because of the greater monetary expense.

This concludes our discussion of hypothesis tests for single parameter values. In the next chapter, we extend this knowledge to consideration of two means, two variances, and two proportions. Chapters 13 and 14 introduce the subject area of "analysis of variance," where techniques of correctly comparing three or more means are presented.

EXERCISES

9.1. Kilgore Auto Supply claims that its special brake shoes last, on the average, 36,000 miles. If $\sigma = 1500$ miles, $\alpha = 0.1$, and β has a desired value 0.05 for $\hat{\mu}_A = 35{,}000$, what value of n is required? If a sample of size $n = 10$ is the maximum size possible, what is the smallest possible value of β?

9.2. Blastron Boats needs a lacquer that will be durable and that will dry in, at most, 17 minutes. Paris Paints, Inc., claims to produce a lacquer that not only meets all other requirements but also dries in 15 minutes, on the average. If the standard deviation for the drying time is $\sigma = 1.75$ minutes and the value of α is set at 0.1, determine the sample size that will ensure that lots with a mean of 17 minutes will be rejected 95% of the time. If that sample is taken and it yields $\bar{x} = 16$, what is your decision in regard to the Paris Paints lacquer?

9.3. In Exercise 9.2, suppose that the value of σ for the drying time is unknown and the value of s from the sample is 1.85 minutes. In that case, what would your decision be?

9.4. A dental molding paste is claimed to harden in 3.5 minutes. From the dentist's viewpoint, this drying time must conform to strict standards. If the paste dries too fast, the patient may be condemned to having the mold form reside permanently in his or her mouth. But if it dries too slowly, the patient may become "impatient" or the dentist may remove the form too early, resulting in a bad impression of the patient's teeth. If the drying time of the paste is known to be normally distributed with $\sigma = 0.4$ minute, α has been set to 0.05, and 16 molds are available, construct and plot the o.c. curve for the implied hypothesis test.

9.5. In using Equation 9.3 to arrive at the correct sample size for a specified α and β, you were instructed to round *up* to obtain the value of the sample size but were instructed to use the *unrounded value* to obtain the value of C. Why is this the appropriate procedure? (*Hint:* Try substituting the rounded value of n into Equation 9.1. Now compute the values of the α and β errors associated with the value of C obtained with the rounded value of n. What does this procedure indicate to you?)

9.6. The o.c. curve of Figure 9.11 has been generated under the assumption that the true value of σ was not known. Suppose that σ was known to be equal to 15 (the same value as s). Construct the o.c. curve associated with this new situation. What are the similarities and differences that you observe between the two o.c. curves?

9.7. In Section 9.4, the large sample approximate distribution for s is said to be $N(\sigma, \sigma/\sqrt{2n})$. At what value of n do you think this approximation becomes valid? Why?

9.8. Show that $\alpha = 0.17$ is the correct p-value for Example 9.10.

9.9. In Example 9.10, suppose that the critical alternate value of $\hat{\sigma}_A^2$ is 35. How do you feel about the current experiment? (*Hint:* An interpretation of Figure 9.14 is very helpful here.)

9.10. Forty-seven ball bearings randomly selected from a vendor's product yield s = 0.003 inch. It is important that the true standard deviation of such bearings not exceed 0.0025 inch. What may be concluded from this sample?

9.11. Foghorn O'Reilly is leading the election at midnight. Based on a random selection of 50% of the 1000 votes that will be cast in the election, Foghorn leads with 57% of the vote. How would you assess the likelihood of Foghorn's winning the election?

9.12. Twenty-two randomly selected samples of super-premium gasoline from the Zeckzon refinery have been tested and have yielded an average octane rating of 89 with $s = 3$. Based on this sample, what can we conclude about Zeckzon's claim that the average octane of their super-premium gasoline is 90.5? Would it be appropriate to take a larger sample? Why or why not?

9.13. In a randomly selected sample of 200 surface plates from a space shuttle, 57% of the plates have suffered enough thermal damage that they cannot be used again. May we conclude that over half of the plates on that shuttle may not be reused because of thermal damage?

9.14. A microcomputer retail outlet claims that it can deliver software packages to any location in the city within 30 minutes of receiving a telephone order. A sample of 28 such deliveries yields an average delivery time of 34.5 minutes with s = 2.3 minutes. What can be said about the outlet's claim? In your opinion, is the sample adequate to make such an assessment?

9.15. A manufacturer of concrete pillars has developed a new additive. The company president claims that pillars made with this new additive have a mean compressive strength of 15 standard units. From past experience, the standard deviation of the pillars is known to be 0.5 standard unit. A sample of 50 pillars yields $\bar{x}$ = 14.8 standard units. Do you think that the president's claim is valid? If it is important to detect a difference of 0.75 unit, is the sample size adequate?

9.16. Ace Cable Company claims that its 0.375-inch cable (Inventory #336129) will support an average load of 5000 pounds. Sixteen such cables were tested to failure. The results yielded $\bar{x}$ = 4925 pounds with s = 190 pounds. What would you conclude about Ace's claim? Is the sample size adequate?

9.17. The liquid crystal displays that have been selected for use on the new Klone portable microcomputers have been advertised as having a functional service life of 6.5 years. The company providing the displays to Klone, Inc., has made available to an independent testing laboratory the data concerning the lifetimes of 120 displays identical to those sold to Klone, Inc. The average, $\bar{x}$, of that lifetime data is 6.3 years with s = 0.95 year. What may we conclude about Klone's claim?

9.18. The average adhesive strength of spot welds in the manufacture of galvanized garbage cans is μ = 300 pounds with a standard deviation, σ = 12 pounds. An additive for the welding material is being considered—one that is said to increase the strengths of the welds. An experiment is run to ascertain the properties of the new material. If the average strength is increased by as much as 30 pounds, such a change should be detected with a probability of 0.97. If there is no change, this should be detected with a probability of 0.95.

(a) How many observations are required to achieve the probabilities just stated?

(b) If $\bar{x}$ obtained in the experiment is 325 psi, should we conclude that the average strength has been increased by the additive?

9.19. A producer of brass rivets claims that its rivet diameters have σ = 0.01 inch. A sample of 10 rivets has s = 0.018. What can you conclude about the rivet maker's claim? Is the sample adequately sized to make any decision?

9.20. PEMCO Oil buys 55-gallon barrels of oil in large lots. PEMCO is interested in whether the average number of gallons in the barrels is adequate. The process that fills the barrels yields a standard deviation of 1.25 gallons. If a lot's average contents for a barrel is as small as 53 gallons, PEMCO would like to reject the lot with a probability of at least 0.7. If the average size is as stipulated (55 gallons), PEMCO would like to accept the lot with a probability of 0.95.

(a) How many barrels should be included in the sample?

(b) If $\bar{x}$ from that sample is 52.5 gallons, should the lot be accepted or rejected?

(c) If the sample size was 10 barrels, instead of the sample you obtained in part a, what would be the value of the β error?

9.21. You work for Acme Levelers, the leading user of purple widgets. Your company uses these widgets as shims in the process of leveling buildings. For your purposes, you are concerned with the thickness of the widgets. If they are too thin, your company must use more, whereas if they are too thick, the appropriate raising distance may be impossible to achieve. The manufacturer claims that the average widget thickness is 0.5 inch. You test a sample of 25 widgets and find $\bar{x} = 0.53$ inch and $s = 0.03$ inch. Do you accept the manufacturer's claim? Is this sample size adequate? Explain.

9.22. After solving Exercise 9.21, you are told that the β error must be 0.05 for detecting differences of 0.005 inch between μ_0 and $\hat{\mu}_A$. You are also told that the true standard deviation is 0.025. What is the appropriate sample size for this test?

9.23. You are concocting a test for your probability and statistics classmates. You want everyone to do equally well. You give the test, and s for the 10 student grades is 2.1 points. You want the true standard deviation overall to be less than 3.5 points. What is your conclusion about this desire given the sample results?

9.24. The manager of your production facility has just come to you with a claim that the reason he didn't meet the production orders last month is that the variance in production times of different employees was too great, creating too many delays. He says that the standard deviation was over 30 minutes for the workers in his plant. You must decide whether he is right, so you sample the production times for 25 of the employees from last month, and find that $s = 23$ minutes. What do you think of his claim? How confident are you in your conclusions? (Use $\alpha = 0.05$.)

9.25. The proportion of people missing a particular class day must be less than 0.25. A sample of 100 days is taken, yielding an average proportion of 0.23. What is your conclusion? Is the sample size sufficient?

9.26. Guns-R-Us, a local shooting school, claims that after graduating from their marksmanship course, "our students will hit the target 95 out of 100 times—anytime, anywhere." After taking a sample of 170 graduates of the course, an average proportion of 0.923 is found. What is your conclusion? How confident are you about your conclusion?

CHAPTER

10

HYPOTHESIS TESTS FOR TWO MEANS, TWO VARIANCES, OR TWO PROPORTIONS

When you understand the following concepts and procedures, you will have mastered the important ideas in this chapter.

- Know why a hypothesis test comparing the means of two normally distributed random variables with known variances may be *reduced* to a hypothesis test about the mean of one normally distributed random variable with a known variance
- Be able to perform a test of hypothesis comparing the means of two normally distributed populations when both population variances are known
- Be able to perform a test of hypothesis comparing two means when the samples are small, the populations are normal, and the variances must be estimated from the samples
- Know what assumptions must be satisfied for both large- and small-sample hypothesis tests comparing two means
- Know what a *pooled estimate of variance* is, how to compute such an estimate, and where to use it
- Be able to obtain the appropriate sample sizes to enforce specified α and β errors for a test of hypothesis comparing two means and be able to construct the o.c. curve associated with a test of hypothesis comparing two means
- Know why the F distribution is used to perform tests of hypotheses comparing *two variances* of normally distributed populations

- Be able to perform tests of hypotheses about two variances taken from normal populations and be able to construct o.c. curves for such tests
- Be able to perform large sample tests of hypotheses about two proportions and to compute β errors for such tests.

In the last chapter, we studied hypothesis tests in general and then considered hypothesis tests about one mean, one proportion, and one variance. We now study three natural extensions of that discussion—hypothesis tests comparing two means, two variances, and two proportions.

10.1 TESTS COMPARING TWO MEANS WHEN σ_1 AND σ_2 ARE KNOWN

Suppose that X_1 and X_2 are random variables representing a common characteristic in two different populations and that we are interested in whether the means of X_1 and X_2 are significantly different. Let us first assume that X_1 and X_2 are normally distributed and possess *known* standard deviations σ_1 and σ_2. If we *define* $Y = X_1 - X_2$, the null and alternate hypotheses are

$$H_0 : \mu_Y = \mu_1 - \mu_2 = \mu_{Y_0} \quad H_A : \mu_Y = \mu_1 - \mu_2 \ (>, <, \neq) \ \mu_{Y_0}$$

We know from Chapter 4 that when you add or subtract normally distributed random variables, the result is also a normally distributed random variable. From Chapter 7 we have learned that if X_1 is distributed as $N(\mu_1, \sigma_1)$ and X_2 is distributed as $N(\mu_2, \sigma_2)$, then $Y = X_1 - X_2$ is distributed as $N(\mu_1 - \mu_2, \sqrt{\sigma_1^2 + \sigma_2^2}) = N(\mu_Y, \sigma_Y)$. Further, $\overline{Y} = \overline{X}_1 - \overline{X}_2$ is distributed as

$$N\left(\mu_1 - \mu_2, \sqrt{\frac{\sigma_1^2}{n_1} + \frac{\sigma_2^2}{n_2}}\right) = N(\mu_{\overline{Y}}, \sigma_{\overline{Y}}) \tag{10.1}$$

where n_1 and n_2 are the sizes of the samples from each of the populations.

Equation 10.1 shows that testing a hypothesis comparing two means when both standard deviations are known is *no different* than testing a single mean when the standard deviation is known. In fact, we can use the same decision rules that were given in Table 9.2. For ease of use and understanding, those rules are restated in Table 10.1 using the notation for the new random variable $Y = X_1 - X_2$.

The values of C, C_1, and C_2 are computed using the procedures presented in Chapter 9. Let us consider an example of the test of a hypothesis concerning two means to illustrate the use of Table 10.1.

TABLE 10.1
Decision table for $H_0 : \mu_Y = \mu_{Y_0}$

H_A	Reject H_0 if		
$\mu_Y < \mu_{Y_0}$	$\overline{y} < C$		
$\mu_Y > \mu_{Y_0}$	$\overline{y} > C$		
$\mu_Y \neq \mu_{Y_0}$	$\overline{y} < C_1$	or	$\overline{y} < C_1$

Example 10.1. Suppose that a new vendor of ball bearings claims that its product has lower frictional resistance under very heavy loading conditions. To check this claim, 36 of the old supplier's bearings and 25 of the new supplier's bearings are placed under identical heavy loads and the frictional resistance of each bearing is measured, yielding $\bar{x}_{\text{new}} = 44$ and $\bar{x}_{\text{old}} = 52$. Further, it is known that $\sigma_{\text{old}} = 8$ and $\sigma_{\text{new}} = 12$. If α is set at 0.10, what can we conclude about the new supplier's claim?

Solution. To answer this question, we first state the null and alternate hypotheses as

$$H_0 : \mu_{\text{new}} - \mu_{\text{old}} = 0 \quad H_A : \mu_{\text{new}} - \mu_{\text{old}} < 0$$

For this H_A, the criterion value is

$$C = \mu_{Y_0} - z_\alpha \sigma_{\bar{Y}} = 0 - 1.282\sqrt{\frac{64}{36} + \frac{144}{25}} = -3.52$$

Since $\bar{y} = \bar{x}_{\text{new}} - \bar{x}_{\text{old}} = -8 < -3.52$, we may reject the null hypothesis of equal means and conclude that, at the 90% level of confidence, the new vendor does produce bearings that have a lower frictional resistance.

The next question to ask in this type of situation is, Under these conditions, how likely are we to fail to reject H_0 when it is false? In order to answer this question, a *critical difference value* between μ_{new} and μ_{old} must be specified.

Example 10.2. Let us suppose that prior in experiment described in Example 10.1, it had been decided that an average difference of 3 units was important to detect.

Referring to Figure 10.1, which pictures the distributions of the null and alternate hypotheses, we see that the probability of a Type II, or β, error is unacceptably

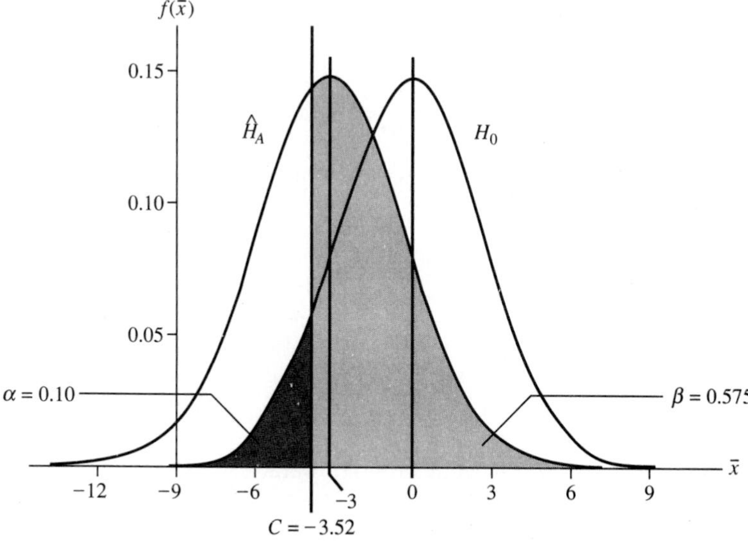

FIGURE 10-1
Distributions of H_0 and $\hat{H}_A$ for Examples 10.1 and 10.2.

large at $\beta = 0.575$. This means that, even if the difference between the means *is* 3 units, we will *fail to reject* H_0 with a probability of 0.575. Indeed, β reaches the acceptable neighborhood of 0.05 only at relatively distant values around $\mu_{Y_A} = -8$.

Example 10.2 has shown how to compute a β error (and therefore an o.c. curve) for this simplest case of comparing two means. The major difference between tests of two means and tests of a single mean is that the variance of the quantity being studied is no longer dependent on a single sample size—it is dependent on two sample sizes. With one mean, we study $X \sim N(\mu, \sigma)$ and $\overline{X} \sim N(\mu, \sigma/\sqrt{n})$, whereas with two means we study

$$Y = X_1 - X_2 \approx N(\mu_1 - \mu_2, \sqrt{\sigma_1^2 + \sigma_2^2}) = N(\mu_Y, \sigma_Y)$$

and

$$\overline{Y} = \overline{X}_1 - \overline{X}_2 = N\left(\mu_1 - \mu_2, \sqrt{\frac{\sigma_1^2}{n_1} + \frac{\sigma_2^2}{n_2}}\right) = N(\mu_{\overline{Y}}, \sigma_{\overline{Y}})$$

There can be a great number of differing pairs of values of n_1 and n_2 that will yield very close to the same value of $\sigma_{\overline{Y}}$ for given values of σ_1 and σ_2. This means that there can be more than one pair of values for n_1 and n_2 that will yield almost identical o.c. curves. This fact can often be used to advantage, especially when samples of one population are more expensive to obtain than samples of the other population. One way to determine how large to set the sample sizes is to make the initial assumption that $n = n_1 = n_2$. This means that

$$\sigma_{\overline{Y}}^2 = \frac{\sigma_1^2}{n_1} + \frac{\sigma_2^2}{n_2} = \frac{\sigma_1^2 + \sigma_2^2}{n}$$

which implies that

$$n = \frac{\sigma_1^2 + \sigma_2^2}{(\sigma_2^2/n_1) + (\sigma_2^2/n_2)} \tag{10.2}$$

From our work with hypothesis tests about a single mean (see Equation 9.3) we know that, if $n = n_1 = n_2$, then

$$n = \frac{(\sigma_1^2 + \sigma_2^2)(z_\alpha + z_\beta)^2}{(\mu_{Y_0} - \hat{\mu}_{Y_A})^2} \tag{10.3}$$

Any pair of values for n_1 and n_2 that causes the values of n given by Equations 10.2 and 10.3 to agree will generate an o.c. curve consistent with the α and β errors and the stated critical difference $\mu_{Y_0} - \hat{\mu}_{Y_A}$. The construction of an o.c. curve for the preceding hypothesis test is easily performed by using the program designed for that purpose in your program library.

Example 10.3. Figure 10.2 presents the o.c. curve constructed by the program for Examples 10.1 and 10.2 when $n_1 = 64$ and $n_2 = 80$.

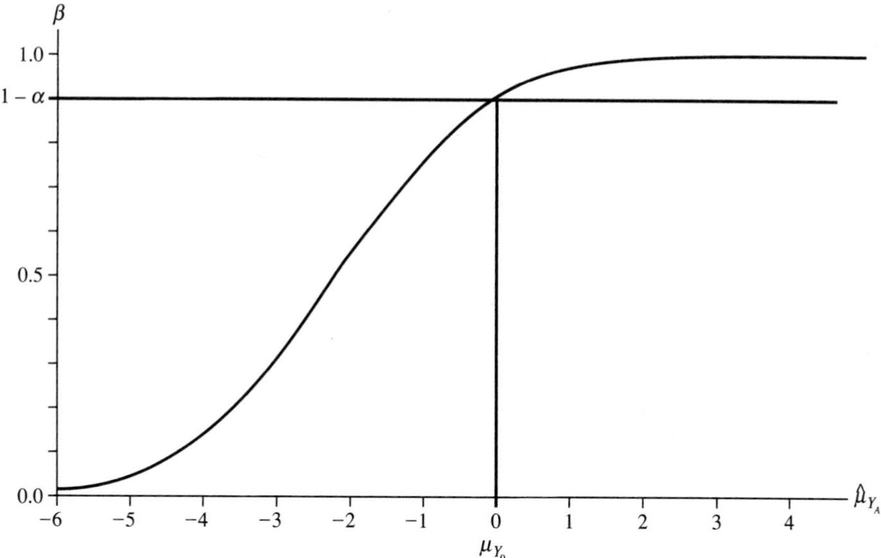

FIGURE 10-2
The o.c. curve for Examples 10.1 and 10.2 with $n_1 = 64$ and $n_2 = 80$.

10.2 TESTS COMPARING TWO MEANS WHEN σ_1 AND σ_2 ARE UNKNOWN

Often the exact values of the standard deviations for X_1 and X_2 are *not known*. When this happens, we must first determine if both sample sizes are large. A good guideline tells us that, if $n_1 \geq 30$ and $n_2 \geq 30$, the samples are large. If the *samples are large,* replace σ_1 and σ_2 everywhere with their sample estimates s_1 and s_2 and proceed as discussed in Section 10.1. Indeed, because of the central limit theorem, we need not even assume that the distributions of X_1 and X_2 are normal.

If the sample sizes are *not large,* we must use the t distribution. In this situation, we both (1) *assume normality* for both populations and (2) make the *added assumption* that $\sigma = \sigma_1 = \sigma_2$. The assumption of equal variance allows the computation of a *pooled* estimate of the variance by combining the results of both samples. As above, $\sigma_{\bar{Y}}^2 = (\sigma_1^2/n_1) + (\sigma_2^2/n_2)$. Since we assume that $\sigma_1 = \sigma_2 = \sigma$ and $\sigma_{\bar{Y}} = \sigma\sqrt{(1/n_1) + (1/n_2)}$, the best estimate of the common value σ^2 is

$$s^2 = \frac{\sum(x_{1i} - \bar{x}_1)^2 + \sum(x_{2i} - \bar{x}_2)^2}{n_1 + n_2 - 2}$$

This is equivalent to

$$s^2 = \frac{(n_1 - 1)s_1^2 + (n_2 - 1)s_2^2}{n_1 + n_2 - 2} \tag{10.4}$$

where s_1^2 and s_2^2 are the sample variances. Notice that the degrees of freedom for this *pooled estimate of the variance* are $n_1 + n_2 - 2$. We subtract 2 from the total of $n_1 + n_2$ degrees of freedom because 1 degree of freedom was lost for computing each of $\bar{x}_1$ and $\bar{x}_2$, a necessary step prior to computing s^2.

With this knowledge, we may perform the above tests of hypotheses using Table 10.1. The only difference is in the computation of the decision criteria, C, C_1, and C_2. Instead of using the standard normal distribution, z, we use the t distribution with $n_1 + n_2 - 2$ degrees of freedom. To clarify these ideas, let us consider the following example.

Example 10.4. Two gunpowder manufacturers are competing for our business. To guarantee a ready supply of gunpowder, we desire to assign half of our purchases to each firm. However, as a preliminary check, we want assurance that their products are not markedly different in quality. One of the primary measures of quality is the muzzle velocity that a gunpowder produces.

Ten cases of powder from each vendor are purchased, and each case is tested to yield a measure of muzzle velocity. The tests have given $\bar{x}_1 = 1210, s_1^2 = 2550, \bar{x}_2 = 1175$, and $s_2^2 = 3600$. What may be concluded, at the 95% level of confidence, from these results?

Solution. The pooled estimate of the variance is $s^2 = [(9)(2550)+(9)(3600)]/18 = 3075$, and the appropriate hypotheses are

$$H_0 : \mu_Y = \mu_{Y_0} = 0 \quad H_A : \mu_Y \neq \mu_{Y_0} = 0$$

Since the alternate hypothesis is two-tailed, we must compute two decision criteria, C_1 and C_2, as follows.

$$C_1 = \mu_{Y_0} - t_{n_1+n_2-2,\alpha/2} s_{\overline{Y}} = 0 - 2.101\sqrt{3075}\sqrt{\frac{1}{10} + \frac{1}{10}} = -52.1$$

and

$$C_2 = \mu_{Y_0} + t_{n_1+n_2-2,\alpha/2} s_{\overline{Y}} = 52.1$$

Since the difference between the sample means of 35 falls between C_1 and C_2, we conclude that there is *insufficient evidence* to cause rejection of the null hypothesis. Until other evidence presents itself, we should feel free to do business with both competing firms.

Once more, it is not sufficient simply to take a sample of an arbitrary size and go through the mechanics just described for a two-sample t test. The conscientious analyst will first identify appropriate values for the α and β errors and the measure of a critical difference and then obtain sample sizes consistent with those values prior to taking the samples and performing the analysis. Just as in the one-sample t test, the mechanics of generating the o.c. curve for our two-sample t test are beyond the scope of this text. However, the computer program for generating o.c. curves for this test is easy to use and you will be given an opportunity to do so in the exercises at the end of this chapter.

10.3 TESTS COMPARING TWO MEANS WHEN THE DATA ARE PAIRED

The hypothesis tests for comparing two means just discussed have all incorporated the assumption that the two samples are taken from two different populations and that all observations are statistically independent from one another, both within the same sample and across the two samples. In certain circumstances, these assumptions may not be satisfied.

In many cases, the experimental environment may have external effects so strong that they mask the differences in the parameter of interest. For example, one such experiment might compare the wearability of two kinds of marine deck paint. The physical location to which the paint is applied may be a major factor influencing wearability. To remove concern about the possible bias introduced by such an external effect, a completely random assignment of locations to each of the two kinds of paint may be done in order to "average out" any adverse effects. Another approach would use both kinds of paint on every selected experimental location.

The latter approach would provide a kind of experimental *control*; each sample application of paint type A would be paired with a sample application of paint type B. In this way, the *difference* in the wearabilities could be studied with the knowledge that both kinds of paint would have been exposed to exactly the same ravages of weather, heat, traffic, and so on. Another common circumstance for which the assumptions of independence are not satisfied occurs when an experimenter conducts "before-and-after" tests on a selected set of items. In this kind of experiment, each item generates two data values, one before some operation is performed on it and one after the operation is performed. The hope in this kind of experiment is that the individual differences between the items are controlled and will not bias the results of the experiment.

Example 10.5. As an example of this kind of experiment, consider a classroom where the students are given a test before they are taught the subject matter covered by the test. The students' scores on this *pretest* are recorded as the first data set. Next, the subject matter is presented to the class. After the instruction is completed, the students are retested on the same material. The scores on the second test, the *posttest*, compose the second data set. It is reasonable to expect that a student that scored high on the pretest will also score high on the posttest (and vice versa). Inherently, a strong dependency exists between the members of a pair of scores generated by each individual. However, there is no reason to believe that a dependency exists between any two individuals.

Suppose that the scores in Table 10.2 have been generated by 15 students under the conditions just described. How would you decide whether the instruction had been effective? Because the scores for each individual are dependent, we do not have the $n_1 + n_2 - 2 = 28$ degrees of freedom that a two-sample t test would generate if the ordinary assumptions of independence are satisfied. Rather, because we have only 15 independent pieces of data, we have $15 - 1 = 14$ degrees of freedom in our paired data set, where our basic unit of data is the *difference* in the scores for each individual, d_i, $i = 1, \ldots, n$. As before, we lose 1 degree of freedom in computing $\overline{d}$.

TABLE 10.2
A data set with paired scores

Student	Pretest	Posttest	d
1	54	66	12
2	79	85	6
3	91	83	−8
4	75	88	13
5	68	93	25
6	43	40	−3
7	33	78	45
8	85	91	6
9	22	44	22
10	56	82	26
11	73	59	−14
12	63	81	18
13	29	64	35
14	75	83	8
15	87	81	−6

Realizing this fact, we may revert to the methods of Chapter 9 for testing a hypothesis about the single mean of a normal distribution using d_i, just as we used the x_i in Chapter 9. For our classroom example, the null hypothesis is that the mean for the differences is zero and, because we want to place the burden of proof on the teacher, the alternate hypothesis is that the mean of the differences is greater than zero.

For paired data, the standard deviation of the difference in the scores is rarely known. Even if the standard deviations associated with each data set are known, it is often difficult to compute the standard deviation of the difference because of the dependencies present. For this reason, we must estimate the standard deviation of the differences in the usual way from the individual d_i values. Thus $\overline{d} = 12.333$, $s = 16.543$, and $C = \mu_0 + t_{14,0.1}(s/\sqrt{n}) = 0 + 1.345(16.543/\sqrt{15}) = 5.745$ (for $\alpha = 0.1$). Since $\overline{d}$ is greater than C, we may conclude that there is sufficient evidence in these data to prove that the instruction did add to the knowledge base of the students.

The construction of the o.c. curve is achieved in exactly the same way as in Chapter 9 for a test about a single mean when the standard deviation was not known. Because of the massive loss of degrees of freedom and the associated degradation of the β error, pairing of this sort should be done only when dependencies are present that would violate the assumptions of the two-sample test.

10.4 TESTS CONCERNING TWO VARIANCES

Often, we need to compare the variances from two normal populations.

Example 10.6. Suppose that we are purchasing cylindrical ball-bearing assemblies from two suppliers. The inside diameter of each bearing must fall between specification limits to function correctly. If the inside diameter is too small, it will not fit on the shaft during assembly. If it is too large, excessive vibration will result while the finished unit is running. Such vibration will not only damage the bearing and shaft but will also cause other parts of the assembly to undergo an early failure.

One measure of how well the bearings fit the shaft is their mean inside diameter. However, even if the two suppliers' bearings do not differ significantly in their average inside diameters, there may still be a noticeable difference in the number of bearings that should be rejected and not used in our assembly process.

As an extreme example, suppose that the nominal inside diameter is 0.500 inch and the specification limits are from 0.495 to 0.505 inch. If one-half of the bearings from supplier B have inside diameters exactly equal to 0.510 inch and the other half of the bearings have inside diameters of 0.490 inch, the overall average of the inside diameters from supplier B will be *precisely* equal to the nominal value of 0.500. Unfortunately, *not even one* of supplier B's bearings could be used in our assemblies. As we recall, our best measure of variability from the mean is the variance, σ^2. The larger the variance of the inside diameters, the fewer bearings that will meet the specification limits. In comparing two suppliers, one essential comparison is to determine whether the variances of the products from the two suppliers are significantly different.

If both sampled populations in Example 10.6 are governed by a normal distribution, the sample variances from both of the populations are related to the χ^2 distribution. The relationship between the χ^2 and F distributions leads to the construction of the test statistic to be used in performing this comparison. As discussed in Chapter 4, the ratio of two χ^2 random variables, each divided by their respective degrees of freedom, forms an F random variable. Therefore, if we set the sample sizes from our two populations as n and m, then

$$F = \frac{\chi^2_{n-1}/(n-1)}{\chi^2_{m-1}/(m-1)} = \frac{[(n-1)S_1^2]/\sigma_1^2}{(n-1)} \bigg/ \frac{[(m-1)S_2^2]/\sigma_2^2}{(m-1)}$$

$$= \frac{S_1^2/\sigma_1^2}{S_2^2/\sigma_2^2} \qquad (10.5)$$

is distributed as an F random variable with $n-1$ and $m-1$ degrees of freedom. If $\sigma_1 = \sigma_2$, i.e., if the null hypothesis of equal variances is true, then

$$F = \frac{S_1^2}{S_2^2} \qquad (10.6)$$

is also distributed as an F random variable with $n-1$ and $m-1$ degrees of freedom.

These facts allow a simple test of hypothesis to be performed using Table 10.3, where $f = s_1^2/s_2^2$.

TABLE 10.3
Decision table for $H_0 : \sigma_1^2 = \sigma_2^2$

H_A	Reject H_0 if
$\sigma_1^2 > \sigma_2^2$	$f > F_{n-1,m-1,\alpha}$
$\sigma_1^2 < \sigma_2^2$	$f < F_{n-1,m-1,1-\alpha}$
$\sigma_1^2 \neq \sigma_2^2$	$f > F_{n-1,m-1,\alpha/2}$ or $f < F_{n-1,m-1,1-\alpha/2}$

Example 10.7. Suppose that we desire to test whether the variances of the two gunpowder populations in Example 10.4 are equal. (This is a good idea when using the two-sample t test to compare means, since equal variances are assumed in the use of that test.) The alternate hypothesis is two-tailed, $H_A : \sigma_1^2 \neq \sigma_2^2$, and we compute $f = s_1^2/s_2^2 = 2550/3600 = 0.708$. To reject $H_0 : \sigma_1^2 = \sigma_2^2$ at the 90% level of confidence, f must lie outside the interval from $F_{9,9,0.95} = 0.31$ to $F_{9,9,0.05} = 3.18$. Since 0.708 is inside the range of these values, the null hypothesis cannot be rejected.

We will not consider the analytical details associated with the construction of an o.c. curve for the comparison of two variances. However, by using the program in your computer package for this purpose, you can arrive at appropriate sample sizes for specified α and β errors and a critical difference in the *ratio* of the two population variances. Figure 10.3 gives the o.c. curve for Example 10.7. The common sample size of $n = m = 10$ is insufficient for most applications. Even as extreme a ratio as 0.4 (or 2.5) implies an unacceptable β error of 0.64. Only at unsupportable ratios in the neighborhood of 0.133 (or 7.5) does the β error reach the more acceptable values of about 0.10.

10.5 HYPOTHESIS TESTS ABOUT TWO PROPORTIONS

Suppose that two populations are governed by binomial distributions with probabilities of success p_1 and p_2, respectively. If we plan to take random samples of size n_1 and n_2 from the populations, then $\hat{D} = (X_1/n_1) - (X_2/n_2) = \hat{P}_1 - \hat{P}_2$

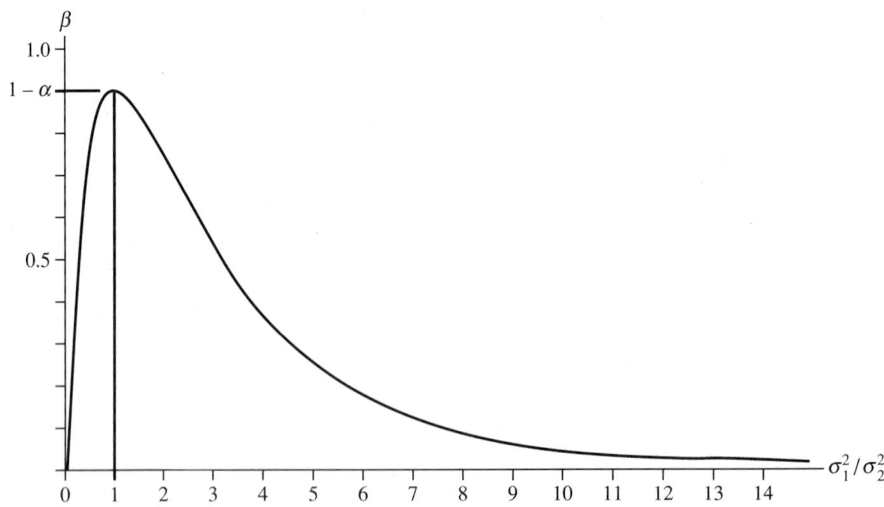

FIGURE 10-3
The o.c. curve for the test comparing the two variances of Example 10.7.

has the mean

$$d = p_1 - p_2 \tag{10.7}$$

and variance

$$\sigma^2_{\hat{D}} = \frac{p_1 q_1}{n_1} + \frac{p_2 q_2}{n_2} \tag{10.8}$$

where $q_1 = 1 - p_1$ and $q_2 = 1 - p_2$. If both n_1 and n_2 are large, $\hat{D}$ is approximately normally distributed. If the null hypothesis is

$$H_0 : d = p_1 - p_2 = d_0 = 0 \tag{10.9}$$

then the best estimate of the common value of the probability of success, $p = p_1 = p_2$, is the *pooled* estimate $\hat{p} = (x_1 + x_2/n_1 + n_2)$, which implies that $d_0 = 0$ and $s^2_{\hat{D}} = \hat{p}\hat{q}\,((1/n_1) + (1/n_2))$ where $\hat{q} = 1 - \hat{p}$ (under the assumption that H_0 is true).

Thus the distribution of the null hypothesis of Equation 10.9 for large n_1 and n_2 is approximately $N(0, \sigma_{\hat{D}})$. The decision table given in Table 10.4 governs the hypothesis test for the null hypothesis of Equation 10.9, where $\hat{d} = \hat{p}_1 - \hat{p}_2$ is the point estimate of $d = p_1 - p_2$.

If the null hypothesis is

$$H_0 : d = p_1 - p_2 = d_0 \neq 0 \tag{10.10}$$

we can no longer use both samples to estimate a common probability of success. Further, there are an infinite number of possible pairs of values of p_1 and p_2 that can yield a specific value of d_0 (except $d_0 = 1$). For each of the possible pairs of values, there is a unique and different value of $\sigma^2_{\hat{D}}$ as given in Equation 10.8. The distribution of the null hypothesis for Equation 10.10 depends both on d_0 and on the values of p_1 and p_2. For the null hypothesis of Equation 10.10, we estimate $\sigma^2_{\hat{D}}$ by substituting the values of the best sample estimates $\hat{p}_1$ and $\hat{p}_2$ for p_1 and p_2 in Equation 10.8; i.e.,

$$s^2_{\hat{D}} = \frac{\hat{p}_1 \hat{q}_1}{n_1} + \frac{\hat{p}_2 \hat{q}_2}{n_2} \tag{10.11}$$

(Unfortunately, this may not be entirely consistent; d_0 may not be precisely equal to $\hat{p}_1 - \hat{p}_2$.) To perform the test of hypothesis associated with Equation 10.10, use the decision table of Table 10.4. Now let us consider the computation of the β error for this kind of hypothesis test.

TABLE 10.4
Decision table for $H_0 : d = d_0$

H_A	Reject H_0 if
$d < d_0$	$\hat{d} < C = d_0 - z_\alpha s_{\hat{D}}$
$d > d_0$	$\hat{d} > C = d_0 + z_\alpha s_{\hat{D}}$
$d \neq d_0$	$\hat{d} < C_1 = d_0 - z_{\alpha/2} s_{\hat{D}}$ or $\hat{d} > C_2 = d_0 + z_{\alpha/2} s_{\hat{D}}$

For a specific alternate hypothesis

$$\hat{H}_A : \hat{d}_A = \hat{p}_{1,A} - \hat{p}_{2,A} \tag{10.12}$$

the distribution of $\hat{H}_A$ is dependent not only on the value of $\hat{d}_A$ but also on the values of $\hat{p}_{1,A}$ and $\hat{p}_{2,A}$. Indeed, the distribution of the alternate hypothesis is approximately normal with the mean $\mu_A = \hat{d}_A = \hat{p}_{1,A} - \hat{p}_{2,A}$ and variance

$$\hat{\sigma}_A^2 = \frac{\hat{p}_{1,A}\hat{q}_{1,A}}{n_1} + \frac{\hat{p}_{2,A}\hat{q}_{2,A}}{n_2} \tag{10.13}$$

Example 10.8. Suppose that, in a specific case, we have $H_0 : d = d_0 = 0.2$ and $\hat{H}_A : d = \hat{d}_A = 0.25$ with $\hat{p}_1 = 0.63$, $\hat{p}_2 = 0.39$, $\hat{p}_{1,A} = 0.95$, $\hat{p}_{2,A} = 0.7$, $n_1 = 120$, and $n_2 = 185$. What is the probability of a β error for this specific alternate hypothesis?

The easiest way to answer this question is to use the program in your software package for computing the β error for tests comparing two proportions. Figure 10.4 shows the distribution of H_0, which is approximately $N(0.2, 0.0568)$ (by Equations 10.10 and 10.11) and the distribution of $\hat{H}_A$, which is approximately $N(0.25, 0.0391)$ (by Equations 10.12 and 10.13).

If $\alpha = 0.05$, then $C = d_0 + z_\alpha s_{\hat{D}} = 0.2 + (1.6452)(0.0568) = 0.293$. This implies that $z_\beta = (C - \hat{d}_A)/\hat{\sigma}_A = (0.293 - 0.25)/0.0391 = 1.1$, which yields $\beta = 0.867$.

Given the values for $\hat{p}_1$, $\hat{p}_2$, $\hat{p}_{1,A}$, $\hat{p}_{2,A}$, n_1, n_2, and α, the β error is obtainable in the manner just described. However, to obtain an o.c. curve, the values of $\hat{p}_{1,A}$ and $\hat{p}_{2,A}$ must be stipulated in such a way that a spectrum of unique values of $\hat{d}_A = \hat{p}_{1,A} - \hat{p}_{2,A}$, ranging from near 0 to near 1 are provided. Because of the infinite number of possibilities of $\hat{p}_{1,A}$ and $\hat{p}_{2,A}$ that each $\hat{d}_A$ may embrace, no attempt has been made to formulate a program that generates an o.c. curve automatically for the comparison of two proportions.

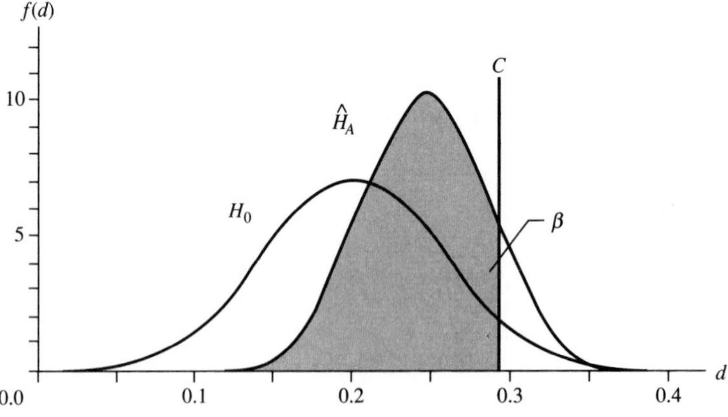

FIGURE 10-4
Distributions of H_0 and $\hat{H}_A$ for the comparison of two proportions.

The next two chapters present a method of *fitting equations to data* and begin the final set of chapters in the text. Each of these *application* chapters is designed to allow you to use and extend the knowledge that you have gained from your study of the earlier chapters of the book.

EXERCISES

10.1. The ductility of metal bar stock is an important property of the raw material for companies engaged in extrusion manufacturing, such as in the making of electrical conduit. It is equally important that the metal be neither too ductile nor not ductile enough. Two suppliers currently provide raw materials to your company. A sample of 14 bars is taken from the inventory of each supplier's product, and the average ductilities from the samples are $\bar{x}_1 = 46$ and $\bar{x}_2 = 53$. The standard deviation of each measure is known to be $\sigma_1 = 5$ and $\sigma_2 = 7$. Based on this information, determine if there is a significant difference in ductility between the two products.

10.2. Two companies sell varnish to the general public. Seventeen cans of vendor A's product and 12 cans of vendor B's product are purchased at random locations. Each can's contents are used to paint a 1-square-foot area on a test board, and the drying time for each sample is measured and recorded. The following results have been obtained: $\bar{x}_A = 20, \bar{x}_B = 24, s_A = 3.5$, and $s_B = 5.0$. If it is important that B's product not take longer than A's to dry, determine if this data set implies a significant difference in drying times between the two products.

10.3. Determine if a significant difference exists between the variabilities of the two populations studied in Exercise 10.2.

10.4. Construct an o.c. curve for Example 10.1 under the assumption that $n_1 = n_2$ and $\beta = 0.05$ for a critical difference of $\mu_{Y_0} - \mu_{Y_A} = -3$.

10.5. Construct an o.c. curve for Example 10.4. (Use the sample sizes of $n_1 = n_2 = 10$.)

10.6. Figure 10.4 shows the distribution of H_0, which is approximately $N(0.2, 0.0568)$ (by Equations 10.10 and 10.11), and $\hat{H}_A$, which is approximately $N(0.25, 0.0391)$ (by Equations 10.12 and 10.13). If $\hat{d}_A = \hat{p}_{1,A} - \hat{p}_{2,A} = 0.25$, determine the values of $\hat{p}_{1,A}$ and $\hat{p}_{2,A}$ that minimize the β error for $n_1 = 120$ and $n_2 = 185$.

10.7. The Field Research Group of the Indianapolis Branch Laboratory of the Day, Tee, and Tee Development Corporation is charged with not allowing any cost-reduced product to be put into the field before it has been shown to be at least as reliable as the product that it is replacing. Because all items that DT&T provides to its customers are on a lease-only basis, it is possible to obtain very good records in regard to the number of products of a particular type that have experienced any difficulty. The Illinois Branch of DT&T agreed to allow a field test of the new "Autocaller 2.0" to be conducted in their state. In order to perform a valid comparison, two samples of 1000 each of the old Autocaller 1.0 and the new Autocaller 2.0 are put into service in Illinois customers' homes. After 6 months, 52 of the old units and 66 of the new units have required repair operations. What may we conclude about the relative reliability of the two products? If we are interested in being able to detect a difference of 0.75%, what is your opinion of the sample sizes that were used?

10.8. The viscosity of oil after extended use in very demanding environments is extremely important to the Borchah Automobile Racing Team. To help determine which of two brands of oil to select, the team decides to try each brand of oil in 20 of the team's racing cars. Each car is run two times under identical competitive racing conditions for the same distance, once with one of the two oil brands and again with the other. To avoid any wear-related bias, the brand of oil used first is randomly assigned so that exactly half of the cars use one of the two brands of oil in their first runs. After standard viscosity tests are run, the following data are obtained from the cars.

	Brand of Oil			Brand of Oil	
Car	**Pegasus Supreme**	**Trekkzon Z-80**	**Car**	**Pegasus Supreme**	**Trekkzon Z-80**
1	27.4	22.6	11	25.4	24.6
2	23.4	25.5	12	26.5	27.2
3	26.1	25.8	13	22.1	22.6
4	24.3	23.3	14	28.8	27.1
5	27.3	25.5	15	29.4	27.2
6	24.2	23.8	16	27.9	28.1
7	30.1	29.4	17	22.5	22.4
8	23.6	22.3	18	25.7	24.5
9	26.6	26.1	19	32.3	27.9
10	24.3	22.9	20	26.2	24.9

Is there a significant difference between the two brands of oil? What difference between the two brands can we expect to be able to detect with this data set?

10.9. The purity of a certain chemical catalyst is very important when used in a gene-splicing procedure. Twenty-two samples of the catalyst from a synthetic production process and 34 samples from the traditional organic process are analyzed for impurities. The following data (total impurities in parts per million) have been generated from this analysis.

Production Process

Synthetic				Organic				
1.5	1.5	1.4	1.6	2.0	2.0	1.8	0.9	1.7
1.1	1.7	1.4	1.7	1.6	1.7	1.5	1.9	2.0
1.4	1.4	1.1	1.7	1.8	1.6	1.8	1.7	1.6
1.5	1.2	2.0	1.6	1.7	2.1	1.5	1.7	2.0
1.1	1.3	1.5	1.7	1.8	1.7	1.5	1.6	1.6
1.9	1.0			1.7	1.7	1.4	1.5	1.7
				1.6	2.0	1.9	2.1	

Is there a significant difference in the amount of impurities found in the two catalysts? Do you feel that the sample sizes are large enough? Why or why not?

10.10. Two bag-filling machines are used by the Bilsdury Flour Company. There are two points of concern in regard to the machines. First, the average quantity of flour that is put into the bag should be the same for each machine, and, second, the amount of variability should not differ between the two machines. As a start, 10 (nominal) 50-pound bags of flour from each machine's output are selected at random, and the amount of flour in each bag is carefully measured.

Filling Machine

No. 1					No. 2				
49.37	49.88	49.91	49.33	49.77	49.68	49.75	50.12	48.99	49.67
50.52	50.14	49.81	49.75	50.50	49.09	50.30	50.14	50.44	49.72

What may we conclude from these data about the relative averages and variabilities of each filling machine? Is the sample adequate? Why or why not?

10.11. One of the expensive and troublesome things associated with providing Millsville City with electrical power is the federal requirement that no excess sulphur emissions are allowed from the municipal coal-burning power plant. The easiest way to deal with this requirement is to purchase coal with a minimum amount of sulphur content. For various economic reasons, only two suppliers of coal can be considered. Over the past nine months, 37 batches from the Great Northern Coal Company and 47 batches from the Ticonderoga Coal Company have been analyzed to determine sulphur content. These analyses have generated the following sample averages and sample standard deviations. Should either coal supplier be preferred over the other?

	Average	Std. Dev.
Great Northern	7.38	0.62
Ticonderoga	7.7	1.03

10.12. The thicknesses of wax coatings, which are applied to the inside and outside of a paper bag used for doughnuts, are known to be statistically independent random variables. We have reason to believe that there is greater variation in the amount of wax applied to the inner surface of a bag than to the outer surface. A sample of 60 bags has yielded the following information.

	Outer Surface	Inner Surface
$\overline{x}$	0.95	0.65
s^2	0.043	0.052

Determine whether the variability in the amount of wax on the inner surface is greater than the variability associated with the outer surface.

10.13. Construct the operating characteristic (o.c.) curve associated with Exercise 10.8. (Use $\alpha = 0.075$.) How likely are you to detect a 1-unit difference between the means of the two oils?

10.14. On occasion, the assumption of Section 10.2 that $\sigma_1 = \sigma_2$ may not be supportable. If you need to compare two means of normally distributed populations when the samples are small and the true values of the variances are not known but are known to be *unequal*, the following procedure may be used for an approximate test of the hypothesis $H_0 : \mu_1 = \mu_2$.

$$T = \frac{\overline{X}_1 - \overline{X}_2}{\sqrt{(\sigma_1^2/n_1) + (\sigma_2^2/n_2)}}$$

is distributed approximately accordingly to a t distribution with

$$n = \frac{[(s_1^2/n_1) + (s_2^2/n_2)]^2}{[(s_1^2/n_1)^2/(n_1 + 1)] + [(s_2^2/n_2)^2/(n_2 + 1)]} - 2 \text{ degrees of freedom}$$

Unfortunately, the distribution of T when H_0 is not true is unknown. This precludes the computation of the β error and thus the construction of an o.c. curve. Rework Exercise 10.2 under the assumption that $\sigma_1 \neq \sigma_2$. Does your conclusion change? How do you feel about the relative accuracy of the two tests of hypotheses?

10.15. Ace Explosive Demolition is understandably concerned with consistent and reliable burn times for the fuses that they use in their work. Their two suppliers of fuses both deliver their fuses on spools that contain 10,000 feet of fuse. Thirty-two fuses of 30 feet in length are prepared from each supplier's spool. The fuses are then lit, and each fuse's elapsed burn time is carefully measured. Supplier A's sample average and sample standard deviation are 30.62 seconds and 0.62 seconds, respectively. Supplier B's sample average and sample standard deviation are 31.37 seconds and 0.43 seconds, respectively. What may we conclude about the relative differences and variabilities of the two suppliers' fuses?

10.16. Two types of corrosion-resistant coatings are compared in a laboratory. Coating A has been in use for quite some time and has given reliable and adequate service. Coating B has not been used outside the laboratory but is cheaper and somewhat easier to apply. Seventeen specimens are coated with A and 25 specimens are coated with B. The normalized wear indices of the specimens give the following results after all specimens have been subjected to identical corrosive conditions: $\overline{x}_A = 5.73$, $\overline{x}_B = 5.58$, $s_A = 0.36$ and $s_B = 0.31$. Based on this evidence, should coating B replace coating A in actual use?

CHAPTER

11

FITTING EQUATIONS TO DATA, PART I: SIMPLE LINEAR REGRESSION ANALYSIS AND CURVILINEAR REGRESSION ANALYSIS

The following concepts and procedures form the major ideas of this chapter.

- Be able to cite four important purposes of fitting equations to data
- Know five properties that a good method of fitting equations to data should possess
- Understand the *assumptions* that are associated with *simple regression analysis* and be able to state the three most important assumptions
- Understand that the *mathematical model* for simple regression analysis is the equation of a straight line augmented by the presence of a *random error term*. Be able to write that model and identify the *parameters*, the *dependent* variable, the *independent* variable, and the error term
- Know how to handle points that appear to be *outliers*. Understand that such points *cannot* be arbitrarily ignored or deleted from a data set
- Understand the graphical interpretation of the two parameters that are estimated in simple linear regression analysis
- Know what is meant when an equation is "linear in the parameters"

- Understand how the *normal equations* are developed
- Be able to use the computer program for simple linear regression to fit a straight line to a given set of data in two variables and be able to interpret the results from that program
- Understand why it is important to identify correctly the independent variable, x, and the dependent variable, Y, in a linear regression analysis
- Understand the meanings of the corrected sum of squares for y, the sum of squares of the residuals, and the sum of squares due to the regression and know how they are related to the *multiple correlation coefficient squared*, r^2
- Be able to interpret the meaning of a specific value of r^2
- Be able to construct a hypothesis test about β_1 using both the F distribution and the t distribution
- Be able to construct a univariate confidence interval on each of the parameters β_0 and β_1, on the expected value of Y at a selected value of x, and on a future value of Y at a specific value of x
- Be able to construct and interpret a *joint confidence region* for β_0 and β_1
- Be able to interpret plots of both the data and the residuals to assist in finding possible inadequacies in the fitted line

11.1 INTRODUCTION

There are many purposes for fitting equations to data sets. Among the more important purposes are the following.

1. To summarize and condense a data set in order to obtain predictive formulas
2. To reject or confirm a proposed mathematical relation
3. To assist in the search for a mathematical relation
4. To perform a quantitative comparison of two or more data sets

A reliable method of fitting equations should do the following five things.

1. Let all the data contribute to the estimation of the parameter values present in the model
2. Require a reasonably small number of parameters
3. Provide estimates of random errors present in the data and indicate how those errors affect the use of the resulting model
4. Provide methods of identifying systematic deviations from the fitted equation
5. Provide a "measure of usefulness" of the fitted equation for predictive purposes

In this chapter, we consider *simple* linear regression analysis (SLR); i.e., we work with one *dependent variable* and one *independent variable*. The independent variable, x, is under our control in the sense that we can set the value of x to whatever we wish (within the physical limitations imposed by the process under study). The dependent variable, Y, is a random variable that attains its realization, y, in response to the selected setting of x. For this reason, Y is also called the *response variable*. We assume that a data set of n pairs of values (x_i, y_i), $i = 1, \ldots, n$, has been obtained from the process that is under study. Let us consider the mathematical approach associated with SLR.

11.2 THE MATHEMATICAL MODEL FOR SIMPLE LINEAR REGRESSION ANALYSIS

In applying SLR, we assume the following.

1. The variables Y and x are theoretically related to one another by the equation of a straight line,

$$E[Y] = \beta_0 + \beta_1 x \tag{11.1}$$

2. The data set is typical of the behavior of the process under study.
3. The Y_i, $i = 1, \ldots, n$, are pairwise statistically independent of one another.
4. The Y_i are random variables possessing the same variance, σ^2.
5. Each data pair is "good"; i.e., there are no "outliers" that have arisen under unusual, accidental, or careless circumstances.
6. The uncontrolled random error associated with the Y_i is governed by a normal distribution.

Although each of these assumptions is important, the first three are more important in practical applications of SLR.

Equation 11.1 defines the relationship that joins x and Y, but it is *not* the mathematical model for SLR because Equation 11.1 does not acknowledge the presence of the uncontrolled and unremovable error that is present in any experimental situation. Let us form the *mathematical model for SLR* by modifying Equation 11.1 with the addition of the uncontrolled error, ε_i:

$$Y_i = \beta_0 + \beta_1 x_i + \varepsilon_i \tag{11.2}$$

where ε_i represents a random variable uniquely associated with Y_i for each $i = 1, \ldots, n$. The ε_i variables are identically distributed $N(0, \sigma)$ random variables and are pairwise statistically independent; i.e., each ε_i is statistically independent of any other ε_j as long as $i \neq j$. The parameters β_0 and β_1 are constants whose "true" values are unknown and must be estimated from the data. From a theoretical view, each x_i is assumed to be a known constant whose value is set to a predetermined value (with *no variability or error*). Once x_i is set, the realization of the dependent (response) variable, y_i, is observed and recorded with the value of x_i.

From a practical viewpoint, it is unreasonable to believe that each x_i is known *exactly*, with no error. However, the SLR approach requires that it be known with reasonable accuracy and that any uncertainty associated with the values of the x_i be much smaller than the inherent variability associated with the Y_i. This restriction *prohibits* the application of SLR to the situation in which the independent and dependent variables are both random variables that are simultaneously observed. The proper analysis of such a situation is beyond the scope of this text.

The right-hand side of Equation 11.2 possesses only one random variable, ε_i. The three other quantities are constants. Since ε_i is distributed $N(0, \sigma)$, both the right-hand side of Equation 11.2 and Y_i (the *left* side of the equation) are random variables distributed as $N(\beta_0 + \beta_1 x_i, \sigma)$.

Figure 11.1 presents a graphical interpretation of the theoretical content of Equation 11.2. Notice that the *expected value* (mean) of Y_i, $E(Y_i)$, varies directly, in a linearly proportional manner, with the value of x_i. However, the variance of Y_i remains constant, no matter what value x_i assumes. From a "three-dimensional" viewpoint, the probability distributions of the Y_i, the $f(y_i)$, form a *contour* of identical normal density functions, differing only in their means, which follow the locus given by Equation 11.1.

Since our model has the form of a straight line in two dimensions, β_0 is the y intercept and β_1 is the slope of the line. Unlike $Y_i = \beta_0 + \beta_1 x_i + \beta_2 x_i^2 + \varepsilon_i$—which is *linear* in the parameters and which is of *second order in the* x_i—and unlike $Y_i = \beta_0 e^{(\beta_1 x_i)} + \varepsilon_i$, which is nonlinear in the parameters, Equation 11.2 is *linear* in the parameters and is *first order in the* x_i.

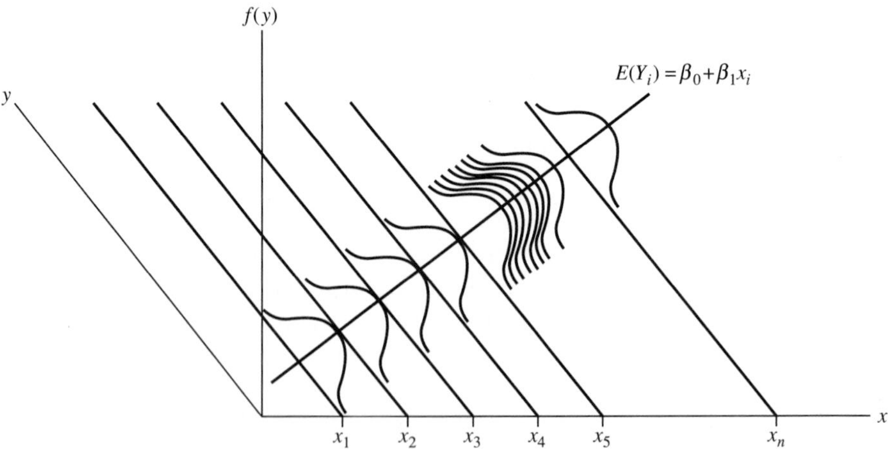

FIGURE 11-1
A graphical interpretation of $Y_i = \beta_0 + \beta_1 x_i + \varepsilon_i$.

11.3 OBTAINING THE BEST ESTIMATES OF β_0 AND β_1

Now that we have a mathematical model for SLR, let us investigate the approach that will yield the best unbiased estimates of β_0 and β_1. Since the errors associated with the Y_i are independent, have a mean of 0, and possess the same variance, it may be shown that these best estimates, b_0 and b_1, are found by the *method of least squares*. Substituting y_i, b_0, and b_1 for Y_i, β_0, and β_1 in Equation 11.2, we may solve for the *residual*, e_i, our estimate of ε_i. This process yields

$$e_i = y_i - (b_0 + b_1 x_i) \tag{11.3}$$

The parameter values that result from the minimization of the sum of the squared residuals, b_0 and b_1, are often called the *least squares estimates* of β_0 and β_1. Thus we want to

$$\text{minimize} = \sum_{i=1}^{n} e_i^2 = \sum_{i=1}^{n} (y_i - b_0 - b_1 x_i)^2 \tag{11.4}$$

Taking the partial derivative of Equation 11.4 with respect to b_0 and b_1 yields, respectively,

$$-2\sum_{i=1}^{n}(y_i - b_0 - b_1 x_i) = 0 \quad \text{and} \quad -2\sum_{i=1}^{n} x_i(yi - b_0 - b_1 x_i) = 0$$

Multiplying both equations by -0.5 and expanding their sums term by term yield $\sum y_i - \sum b_0 - \sum b_1 x_i = 0$ and $\sum x_i y_i - \sum b_0 x_i - \sum b_1 x_i^2 = 0$. With expansion and rearrangement, we find

$$\sum_{i=1}^{n} y_i - n b_0 - b_1 \sum_{i=1}^{n} x_i = 0 \quad \text{and} \quad \sum_{i=1}^{n} x_i y_i - b_0 \sum_{i=1}^{n} x_i - b_1 \sum_{i=1}^{n} x_i^2 = 0$$

These last two equations are the *normal equations*, which, on simultaneous solution, yield the least squares estimates. Solving for b_0 in the first normal equation yields

$$b_0 = \sum \frac{y_i}{n} - b_1 \sum \frac{x_i}{n} = \overline{y} - b_1 \overline{x} \tag{11.5}$$

Substituting this result in the second normal equation and solving for b_1 yield

$$b_1 = \frac{\sum x_i y_i - (\sum x_i \sum y_i)/n}{\sum x_i^2 - (\sum x_i)^2/n} = \frac{\sum (x_i - \overline{x})(y_i - \overline{y})}{\sum (x_i - \overline{x})^2} \tag{11.6}$$

Equations 11.5 and 11.6 give the best unbiased estimates of β_0 and β_1. From that viewpoint, Equation 11.7 gives us the best value of $\hat{y}_i$, the predicted value of Y at $x = x_i$:

$$\hat{y}_i = b_0 + b_1 x_i \tag{11.7}$$

It is useful in later sections to note that substituting Y_i, the random variable, everywhere for the realized value, y_i, in the *estimates* given in Equations 11.5, 11.6, and 11.7 also provides us with the following minimum variance unbiased *estimators* (random variables) associated with b_0, b_1, and $\hat{y}_i$.

$$\hat{\beta}_0 = \overline{Y} - \hat{\beta}_1\overline{x} \tag{11.5a}$$

$$\hat{\beta}_1 = \frac{\sum(x_i - \overline{x})(Y_i - \overline{Y})}{\sum(x_i - \overline{x})^2} \tag{11.6a}$$

$$\hat{Y}_i = \hat{\beta}_0 + \hat{\beta}_1 x_i \tag{11.7a}$$

Example 11.1. As an illustration of the use of these equations and the program that performs SLR, consider the data set given in Table 11.1.

Applying Equations 11.5 and 11.6 to the data of Table 11.1, the program obtains (with $\overline{x} = 1196.52$ and $\overline{y} = 852.48$)

$$b_1 = \frac{(1148 - 1196.52)(724 - 852.48) + \cdots + (875 - 1196.52)(580 - 852.48)}{(1148 - 1196.52)^2 + \cdots + (875 - 1196.52)^2}$$

$$= 0.97$$

and

$$b_0 = 852.48 - (0.96982743)(1196.52) = -307.94$$

A *scatter plot* of the data in Table 11.1 is presented in Figure 11.2. The program provides a scatter plot with the fitted line superimposed, as shown in Figure 11.3. As we see later in the chapter, it is always important to construct a scatter plot when there is only one independent variable. In this special case, we may actually *observe* the relation that joins x and Y and determine graphically if our assumption of a straight-line relation has any practical validity. Figures 11.2 and 11.3 show clearly that the data of Table 11.1 do have an underlying linear relation.

When we perform a fit using x as the independent variable, we are performing a *regression of Y on x*; i.e., we are predicting the mean of Y *given* the value of x. It is important to identify the independent and dependent variables correctly, especially when using a computer program to perform your analysis.

Example 11.2. Suppose that you reverse your data set and input the y_i as the x_i values and the x_i as the y_i. If you make this mistake for the data of Table 11.1, the computer treats the y_i as the independent variable and x_i as the dependent variable.

TABLE 11.1
An example data set

Datum	x	y	Datum	x	y	Datum	x	y
1	1148	724	9	867	550	17	1357	1007
2	1638	1293	10	1158	870	18	1405	978
3	1678	1296	11	1082	669	19	1127	849
4	1292	925	12	907	517	20	1073	670
5	1422	1078	13	752	495	21	1308	953
6	1285	948	14	1115	692	22	812	497
7	1152	893	15	1307	1014	23	1260	798
8	1357	1077	16	1528	1282	24	1008	657
						25	875	580

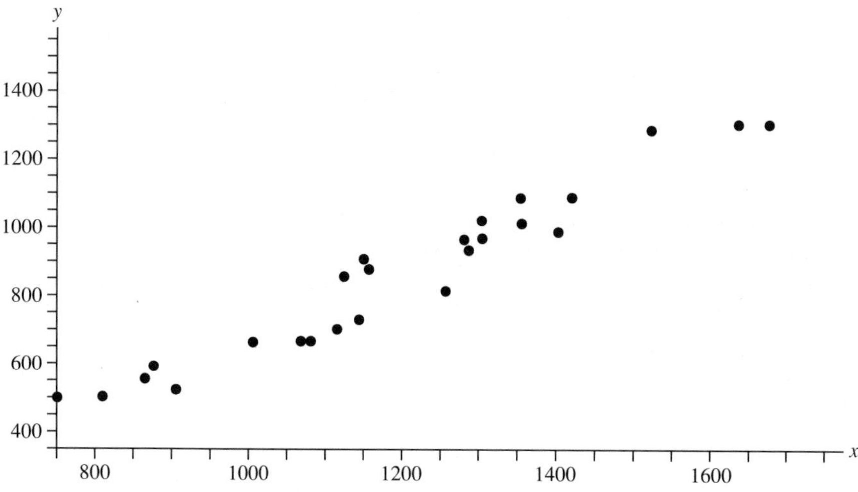

FIGURE 11-2
A scatter plot of the data of Table 11.1.

As a consequence of this error, the program yields

$$\hat{b}_0 = \sum \frac{x_i}{n} - \hat{b}_1 \sum \frac{y_i}{n} = \overline{x} - \hat{b}_1 \overline{y} = 370.37$$

and

$$\hat{b}_1 = \frac{\sum (x_i - \overline{x})(y_i - \overline{y})}{\sum (y_i - \overline{y})^2} = 0.97$$

These values imply that

$$x_i = \hat{b}_0 + \hat{b}_1 y_i = 370.37 + 0.97 y_i \tag{11.8}$$

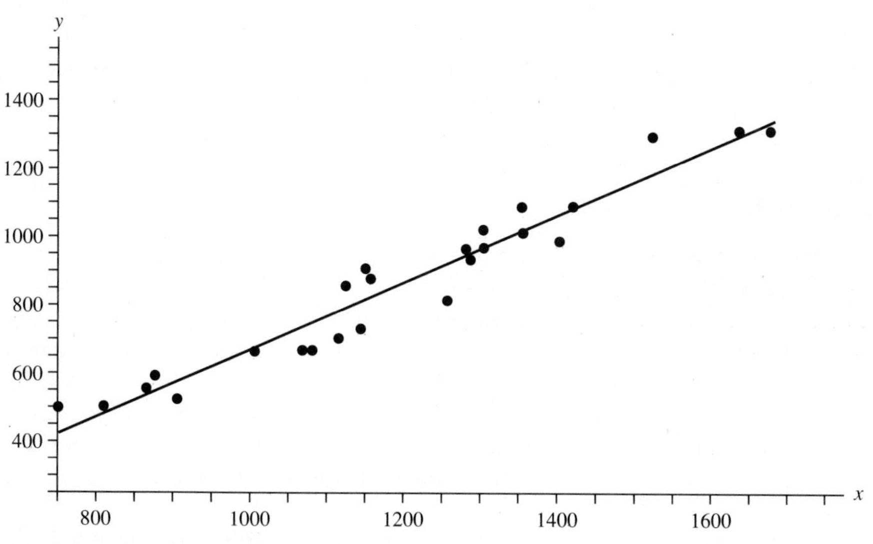

FIGURE 11-3
The scatter plot with the fitted line superimposed.

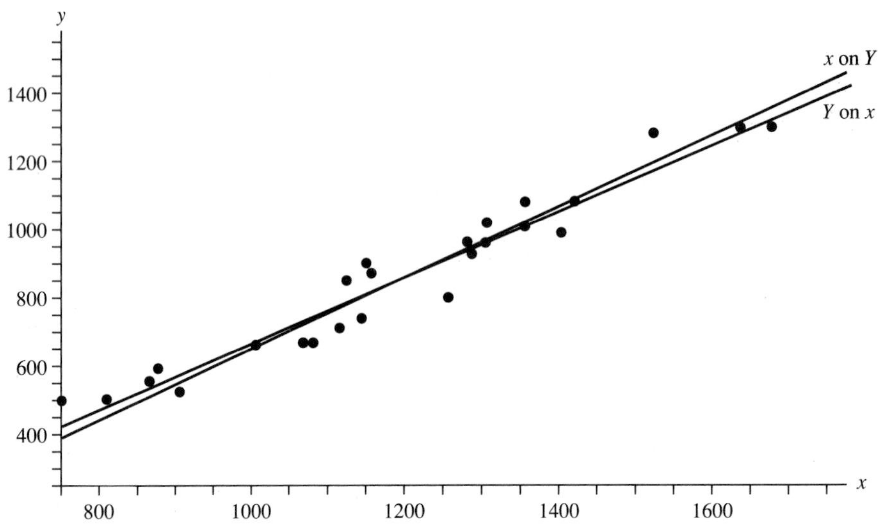

FIGURE 11-4
The regressions of Y on x and x on Y for the data of Table 11.1.

In order to compare with our earlier result, when the variables were correctly identified, we solve for y in Equation 11.8 and obtain $\hat{y}_i = -381.825 + 1.0309x_i$. This is markedly different from the correct equation, $\overline{y}_i = -307.94 + 0.97x_i$.

These results imply that the regression of "Y on x" is not necessarily the same as the regression of "x on Y." Indeed, the *only* time when the two regressions will yield the same equation for a line is when the data lie *perfectly* on a straight line.

Figure 11.4 shows both lines superimposed on the scatter plot to illustrate the difference. The reason that the two lines are so close together is that the data lie so close to a straight line. In general, the only thing of which we can be assured is that *both* lines will pass through the *centroid* of the data, $(\overline{x}, \overline{y})$.

11.4 THE MULTIPLE CORRELATION COEFFICIENT SQUARED, r^2

Up to this point we have been considering only *point estimates* of the true but unknown parameter values β_0 and β_1. Further, we have considered only a point estimate of the "true line." With our discussions in Chapter 8, we are aware of the shortcomings of point estimates as opposed to interval estimates. Interval estimates of any parameter assist us in determining how *precise* our knowledge is about that parameter. One approach to overcoming the implicit objections to providing only point estimates of the parameters of the fitted line is to develop the general form of the $1 - \alpha$ confidence intervals for both β_0 and β_1. We do that later in the chapter.

Another approach is to develop a measure of goodness for the fitted line viewed as a single entity. Indeed, this might be a superior approach, since our fitted equation embraces both parameters, β_0 or β_1, as components of a greater whole. With this in mind, consider the partition in Equation 11.9 of the *corrected sum of squares of the* y_i, also known as the *total sum of squared deviations of the* y_i *about* $\bar{y}$ (SST). (Although this equation may be proven to be true, the proof is cumbersome, adds nothing to the content of the discussion, and will not be presented here.)

$$\sum(y_i - \bar{y})^2 = \sum(y_i - \hat{y}_i)^2 + \sum(\hat{y}_i - \bar{y})^2 \quad \text{or} \quad \text{SST} = \text{SSE} + \text{SSR} \qquad (11.9)$$

Let us consider the two terms on the right-hand side of the equation. The first term on the right is the sum of the e_i^2. This first term is known by three synonymous names: the *sum of squares about the regression* (line), *sum of squares of the residuals*, and *sum of squares of the errors* (SSE). Figure 11.5 pictures the general relationships between the fitted regression line, the y_i, and the residual. Since $e_i = y_i - \hat{y}_i$, the specific residual selected in Figure 11.5 would have a value less than zero. Notice further that the distance from the observed y_i to the regression line is measured *parallel to the y axis*. The *second term* on the right side of Equation 11.9 is the sum of squared deviations of the $\hat{y}_i$ about $\bar{y}$. It is most often called the *sum of squares due to the regression* (SSR). Notice that this second term is identical to the left-hand side except that the deviations are not measured relative to the observed values but rather to the fitted values.

If we divide SSR by SST, we obtain the *coefficient of multiple determination*, which is also known as the *multiple correlation coefficient squared*, r^2:

$$r^2 = \frac{\sum(\hat{y}_i - \bar{y})^2}{\sum(y_i - \bar{y})^2} = \frac{\text{SSR}}{\text{SST}} = 1 - \frac{\text{SSE}}{\text{SST}} \qquad (11.10)$$

Suppose that $r^2 = 1$. By Equation 11.9, this would imply that $\sum(y_i - \hat{y}_i)^2 = 0$, which would mean that, *in every case*, $y_i = \hat{y}_i$—i.e., *a perfect fit*. In general, r^2 tells us the proportion of SST that is *explained* by the fitted equation. From Equation 11.9, this means that the SSE is the proportion of SST that is *not explained* by the model.

$$s_Y^2 = \frac{\sum(y_i - \bar{y})^2}{n - 1} = \frac{\text{SST}}{n - 1} \qquad (11.11)$$

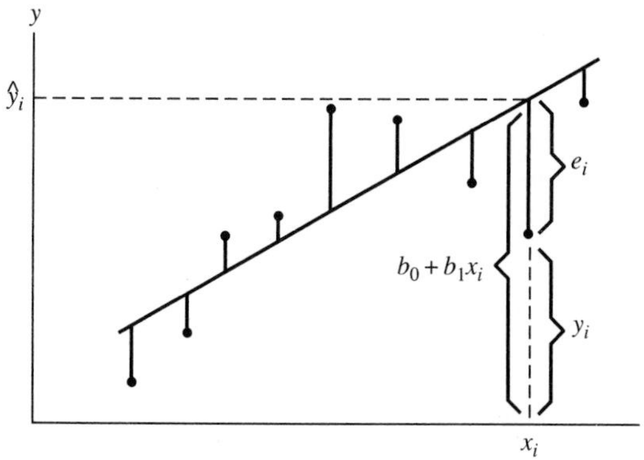

FIGURE 11-5
The relationships between the fitted line, the y_i, and the residual.

is the *unconditional estimate of the variance of* Y; i.e., s_Y^2 is the appropriate estimate of the variance if you are given only the y_i values and use the methods of Chapter 6.

$$\frac{\sum(y_i - \hat{y}_i)^2}{n-2} = \frac{\text{SSE}}{n-2} = s_{Y|x}^2 \tag{11.12}$$

is the *conditional estimate of the variance of* Y *given knowledge of the associated paired values of* x. It is also our best estimate of the true but unknown value of σ^2, the variance of our uncontrolled error as defined in Equation 11.2 and pictured in Figure 11.1.

Example 11.3. In obtaining the above items for the data of Table 11.1, the program first uses Equation 11.7 to compute the values of $\hat{y}_i$ for each value of x_i. It then uses Equation 11.9 to obtain SST = 1,451,542, SSR = 1,364,264, and SSE = 87,278.47. These values then are used in Equations 11.11 and 11.12 to obtain s_Y^2 = 60,480.93, $s_{Y|x}^2$ = 3794.72, and r^2 = 0.939. Thus, for our example, we see that about 94% of SST and thus of the variation, s_Y^2, is explained by the fitted model.

11.5 A HYPOTHESIS TEST FOR THE SIGNIFICANCE OF THE FITTED LINE

From Equation 11.9, we see that the effect of *successfully* fitting the model of Equation 11.2 to a data set has the effect of explaining, or "filtering," much of the unexplained variability of the Y_i. This reduces the unexplained error to a minimum and allows us to predict and estimate values of Y (given an x_i) in a much more accurate and precise fashion. Figure 11.6 presents the idea of this *linear filter model* in a schematic fashion.

The partition of Equation 11.9 has given us one *measure of goodness* for our fitted equation. Let us see how it can be used to construct a proper statistical hypothesis test for whether there is a relation between x and Y; i.e., Is there any reason to perform this analysis?

Substituting Equation 11.5 into Equation 11.7, we obtain $\hat{y}_i = \bar{y} - b_1\bar{x} + b_1 x_i$, which yields $\hat{y}_i - \bar{y} = b_1(x_i - \bar{x})$. Squaring both sides and summing over $i = 1, \ldots, n$ yield

$$\sum(\hat{y}_i - \bar{y})^2 = b_1^2 \sum(x_i - \bar{x})^2 \tag{11.13}$$

So we see that SSR may be reexpressed in terms of a single function of the y_i, namely, b_1 [times the *constant* $\sum(x_i - \bar{x})^2$]. For this reason, SSR possesses only a single degree of freedom. Since we know that SST has $n - 1$ degrees of freedom, we may infer (by subtraction) that SSE has $n - 2$ degrees of freedom. It may be shown that

$$E\left(S_{y|x}^2 = \frac{\sum(Y_i - \hat{Y}_i)^2}{n-2}\right) = \sigma^2$$

and that

$$E\left(\sum(\hat{Y}_i - \bar{Y})^2\right) = \sigma^2 + \beta_1^2 \sum(x_i - \bar{x})^2$$

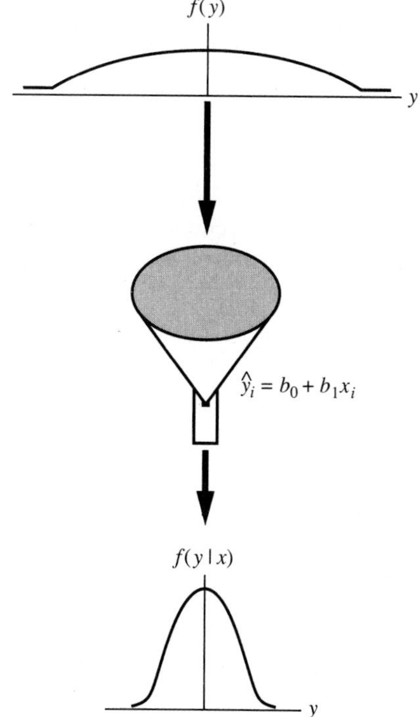

FIGURE 11-6
The linear filter model for simple linear regression analysis.

where $S^2_{Y|x}$ and $\sum(\hat{Y}_i - \overline{Y})^2$ are the estimators associated with $s^2_{Y|x}$ and SSR, respectively.

Further, because of the normality assumptions that we have made about the Y_i, it may be shown that

$$\frac{(n-2)S^2_{Y|x}}{\sigma^2} \sim \chi^2_{n-2} \qquad \text{and} \qquad \frac{\sum(\hat{Y}_i - \overline{Y})^2}{\sigma^2} \sim \chi^2_1 \qquad \text{if} \qquad \beta_1 = 0$$

Therefore, the random variable

$$F = \frac{\sum(\hat{Y}_i - \overline{Y})^2}{S^2_{Y|x}} \tag{11.14}$$

is distributed according to an $F_{1,n-2}$ distribution *if* $\beta_1 = 0$. We may use the point estimate $f = \text{SSR}/s^2_{Y|x}$ to test the following hypotheses.

H_0—There is no significant relationship between x and $Y(\beta_1 = 0)$.

H_A—There is a significant relationship between x and $Y(\beta_1 \neq 0)$.

Because the numerator of f contains b_1^2, the value of f will tend to be *large* when a significant relationship ($\beta_1 \neq 0$) is present. Therefore, we should reject H_0 when $f > F_{1,n-2,\alpha}$.

Using the relationship between the F and t distributions that was first presented in Chapter 4, an equivalent hypothesis test can easily be formed by using the relation $T_{n-2} = \sqrt{F_{1,n-2}}$; i.e., $t = \sqrt{f} = \sqrt{\text{SSR}/s^2_{Y|x}}$. We should reject H_0 if $t > t_{n-2,\alpha/2}$.

Example 11.4. Using the values of SSR and $s^2_{Y|x}$ that were obtained earlier for the data of Table 11.1, the program obtains $f = 1364264/3794.72 = 359.52$ and presents the fact that such an inflated f would be significant even if the value of α is set as small as 0.00005. Thus we may confidently reject the null hypothesis and conclude that a significant relation exists between x and Y.

Since $b_0 = \overline{y} - b_1\overline{x}$, if we conclude that $b_1 = 0$, this implies that $b_0 = \overline{y}$. This is consistent with our earlier work because, if no relation between x and Y exists, the best estimate of Y at *any* value of x is $\overline{y}$.

11.6 THE CONSTRUCTION OF CONFIDENCE INTERVALS ABOUT β_0, β_1, THE MEAN OF Y, AND A PREDICTED VALUE OF Y

From Equation 11.6(a), we know that

$$\hat{\beta}_1 = \frac{\sum(x_i - \overline{x})(Y_i - \overline{Y})}{\sum(x_i - \overline{x})^2} \tag{11.6a}$$

Expanding the numerator about the second term, we obtain

$$\hat{\beta}_1 = \frac{\sum(x_i - \overline{x})Y_i - \sum(x_i - \overline{x})\overline{Y}}{\sum(x_i - \overline{x})^2}$$

However, the second term in our expansion, $\sum(x_i - \overline{x})\overline{Y} = \overline{Y}\sum(x_i - \overline{x})$ is zero because $\sum(x_i - \overline{x})$ is always zero. This allows the following simpler expression to be written.

$$\hat{\beta}_1 = \frac{\sum(x_i - \overline{x})Y_i}{\sum(x_i - \overline{x})^2} \tag{11.15}$$

Earlier, we made the assumptions that all the Y_i are statistically independent and that the variance of Y_i at any value of x_i is σ_2. The following two facts are well known.

1. The variance of a sum of independent random variables is equal to the sum of the variances of the random variables that contribute to the sum; i.e., if $Q = R + S + T$, then $\sigma^2_Q = \sigma^2_R + \sigma^2_S + \sigma^2_T$ if R, S, and T are statistically independent random variables.
2. The variance of a constant times a random variable is equal to the constant squared times the variance of the random variable; i.e., if V is a random variable with variance equal to σ^2_V and k is a constant, the variance of the random variable $W = kV$ is $\sigma^2_W = k^2\sigma^2_V$.

From these assumptions and facts, we see that

$$\begin{aligned}\sigma^2(\hat{\beta}_1) &= \text{Var}(\hat{\beta}_1) = \sum \text{Var}\left(\frac{\sum(x_i - \overline{x})Y_i}{\sum(x_i - \overline{x})^2}\right) \\ &= \sum \frac{(x_i - \overline{x})^2\sigma^2}{(\sum(x_i - \overline{x})^2)^2} = \frac{\sigma^2}{\sum(x_i - \overline{x})^2}\end{aligned} \tag{11.16}$$

Taking the square root of the rightmost term in Equation 11.16, we obtain the standard deviation of $\hat{\beta}_1$,

$$\sigma(\hat{\beta}_1) = \frac{\sigma}{\sqrt{\sum(x_i - \overline{x})^2}} \tag{11.17}$$

The *estimated* standard deviation of $\hat{\beta}_1$, $s(\hat{\beta}_1)$, is obtained by substituting $s_{Y|x}$ for σ in Equation 11.17.

Looking at Equation 11.15 and realizing that the x_i and *any function* of the x_i are constants, we see that $\hat{\beta}_1$ is a sum of constants multiplied with normally distributed random variables, the Y_i. Since any constant multiplied by a normal random variable yields another normal random variable and any sum of normal random variables is a normal random variable, $\hat{\beta}_1$ is $N(\beta_1, \sigma(\hat{\beta}_1))$. From this, the $1 - \alpha$ confidence limits on β_1 are

$$b_1 \pm t_{n-2,\alpha/2} \frac{s_{Y|x}}{\sqrt{\sum(x_i - \overline{x})^2}} \tag{11.18}$$

If $(n - 2) \geq 30$, we may replace $t_{n-2,\alpha/2}$ with $z_{\alpha/2}$ without serious loss of accuracy.

From Equation 11.5(a), we know that $\hat{\beta}_0 = \overline{Y} - \hat{\beta}_1\overline{x}$. Therefore, $\text{Var}(\hat{\beta}_0) = \text{Var}(\overline{Y} - \hat{\beta}_1\overline{x})$. We also know that $\text{Var}(\overline{Y}) = \text{Var}(Y_i)/n = \sigma^2/n$ and $\text{Var}(\hat{\beta}_1) = \sigma^2/\sum(x_i - \overline{x})^2$. This implies that $\text{Var}(b_1\overline{x}) = \overline{x}^2\sigma^2/\sum(x_i - \overline{x})^2$. Since it may be shown that $\overline{Y}$ and $\hat{\beta}_1$ are statistically independent,

$$\text{Var}(\hat{\beta}_0) = \text{Var}(\overline{Y} - \hat{\beta}_1\overline{x}) = \frac{\sigma^2}{n} + \frac{x^2\sigma^2}{\sum(x_i - \overline{x})^2}$$

After several algebraic manipulations, this expression can be reduced to

$$\sigma^2(\hat{\beta}_0) = \text{Var}(\hat{\beta}_0) = \frac{\sigma^2 \sum x_i^2}{n\sum(x_i - \overline{x})^2} \tag{11.19}$$

Substituting $s_{Y|x}$ for σ in Equation 11.19 yields the sample estimate of $\text{Var}(\hat{\beta}_0)$, and the estimated standard deviation of $\hat{\beta}_0$, $s(\hat{\beta}_0)$, is obtained by taking the square root of Equation 11.19 after that substitution is made. The $1 - \alpha$ confidence limits on β_0 are

$$\beta_0 \pm t_{n-2,\alpha/2} s_{Y|x} \frac{\sqrt{\sum x_i^2}}{\sqrt{n\sum(x_i - \overline{x})^2}} \tag{11.20}$$

Example 11.5. Using Equations 11.17 and 11.19, the program finds, for the data of Table 11.1, $s^2(\hat{\beta}_0) = 3897.288$ and $s^2(\hat{\beta}_1) = 0.003$. Equations 11.18 and 11.20 imply the 95% confidence limits on β_0 and β_1 are, respectively, $(-437.1, -178.8)$ and $(0.86, 1.08)$.

Using techniques similar to those used above, the following $1 - \alpha$ confidence limits may be stated. For the mean of Y given a specific paired value x_k,

$$\hat{y}_k \pm t_{n-2,\alpha/2} s_{Y|x} \sqrt{\frac{1}{n} + \frac{(x_k - \overline{x})^2}{\sum(x_i - \overline{x})^2}} \tag{11.21}$$

For a future (or predicted) value of Y given a specific paired value x_k,

$$\hat{y}_k \pm t_{n-2,\alpha/2} s_{Y|x} \sqrt{1 + \frac{1}{n} + \frac{(x_k - \overline{x})^2}{\sum(x_i - \overline{x})^2}} \tag{11.22}$$

Equation 11.21 places a confidence interval on the *long-term average* of Y when x is set to a specific value, x_k, and Equation 11.22 places a confidence interval on a single *future* realization for a given x_k. The second kind of interval is appropriate when we are interested in whether a single future "trial" will fall between certain specification limits. Note that the limits of Equation 11.22 are wider than those of Equation 11.21. The single difference in the two expressions is the presence of an additional "1" in the square root term. This additional factor reflects the uncertainty associated with this single future trial and adds $s_{Y|x}^2$ to the variance associated with the interval.

Example 11.6. It is often informative to construct a *confidence band* on a future value of Y using Equation 11.22. Figure 11.7 shows the two-tailed confidence band on a single future value of Y at $\alpha = 0.05$ (constructed by the SLR program) as superimposed on Figure 11.3.

11.7 THE CORRELATION COEFFICIENT AND A JOINT CONFIDENCE REGION FOR β_0 AND β_1

Since we have the confidence intervals of both β_0 and β_1, it is tempting to use them as independent pieces of information. However, because $\hat{\beta}_0$ and $\hat{\beta}_1$ are *not necessarily* statistically independent, this may be *incorrect*.

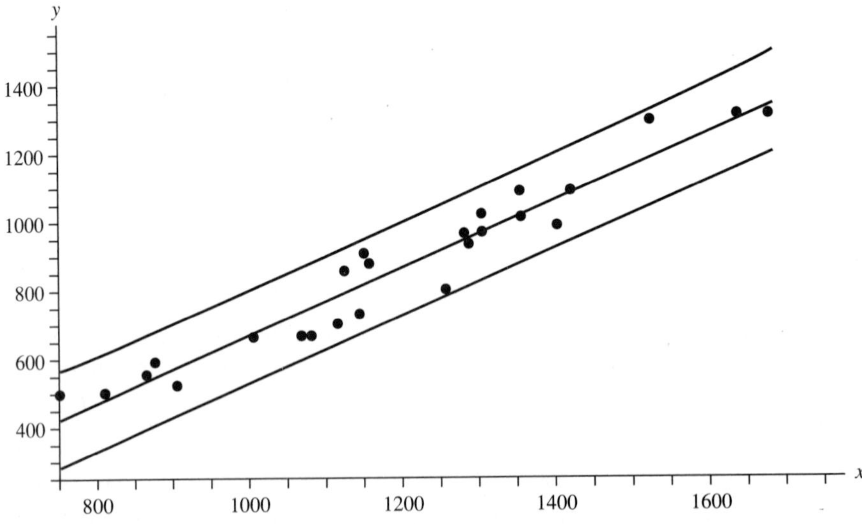

FIGURE 11-7
A confidence band on a future value of y.

As shown in Chapter 5, one way to determine if two *normal* random variables, V and W, are statistically independent is to compute their *correlation coefficient*, which is defined theoretically as

$$\rho = \frac{\sigma_{V,W}}{\sigma_V \sigma_W} = \frac{E((V - \mu_V)(W - \mu_W))}{\sqrt{\mathrm{Var}(V)\,\mathrm{Var}(W)}} \tag{11.23}$$

where $\sigma_{V,W} = E((V - \mu_V)(W - \mu_W))$ is the *covariance* of V and W. The sample estimate of ρ, r, is obtained by substituting the sample estimates $s_{V,W}$, s_V, and s_W of the theoretical quantities contained in Equation 11.23, where

$$s_{V,W} = \frac{\sum(v - \overline{v})(w - \overline{w})}{n - 1} \tag{11.24}$$

It may be shown that both ρ and r may assume any value from -1 to 1. If $\rho = 0$, there is *no correlation* between V and W. The farther ρ departs from 0, the more correlated V and W become, until V and W become *perfectly correlated* at either -1 or $+1$. If r is equal to -1 or $+1$, the paired values $(v_i,\ w_i)$ will lie exactly on a straight line. If $r = -1$ $(+1)$, the line on which the data pairs lay has a negative (positive) slope. (It is important to note that ρ and r are only linear measures of correlation. It is possible that V and W could be perfectly correlated in some nonlinear relationship and have $\rho = 0$.)

Example 11.7. For the data set of Table 11.1, the sample correlation between x and y is $r_{x,y} = \sqrt{r^2} = 0.969$. We would have expected the value of r to be positive and near 1, since the slope of the scatter diagram as given in Figure 11.2 is generally positive and is tightly bunched about the fitted line, as seen in Figure 11.3.

It may be shown that the covariance of $\hat{\beta}_0$ and $\hat{\beta}_1$ is

$$\mathrm{Cov}(\hat{\beta}_0, \hat{\beta}_1) = -\frac{\overline{x}\sigma^2}{\sum(x_i - \overline{x})^2} \tag{11.25}$$

By substituting $s_{Y|x}$ for σ, we obtain the sample estimate of the covariance between $\hat{\beta}_0$ and $\hat{\beta}_1$. By applying Equation 11.23 with the appropriate sample estimators for the variances of $\hat{\beta}_0$ and $\hat{\beta}_1$, the SLR program yields a sample estimate of the correlation between $\hat{\beta}_0$ and $\hat{\beta}_1$ equal to -0.98. What does this mean, in a practical sense, to us in the analysis of this data? The high negative correlation means that a low (high) value of $\hat{\beta}_1$ implies that it is very likely that the value of $\hat{\beta}_0$ will be high (low). This is best understood in the context of a *joint confidence region*. It may be shown that the joint confidence region for (β_0, β_1) is the interior of an *ellipse*, defined mathematically as

$$n(\beta_0 - b_0)^2 + 2n\overline{x}(\beta_0 - b_0)(\beta_1 - b_1) + \sum x_i^2(\beta_1 - b_1)^2 \le 2s^2_{Y|x}F_{2,n-2,\alpha} \tag{11.26}$$

Except for β_0 and β_1, we may obtain the values of everything in Equation 11.26. Thus, we can solve for a sufficient number of (β_0, β_1) points to draw the ellipse that forms the joint confidence region for β_0 and β_1.

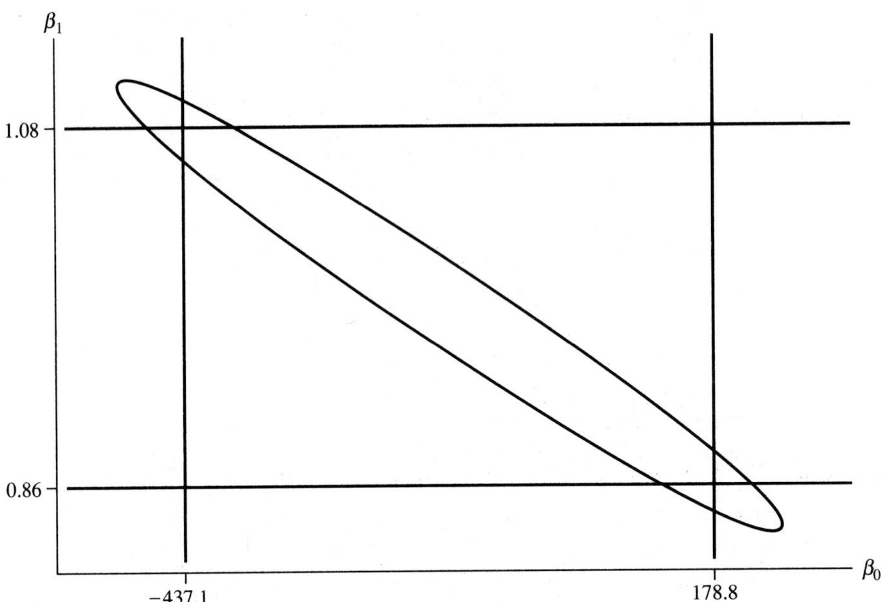

FIGURE 11-8
The joint confidence region for β_0 and β_1 with $\alpha = 0.05$ for the data of Table 11.1.

Example 11.8. Figure 11.8 presents the 95% confidence region for β_0 and β_1 for the data of Table 11.1 as constructed by the SLR program. If we constructed a large number of confidence regions, using Equation 11.26, we would expect that about 95% of them would contain the true values of β_0 and β_1. Notice that the ellipse is elongated in the direction of a negative slope in the $\beta_0 - \beta_1$ plane. The lines emanating from the β_0 and β_1 axes at the respective univariate confidence limits on β_0 and β_1 form the rectangle in the $\beta_0 - \beta_1$ plane that would be implied if one were to interpret incorrectly those univariate confidence intervals simultaneously.

From Figure 11.8 we see that, even though the simultaneous values of $\beta_0 = -440$ and $\beta_1 = 1.08$ are reasonably consistent with the data, simultaneous values of $\beta_0 = 160$ and $\beta_1 = 1.00$ are *extremely inconsistent.*

It is possible to structure your data at the outset to ensure that the covariance, and thus the correlation, between $\hat{\beta}_0$ and $\hat{\beta}_1$ is zero. From Equation 11.25, we see that all that needs to be done is to ensure that $\overline{x} = 0$. If the correlation between $\hat{\beta}_0$ and $\hat{\beta}_1$ is zero, the confidence region on β_0 and β_1 becomes an ellipse with its major axes parallel to the β_0 and β_1 axes.

If the confidence limits on β_0 and β_1 are of equal width, the confidence region becomes a circle. Notice that, even when the correlation between $\hat{\beta}_0$ and $\hat{\beta}_1$ is zero, the confidence region is not a rectangle. However, the approximation of the rectangle formed by the univariate confidence intervals is certainly closer when the correlation is zero. From this discussion, we see that knowledge of the correlation between the estimators $\hat{\beta}_0$ and $\hat{\beta}_1$ may be critical in some uses of simple linear regression.

One additional interpretation of the confidence ellipse can be made. If one takes each possible pair of values of the parameters as defined by the boundary of a confidence ellipse and plots (on the same graph) each of the lines associated with each parameter pair, then the part of the graph darkened by the lines would form two *envelopes*. These envelopes constitute the *confidence region for the line*. It may be shown that the boundaries of the envelopes forming the confidence region for the line are given by a pair of hyperbolas whose mathematical expression is

$$b_0 + b_1 x \pm \sqrt{2s_{Y|x}^2 F_{2,n-2,\alpha}\left(\frac{1}{n} + \frac{(x-\overline{x})^2}{\sum(x-\overline{x})^2}\right)} \tag{11.27}$$

Any line that does not lie wholly in the region defined by Equation 11.27 is improbable in the sense of the selected level of confidence.

11.8 GRAPHICAL METHODS OF INVESTIGATING DATA STRUCTURE IN SLR

Now that we have covered most of the mathematical material associated with simple linear regression, let us consider some graphical methods both to detect pathological structures in the data set and to detect if there is significant "lack of fit" in the model that has been fitted to the data. As mentioned earlier in the chapter, *before* any great effort is expended in trying to fit an equation to a data set that has only a single independent variable, one should always examine a plot, or *scatter diagram*, of the data. Here are some compelling reasons why this is so.

Consider the scatter diagram in Figure 11.9. Data of this form could easily have been taken in five runs, or "batches," such that all y values at each particular x are taken without resetting the equipment and without going to a different value

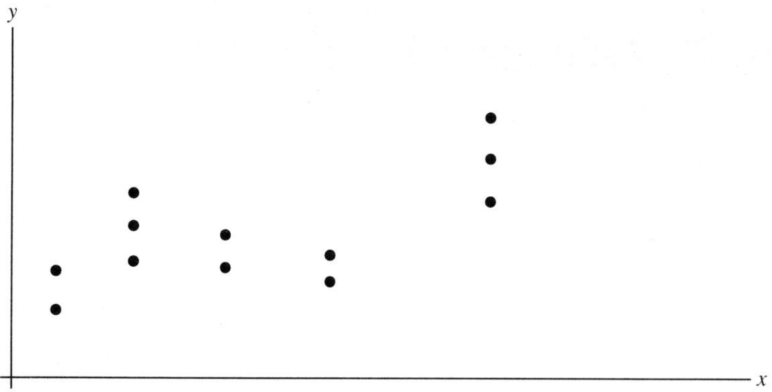

FIGURE 11-9
An example scatter plot showing batches at selected x values.

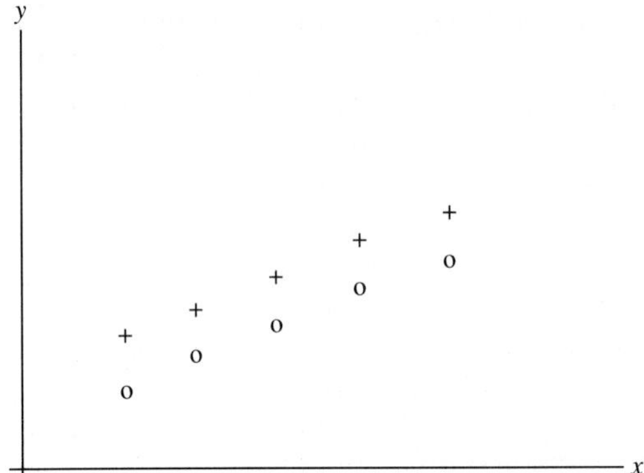

FIGURE 11-10
An example scatter plot showing runs on a sequence of x values.

of x. This obviously would cause dependencies among the y values at each x value, a clear violation of our assumption in regard to the pairwise statistical independence of the Y_i. If the data are taken in runs, we would *not* have $n - 2 = 10$ degrees of freedom for our estimation of β_0 and β_1. Rather, because our overall data contains only 5 independent sets of data, we would have only $5 - 2 = 3$ degrees of freedom for our estimation procedures.

Note that there is nothing wrong, when you have the luxury of taking extra data values, to take one or more "repeats" at the same value of x. Indeed, such repeats can be used to construct a *model-independent* "pooled" estimate of $\sigma^2_{Y|x}$. However, you can do this only when steps are taken to ensure that the repeats are statistically independent of other observations in the data set.

Consider Figure 11.10. In obtaining this data set, it is very possible that the data taker made a run through a sequence of values of x to get the +'s and then another run of exactly the same form to get the o's. If that incorrect data-taking method was actually used, there is only 1 degree of freedom available to estimate the y intercept, β_0.

In general, the experimenter may avoid problems like those illustrated in Figures 11.9 and 11.10 by randomizing the order that the data are taken. Consider the example scatter plot in Figure 11.11. What point will weigh most heavily on the results of the regression analysis?

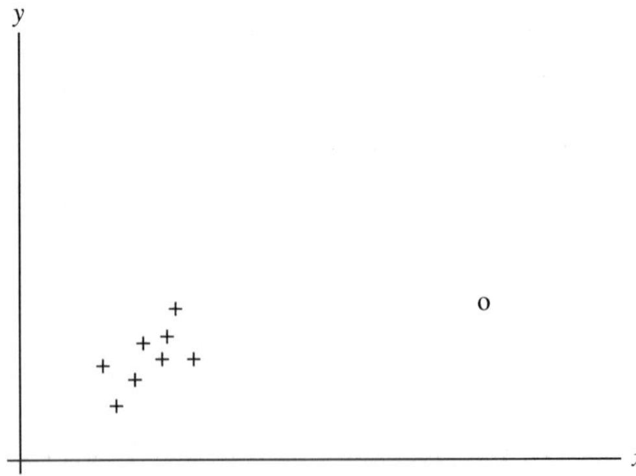

FIGURE 11-11
A scatter plot with an outlying observation.

The single "outermost" data point, o, at the higher value of x is *controlling the regression*. It probably has as much influence on the fitted line as all eight of the points at the lower values of x; the reason is that our "best" line minimizes the sum of the squared distances from the line to the data points. Because the eight data values are so close together, several lines can be drawn through them and about the same penalty, in the form of the sum of the squared differences, would be incurred. However, few lines can be drawn that would not incur a massive penalty associated with point o.

Indeed, we would expect in general that outermost points like the one in Figure 11.11 would have very small residuals and would therefore not stand out in the analysis unless the scatter plot is examined. This possible, bad data structure emphasizes the need to design the data-gathering procedure prior to taking the data so that such an unbalanced situation is avoided. Whenever possible, data should be taken in a random way and data points should be taken over the entire relevant range of x.

Another kind of data problem should be considered prior to and during the analysis. There is a possibility, despite our assumption to the contrary, that a "bad" data point, or *outlier*, will become part of our data set. There are a great number of ways such a point can be generated, ranging from a simple clerical error to a contaminated physical sample or an uncalibrated measuring device. Consider the scatter plots in Figure 11.12. If the circled data points are bad data and they are not deleted, one can see that they will cause markedly different fits to result with an increase in $s_{Y|x}$ far above the true value of $\sigma_{Y|x}$. If the circled data points are *not* bad, their presence certainly merits additional investigation to explain why they have assumed values so radically different from those that might have been expected.

One *strong point* must be made: A data point should not be deleted simply because it does not fit into the general scheme. There should be a verified reason to call the sample point a bad datum before deleting it. There is an additional reason for investigating apparent outliers: Many discoveries have been the direct results of new knowledge and new factors that were discovered during the investigation of outliers.

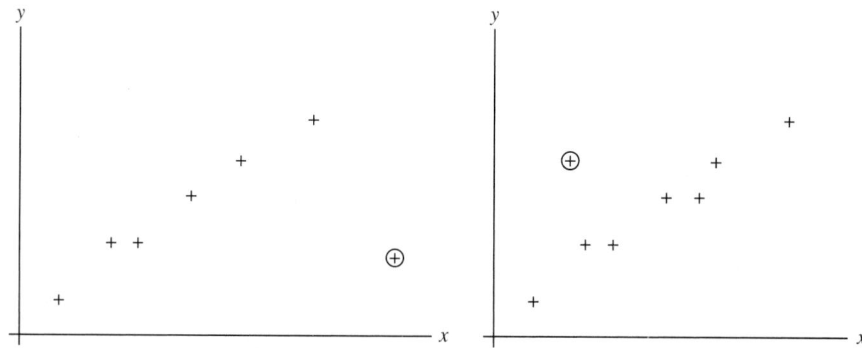

FIGURE 11-12
Two scatter plots showing possible outliers.

A pragmatic method that is often employed when the presence of an outlier is suspected is to perform identical analyses, first with the suspect point included in the data and then with it excluded. Such an approach often lends insight into the importance, or lack thereof, of the questioned datum.

11.9 THE STUDY OF SAMPLE RESIDUALS IN SLR

In the context of SLR, a residual has been defined in Equation 11.3 as $e_i = y_i - \hat{y}_i$. One purpose of studying the e_i is to determine whether our assumption that the ε_i are distributed $N(0, \sigma_{Y|x})$ is satisfied. If this assumption is not satisfied, at least in a "robust" sense, then all the confidence intervals, the various hypothesis tests on the parameters and on the validity of the regression, and the joint confidence region discussed earlier are no longer valid. If the assumption of normality is true, the random variables $\gamma_i = \varepsilon_i/\sigma_{Y|x}$, which are estimated by the $g_i = e_i/s_{Y|x}$, are distributed $N(0, 1)$—where γ_i is the ith *standardized residual* $[P(-2 < \gamma_i < 2) \cong 0.95$ and $P(-3 < \gamma_i < 3) \cong 0.997]$. If the sample size is small, we must use the appropriate t statistic.

Example 11.9. For the data of Table 11.1, the SLR program uses Equation 11.3 to compute the e_i. It then forms the g_i and provides Table 11.2. Since the standardized

TABLE 11.2
Sample residuals for the data of Table 11.1

IND VAR	OBS	EST	RES	STD RES
1148	724	805.42	−81.42	−1.32
1638	1293	1280.64	12.36	0.20
1678	1296	1319.43	−23.43	−0.38
1292	925	945.08	−20.08	−0.33
1422	1078	1071.16	6.84	0.11
1285	948	938.29	9.71	0.16
1152	893	809.30	83.70	1.36
1357	1077	1008.12	68.88	1.12
867	550	532.90	17.10	0.28
1158	870	815.12	54.88	0.89
1082	669	741.42	−72.42	−1.18
907	517	571.70	−54.70	−0.89
752	495	421.37	73.63	1.20
1115	692	773.42	−81.42	−1.32
1307	1014	959.63	54.37	0.88
1528	1282	1173.96	108.04	1.75
1357	1007	1008.12	−1.12	−0.02
1405	978	1054.67	−76.67	−1.24
1127	849	785.06	63.94	1.04
1073	670	732.69	−62.69	−1.02
1308	953	960.60	−7.60	−0.12
812	497	479.56	17.44	0.28
1260	798	914.04	−116.04	−1.88
1008	657	669.65	−12.65	−0.21
875	580	540.66	39.34	0.64

sample residuals extend only from -1.88 to 1.75, the residuals give no immediate cause to question the assumption of normality. This observation stems directly from the distribution theory discussed in Chapter 4.

However, because SLR can tolerate moderate departures from the normality assumptions, graphical methods of studying the distribution of the residuals are just as good, if not better, for this particular application. Indeed, any departure from normality that does not appear in an obvious way while using the techniques described next will probably have no real effect on the analysis or the correctness of the results.

The SLR program in your software library provides a variety of different residual plots to assist you not only in detecting departures from the normality assumption but also in determining the cause of such departures. Figure 11.13 shows the menu that is presented at this point in the analysis by the SLR program.

Example 11.10. Selection 1 on the menu requests a plot of the *empirical cumulative distribution function* (ECDF) on a "probability" scale. Figure 11.14 gives the ECDF for the data of Table 11.1 and our fitted line. The plotted points are obtained by first ordering the g_i from the smallest to largest values. These ordered g_i are then assigned, in that ascending order—the indices $i = 1, 2, \ldots, n$. The associated horizontal axis value for g_i is $p_i = (i - 0.5)/n$, and the horizontal axis is scaled according to a standard normal probability distribution.

If the residuals were *exactly* distributed according to a normal distribution, the plot of the ECDF would be *precisely* a straight line. However, the ECDF is a plot of the realizations of random variables and thus will show variability about the hypothetical line even when the underlying distribution of the residuals is normal. For this reason, it is advantageous to have a feel for how normal deviates will appear in an ECDF. To assist you in obtaining this understanding, Figure 11.15 contains the ECDF plots of standardized normal deviates for various sample sizes. You should study these plots until you have a good idea of the relation between the sample size and the kind and amount of departures from the theoretical line. Remember, these plots are *known* to be constructed from standard normal deviates, whereas you will not have that assurance in your attempts to perform a simple linear regression analysis on a set of data.

```
MENU FOR RESIDUAL PLOTS

SELECT:
1. EMPIRICAL CUMULATIVE DIST PLOT
2. RESID. AGAINST INPUT ORDER
3. RESID. AGAINST INDEPENDENT VARIABLE, X
4. RESID. AGAINST DEPENDENT VARIABLE, Y
5. RESID. AGAINST FITTED Y
6. NO RESIDUAL PLOTS
```

FIGURE 11-13
The SLR program's menu for residual plots.

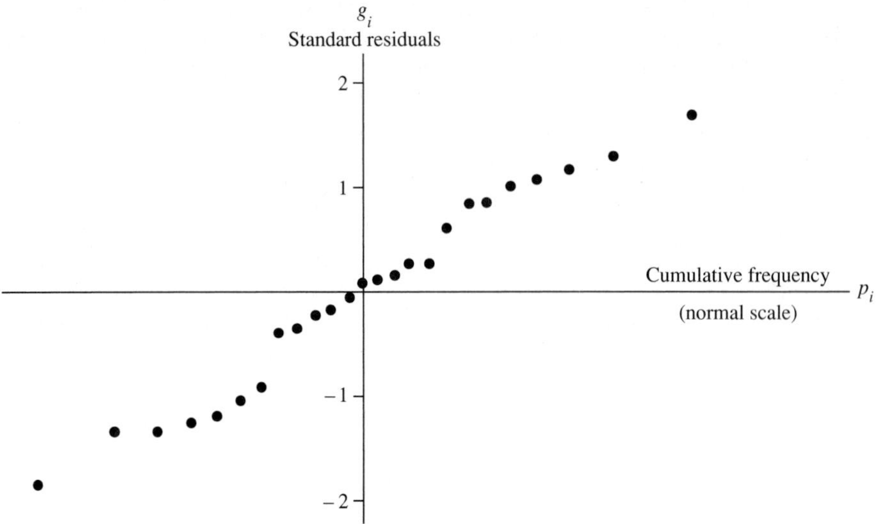

FIGURE 11-14
ECDF plot for Table 11.1.

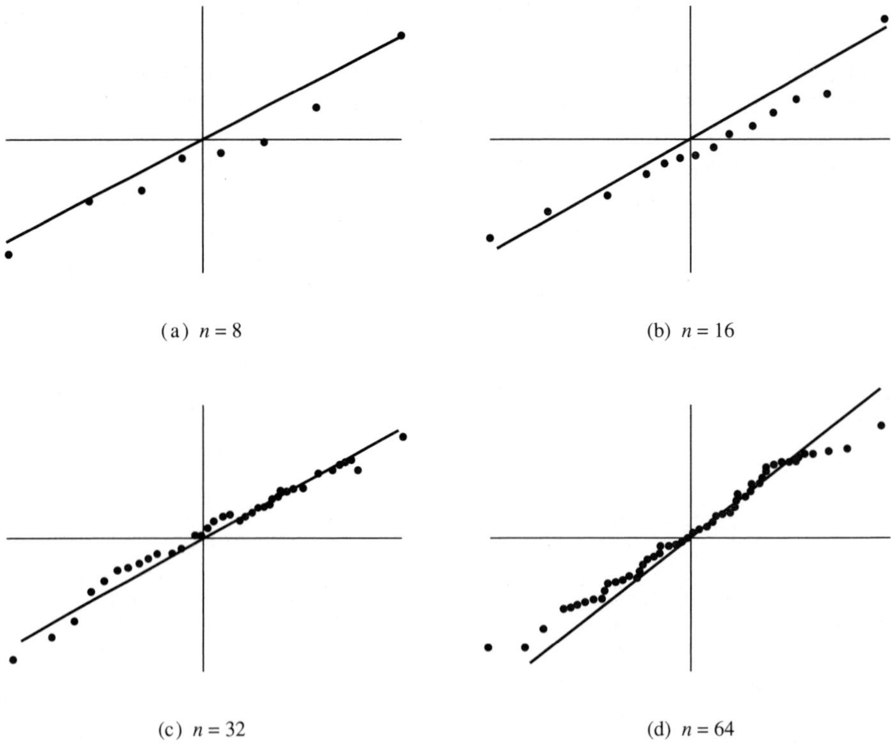

FIGURE 11-15
Plots of the ECDF of $N(0, 1)$ deviates for four sample sizes.

The departures from the theoretical line at the smaller values of n take place both in the middle of the plot and at the extremes. Only when the sample size reaches a value of 32 does any adherence to the line becomes obvious, and this adherence is in the center of the plot. As the sample size becomes larger, we may observe that the departures from the line are restricted to the extremes of the line and correspond to the low-probability tails of the underlying normal distribution. With this information and understanding, we may safely conclude that Figure 11.14 provides no evidence of nonnormality in the residuals associated with our current example.

In addition to the ECDF, the residuals from a regression fit should be plotted, on an ordinary scale, against various quantities relevant to the phenomenon and the data. A nonexhaustive set of such plots might be to plot the residuals against input order, against the independent variable, against the dependent variable, and against the fitted values of the dependent variable as yielded from the model. All of these residual plots are provided automatically by the SLR program through selection from the menu of Figure 11.13.

The idea behind this type of plot is to search for evidence of nonrandom trends or tendencies in the residuals. If the fit is good and the assumptions that we have made earlier are satisfied, we would expect an *even* band to be exhibited by the plot; an example is given in Figure 11.16. Four common departures from this result are shown in Figure 11.17. In Figure 11.17*a,* the variance of Y is increasing as the horizontal axis increases, whereas in Figure 11.17*b* the converse is true. Both cases are clear indications of a violation of the assumption of constant variance as given in Section 11.2. Figure 11.17*c* shows a marked curvature, which is an indication that a second-order effect present in the data has not been included in the model. In Figure 11.17*d,* a linear trend is present. This means that a strong first-order effect has not been explained by the model.

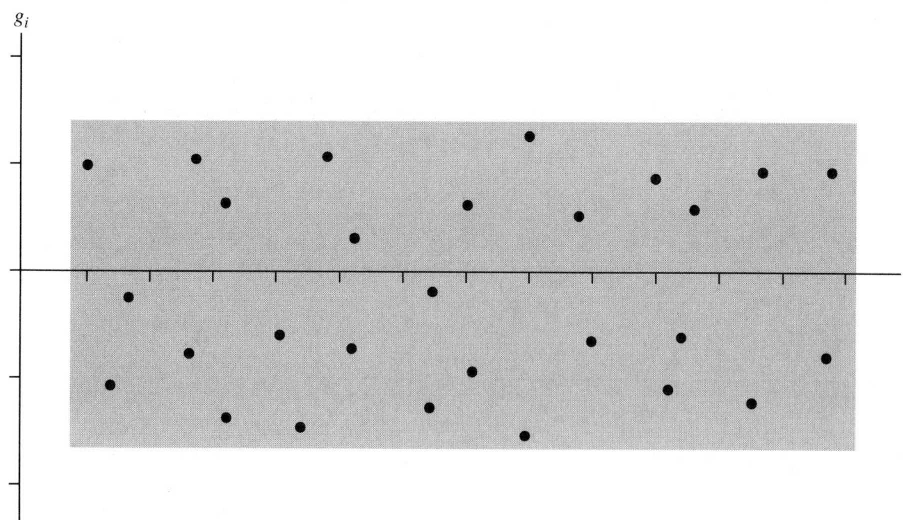

FIGURE 11-16
A residual plot with an even random band.

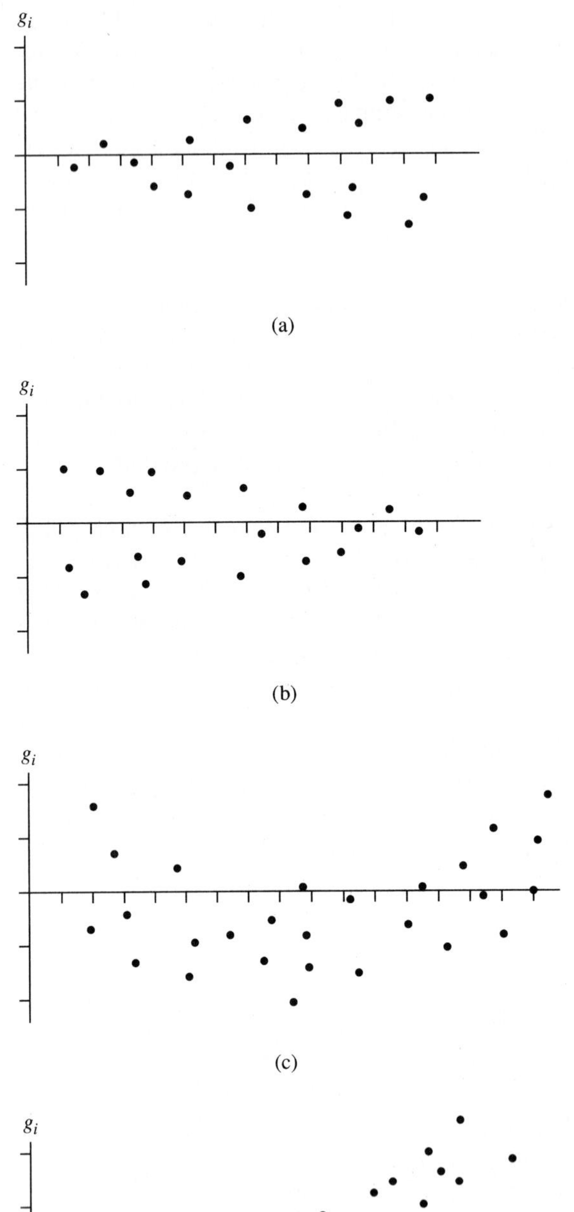

FIGURE 11-17
Common departures from a random residual plot.

If any of these kinds of tendencies are present in the residual plots from one of your analyses, you have not achieved a satisfactory fit. A number of ways exist to attempt correcting such situations. In the cases of Figure 11.7*a* and *b*, the method of *weighted least squares* can be used (see Draper and Smith [1966]). In the cases of Figure 11.17*c* and *d*, we must resort to adding more terms to our fitted equation. We discuss methods of adding variables in the next chapter when we address the topic of *multivariate linear regression analysis* (MLR).

Example 11.11. The SLR program's residual plots for the data of Table 11.1 for menu selections 2 through 5 in the residual plotting menu are presented in Figure 11.18. As may be seen, there is no evidence of nonrandom effects in those residual plots.

Before we consider models with more than one independent variable, let us consider a method that extends the application of SLR to cases in which the relation between the dependent and independent variable may be nonlinear.

11.10 CURVILINEAR REGRESSION

Many possible nonlinear mathematical relationships between two variables can be transformed to linear relationships in one or two *new* variables by applying relatively simple mathematical operations to the original nonlinear form.

Example 11.12. Consider the exponential relation

$$y = \beta_0 e^{\beta_1 x} \tag{11.28}$$

This *intrinsically* linear relationship is transformed to a linear relation by taking the natural logarithm of both sides of the equation. This mathematical operation yields

$$\ln y = \ln \beta_0 + \beta_1 x \qquad \text{or} \qquad w = \beta_0' + \beta_1 x \tag{11.29}$$

Therefore, if your data appears to follow an exponential relationship instead of a linear relationship, all you need to do is transform your data set by taking the natural logarithm of each y_i (nothing needs to be done to the x_i) and perform the simple linear regression analysis on the transformed data set. However, a *word of caution* is appropriate here. Anytime you perform a transformation on the dependent variable, Y, you are implicitly modifying the error structure that underlies your analysis. As an example of such a modification, let us consider what happens to the error structure in the untransformed "space" when the logarithmic transform is applied, as in Equation 11.29. Because we are performing the analysis in the transformed space, our model is

$$W_i = \beta_0' + \beta_1 x + \varepsilon_i \tag{11.30}$$

Note that the *additive* error structure and all the assumptions that are made about linear regression analysis at the start of the chapter apply to the transformed space.

Now, let us perform the *inverse transformation*, i.e., raise both sides of Equation 11.30 as powers of e, the base of the natural logarithms. Doing so

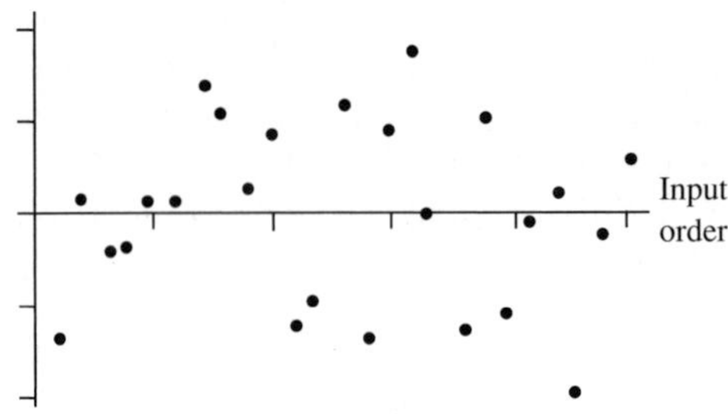

(a) e_i versus input order

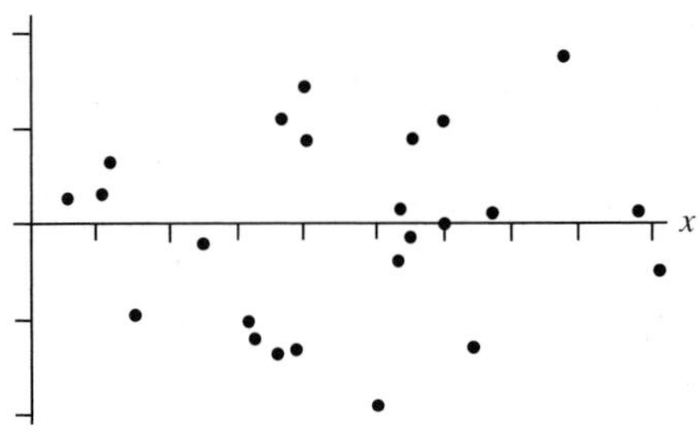

(b) e_i versus x_i

(c) e versus y_i

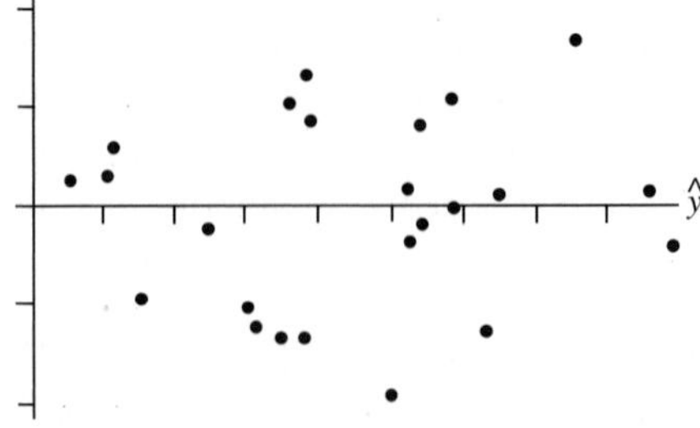

(d) e_i versus $\hat{y}_i$

FIGURE 11-18
Residual plots for selections 2 through 5.

yields

$$e^{W_i} = e^{\beta_0 + \beta_1 x_i + \varepsilon_i} \qquad \text{or} \qquad Y_i = \beta_0 e^{\beta_1 x_i} e^{\varepsilon_i}$$

Thus, a normally distributed additive error structure in the transformed space of an exponential relationship implies a *multiplicative* "exponential-normal" error structure in the untransformed space. In many applications such an unusual error structure cannot be justified or tolerated. Fortunately, there are ways to overcome this problem. They reside mainly in the area of nonlinear regression analysis, whose study is beyond the scope of this text. Often a curvilinear fit of the data as above is used as a starting point for the nonlinear regression analysis.

Of course, not all transformations of a data set require transforming the dependent variable. When this is true, only the independent variable x is transformed, and no transformation of the error structure is implied because x is assumed to be a known constant with *no error*.

Table 11.3 joined with Figure 11.19 give only a representative set of the great many transformations that can be entertained when the scatter plot of the data indicates that a nonlinear relationship between x and y might be present.

TABLE 11.3
Curvilinear functions and their SLR transforms

Function	Equation	Transformed Equation
Hyperbolic	$y = \dfrac{x}{\beta_0 x + \beta_1}$	$\dfrac{1}{y} = \beta_0 + \beta_1 \dfrac{1}{x}$
Exponential	$y = \beta_0 \exp(\beta_1 x)$	$\ln y = \ln \beta_0 + \beta_1 x$
Power	$y = \beta_0 x^{\beta_i}$	$\ln y = \ln \beta_0 + \beta_1 \ln x$
Logarithmic	$y = \beta_0 + \beta_1 \ln x$	$y = \beta_0 + \beta_1 \ln x$
Inverse exponential	$y = \beta_0 \exp(\beta_1 / x)$	$\ln y = \ln \beta_0 + \dfrac{\beta_1}{x}$
Pseudo-exponential	$y = \dfrac{1}{\beta_0 + \beta_1 e^{-x}}$	$\dfrac{1}{y} = \beta_0 + \beta_1 e^{-x}$

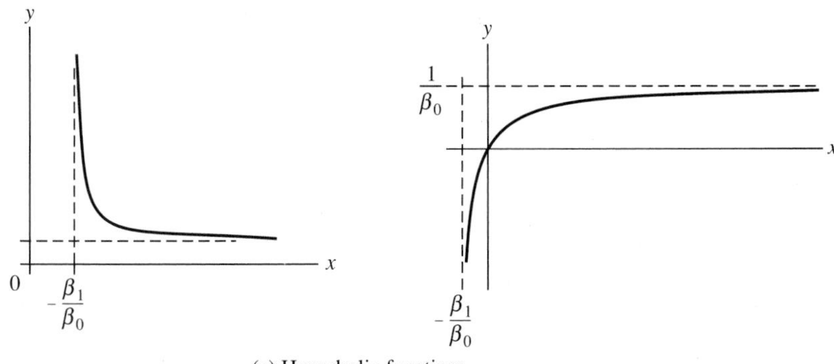

(a) Hyperbolic functions

FIGURE 11-19
A set of curvilinear functions for SLR.

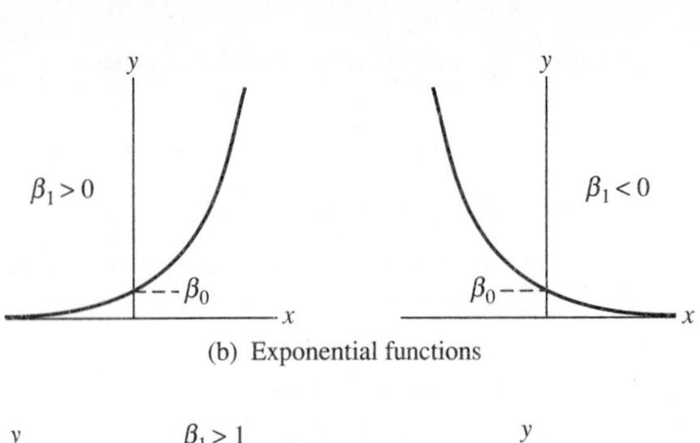

(b) Exponential functions

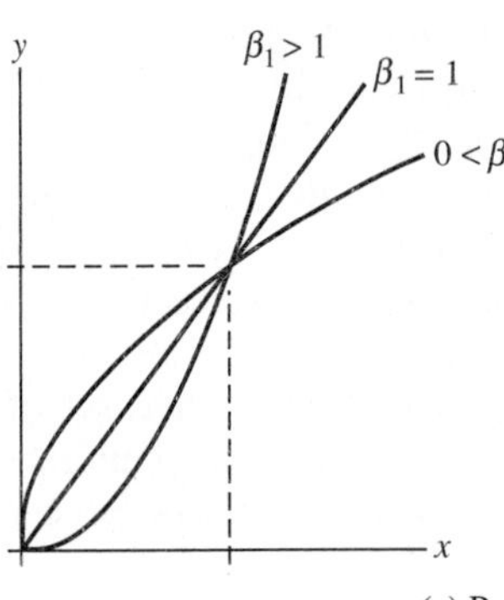

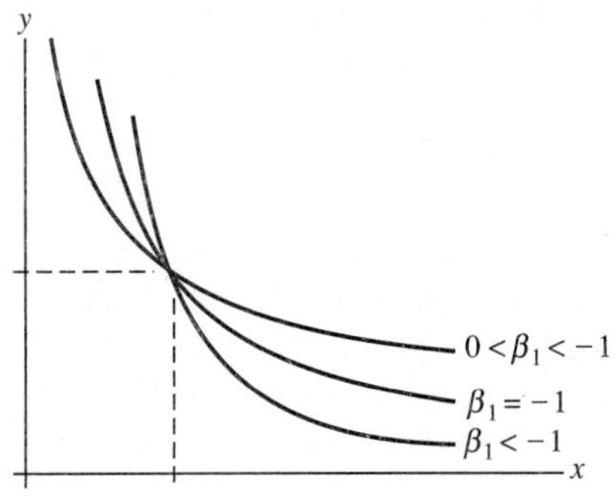

(c) Power functions

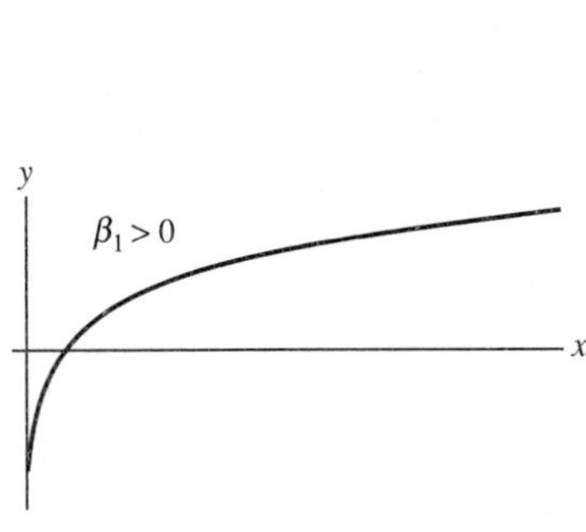

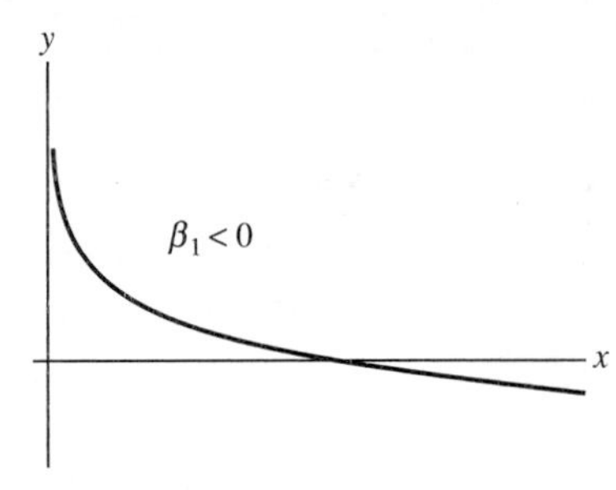

(d) Logarithmic functions

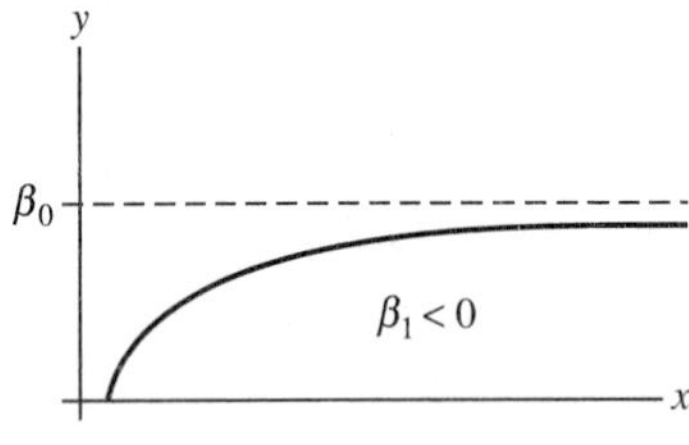

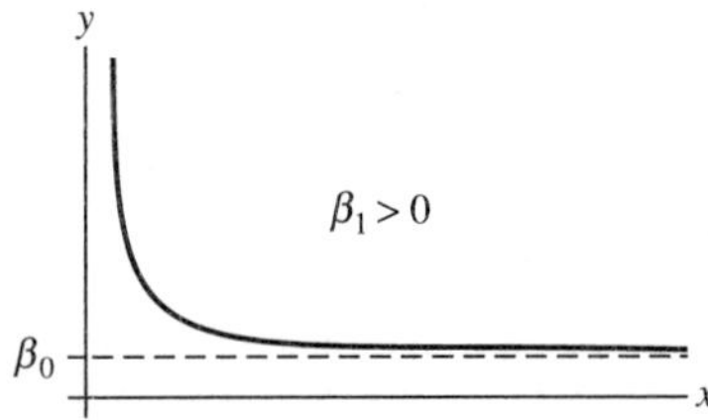

(e) Inverse exponential functions

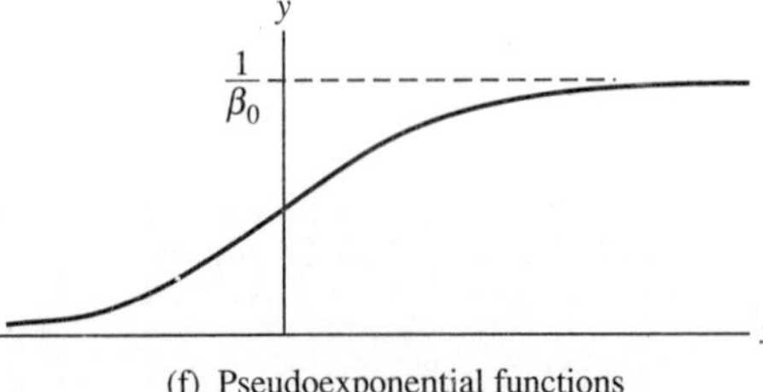

(f) Pseudoexponential functions

FIGURE 11-19
(Continued)

Example 11.13. As an example of a curvilinear regression analysis, consider the data set of Table 11.4. By first applying the SLR program without any transformations, we achieve the scatter plot given in Figure 11.20. Although several *possibilities* exist (such as a hyperbola, a power function, a logarithmic function, or one of the pseudoexponential forms), a promising one is that the variables are related through a logarithmic function, $y = b_0 + b_1 \log(x)$.

Since we consider methods of choosing between competing models of differing mathematical forms in the next chapter, we examine only the logarithmic form at this point. In order to test the conjecture that the variables are related by a logarithmic function, we first use the program in your software library for transforming regression data to replace the current values of the independent variable with their common logarithms (log base 10). This operation yields the new data set of Table 11.5. The transformation program is a general program for this purpose and will allow not only transformation of variables but also deletion and addition of new variables to a set of data. This will prove quite useful in our work in the next chapter.

Having completed the transformation of the x values, we need only run the SLR program on the "new" data set. Leaving the details to the exercises, the resulting fit is quite good, as evidenced in Figure 11.21. In addition, the result would have a

TABLE 11.4
Data set for curvilinear regression

x	y	x	y	x	y	x	y	x	y	x	y
1.122	0.902	1.259	0.853	1.413	0.674	1.584	0.859	1.778	1.230	1.995	0.961
2.239	1.143	2.512	1.334	2.818	1.746	3.162	1.286	3.548	1.512	3.981	1.839
4.467	1.433	5.012	1.702	5.623	1.289	6.310	2.065	7.080	1.773	7.940	1.703
8.913	2.132										

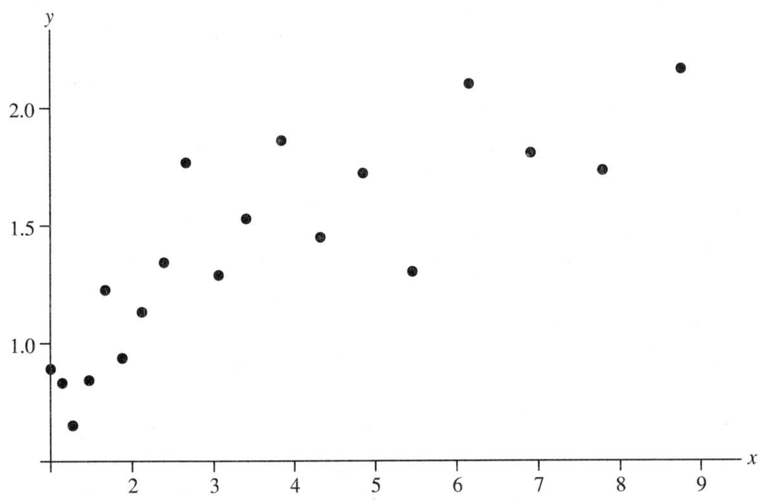

FIGURE 11-20
Untransformed scatter plot of Table 11.4.

TABLE 11.5
Transformed data set from Table 11.4

x	y	x	y	x	y	x	y
0.05	0.902	0.1	0.853	0.15	0.674	0.2	0.859
0.25	1.230	0.3	0.961	0.35	1.143	0.4	1.334
0.45	1.746	0.5	1.286	0.55	1.512	0.6	1.839
0.65	1.433	0.7	1.702	0.75	1.289	0.8	2.065
0.85	1.773	0.9	1.703	0.95	2.132		

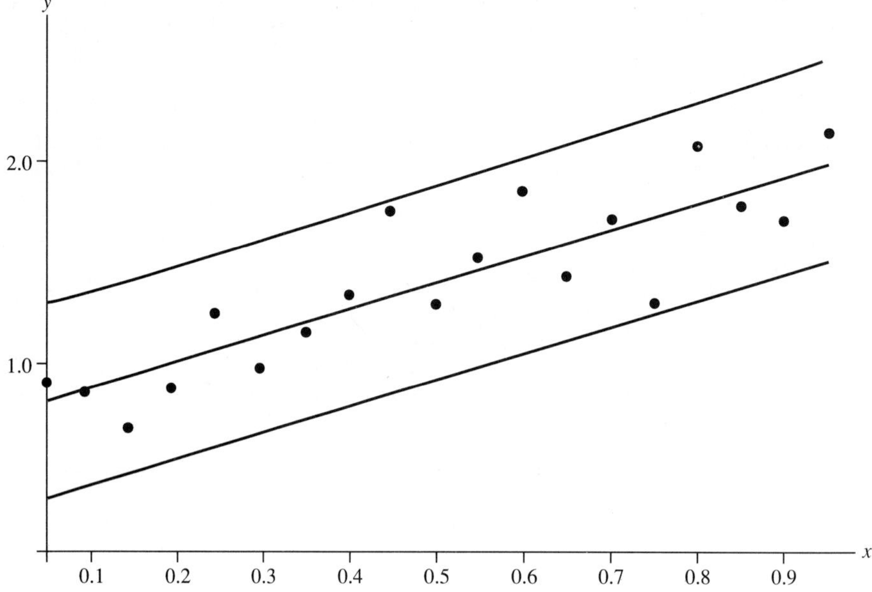

FIGURE 11-21
Transformed scatter plot of Table 11.4 with fitted line and confidence band.

very good chance of acceptance in a real application because the commonly accepted additive error structure has not been changed.

In the next chapter we turn our attention to fitting equations to data sets that have more than one independent variable, multiple linear regression (MLR). Although the mathematical structure associated with MLR is more cumbersome and sophisticated, the conceptual framework for MLR is very similar to that for SLR.

EXERCISES

11.1. The following data set has been taken from a laboratory experiment in which the weld area (x) and the shear load at failure (y) were studied. Perform a regression analysis on the data and state your conclusions about the goodness of your fit.

60 Data Points

x	y	x	y	x	y	x	y	x	y
1	7.2104	2	8.3691	3	10.4150	4	12.4242	5	15.5456
6	17.0350	7	19.6290	8	19.3004	9	22.7433	10	23.7436
11	26.3269	12	29.2092	13	32.8558	14	31.2383	15	35.6833
16	39.3606	17	39.2260	18	40.1267	19	43.9814	20	43.4162
21	45.1042	22	50.0318	23	50.2253	24	55.5276	25	54.3018
26	59.0039	27	56.8852	28	59.2104	29	62.4211	30	66.4288
31	69.2883	32	72.2772	33	72.3958	34	75.2927	35	77.5615
36	78.3066	37	75.7109	38	79.8616	39	78.9663	40	84.7872
41	86.5572	42	86.7927	43	95.1320	44	92.3665	45	99.3675
46	101.4774	47	94.8469	48	99.9411	49	100.9209	50	108.5388
51	104.3818	52	107.5911	53	109.2293	54	110.1580	55	113.4751
56	113.3834	57	118.4172	58	122.3659	59	125.2835	60	129.5720

11.2. In a study of the particle emission rate of indium 115, the average rate (y) was recorded for 20 contiguous 3-week periods (x). Fit the best equation possible to the following data.

20 Data Points

x	y	x	y	x	y	x	y	x	y
3	673	6	642	9	618	12	598	15	587
18	555	21	525	24	508	27	493	30	477
33	459	36	446	39	431	42	414	45	398
48	379	51	363	54	353	57	344	60	327

11.3. A departmental study has generated the following data in regard to years of service (x) and value of equipment (y) for forklifts used by the loading-dock personnel. Fit an equation to this data set. Discuss the strong and weak points of the fit.

26 Data Points

x	y	x	y	x	y	x	y
3.2	52.2105	3.6	45.8395	4.0	23.9653	4.4	17.6200
4.8	12.3837	5.2	11.7717	5.6	10.1653	6.0	11.6114
6.4	9.5895	6.8	9.7419	7.2	8.9268	7.6	8.1678
8.0	7.3133	8.4	8.4854	8.8	7.3621	9.2	6.7481
9.6	7.2075	10.0	7.3892	10.4	6.7838	10.8	7.3301
11.2	7.2958	11.6	6.5252	12.0	6.8330	12.4	6.1248
12.8	6.6680	13.2	6.1193				

11.4. With regard to curvilinear regression, investigate and discuss the error "structure" in the *untransformed* space for each of the following.
(a) the hyperbolic transformation
(b) the power transformation

11.5. Outline the method for obtaining the least squares estimates b_0 and b_1 for $y = \beta_0 \exp(\beta_1 x)$ *without* transforming the equation. State the general form of the normal equations, and indicate how you would solve the normal equations in a specific case.

11.6. In an automobile-collision study, one item of interest was the dollar value of collision damage (Y) relative to the *normalized* speed of the vehicle at impact (x). Perform a regression analysis on the following data from that experiment, and comment on the conclusions that might be drawn from your analysis.

19 Data Points

x	y	x	y	x	y	x	y	x	y
1	41	2	44	3	48	4	50	5	58
6	61	7	66	8	70	9	75	10	81
11	89	12	94	13	96	14	106	15	118
16	129	17	134	18	142	19	147		

11.7. Consider the joint 95% confidence region on β_0 and β_1 given here. Take a suitable number of pairs of (β_0, β_1) values from the boundary of the ellipse, and plot the associated regression lines on a separate graph. Use the locus formed by the area that is exterior to the lines to sketch the 95% confidence band *on the line*. (This is the same band that you would obtain using Equation 11.27.)

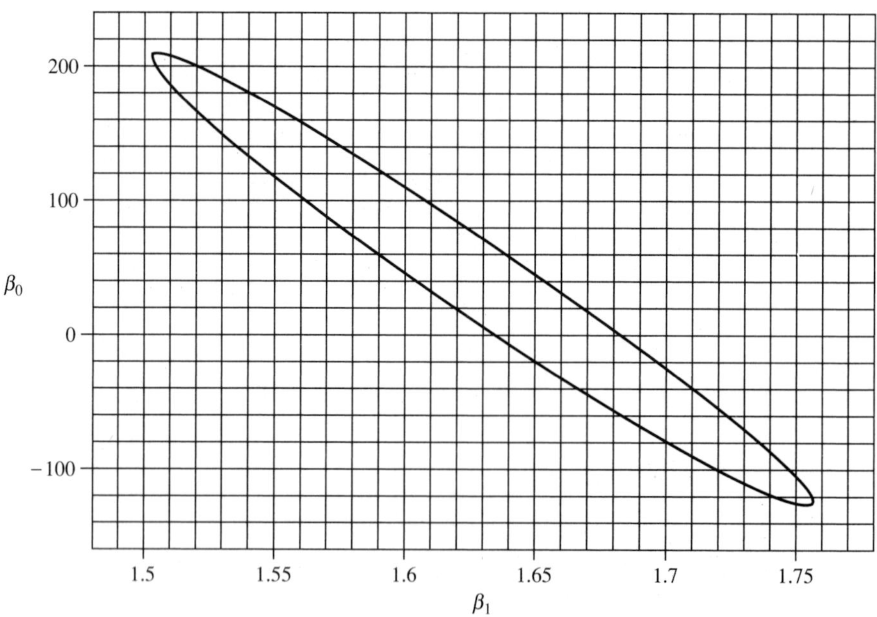

11.8. Tensile members to be used in a highway pedestrian overpass are tested as part of a quality assurance program. The 22 data points listed here give the amount of elongation (y) that occurred in response to a specified stress level (x), relative to a basic stress level. Fit an equation to the data, and comment on the characteristics of the fit you have achieved.

22 Data Points

x	y	x	y	x	y
−1	0.0360	0	0.0969	1	0.2519
2	0.5090	3	1.0915	4	1.0595
5	2.0127	6	1.3827	7	3.2926
8	3.0055	9	2.2371	10	1.5823
11	1.4138	12	1.2626	13	1.6066
14	1.4325	15	1.3927	16	1.6403
17	4.3900	18	2.4567	19	5.7158
20	2.0701				

11.9. Perform a regression analysis on the data of Table 11.4 *after* the independent variable has been transformed; i.e., use $x' = \log_{10} x$. How well does the transformation work in regard to the fitted model?

11.10. In a study of the heating requirements of a small organic chemistry laboratory, the following data were taken (over a period of 25 days). Use a regression analysis to determine if there is a valid linear relationship between the outside temperature (x) and the laboratory heating requirements (y).

25 Data Points

x	y	x	y
35.30	10.98	29.70	11.13
30.80	12.51	58.80	8.40
61.40	9.27	71.30	8.73
74.40	6.36	76.70	8.50
70.70	7.82	57.50	9.14
46.40	8.24	28.90	12.19
28.10	11.88	39.10	9.57
46.80	10.94	48.50	9.58
59.30	10.09	70.00	8.11
70.00	6.83	74.50	8.88
72.10	7.68	58.10	8.47
44.60	8.86	33.40	10.36
28.60	11.08		

11.11. An experiment was conducted to determine if adding a certain chemical (x) to a plastic forming process would have an effect on the heat (y) produced by the chemical reaction. The following data have been obtained. Does the additive have any effect on the heat produced?

52 Data Points

x	y	x	y	x	y
3.1	5.1	3.5	4.2	3.9	3.9
3.9	4.9	3.9	6.9	4.1	5.7
4.5	3.9	4.5	5.1	4.6	6.5
5.0	4.5	5.2	5.5	5.5	3.0
5.5	6.4	5.5	7.4	5.7	4.0
5.7	5.0	6.1	6.5	6.1	8.1
6.4	9.9	7.0	2.9	7.0	4.9
7.0	5.7	7.0	7.0	7.7	4.0
7.8	6.8	7.9	8.1	7.9	9.1
8.2	1.7	8.4	5.0	8.2	5.8
8.5	7.5	8.6	8.3	8.9	2.8
8.9	4.5	8.9	6.7	9.3	7.7
9.5	3.6	9.8	5.0	10.0	9.0
10.1	6.8	10.5	3.3	10.6	4.9
10.7	6.1	10.8	8.2	11.2	7.2
11.8	6.0	12.1	8.2	12.2	7.0
12.4	4.7	13.0	3.1	14.0	4.0
14.1	5.7				

11.12. In a study of a water supply pipe network in an office building, the hydraulic pressure was changed at point A by a specific amount (x) relative to a nominal pressure value. The change in the pressure at point B was measured after each change at point A. For the accompanying data set, is there a linear relation between the pressure changes at points A and B? Why or why not?

21 Data Points

x	y	x	y	x	y
−4	−7.0	−3	−3.0	−2	−3.0
−1	1.0	0	3.0	1	3.0
2	6.0	3	8.0	4	9.0
−4	−7.5	−5	−8.0	−6	−9.3
−8	−12.5	4	11.2	5	11.0
6	15.0	8	19.5	−8	15.4
8	21.0	−9	−14.5	9	23.1

11.13. As stated in Section 11.7, r applies only to the possible linear correlation between two variables. It is possible that $r = 0$ even in the presence of a strong nonlinear relation between the two variables. As an example of this fact, use the simple linear regression program to fit a straight line to the following data set.

x	y	x	y
−4	16	1	1
−3	9	2	4
−2	4	3	9
−1	1	4	16
0	0		

What observations can you make from this analysis? If you were to use the transformation program to transform the independent variable, x, to the square of its previous value in each of the 9 data pairs, what do you think would happen? Why?

CHAPTER 12

FITTING EQUATIONS TO DATA, PART II: MULTIVARIATE REGRESSION ANALYSIS

An introductory background in matrices, vectors, and their operational rules is helpful in understanding the material in this chapter. Vectors and matrices will be represented by **boldface** characters.

When you understand the following concepts and procedures, you will have mastered the important ideas of this chapter.

- Be able to interpret the *variance-covariance* and *correlation matrices* of the variables contained in your model
- Understand the *linear model*, $\mathbf{Y} = \mathbf{X}\boldsymbol{\beta} + \boldsymbol{\varepsilon}$, and the meaning of each of the symbols in the linear model
- Be able to state the assumptions associated with *multivariate linear regression* analysis (MLR)
- Know the steps to follow in obtaining the least squares estimates of the elements of $\boldsymbol{\beta}$
- Understand the meaning of the *sum of squares for error, mean square of errors, sum of squares for regression, mean square of regression,* and *multiple correlation coefficient squared*

- Be able to interpret the *variance-covariance* and *correlation matrices* of $\hat{\boldsymbol{\beta}}$
- Be able to construct univariate confidence intervals on any of the β_j and on the mean and the predicted value of Y at a specific $\mathbf{x}_k$
- Understand why a confidence ellipse on two parameters is not as helpful when there are more than two parameters in the model
- Have a conceptual understanding of a joint confidence region in three or more dimensions (parameters)
- Understand the MLR program's output for the estimates of the parameters
- Be able to state and explain the primary measures of goodness of fit for an MLR model
- Know the different kinds of residual plots that should be made as part of the diagnostic analyses of an MLR model and how to interpret them
- Know how to determine which of a competing set of models best fits a particular data set
- Understand that the only way to be sure that the "best" model has been identified is to examine all 2^{p-1} possible models
- Be able to obtain new independent variables by forming algebraic combinations of the original independent variables
- Be aware that more advanced texts and software packages exist to help you perform more sophisticated analyses than would be possible using only this chapter and its associated MLR program

12.1 INTRODUCTION

The previous chapter has covered the cases in which only one independent variable is allowed in the regression model. Suppose that two or more independent variables are allowed. As an example of such a multivariate model, consider

$$Y_i = \beta_0 + \beta_1 x_{1i} + \beta_2 x_{2i} + \varepsilon_i \tag{12.1}$$

which contains two independent variables, x_1 and x_2.

Example 12.1. Suppose that we wish to fit the model of Equation 12.1 to the data set in Table 12.1. These data come from a larger data set concerned with the study of the amount of heat produced by portland cement during hardening. Variables x_1 and x_2 measure the amounts of two chemicals that have an effect on that exothermic reaction. Initially, the multivariate linear regression program in your package performs some basic analysis on the input data. It first computes estimates of the means and standard deviations of the input variables:

	x_1	x_2	y
Sample mean	7.462	48.154	95.423
Sample standard deviation	5.882	15.561	15.044

TABLE 12.1
An example multivariate regression analysis data set

i	x_{1i}	x_{2i}	y_i	i	x_{1i}	x_{2i}	y_i
1	7	26	78.5	7	3	71	102.7
2	1	29	74.3	8	1	31	72.5
3	11	56	104.3	9	2	54	93.1
4	11	31	87.6	10	21	47	115.9
5	7	52	95.9	11	1	40	83.8
6	11	55	109.2	12	11	66	113.3
				13	10	68	109.4

Next the estimate of the variance-covariance matrix of the input variables is displayed.

Variance-covariance matrix of the input variables

34.603	20.923	64.663
20.923	242.141	191.079
64.663	191.079	226.314

This matrix is formed by using Equation 11.24, where the matrix element at row i and column j contains the sample covariance of variables i and j. (In the above matrix, the order of the variables is x_1, x_2, and y; i.e., the covariance of x_1 with x_2 is 20.923, and the covariance of x_2 with y is 191.079.) Since the elements on the major diagonal give the covariance of each variable with itself, the element at position (i, i) gives the variance of each input variable i; i.e., $s^2_{x_1} = 34.603$, $s^2_{x_2} = 242.141$, $s^2_Y = 226.313$.

The information contained in the covariance matrix is then used in Equation 11.23 to obtain the estimated correlation matrix of the input variables.

Simple correlation coefficients

1.0	0.229	0.731
0.229	1.0	0.816
0.731	0.816	1.0

From this matrix we may observe that the independent variables, x_1 and x_2, are *not* highly correlated with one another, whereas both independent variables have a *strong* correlation with Y—x_2 being somewhat more strongly related to Y. The final result of the preliminary analysis is range information on the input data.

	Min	Max	Range
x_1	1	21	20
x_2	26	71	45
y	72.5	115.9	43.4

A scatter plot of the data, as in SLR, is no longer possible in MLR. Indeed, when more than two independent variables are present, no comprehensive plot is available for the entire data set, since four or more dimensions would be required. When only two independent variables are present, contour plots and perspective plots of the data can offer additional insight into the structure of the data.

12.2 ESTIMATING THE PARAMETER VALUES IN MLR

A greater amount of data must be considered in performing multivariate linear regression analysis (MLR), particularly when several independent variables are present. For this reason, it is much more convenient to work with *matrices* to represent specific sets of numbers rather than laboriously writing out the numbers each time we refer to them. Further, using the mathematical methods of *linear algebra* makes the necessary numerical operations easier to describe and understand.

Therefore, let us represent all observed values of the dependent variable with $\mathbf{y}$, all corresponding values of the independent variables with $\mathbf{X}$, all parameters with $\boldsymbol{\beta}$, and all errors with $\boldsymbol{\varepsilon}$. The vector of random variables corresponding to $\mathbf{y}$ will be denoted as $\mathbf{Y}$.

Example 12.2. For the data of Table 12.1 and for Equation 12.1, $\mathbf{Y}$, $\mathbf{y}$, $\mathbf{X}$, $\boldsymbol{\beta}$, and $\boldsymbol{\varepsilon}$ take the specific forms given here, where $\mathbf{y}$ is an n (13) element column vector, $\mathbf{X}$ is an n by p (13 by 3) matrix, $\boldsymbol{\beta}$ is a p (3) element column vector, and $\boldsymbol{\varepsilon}$ is an n (13) element column vector. (Notice that β_0 has an associated *dummy variable* in the first column of $\mathbf{X}$, which has a value of 1 for all observations.)

$$\mathbf{Y} = \begin{bmatrix} Y_1 \\ Y_2 \\ Y_3 \\ Y_4 \\ Y_5 \\ Y_6 \\ Y_7 \\ Y_8 \\ Y_9 \\ Y_{10} \\ Y_{11} \\ Y_{12} \\ Y_{13} \end{bmatrix} \quad \mathbf{y} = \begin{bmatrix} 78.5 \\ 74.3 \\ 104.3 \\ 87.6 \\ 95.9 \\ 109.2 \\ 102.7 \\ 72.5 \\ 93.1 \\ 115.9 \\ 83.8 \\ 113.3 \\ 109.4 \end{bmatrix} \quad \mathbf{X} = \begin{bmatrix} 1 & 7 & 26 \\ 1 & 1 & 29 \\ 1 & 11 & 56 \\ 1 & 11 & 31 \\ 1 & 7 & 52 \\ 1 & 11 & 55 \\ 1 & 3 & 71 \\ 1 & 1 & 31 \\ 1 & 2 & 54 \\ 1 & 21 & 47 \\ 1 & 1 & 40 \\ 1 & 11 & 66 \\ 1 & 10 & 68 \end{bmatrix} \quad \boldsymbol{\beta} = \begin{bmatrix} \beta_0 \\ \beta_1 \\ \beta_2 \end{bmatrix} \quad \boldsymbol{\varepsilon} = \begin{bmatrix} \varepsilon_1 \\ \varepsilon_2 \\ \varepsilon_3 \\ \varepsilon_4 \\ \varepsilon_5 \\ \varepsilon_6 \\ \varepsilon_7 \\ \varepsilon_8 \\ \varepsilon_9 \\ \varepsilon_{10} \\ \varepsilon_{11} \\ \varepsilon_{12} \\ \varepsilon_{13} \end{bmatrix}$$

Thus, the matrix representation of linear models like Equation 12.1 is

$$\mathbf{Y} = \mathbf{X}\boldsymbol{\beta} + \boldsymbol{\varepsilon} \tag{12.2}$$

The assumptions that applied to the SLR model (Section 11.2) apply equally well to the MLR model of Equation 12.2. In the context of Equation 12.2, they may be restated as the following.

1. The elements of $\boldsymbol{\varepsilon}$, the ε_i, $i = 1, \ldots, n$, are NID$(0, \sigma^2)$, where NID is an abbreviation for *normal and independently distributed.*
2. The p elements of $\boldsymbol{\beta}$, the β_j, $j = 0, 1, \ldots, p-1$, are constant parameters whose true values are unknown and must be estimated from the data.
3. The elements of $\mathbf{X}$ are known constants that are set to predetermined values. They embody, as a group, all the values of the p independent variables, x_{ji}, $j = 0, \ldots, p-1$ and $i = 1, \ldots, n$, associated with the p parameters, β_j, $j = 0, 1, \ldots, p-1$.
4. The data are typical of the process and are good and valid.

If we perform the product $\mathbf{X\beta}$ for the matrices of Example 12.2 and write out the result in expanded form for our current example, we would obtain 13 equations just like Equation 12.1, i.e.,

$$\begin{aligned} Y_1 &= \beta_0 + \beta_1(7) + \beta_2(26) + \varepsilon_1 \\ Y_2 &= \beta_0 + \beta_1(1) + \beta_2(29) + \varepsilon_2 \\ &\vdots \\ Y_{13} &= \beta_0 + \beta_1(10) + \beta_2(68) + \varepsilon_{13} \end{aligned}$$

Just as in SLR, we need the least squares estimates of the elements of $\boldsymbol{\beta}$, i.e., β_0, β_1, and β_2. This means that we want the values of b_0, b_1, and b_2 that minimize

$$\sum e_i^2 = \sum (y_i - b_0 - b_1 x_{1i} - b_2 x_{2i})^2 = \mathbf{e}^{\mathrm{T}}\mathbf{e} = (\mathbf{y} - \mathbf{Xb})^{\mathrm{T}}(\mathbf{y} - \mathbf{Xb})$$

where $\mathbf{e}^{\mathrm{T}}$ is the transpose of $\mathbf{e}$.

It may be shown that by applying the same approach that was used in developing Equations 11.5 and 11.6, i.e., taking the partial derivative with respect to b_0, b_1, and b_2, and setting each result equal to zero, we arrive at the following expression for the solution of the *normal equations for MLR.*

$$\mathbf{b} = (\mathbf{X}^{\mathrm{T}}\mathbf{X})^{-1}\mathbf{X}^{\mathrm{T}}\mathbf{y} \tag{12.3}$$

Just as in Chapter 11, let us define $\hat{\boldsymbol{\beta}}$ to be the vector of estimators (random variables) for $\boldsymbol{\beta}$, i.e.,

$$\hat{\boldsymbol{\beta}} = (\mathbf{X}^{\mathrm{T}}\mathbf{X})^{-1}\mathbf{X}^{\mathrm{T}}\mathbf{Y} \tag{12.3a}$$

where $\mathbf{Y}$ is substituted for $\mathbf{y}$.

Example 12.3. For the matrices of Example 12.1, Equation 12.3 implies that

$$\mathbf{b} = \begin{bmatrix} b_0 \\ b_1 \\ b_2 \end{bmatrix} = \left[\begin{bmatrix} 1 & 1 & \cdots & 1 \\ 7 & 1 & \cdots & 10 \\ 26 & 29 & \cdots & 68 \end{bmatrix} \begin{bmatrix} 1 & 7 & 26 \\ 1 & 1 & 24 \\ \vdots & \vdots & \vdots \\ 1 & 10 & 68 \end{bmatrix} \right]^{-1} \begin{bmatrix} 1 & 1 & \cdots & 1 \\ 7 & 1 & \cdots & 10 \\ 26 & 29 & \cdots & 68 \end{bmatrix} \begin{bmatrix} 78.5 \\ 74.3 \\ \vdots \\ 109.4 \end{bmatrix}$$

The multivariate linear regression program in your software package uses the data of Table 12.1 and Equation 12.3 to obtain

$$\mathbf{b} = \begin{bmatrix} b_0 \\ b_1 \\ b_2 \end{bmatrix} = \begin{bmatrix} 52.57 \\ 1.468 \\ 0.662 \end{bmatrix}$$

12.3 SOME USEFUL THEORETICAL PROPERTIES OF MLR

Just as in SLR, several things of interest are associated with MLR; the following pages list them.

1. $\hat{\boldsymbol{\beta}}$ is an unbiased estimator of $\boldsymbol{\beta}$.

$$E\{\hat{\boldsymbol{\beta}}\} = E\{(\mathbf{X}^T\mathbf{X})^{-1}\mathbf{X}^T\mathbf{Y}\} = (\mathbf{X}^T\mathbf{X})^{-1}\mathbf{X}^T E\{\mathbf{Y}\} = (\mathbf{X}^T\mathbf{X})^{-1}\mathbf{X}^T\mathbf{X}\boldsymbol{\beta} = \boldsymbol{\beta}$$

2. The variance-covariance matrix of $\hat{\boldsymbol{\beta}}$ is $\sigma^2(\hat{\boldsymbol{\beta}}) = (\mathbf{X}^T\mathbf{X})^{-1}\sigma^2$.

$$\begin{aligned} \boldsymbol{\sigma}^2(\hat{\boldsymbol{\beta}}) &= \text{var}\{(\mathbf{X}^T\mathbf{X})^{-1}\mathbf{X}^T\mathbf{Y}\} = (\mathbf{X}^T\mathbf{X})^{-1}\mathbf{X}^T\text{var}(\mathbf{Y})\mathbf{X}(\mathbf{X}^T\mathbf{X})^{-1} \\ &= (\mathbf{X}^T\mathbf{X})^{-1}\mathbf{X}^T\boldsymbol{\sigma}^2\mathbf{X}(\mathbf{X}^T\mathbf{X})^{-1} = (\mathbf{X}^T\mathbf{X})^{-1}\mathbf{X}^T\mathbf{X}(\mathbf{X}^T\mathbf{X})^{-1}\sigma^2 \\ &= (\mathbf{X}^T\mathbf{X})^{-1}\sigma^2 \end{aligned} \tag{12.4}$$

The specific form and interpretation of $\boldsymbol{\sigma}^2(\hat{\boldsymbol{\beta}})$ is discussed in detail in the next section of this chapter.

3. The sum of squares for error is

$$\begin{aligned} \text{SSE} &= \text{SST} - \text{SSR} \qquad (12.5) \\ &= \sum (y_i - \hat{y}_i)^2) \\ &= \sum (y_i - \bar{y})^2 - \sum (\hat{y}_i - \bar{y})^2 \\ &= (\mathbf{y}^T\mathbf{y} - n\bar{y}^2) - (\mathbf{b}^T\mathbf{X}^T\mathbf{y} - n\bar{y}^2) \end{aligned}$$

In words, the sum of squares for error is equal to the *total sum of squares minus the sum of squares due to regression*—where SSR is the reduction of the total "corrected" sum of squares of the y_i about $\bar{y}$, SST, due to the "fitted" model.

$$\begin{aligned} \text{SSE} &= (\mathbf{y} - \mathbf{X}\mathbf{b})^T(\mathbf{y} - \mathbf{X}\mathbf{b}) = \mathbf{y}^T\mathbf{y} - \mathbf{b}^T\mathbf{X}^T\mathbf{y} - \mathbf{y}^T\mathbf{X}\mathbf{b} + \mathbf{b}^T\mathbf{X}^T\mathbf{X}\mathbf{b} \\ &= \mathbf{y}^T\mathbf{y} - 2\mathbf{b}^T\mathbf{X}^T\mathbf{y} + \mathbf{b}^T\mathbf{X}^T\mathbf{X}\mathbf{b} \qquad (12.6) \\ &= \mathbf{y}^T\mathbf{y} - 2\mathbf{b}^T\mathbf{X}^T\mathbf{y} + \mathbf{b}^T\mathbf{X}^T\mathbf{y} = \mathbf{y}^T\mathbf{y} - \mathbf{b}^T\mathbf{X}^T\mathbf{y} \\ &= (\mathbf{y}^T\mathbf{y} - n\bar{y}^2) - (\mathbf{b}^T\mathbf{X}^T\mathbf{y} - n\bar{y}^2) \end{aligned}$$

For the data of Table 12.1, the MLR program obtains SST $= 2715.76$ and SSR $= 2657.86$, which yields SSE $= 57.90$. (Note that $\mathbf{y}^T\mathbf{y} = \sum y_i^2$ is the "uncorrected" sum of squares of the y_i and $n\bar{y}^2$ is the "correction factor," which is also the sum of squares due to the presence of b_0 in the model. This implies that $\mathbf{b}^T\mathbf{X}^T\mathbf{y}$ is the sum of squares due to the presence of *all* parameters in the model. Thus SSR $= \mathbf{b}^T\mathbf{X}^T\mathbf{y} - n\bar{y}^2$ is the sum of squares due to all parameters excluding b_0, i.e., due to the fitted model.)

4. The *mean square of errors,*

$$s^2 = s^2_{Y|x} = \frac{\text{SSE}}{(n - p)} = \text{MSE} \tag{12.7}$$

is the best unbiased estimate of σ^2 where p is the number of parameters in the MLR model. (In Equation 12.1, p = 3.) For Table 12.1, $s^2 = 57.90/10 = 5.79$ ($s = 2.406$). This fact is very useful, since we rarely know the true value of σ^2 and are forced to use an estimate from the data in any confidence interval computations or predictions based on the final equation that is fitted to the data. Further, it may be shown that

$$\frac{(n - p)S^2}{\sigma^2} \tag{12.8}$$

is distributed according to a χ^2_{n-p} distribution (where S^2 is the estimator associated with the sample estimate, s^2 = MSE).

5. The *mean square of regression* is

$$\text{MSR} = \frac{\text{SSR}}{(p - 1)} \tag{12.9}$$

For Table 12.1, MSR = 2657.86/2 = 1328.93. Since the degrees of freedom for MSR are $p - 1$ and the degrees of freedom for s^2 are $n - p$, MSR and s^2 account for all $n - 1$ degrees of freedom in the data set. In addition, it may be shown that

$$\frac{(p - 1)\sum(\hat{Y}_i - \overline{Y})^2}{\sigma^2} \tag{12.10}$$

is distributed according to a χ^2_{p-1} distribution if and only if the β_j, $j = 1, \ldots, p - 1$ are all equal to zero [where $\sum(\hat{Y}_i - \overline{Y})^2$ is the estimator associated with the estimate SSR]. This fact is important in item 7.

6. The *multiple correlation coefficient squared,*

$$r^2 = \frac{\text{SSR}}{\text{SST}} \tag{12.11}$$

gives us the proportion of SST that is explained by the fitted model. For Table 12.1, $r^2 = 2657.86/2715.76 = 0.9787$. (Notice that in MLR the simple interpretation of r^2 that existed in SLR—namely, $r_{x,y} = \sqrt{r^2}$—is no longer valid. Some authors strengthen this distinction by denoting the multiple correlation coefficient squared by R^2 when $p > 2$.)

7. In Equation 11.14 we state the form of the F ratio that can be used to test whether $\beta_1 = 0$ for SLR. In MLR, the same ratio can be used to test the "composite" hypothesis,

$$\beta_1 = \beta_2 = \cdots = \beta_{p-1} = 0$$

If this hypothesis is true, then the expressions in Equations 12.8 and 12.10 are both distributed according to independent χ^2 distribution functions. Putting those expressions in ratio and dividing them by their respective degrees of freedom yield

$$F = \frac{(p-1)\sum(\hat{Y}_i - \overline{Y})^2/[\sigma^2(p-1)]}{(n-p)S^2/[\sigma^2(n-p)]} = \frac{\sum(\hat{Y}_i - \overline{Y})^2}{S^2} \tag{12.12}$$

which is distributed according to $F_{p-1,n-p}$ *if* $\beta_1 = \beta_2 = \cdots = \beta_{p-1} = 0$. (Note that in simple linear regression, SSR = MSR, since $p - 1 = 1$.) The realization of F from the data is

$$f = \frac{(p-1)\text{MSR}/[\sigma^2(p-1)]}{(n-p)s^2/[\sigma^2(n-p)]} = \frac{\text{MSR}}{s^2} = \frac{\text{MSR}}{\text{MSE}} \tag{12.12a}$$

If $f > F_{p-1,n-p,\alpha}$, we conclude that at least one of the β_j is not equal to zero and that some significant relation exists between the independent and dependent variables. Otherwise, we may conclude that the regression fails and no relation exists. For Table 12.1 and the fitted model, $f = 229.505$. Comparing this value with the $F_{2,10}$ distribution yields the conclusion that the p-value associated with the hypothesis of $\beta_1 = \beta = \cdots = \beta_{p-1} = 0$ is less than 0.00005.

12.4 THE VARIANCE-COVARIANCE AND CORRELATION MATRICES OF $\hat{\boldsymbol{\beta}}$

The previous section has shown that $\boldsymbol{\sigma}^2(\hat{\boldsymbol{\beta}}) = (\mathbf{X}^T\mathbf{X})^{-1}\boldsymbol{\sigma}^2$. Since we do not know the true value of σ^2, we must use s^2 in its place. Thus our *estimate* of the variance-covariance matrix of $\boldsymbol{\beta}$ is $\mathbf{s}^2(\hat{\boldsymbol{\beta}}) = (X^TX)^{-1}\mathbf{s}^2$. The element of $s^2(\hat{\boldsymbol{\beta}})$ at row i and column j is an estimate of the covariance between $\hat{\beta}_i$ and $\hat{\beta}_j$. (Remember that i and j both start at a value of 0.) Since the kth diagonal element of $\mathbf{s}^2(\hat{\boldsymbol{\beta}})$ is the estimated covariance of $\hat{\beta}_k$ with itself, it is the estimated variance of $\hat{\beta}_k$, $s^2(\hat{\beta}_k)$.

Example 12.4. For the data of Table 12.1 and our current fitted model from Equation 12.1, the MLR program obtains

$$\mathbf{s}(\hat{\boldsymbol{\beta}}) = \begin{bmatrix} 5.227 & -0.049 & -0.092 \\ -0.049 & 0.015 & -0.001 \\ -0.092 & -0.001 & 0.002 \end{bmatrix} = \overbrace{\begin{bmatrix} 0.9026 & -0.0084 & -0.0158 \\ -0.0084 & 0.0025 & -0.0002 \\ -0.0158 & -0.0002 & 0.0004 \end{bmatrix}}^{(\mathbf{X}^T\mathbf{X})^{-1}} s^2$$

Therefore, the estimate of the covariance between $\hat{\beta}_1$ and $\hat{\beta}_2$ is -0.001, and the estimate of the covariance between $\hat{\beta}_0$ and $\hat{\beta}_1$ is -0.049. The estimate of the variance of $\hat{\beta}_0$ is $s^2(\hat{\beta}_0) = 5.227$, and the estimate of the variance of $\hat{\beta}_1$ is $s^2(\hat{\beta}_1) = 0.015$.

The $\mathbf{s}^2(\hat{\boldsymbol{\beta}})$ matrix is always *symmetric*—a matrix in which the elements above the main diagonal are a reflection of the elements below the diagonal. For this reason, the covariance matrix is often presented with only the elements above and in the main diagonal.

The estimated variance of any parameter still retains its usual interpretation, and the estimated standard deviation of any parameter can still be used to set a univariate confidence interval about that parameter's mean. The covariance between any two parameter estimators, $\hat{\beta}_i$ and $\hat{\beta}_j$, may be interpreted in exactly the same way that covariance between $\hat{\beta}_0$ and $\hat{\beta}_1$ is interpreted in Chapter 11. Indeed, a joint confidence ellipse can be constructed for any pair of parameters in the same manner that a confidence ellipse is constructed for β_0 and β_1 in Chapter 11.

Unfortunately, in the presence of three or more parameters, a bivariate confidence ellipse no longer contains as much information as it does when there are only two parameters. In our current example with three parameters, we would have to construct a trivariate confidence ellipsoid (like a football or watermelon) to present the same simultaneous amount of information that is presented by the confidence ellipse in SLR. As we consider models with four or more parameters, it is still possible mathematically to characterize the joint confidence region for the parameters. However, because we are three-dimensional beings, it is no longer possible to draw or even to conceptualize the "p-dimensional hyperellipsoid" $(p > 3)$ that would be the physical realization of that joint confidence region.

Realizing that limitation, we may still gain some insight into the dependency structure that exists among the parameter estimates in MLR by forming an *estimate* of the correlation matrix for $\hat{\boldsymbol{\beta}}$, $\mathbf{r}(\hat{\boldsymbol{\beta}})$, by substituting the appropriate information contained in $\mathbf{s}^2(\hat{\boldsymbol{\beta}})$ into the formula given in Equation 11.23.

Example 12.5. For Table 12.1,

$$\mathbf{r}(\hat{\boldsymbol{\beta}}) = \begin{bmatrix} 1.000 & -0.175 & -0.875 \\ -0.175 & 1.000 & -0.229 \\ -0.875 & -0.229 & 1.000 \end{bmatrix}$$

From examination of $\mathbf{r}(\hat{\boldsymbol{\beta}})$, we may state that $\hat{\beta}_0$ and $\hat{\beta}_2$ possess a *high* negative correlation, whereas the other two possible pairings have no appreciable interdependence.

12.5 UNIVARIATE CONFIDENCE INTERVALS ON THE β_j AND ON PREDICTED VALUES OF Y

If the assumptions stated in Section 11.2 are reasonably well satisfied, it may be shown that each estimator $\hat{\beta}_j$ is approximately distributed according to an $N(\beta_j, \sigma(\hat{\beta}_j))$ distribution and that the $\hat{\beta}_j$ are statistically independent of S^2. Since the $\hat{\beta}_j$ are normally distributed random variables, we may use the methods of Chapters 8 and 9 to place univariate confidence intervals on any β_j. The general expression for these $1 - \alpha$ confidence intervals is

$$\left(b_j - t_{n-p,\alpha/2}s(\hat{\beta}_j),\, b_j + t_{n-p,\alpha/2}s(\hat{\beta}_j)\right) \tag{12.13}$$

Another way to use the fact that $\hat{\beta}_j$ is $N(\beta_j, \sigma(\hat{\beta}_j))$ is to form a univariate hypothesis test for any selected β_j.

Suppose that $H_0 : \beta_j = \beta_j^0$ and $H_A : \beta_j \neq \beta_j^0$. It may be shown that

$$T = \frac{\hat{\beta}_j - \beta_j^0}{S(\hat{\beta}_j)} \tag{12.14}$$

is distributed according to a t_{n-p} distribution [where $S(\hat{\beta}_j)$ is the estimator of the standard deviation of $\hat{\beta}_j$]. Thus, if the absolute value of $t = (b_j - \beta_j^0)/s(\hat{\beta}_j)$ is greater than $t_{n-p,\alpha/2}$, we may reject H_0 in favor of H_A. The t estimates and the right-hand-tail probability associated with each are a standard part of the output of the MLR program (where $\beta_j^0 = 0$).

Example 12.6. Table 12.2 gives the MLR program output for the parameter estimates associated with the data of Table 12.1. It succinctly displays the parameter index, estimated value, sample standard deviation, t estimate, right-hand-tail probability of the t value, and the value of the standardized b_i. Since b_0, b_1, and b_2 are all greater than zero, the fifth column of Table 12.2 also gives the p-value of $\alpha/2$ that would just barely cause the rejection of $H_0 : \beta_i = \beta_i^0 = 0$.

We may feel quite comfortable with rejecting H_0 for β_0, β_1, and β_2 since the associated p-values are less than 0.00005.

Although the theoretical development of the values of the standardized b_i (STD-B(I)), given in the sixth column, is beyond the scope of this book, their interpretation and use are not. In essence, the standardized b_i give the number of standard deviations that Y will change for each standard deviation change in the value of x_i.

Thus, the standardized b_i is a measure of the effect that a change in independent variable i has on the change of the dependent variable. No value is given for b_0 because its variable never changes from its constant value of 1. Since s_{x_1} is small compared to s_Y, a relatively small change in x_1 will cause a relatively large change in Y.

Suppose that we are concerned with the accuracy of the estimate of the mean of Y at a specified value of $\mathbf{x}_k^T = [1 \quad x_{1k} \quad x_{2k} \quad \ldots \quad x_{p-1,k}]$. Using the replicative property of the normal distribution, it may be shown that the random variable $Y_k = \mathbf{x}_k^T\hat{\boldsymbol{\beta}} = \sum \hat{\beta}_i x_{ik}$ is distributed as

$$N\left(\mathbf{x}_k^T\boldsymbol{\beta}, \sigma\sqrt{\mathbf{x}_k^T(\mathbf{X}^T\mathbf{X})^{-1}\mathbf{x}_k}\right) = N(\mu_{Yk}, \sigma_{Yk}) \tag{12.15}$$

Substituting the values of $\mathbf{b}$ and s (for $\boldsymbol{\beta}$ and σ) in Equation 12.15, we may use the results to obtain the following *confidence interval on the long-term mean of Y_k for a specified value of $\mathbf{x_k}$*.

$$(\mathbf{x}_k^T\mathbf{b} - t_{n-p,\alpha/2}s_{Yk}, \quad \mathbf{x}_k^T\mathbf{b} + t_{n-p,\alpha/2}s_{Yk}) \tag{12.16}$$

Suppose that we are interested in a single predicted or future value of Y for a specific value of $\mathbf{x}_k$. This results in only one change in the considerations

TABLE 12.2
Output for the parameter estimates

I	B(I)	SD(I)	t	PR(I)	STD-B(I)
0	52.57	2.286	22.998	0.0000	
1	1.468	0.121	12.105	0.0000	0.5741
2	0.662	0.046	14.442	0.0000	0.6850

contained in Equations 12.15 and 12.16. We must replace s_{Yk}, the estimate of the standard deviation of Y_k, with the sample standard deviation of the future value of Y, $\sqrt{s_{Yk}^2 + s^2}$. The additional s^2 simply reflects the uncertainty added to the process because of the presence of the single new observation of Y.

12.6 DETERMINING WHETHER A FIT IS ADEQUATE AND COMPARING COMPETING MODELS

Essentially, the measures of goodness of fit that are used in SLR are also used in MLR. The primary measures are the values of r^2, s^2, and the indications from the residual plots. We desire a high value of r^2, a low value of s^2, an acceptably straight ECDF plot, and nice wide horizontal random bands for the plots of the e_i against any selected measure. If these desires are met, we will have explained most of the variability contained in the data and will be able to set reasonably small confidence intervals on items of interest. We will also know that our basic assumptions about the data are correct.

Lack of fit will usually make itself known in a diminished value of r^2, an inflated value of s^2, and in residual plots that indicate departures from the primary assumptions of MLR. (Some statisticians make use of an additional measure, the C_P statistic; it combines much of the information in r^2 and s^2 into one value. Readers who desire more information on the C_P statistic are referred to the elegant development by Daniel and Wood [1971].)

In MLR, a few more residual plots are available, because we will want to plot the residuals against each of the independent variables. After the e_i are computed for the fitted model, they are displayed by the MLR program in a table very similar to the table for the SLR program. Next, the MLR program presents the following menu.

```
MENU FOR RESIDUAL PLOTS
SELECT:
1. EMPIRICAL CUMULATIVE DIST PLOT
2. RESID. AGAINST INPUT ORDER
3. RESID. AGAINST AN INDEPENDENT VAR
4. RESID. AGAINST DEPENDENT VARIABLE, Y
5. RESID. AGAINST FITTED Y
6. NO RESIDUAL PLOTS
```

Selection of options 1, 2, 4, or 5 causes responses similar to those in SLR. The only major difference from SLR is in the selection of option 3, which causes the program to request which independent variable is to be used.

All of the residual plots associated with the current model as given in Equation 12.1 are adequate. (With only 13 data points, any other result would be surprising.) The only remaining question to ask is whether there is a better model to fit the data set of Table 12.1. In general, with a "full" model of p parameters, there are 2^{p-1} different models that can be immediately considered for any data set. These different models correspond to the various possibilities that arise when each independent variable is considered to be either a member of the model or not.

Example 12.7. For our current full model, $p = 3$. The $2^2 = 4$ different models to be considered, with their associated values of r^2 and s^2 and residual indications, are presented in Table 12.3. If we limit our considerations to these four models alone, it is clear that model 1 is superior to the other three. This conclusion is not surprising, since the parameters associated with both independent variables, β_1 and β_2, were found to be significantly different from zero in the implied t tests presented in Table 12.2.

We are not limited to the four models in Table 12.3. Other variables may be formed from algebraic combinations of our two basic variables. Examples of such new variables are $x_3 = x_1 x_2$, $x_4 = x_1^2$, $x_5 = x_2^2$, and $x_6 = \log x_1$. Of course, many other possibilities for the formation of new variables exist, and their form will often be suggested by the data and the physical phenomenon under study. Indeed, on occasion, it may be appropriate to transform Y as was done in some of the cases of curvilinear regression in Chapter 11. When this is done, however, it should always be remembered that the error structure of the model for the untransformed Y may be different from what was intended or appropriate.

As part of your software package, a program for performing transformations on MLR and SLR data is provided. Use of this program will readily allow you to add or delete variables from the data set and to perform a great many types of transformations on the variables that are present in the model. (All transformations, additions, and deletions of variables cited in Tables 12.3 and 12.4 have been performed using that program.)

The only stringent restriction on the formation of new variables is that they *cannot* be simple linear functions of one or more of the variables in the model. For example, $x_9 = x_1 + x_2$ and $x_{10} = 0.65x_3$ are not allowed because they are linear functions of x_1, x_2, and x_3. To introduce such variables in the model would confound the variable definitions of the model and would cause insurmountable numerical difficulties in solving for the estimates of the parameter values.

Let us consider seven more models, as given in Table 12.4; they are modifications of model 1 in Table 12.3. By studying Table 12.4, we can observe the empirical verification of certain facts associated with adding a variable to an MLR model. First, no matter what variable is added to an MLR model, r^2 will never decrease and will almost always increase. Second, the residual plots may

TABLE 12.3
The four basic models from equation 12.1

Model Number	Model	r^2	s^2	e_i Plots
1	$Y_i = \beta_0 + \beta_1 x_{1i} + \beta_2 x_{2i} + \varepsilon_i$	0.9787	5.79	Okay
2	$Y_i = \beta_0 + \beta_1 x_{1i} \quad + \varepsilon_i$	0.5339	115.06	Variable missing
3	$Y_i = \beta_0 \quad + \beta_2 x_{2i} + \varepsilon_i$	0.6663	82.39	ECDF skewed (variable missing)
4	$Y_i = \beta_0 \quad + \varepsilon_i$	—	226.31	—

TABLE 12.4
Seven augmented models from Equation 12.1

Model Number	Model	r^2	s^2	e_i **Plots**
1	$Y_i = \beta_0 + \beta_1 x_{1i} + \beta_2 x_{2i} + \varepsilon_i$	0.9787	5.79	Acceptable
5	$Y_i = \beta_0 + \beta_1 x_{1i} + \beta_2 x_{2i} + \beta_3 x_{1i} x_{2i} + \varepsilon_i$	0.9790	6.35	Marginal
6	$Y_i = \beta_0 + \beta_1 x_{1i} + \beta_2 x_{2i} + \beta_4 x_{1i}^2 + \varepsilon_i$	0.9813	5.64	*Good*
7	$Y_i = \beta_0 + \beta_1 x_{1i} + \beta_2 x_{2i} + \beta_5 x_{2i}^2 + \varepsilon_i$	0.9852	4.47	*Good*
8	$Y_i = \beta_0 + \beta_1 x_{1i} + \beta_2 x_{2i} + \beta_3 x_{1i} x_{2i} + \beta_4 x_{1i}^2 + \varepsilon_i$	0.9815	6.27	Acceptable
9	$Y_i = \beta_0 + \beta_1 x_{1i} + \beta_2 x_{2i} + \beta_3 x_{1i} x_{2i} + \beta_5 x_{2i}^2 + \varepsilon_i$	0.9853	4.98	Acceptable
10	$Y_i = \beta_0 + \beta_1 x_{1i} + \beta_2 x_{2i} + \beta_4 x_{1i}^2 + \beta_5 x_{2i}^2 + \varepsilon_i$	0.9861	4.70	Acceptable
11	$Y_i = \beta_0 + \beta_1 x_{1i} + \beta_2 x_{2i} + \beta_3 x_{1i} x_{2i} + \beta_4 x_{1i}^2 + \beta_5 x_{2i}^2 + \varepsilon_i$	0.9862	5.33	Acceptable

improve or degrade. Third, the value of the mean square error, s^2, may increase or decrease depending on how well the additional variable relates to the dependent variable and how much of that relation has been explained by the variables that are already in the model.

Clearly, the objective of obtaining a large value of r^2 cannot be considered by itself, since the addition of enough new independent variables will almost certainly drive r^2 as close as desired to its maximum value of 1.0. We must also consider the effect that each new variable has on the residual plots and on the value of s^2. For example, creating model 5 in Table 12.4 by adding the variable $x_1 x_2$ to model 1 causes only a small increase in r^2 while causing a large increase in s^2 and damaging the residual plots. The reason that s^2 increases is that the additional amount of the total sum of squares that is explained by the new variable is not sufficient to offset the loss of an additional degree of freedom associated with s^2.

To clarify, recall that

$$\text{SSE} = \text{SST} - \text{SSR} \qquad (12.5)$$

By adding a variable to the model, SSR will almost always be increased to a new value, $\mathbb{SSR}$, yielding the new SSE: $\mathbb{SSE} = \text{SST} - \mathbb{SSR}$, where $\mathbb{SSE} < \text{SSE}$. However, the new value of p, the number of parameters, is $\mathbb{p} = p + 1$. Thus the new value of s^2, $\mathbb{s}^2$, is

$$\mathbb{s}^2 = \frac{\mathbb{SSE}}{(n - p - 1)} \qquad (12.7)$$

Unless the decrease in SSE is proportionally greater than the decrease in the denominator of Equation 12.7, $\mathbb{s}^2$ will be greater than s^2. A new variable that causes this type of behavior simply does not have the "membership credentials" to join the model.

On the other hand, model 7, which was created by adding x_2^2 to model 1, is the best of all of the models considered in Tables 12.3 and 12.4. Model 7 has a value of r^2 that is only a little less than the maximal r^2 of the new "full model" given by model 11, whereas model 7 has the smallest s^2 of all the models and has good residual plots.

As a caution, we note that the decision as to what model is "best" is not always as obvious as in the preceding example. Often, there will be two or more models with about the same r^2 and s^2 and with comparable residual plots. When such is the case, the decision is made on grounds that have not been quantified by the model, i.e., the simplest model, the model with the fewest terms, the model that most nearly agrees with the theoretical formula that was proposed before the experiment, the model that the boss likes best, and so on.

We have not considered all $2^5 = 32$ models that might be formed from subsets of model 11 (Table 12.4). Thus there may be a superior model in the remaining 21 models that we have not examined. In general, the unfortunate fact is that the *only* way we can be *sure* to have found the "best" model is to examine all 2^{p-1} possibilities. To do so without the assistance of a computer, however, quickly becomes an impossible task. Consider the case where p has a relatively small value of 13. This would imply $2^{12} = 4096$ possible models to examine!

Fortunately, other methods, which are beyond the scope of this book, have been developed to attack this problem and find good, but not necessarily the best, fitting models with greatly reduced computational effort. In essence, there are two approaches. The first approach starts with a *full model* with all of the independent variables that might be related to the dependent variable included. Then, based on a selected statistical measure, such as the lowest t value, variables are thrown out until the "membership credentials" of the remaining variables cannot be challenged. The second of the approaches starts with $Y_i = \beta_0$ and adds variables according to a selected statistical measure until no variable possesses suitable credentials. A third method is simply a hybrid of the two just described. At each step, a variable is either added to or deleted from the model until no variable is selected to leave or join the model. The flexibility implied in considering any variable for removal from or addition to the model at each step often leads to superior results over the more basic approaches.

In addition to more sophisticated methods for performing MLR, there exist more sophisticated methods of fitting equations to data. Two prominent methods are *ridge regression* and *nonlinear MLR*. Ridge regression has been especially developed to combat the problems that arise when *multicollinearity* is present in the data set. Multicollinearity is present when the independent variables are excessively intercorrelated, i.e., most or all possible pairs of the independent variables have high correlation. When this condition is present, all or most of the effect of any selected independent variable will be accounted for by the other independent variables as a group. This naturally gives rise to the question of which of the variables should remain in the model, since all variables seem to have about the same importance. Ridge regression introduces a small amount of *bias* or *error* into the estimation procedure in such a way that much of the problem of multicollinearity is removed while hopefully not seriously damaging the accuracy and value of the fitted model. Nonlinear MLR allows the model to contain nonlinear functions of the elements of $\boldsymbol{\beta}$. As pointed out in Exercise 11.5, this can lead to very complicated problems in deriving the estimates of the model parameters.

For those readers interested in learning more about these methods and about MLR in general, consult Daniel and Wood [1971], Draper and Smith [1981], Neter, Wasserman, and Kutner [1990], Nie et al., [1975], IMSL Library [1984], and SAS [1984]. For the reader interested in access to statistical analysis packages that take full advantage of the powers of modern computers, the latter three references contain complete user instructions to the analysis software with which they are associated. Indeed, their emphasis is on the use of the software, and the documentation often leans heavily on other more theoretical sources.

12.7 THE GENERAL LINEAR MODEL AND "THE OTHER SIDE OF THE COIN"

Equation 12.2 is an example of what users of statistics have come to call the *general linear model*. We have seen but one use of this model in the last two chapters, fitting equations to data. There are many other applications of this remarkably versatile model, among which are factor analysis, discriminant analysis, and time series analysis. However, in Equation 12.12, we had a glimpse at one of the several "other sides of the coin": composite hypothesis testing. In the next two chapters, we address a specific type of composite hypothesis test—the analysis of variance—in which we develop methods to test whether the means of several different normally distributed populations are equal.

EXERCISES

12.1. In a study of the possible polluting effects of dispensing large quantities of gasoline from stationary storage tanks to identical tank trucks, data were gathered on x_1 = tank temperature, x_2 = temperature of dispensed gasoline, x_3 = tank vapor pressure, x_4 = vapor pressure of dispensed gasoline, and y = quantity of hydrocarbons released into the atmosphere. Given the following data set, perform an MLR analysis and find the best-fitting equation for the data.

x_1	x_2	x_3	x_4	y	x_1	x_2	x_3	x_4	y	x_1	x_2	x_3	x_4	y	x_1	x_2	x_3	x_4	y
33	53	3.32	3.42	29	31	36	3.10	3.26	24	33	51	3.18	3.18	26	37	51	3.39	3.08	22
36	54	3.20	3.41	27	35	35	3.03	3.03	21	59	56	4.78	4.57	33	60	60	4.72	4.72	34
59	60	4.60	4.41	32	60	60	4.53	4.53	34	34	35	2.90	2.95	20	60	59	4.40	4.36	36
60	62	4.31	4.42	34	60	36	4.27	3.94	23	62	38	4.41	3.49	24	62	61	4.39	4.39	32
90	64	7.32	6.70	40	90	60	7.32	7.20	46	92	92	7.45	7.45	55	91	92	7.27	7.26	52
61	62	3.91	4.08	29	59	42	3.75	3.45	22	88	65	6.48	5.80	31	91	89	6.70	6.60	45
63	62	4.30	4.30	37	60	61	4.02	4.10	37	60	62	4.02	3.89	33	59	62	3.98	4.02	27
59	62	4.39	4.53	34	37	35	2.75	2.64	19	35	35	2.59	2.59	16	37	37	2.73	2.59	22

12.2. Use Equation 12.13 to set 95% confidence intervals on the β_j given in Table 12.2. Do you think that the "box" that would be formed by the simultaneous use of those three univariate confidence intervals would be a reliable measure of the joint behavior of $\hat{\beta}_0$, $\hat{\beta}_1$, and $\hat{\beta}_2$? Why or why not? If not, what would you use instead?

12.3. Because of a decline in the health of many of the older trees, the Forestry Service is conducting an experiment to determine if there is a relationship between the height, y, of a certain conifer and the chemical content of the tree's needles. Toward this end, 26 trees are identified and a sample of each tree's needles are collected. Each sample is then incinerated, and the following percentages of the total needle weight are measured: x_1 = nitrogen, x_2 = phosphorus, x_3 = potassium, x_4 = residual ash. Fit your best MLR model to the following data set. How useful do you think your fitted equation is? Why?

x_1	x_2	x_3	x_4	y	x_1	x_2	x_3	x_4	y	x_1	x_2	x_3	x_4	y
2.20	0.417	1.35	1.79	351	2.10	0.354	0.90	1.08	249	1.52	0.208	0.71	0.47	171
2.88	0.335	0.90	1.48	373	2.18	0.314	1.26	1.09	321	1.87	0.271	1.15	0.99	191
1.52	0.164	0.83	0.85	225	2.37	0.302	0.89	0.94	291	2.06	0.373	0.79	0.80	284
1.84	0.265	0.72	0.77	213	1.89	0.192	0.46	0.46	138	2.45	0.221	0.76	0.95	213
1.88	0.186	0.52	0.95	151	1.93	0.207	0.60	0.92	130	1.80	0.157	0.67	0.60	93
1.81	0.195	0.47	0.57	95	1.59	0.165	0.66	0.80	147	1.53	0.226	0.68	0.66	88
1.43	0.227	0.44	0.45	65	1.54	0.271	0.51	0.95	120	1.13	0.187	0.38	0.63	72
1.63	0.200	0.62	1.10	160	1.36	0.211	0.71	0.47	72	1.76	0.283	0.96	0.96	252
2.53	0.284	0.85	1.39	310	2.59	0.303	1.02	0.95	336					

12.4. An independent testing agency has been retained to determine if there is a significant relationship between the percent of nicotine, y, and the percentages of nitrogen, chlorine, potassium, phosphorus, calcium, and magnesium, x_1 through x_6, in a sample of tobacco. Fit your best MLR model to the data given here and comment on the usefulness of your fit.

x_1	x_2	x_3	x_4	x_5	x_6	y	x_1	x_2	x_3	x_4	x_5	x_6	y
2.02	2.90	2.17	0.51	3.47	0.91	1.38	2.62	2.78	1.72	0.50	4.57	1.25	2.64
2.08	2.68	2.40	0.43	3.52	0.82	1.56	2.20	3.17	2.06	0.52	3.69	0.97	2.22
2.38	2.52	2.18	0.42	4.01	1.12	2.85	2.03	2.56	2.57	0.44	2.79	0.82	1.24
2.87	2.67	2.64	0.50	3.92	1.06	2.86	1.88	2.58	2.22	0.49	3.58	1.01	2.18
1.93	2.26	2.15	0.56	3.57	0.92	1.65	2.57	1.74	1.64	0.51	4.38	1.22	3.28
1.95	2.15	2.48	0.48	3.28	0.81	1.56	2.03	2.00	2.38	0.50	3.31	0.98	2.00
2.50	2.07	2.32	0.48	3.72	1.04	2.88	1.72	2.24	2.25	0.52	3.10	0.78	1.36
2.53	1.74	2.64	0.50	3.48	0.93	2.66	1.90	1.46	1.97	0.46	3.48	0.90	2.43
2.18	0.74	2.46	0.48	3.16	0.86	2.42	2.16	2.84	2.36	0.49	3.68	0.95	2.16
2.14	3.30	2.04	0.48	3.28	1.06	2.12	1.98	2.90	2.16	0.48	3.56	0.84	1.87
1.89	2.82	2.04	0.53	3.56	1.02	2.10	2.07	2.79	2.15	0.52	3.49	1.04	2.21
2.08	3.14	2.60	0.50	3.30	0.80	2.00	2.21	2.81	2.18	0.44	4.16	0.92	2.26
2.00	3.16	2.22	0.51	3.73	1.07	2.14							

12.5. Use Equation 12.16 to set a 95% confidence interval on the mean value of the dependent variable for $\mathbf{x}_k^{\mathrm{T}} = [6, 41]$ for the model of Table 12.2. In addition, set a 95% confidence interval on the single future or predicted value of the dependent variable for $\mathbf{x}_k^{\mathrm{T}} = [6, 41]$.

12.6. The measurements of two additional independent variables have been found for the data set of Table 12.1. They are presented below as x_3 and x_4. Perform an

MLR analysis on this augmented data set. Do the two new variables help any? Why or why not?

i	1	2	3	4	5	6	7	8	9	10	11	12	13
x_{3i}	6	15	8	8	6	9	17	22	18	4	23	9	8
x_{4i}	60	52	20	47	33	22	6	44	22	26	34	12	12

12.7. The following data set is from a chemical plant that produces nitric acid through the oxidation of nitrogen: x_1 is the amount of air flowing into the reaction, x_2 is the temperature of the cooling water that is entering the nitric oxide absorption tower of the plant, x_3 is the coded concentration of nitric acid in the absorbing liquid, and y is an indirect quantitative measure of the yield of the nitric acid from the plant. Given the following data set from the plant, fit your best MLR model to the data and comment on how useful you think your model is. Why?

x_1	x_2	x_3	y	x_1	x_2	x_3	y	x_1	x_2	x_3	y	x_1	x_2	x_3	y	x_1	x_2	x_3	y
80	27	89	42	80	27	88	37	75	25	90	37	62	24	87	28	62	22	87	18
62	23	87	18	62	24	93	19	62	24	93	20	58	23	87	15	58	18	80	14
58	18	89	14	58	17	88	13	58	18	82	11	58	19	93	12	50	18	89	8
50	18	86	7	50	19	72	8	50	19	79	8	50	20	80	9	56	20	82	15
70	20	91	15																

12.8. Given next is a part of the MLR output from a full model, $Y = \beta_0 + \beta_1 x_1 + \beta_2 x_2 + \beta_3 x_3 + \beta_4 x_4 + \varepsilon$, for a set of 30 data points. Comment on how well the model fits the data and what you would do to progress to a better fit. Fully explain your reasons for the steps you would take.

x_1	x_2	x_3	x_4	y	x_1	x_2	x_3	x_4	y
13.920	9.560	7.290	1.020	3.25	17.760	13.560	20.840	3.270	0.31
15.260	11.960	8.890	1.470	1.32	17.890	13.120	9.850	1.340	1.00
15.130	11.010	7.220	.500	1.23	13.650	9.160	5.950	.570	2.83
17.400	10.880	5.910	.420	1.87	14.320	10.770	9.240	.940	2.30
19.590	15.340	22.560	4.150	.31	14.410	10.980	7.470	.700	1.23
14.150	10.520	7.280	.660	1.32	15.450	11.960	8.850	.970	.93
13.040	9.800	6.590	.560	1.32	18.590	13.900	22.630	4.400	.50
14.100	9.000	5.240	.200	5.28	10.560	6.850	5.060	.530	9.19
12.280	8.470	4.440	.200	3.25	17.190	12.030	9.070	.960	.71
15.060	10.800	6.420	.530	1.07	13.640	9.920	6.570	.520	3.48
17.510	12.150	8.660	.640	1.00	14.320	10.900	8.800	1.210	1.52
16.020	12.290	13.520	1.840	.71	12.530	9.400	6.400	.440	1.32
16.460	11.860	7.910	.770	1.52	12.510	8.710	4.740	.380	4.00
12.650	9.650	8.170	1.830	2.46	18.330	13.660	13.090	1.510	.41
16.980	12.930	9.140	.910	1.62	19.540	14.420	14.020	1.770	.62

```
MEANS OF THE VARIABLES                         STD. DEV. OF THE  VARIABLES
15.341 11.185 9.394 1.174 1.93                 2.34  1.966  4.893 1.057 1.82

SIMPLE CORRELATION COEFFICIENTS                VAR MIN    MAX    RANGE
1      .9473 .7287  .6095 -.6918               1    10.56 19.59 9.03
       1     .8126  .7084 -.7802               2    6.85  15.34 8.49
             1      .9639 -.5234               3    4.44  22.63 18.19
                    1     -.4287               4    .2    4.4   4.2
                           1                   5    .31   9.19  8.88


CORREL MATRIX  FOR REGRESSION COEFF      I B(I)  SD(I)    T     PR>T  STD-B(I)
1  -0.3619 -0.1175  0.3189  -0.1627      0 10.59 1.634 6.479  5E-05      0
        1 -0.8545 -0.1162   0.2004       1 .4985 .2859 1.743 .0450  .6412
                1 -0.2571   0.0745       2 -1.56 .3928 -3.97 .9996  -1.68
                        1  -0.9418       3 .0967 .2188 .4419 .3328  .2601
                                 1       4 .2157 .8485 .2542 .3983  .1252


MULTIPLE CORRELATION COEFFICIENT R-SQUARED = .683457516
F STATISTIC = 13.4945851 WITH 4 AND 25 DEGREES OF FREEDOM (PR(X<F) BY
CHANCE=0.99995) STANDARD DEVIATION OF ERROR = 1.10251218 (VARIANCE OF
ERROR = 1.2155331 WITH 25 D.F.)
```

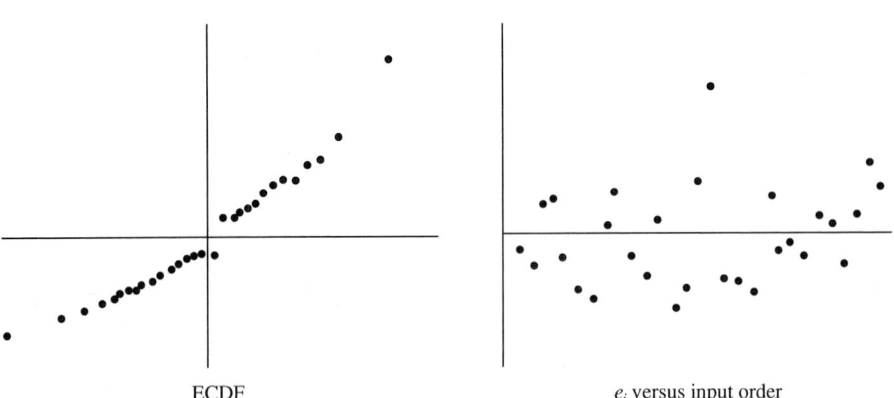

ECDF　　　　e_i versus input order

***12.9.** A physical process has a very strong synergistic effect in the presence of two chemicals. This effect dominates the response except when either of the chemicals is present in an apparent "damping" concentration. When either chemical is at its damping concentration, the response is greatly reduced. Using the following data set, obtain an estimate of the damping concentration of each chemical, x_1 and x_2, and fit your best MLR model to the data set.

x_1	x_2	y	x_1	x_2	y	x_1	x_2	y
−4.729	11.247	8092.4737	−3.405	11.625	6541.3889	−1.376	10.799	3146.8172
−4.288	11.443	7689.9758	−0.045	8.321	1014.8270	2.285	8.148	280.0843
1.07	9.898	967.0259	−0.011	10.509	1819.340	3.58	10.283	139.611
3.202	10.25	220.1754	3.061	8.819	177.3434	4.88	9.686	−2.4922
7.025	10.531	302.5477	10.488	10.913	2395.4059	10.624	11.355	2766.8198
10.825	10.531	2471.71	12.902	8.783	2871.3727	12.554	11.658	5324.589
13.085	9.486	3661.6622	12.363	9.197	2808.566	−1.757	6.554	947.7396
−1.494	5.886	637.070	−2.281	5.251	557.2	−3.245	4.073	291.3258
−0.04	4.673	179.237	2.722	5.992	79.83	2.763	6.83	116.6456
0.787	5.107	170.806	3.509	4.116	10	4.406	4.473	5.0375
0.22	7.552	709	5.236	5.4	−0.1131	3.051	4.485	25.4073
8.54	7.696	407.9	8.112	4.189	44.4	7.263	6.053	86.21
8.984	6.46	311.485	13.834	4.365	439.4	13.202	5.632	889.2716
14.46	4.45	538.7258	5.272	−2.388	2.4462	11.848	4.303	251.4169
−1.05	3.291	58.2085	−2.934	0.385	167.2645	−3.236	0.712	109.1734
−1.82	0.377	124.7005	0.615	0.49	44.1213	1.28	1.429	6.563
0.337	1.4	3.907	2.6	1.574	1.76	−10.04	−13.48	54243.3219
6.976	3.404	9.6382	5.153	0.982	2.6855	5.595	3.781	2.3417
5.572	2.061	−4.0826	7.895	0.931	6.0281	10.584	2.127	−0.3076
8.012	2.287	−1.0113	10.538	0.436	78.1771	13.592	0.274	222.359
19.854	2.7674	134.916	14.578	1.517	26.0449	14.36	2.351	6.5875
11.036	1.577	3.4189	−3.741	−2.358	1448.33	−1.998	−2.572	1022.268
−3.87	−0.667	560.8619	1.945	−0.991	81.9759	2.436	−2.21	118.3477
19.854	2.674	99.3162	−10.04	−13.482	54243.0	0.473	−3.085	528.7297
2.6	−3.419	168.8639	4.308	−0.56	−0.7144	5.289	−1.472	−2.546
5.97	−1.177	8.797	5.72	−0.957	2.0138	9.164	−3.976	616.3603
9.489	−3.468	602.2114	10.728	−3.218	891.6835	9.645	−0.165	99.1746
11.857	−2.561	978.8056	12.384	−0.702	400.6336	14.117	−2.474	1660.5298
11.667	−0.281	235.4217	−4.501	−6.826	7031.375	−1.469	−4.737	1903.2905
−1.848	−6.669	3526.5958	−4.964	−4.414	4082.23	0.478	−7.894	2000.545
0.575	−6.116	1291.983	−0.623	−4.267	1243.0344	2.181	−7.046	645.6975
4.711	−4.305	1.4799	5.188	−5.257	3.1258	5.531	−5.557	13.6521
6.69	−6.443	200.8104	8.271	−6.5040	774.5073	10.389	−7.304	2514.2817
7.163	11.592	431.3946						

*Problems 9 and 10 require mathematical transformations of the data prior to finding a good model.

***12.10.** The following data set consists of (x_1, x_2) points taken from a theoretical surface, $y' = f(x_1, x_2)$, where the values given in the data set for the dependent variable are given in terms of $y' = \ln(y)$. Fit your best MLR model to the data, and comment on the apparent geometric appearance of the theoretical surface, $y' = f(x_1, x_2)$.

x_1	x_2	y	x_1	x_2	y	x_1	x_2	y
−6.729	5.247	4.4909	−5.405	5.625	4.4143	−3.376	4.799	4.0954
−6.288	5.443	4.489	−2.045	2.321	3.4535	−0.285	2.148	3.3497
−0.93	3.898	3.8417	−2.011	4.509	3.9694	1.580	4.283	4.0782
1.202	4.25	4.0341	1.061	2.819	3.5419	2.88	3.686	4.0615
5.025	4.531	4.5363	8.488	4.913	5.0883	8.624	5.355	5.1766
8.825	4.531	5.0482	10.902	2.783	5.0972	10.554	5.658	5.4219
11.085	3.486	5.1952	10.363	3.197	5.0782	−3.757	0.554	3.5663
−3.494	−0.114	3.5032	−4.281	−0.749	3.6167	−5.245	−1.927	4.0572
−2.044	−1.327	2.8685	0.722	−0.008	2.3241	0.763	0.83	2.8166
−1.213	−0.893	2.6346	1.509	−1.884	1.4404	2.406	−1.527	2.0027
−1.78	1.552	3.3956	3.236	−0.522	2.3677	1.051	−1.515	1.8982
6.54	1.696	4.1847	6.112	−1.811	3.0647	5.263	0.053	3.3553
6.984	0.46	3.9824	11.834	−1.635	4.6585	11.202	−0.368	4.714
12.46	−1.55	4.7601	3.272	−8.388	2.827	9.848	−1.697	4.2639
−3.058	−2.709	3.425	−4.934	−5.615	4.3397	−5.236	−5.288	4.3646
−3.823	−5.623	4.1345	−1.385	−5.51	3.2965	−0.72	−4.571	2.9061
−1.663	−4.533	3.0613	0.6	−4.426	2.3086	−12.043	−19.482	6.552
4.976	−2.596	2.1627	3.153	−5.018	0.8386	3.595	−2.219	1.575
5.895	−5.069	2.172	8.584	−3.873	3.3997	6.012	−3.713	2.3005
8.538	−5.564	3.1366	11.592	−5.726	4.1409	17.854	−3.233	5.4514
12.578	−4.483	4.4568	12.36	−3.649	4.5384	9.036	−4.423	3.5573
−5.741	−8.358	4.9053	−3.998	−8.572	4.6097	−5.87	−6.667	4.6685
−0.055	−6.991	3.1178	0.436	−8.211	3.5016	17.854	−3.326	5.4403
−12.043	−19.482	6.5458	−1.527	−9.085	4.2746	0.6	−9.419	3.9371
2.308	−6.56	2.3648	3.289	−7.472	2.4362	3.97	−7.177	2.1346
3.722	−6.957	1.9481	7.164	−9.976	3.467	7.489	−9.468	3.2702
8.728	−9.218	3.421	7.645	−6.165	2.7311	9.857	−8.561	3.6123
10.384	−6.702	3.711	12.117	−8.474	4.0942	9.667	−6.281	3.4475
−6.501	−12.826	5.5403	−3.469	−10.737	4.8627	−3.848	−12.669	5.1995
−6.964	−10.414	5.3095	−1.522	−13.894	5.0949	−1.425	−12.116	4.8069
−2.623	−10.267	4.6768	0.181	−13.046	4.7356	2.711	−10.305	3.7193
3.188	−11.257	4.0091	3.531	−11.557	3.9522	4.69	−12.443	4.0939
6.271	−12.504	4.045	8.389	−13.304	4.2078	5.163	5.592	4.8012

CHAPTER 13

HYPOTHESIS TESTS FOR TWO OR MORE MEANS: ANALYSIS OF VARIANCE—SINGLE-FACTOR DESIGNS

The following are the important concepts and ideas contained in this chapter.

- Know why *analysis of variance, ANOVA,* was developed and know its primary application
- Know what is meant by a *completely randomized single-factor experiment;* understand its mathematical model and its associated null and alternate hypotheses
- Know what is meant by the terms *treatment, treatment effect,* and *factor level*
- Understand why any ANOVA design may be viewed as an MLR model
- Understand why the construction of specific independent estimates of σ^2 allows us to test the hypothesis of equal means through the use of the F distribution
- Be able to interpret an ANOVA table, generated from the ANOVA program, to decide whether the associated hypotheses may be rejected
- Know how to determine which of the means are statistically different when a significant difference is obtained from the ANOVA table

- Be able to use the o.c. curve program for the ANOVA designs covered in this chapter and be able to size samples adequately so that acceptable β errors are associated with the ANOVA test of hypotheses
- Understand the differences between fixed and random models for ANOVA
- Know what is meant by a *randomized block single-factor design*; understand its associated mathematical model and its null and alternate hypotheses

13.1 INTRODUCTION

In Chapter 10, statistical techniques are developed to perform hypothesis tests comparing the means, μ_1 and μ_2, from two different populations. Let us suppose that, instead of two populations, our interest centers on eight populations and that we desire to compare the means of these populations, μ_1, μ_2, μ_3, μ_4, μ_5, μ_6, μ_7, and μ_8. One way to approach such a comparison would be to use the methods of Chapter 10 and to perform the ${}_8C_2 = 28$ possible two-sample comparisons. If we select the appropriate sample size for acceptable α and β errors for each of these 28 tests, what can we say about the conclusions that might be reached from such a group of tests?

Suppose that, in fact, all of the means possess the *same* true but unknown value and $\alpha = 0.05$. To illustrate the kind of problem that is caused by these repeated two-sample tests, let us suppose that all 28 tests are pairwise statistically independent. (They are *not*.) From the basic rules of probability presented in Chapter 2, if all of the tests are statistically independent, the probability that we would conclude *correctly* that $\mu_i = \mu_j$ in each of the 28 two-sample tests would be equal to $(1 - \alpha)^{28} = 0.95^{28} = 0.2378$. This means that even with $\alpha = 0.05$, the probability of our reaching the correct conclusion in all 28 tests would be less than 24%! Indeed, only when α was set to the ridiculously small value of 0.003756 would the probability of reaching the correct conclusion in all 28 tests be at a "reasonable" value of 0.9.

This same kind of catastrophic growth of error is also present when the members of the set of two-sample tests are not statistically independent. The probability of reaching the correct conclusion in all 28 tests is just much more *difficult* to compute. Clearly, some better method of comparing the means of a set of populations must be obtained and used. In this chapter we develop sound statistical methods that will allow the comparison of an arbitrary number of means.

13.2 COMPLETELY RANDOMIZED SINGLE-FACTOR EXPERIMENTS

Example 13.1. The data set given in Table 13.1 presents the results of *randomly sampling* from different, statistically independent, job batches from three different laboratories. Each data value is the amount of a chemical by-product that was present in the leaching-process cooling water from a job batch. We wish to determine whether a statistically significant difference exists between the three laboratories in their average concentrations of the chemical by-product. Since a series of two-sample

TABLE 13.1
Concentrations of a chemical by-product (mg/liter)

Laboratory		
1	**2**	**3**
19	19	15
22	22	16
21	26	19
22	24	15
18	26	19
19	19	16
18	22	24
25	23	20
19	27	17
19	23	18
16	21	19
22	19	24
20	22	22
19	24	26
18	24	16
18	22	24
19	26	20
21	22	19
22	25	19
25	22	24
19	25	14
17	23	16
23	19	20

tests of hypothesis is not a correct approach, we must develop an approach that does not have the disadvantages cited in the previous section.

In Table 13.1, there is only one *factor* of interest, the laboratories, and that factor occurs at three levels, $j = 1, 2$, and 3. The "factor" in a single-factor design may also be called a *treatment*. In general, we will use the symbol T_j to represent the treatment effect of factor level j, where the index j will assume values from 1 to k, i.e., T_j, $j = 1, 2, \ldots, k$. Each treatment level has a selected number of observations, or data values, which we symbolize by y_{ij}, $j = 1, 2, \ldots, k$ and $i = 1, 2, \ldots, n$. In Table 13.1, $n = 23$ and $k = 3$.

In general, the data set for a completely randomized single-factor experiment may be pictured schematically as shown in Table 13.2.

For the present discussion, we will assume that the three laboratories of Example 13.1 are the only ones of interest and that the results of our analysis are not be extendible to any other laboratories. This assumption means that we are considering the treatment effects to be *fixed*. At the end of Section 13.4, we briefly discuss the *random* effect model, in which the laboratories would represent a random sample of all such laboratories and the results of the analysis could be extended to all laboratories in the sampled population.

TABLE 13.2
The general data layout for a completely randomized single-factor design

Factor (treatment)					
1	**2**	...	**j**	...	**k**
y_{11}	y_{12}	...	y_{1j}	...	y_{1k}
y_{21}	y_{22}	...	y_{2j}	...	y_{2k}
⋮	⋮		⋮		⋮
y_{i1}	y_{i2}	...	y_{ij}	...	y_{ik}
⋮	⋮		⋮		⋮
y_{n1}	y_{n2}	...	y_{nj}	...	y_{nk}

As indicated at the end of Chapter 12, we may use the general linear model to assist us with the current analysis. Let us consider how the general linear model is connected to ANOVA. The null hypothesis that we are interested in testing is whether the means of the different populations are the same against the alternate hypothesis that at least one population's mean is different from the others. In mathematical notation, this may be expressed as

$$\begin{aligned} H_0 &: \mu = \mu_1 = \mu_2 = \cdots = \mu_k \\ H_A &: \mu_p \neq \mu_q \qquad \text{for some } p \text{ and } q \end{aligned} \tag{13.1}$$

Further, each population mean, μ_j, may be viewed as the sum of the hypothesized common mean, μ, and the "treatment effect"* for the population, T_j; i.e.,

$$\mu_j = \mu + T_j \tag{13.2}$$

We will assume, without loss of generality, that $\sum_{j=1}^{k} T_j = 0$.

Clearly, if H_0 is *true*, $T_j \equiv 0$ for all $j = 1, 2, \ldots, k$. Indeed, using Equation 13.2, the hypotheses of Equation 13.1 may be reexpressed as

$$\begin{aligned} H_0 &: T_1 = T_2 = \cdots = T_k = 0 \\ H_A &: T_p \neq 0 \qquad \text{for some } p \end{aligned} \tag{13.3}$$

Thus, the value of each observation in the data set, y_{ij}, may be viewed as the *realization* of the additive result of the hypothesized common mean, μ, the treatment effect for population j, T_j, and a random error associated with that observation, ε_{ij}; i.e.,

$$Y_{ij} = \mu_j + \varepsilon_{ij} = \mu + T_j + \varepsilon_{ij} \tag{13.4}$$

*In this chapter, in accordance with traditional notation for analysis of variance, we use T_j to denote the parameter associated with the treatment effect for population j. Other parameters introduced later in this chapter and in Chapter 14 will be similarly denoted.

Recalling the dummy variable that is associated with β_0 in the MLR model, suppose that we associate a similar dummy variable, δ_{i0} (always equal to 1), with μ. Further, let us associate a δ_{ij} with each T_j so that $\delta_{ij} = 1$ if Y_{ij} is from population j and that $\delta_{ij} = 0$, otherwise. With this additional set of variables, Equation 13.4 can be expanded to the following more general form.

$$Y_{ij} = \mu\delta_{i0} + T_1\delta_{i1} + T_2\delta_{i2} + \cdots + T_j\delta_{ij} + \cdots + T_k\delta_{ik} + \varepsilon_{ij} \tag{13.5}$$

Viewing the δ_{ij} as the independent variables and μ and the T_j as parameters of an MLR model (where $p = k + 1$, $\beta_0 = \mu$, and $\beta_j = T_j$, for $j = 1, 2, \ldots, k$) makes the relationship between ANOVA and regression analysis clearer. Indeed, using the techniques of multivariate regression analysis, the T_j may be estimated just as the values of the β_j are estimated in Chapter 12. Substituting the appropriate values of δ_{ij} into Equation 13.5 yields Equation 13.4. (An excellent development of how the T_j are estimated is found in Neter et al. [1990].)

At this point we can immediately impose all the basic assumptions of the MLR model on our model for ANOVA. Principal among these, for our current purposes, is the assumption that each of the populations being compared is governed by a normal distribution function with the same value of the variance. This implies that the ε_{ij} are NID(0, σ^2). Although advanced methods exist to test whether the assumptions of normality and equality of variance are indeed true, it is fortunate that the methods developed next yield usable results even in the presence of moderate departures from these assumptions. (Statistical methods that possess this relative insensitivity to the above assumptions are often referred to as *robust* methods of analysis.)

Since, by assumption, the variance of each population, σ_j^2, is equal to the common value of the variance, σ^2, we may form two estimates of σ^2 from the data set. The first is the pooled estimate of σ^2 constructed from the variance estimates of the k different populations of treatment levels; i.e., an unbiased estimate of σ^2 is provided by each of the variance estimates for population j, ($j = 1, \ldots, k$):

$$s_j^2 = \frac{\sum_{i=1}^{n} (y_{ij} - \overline{y}_j)^2}{n - 1} \tag{13.6}$$

where $\overline{y}_j$ is the sample mean of the data from population j. To obtain the pooled *within-treatment-levels* unbiased estimate of σ^2, we form

$$s_e^2 = \frac{\sum_{j=1}^{k}\sum_{i=1}^{n} (y_{ij} - \overline{y}_j)^2}{k(n - 1)} = \frac{\sum_{j=1}^{k} s_j^2}{k} \tag{13.7}$$

and we know from our work in Chapters 7 through 10 that $k(n - 1)S_e^2/\sigma^2$ is distributed according to a $\chi^2_{k(n-1)}$ distribution, where S_e^2 is the estimator associated with s_e^2. This relationship is *not* dependent on whether H_0 is true.

The second estimate of σ^2 is formed using the fact that an estimate of $\sigma^2_{\overline{Y}}$ is

$$s^2_{\overline{Y}} = \frac{\sum_{j=1}^{k} (\overline{y}_j - \overline{y}_\bullet)^2}{k - 1}$$

where $\bar{y}_\bullet$ is the grand mean of all observations in the data set. Notice that this estimate uses the differences between the $\bar{y}_j$ and the grand mean. We recall that $s_{\bar{Y}}^2$ is an estimate of $\sigma_{\bar{Y}}^2/n$, and therefore

$$s_T^2 = n\frac{\sum_{j=1}^{k}(\bar{y}_j - \bar{y}_\bullet)^2}{k-1} = ns_{\bar{Y}}^2 \tag{13.8}$$

is an estimate of σ^2. The second estimate, s_T^2 is an unbiased estimate of σ^2 if and only if H_0 is true.

It can be shown that not only is the estimator, S_T^2, statistically independent of S_e^2 but also that $[(k-1)S_T^2]/\sigma^2$ is distributed according to a χ_{k-1}^2 distribution *if and only if* the null hypothesis is true. If $H_0 : T_1 = T_2 = \cdots = T_k = 0$ *is not true,* there is no valid estimate of a common mean, μ, on which to base the computation of s_T^2, and s_T^2 is *not* an unbiased estimate of σ^2. [If H_0 is not true, it may be shown that $E(S_T^2) = (n\sum T_j^2/(k-1)) + \sigma^2$.]

From this development, we may state that if H_0 is true, then

$$F = \frac{[(k-1)S_T^2]/[\sigma^2(k-1)]}{[k(n-1)S_e^2]/[\sigma^2 k(n-1)]} = \frac{S_T^2}{S_e^2} \tag{13.9}$$

is distributed as $F_{k-1,k(n-1)}$. Equation 13.9 is analogous to Equation 12.12, in which we are testing whether the T_j (all of the parameters except μ) are each identically equal to zero. When H_0 is *not* true, we expect that s_T^2, the between-treatment estimate of σ^2, will give a larger value than s_e^2, the within-treatment estimate of σ^2. Indeed, the expected value of F in Equation 13.9 is

$$E(F) = \frac{(n\sum T_j^2)/(k-1) + \sigma^2}{\sigma^2} = 1 + \frac{(n\sum T_j^2)/(k-1)}{\sigma^2} \tag{13.10}$$

If $f = s_T^2/s_e^2$ *exceeds* the tabled value of $F_{k-1,k(n-1),\alpha}$, the null hypothesis may be rejected at the $1-\alpha$ level of confidence.

It is common practice to present the preceding results in an *ANOVA table,* as exemplified in Table 13.3. Such tables also assist in understanding the more complex designs to be considered later. Table 13.3 illustrates even more clearly

TABLE 13.3
ANOVA table for a completely randomized single-factor design

Source of Variation	Degrees of Freedom	Sum of Squares	Mean Square	f
Between Treatments (regression)	$k-1$	$n\sum_{j=1}^{k}(\bar{y}_j - \bar{y}_\bullet)^2$	s_T^2	$\frac{s_T^2}{s_e^2}$
Within Treatments (error)	$k(n-1)$	$\sum_{j=1}^{k}\sum_{i=1}^{n}(y_{ij} - \bar{y}_j)^2$	s_e^2	
Totals	$kn-1$	$\sum_{j=1}^{k}\sum_{i=1}^{n}(y_{ij} - \bar{y}_\bullet)^2$		

TABLE 13.4
The ANOVA table for Table 13.1

ANALYSIS OF VARIANCE TABLE

SOURCE	DF	SS	MS	f
TREAT	2	164.435	82.217	10.636
ERROR	66	510.174	7.730	
TOTAL	68	674.609		

the relationship between ANOVA and regression analysis. If f is not large enough to conclude that H_0 should be rejected, the analysis is complete. If, on the other hand, f does indicate that H_0 should be rejected, more analysis will probably be necessary.

Example 13.2. Table 13.4 presents the table that the ANOVA program for completely randomized single-factor designs has produced when it was applied to the data of Table 13.1. The program also assists the user in interpreting the f value by displaying the following information about the p-value of α.

```
THE f VALUE OF 10.636
WITH 2 AND 66
DEGREES OF FREEDOM WOULD BE CAUSE FOR
REJECTION OF THE NULL HYPOTHESIS AT AN
ALPHA LEVEL OF 0.0001
```

This means that there is very strong evidence for rejecting the null hypothesis. Indeed, we should reject H_0 if we have set the value of α at any value greater than or equal to 0.0001.

Table 13.4 reaffirms that not all of the sums of squares in Table 13.3 must be explicitly computed. Rather, once the total sum of squares and the between-treatment sum of squares are computed, the error sum of squares may be obtained by subtraction. As we see later for more complex designs, this general idea remains valid; i.e., we will always be able to obtain the sum of squares for error by subtracting all other sums of squares from the total sum of squares.

The preceding development assumes that the sample sizes for each factor level are identical, i.e., $n_1 = n_2 = \cdots = n_k = n$. The necessary analysis and computations are only marginally changed, as illustrated in Table 13.5, when one or more of the n_j differ. Your computer program handles the analysis of this case just as easily as when all the sample sizes are the same.

13.3 HOW TO DETERMINE WHICH MEANS DIFFER WHEN H_0 IS REJECTED

At this point in the analysis of the example of Table 13.1, we can conclude only that one or more of the T_j's differ from zero; i.e., two or more of the μ_j's differ from one another. Unfortunately, we do not know which populations differ.

TABLE 13.5
ANOVA Table for a completely randomized single-factor design when the numbers of observations for the treatments differ

Source of Variation	Degrees of Freedom	Sum of Squares	Mean Square	f
Between Treatments (regression)	$k-1$	$\sum_{j=1}^{k} n_j(\bar{y}_j - \bar{y}_\bullet)^2$	s_T^2	$\frac{s_T^2}{s_e^2}$
Within Treatments (error)	$\sum n_j - k$	$\sum_{j=1}^{k}\sum_{i=1}^{n_j}(y_{ij} - \bar{y}_j)^2$	s_e^2	
Totals	$\sum n_j - 1$	$\sum_{j=1}^{k}\sum_{i=1}^{n_j}(y_{ij} - \bar{y}_\bullet)^2$		

Example 13.3. As a start to determining which μ_j differ, let us consider the following summary information presented in Table 13.6 (provided, on request, by the program). The final column of Table 13.6 gives the estimates of the treatment effects, the T_j. These are obtained by using the estimates of μ and μ_j in Equation 13.2, i.e., by computing $T_j' = \bar{y}_j - \bar{y}_\bullet$.

Since we know that at least one of the means *is* statistically different, we could conclude, on the basis of the sample means given in the third column, that the mean of population 2 is significantly different from population 3. However, we do not know, from a purely statistical viewpoint, if other significant differences exist. Does μ_2 also differ from μ_1? Does μ_1 differ from μ_3?

We have already shown that a sequence of two-sample tests is incorrect, so there must be another way to use the sample means and the results of the ANOVA to attack questions like these. The Tukey-Kramer procedure [Tukey, 1953; Kramer, 1956; and Hayter 1984], presented in the following pages, is the best of the several methods that have been developed.

Because we have waited until after the data have been taken and analyzed, we are able to order the sample means and possibly (almost certainly) perform

TABLE 13.6
Summary information for Table 13.1

POP	CNT	MEAN	SD	MIN	MAX	TJ′
1	23	20.043	0.497	16	25	− 0.652
2	23	22.826	0.502	19	27	2.130
3	23	19.217	0.714	14	26	− 1.478
Totals	69	20.696				

fewer pairwise tests in determining which means differ. However, because we have used the results of our ANOVA to gain this added knowledge, the error structure of the subsequent analysis is biased and changed. Fortunately, these additional complications have been overcome and we need only make use of the *Studentized range table*, Table A.7. (Because of the ease of implementation using Table A.7 and because of the complexity of evaluating the Studentized range for general parameters, no program for this application exists in your software package.)

The Tukey-Kramer procedure is performed in the following way (and is used only if a significant f has been found).

1. Order the $\overline{y}_j$ from smallest to largest, in a schematic fashion, along a horizontal line.

$$\overline{y}_A \quad \overline{y}_B \quad \overline{y}_C \quad \overline{y}_D \quad \overline{y}_E$$

2. From the ANOVA table that yielded the significant f value, obtain the value of s_e^2, the unbiased estimate of σ^2, and the degrees of freedom, ν, associated with s_e^2. (For a completely randomized single-factor design, $\nu = k(n-1)$. In other uses of the Tukey-Kramer procedure, ν may be different.)

3. Obtain the required values of the Studentized range, $R_{\alpha,\nu,d}$, for $d = 2, 3, \ldots, k$ (from Table A.7).

4. Begin by comparing the largest and smallest $\overline{y}_j$. In subsequent comparisons, consider the uncompared $\overline{y}_j$ pair that exhibits the largest difference in value. If, at any time, you encounter a pair of $\overline{y}_j$, say $\overline{y}_C$ and $\overline{y}_G$ that imply that their associated means, μ_C and μ_G, are not significantly different, you may ignore the $\overline{y}_j$ that are between $\overline{y}_C$ and $\overline{y}_G$ for the purpose of further comparisons. Those "interior" μ_j logically cannot be different, in view of the fact that μ_C and μ_G are not different.

The comparison of any two treatment means, μ_j and μ_i, is based on a $1-\alpha$ confidence interval about their theoretical difference, $\mu_j - \mu_i$. If the number of observations in either treatment is the same; i.e., if $n_j = n_i = n$, then the *exact* $1-\alpha$ confidence limits are $\overline{y}_j - \overline{y}_i \pm R_{\alpha,\nu,d}s_e/\sqrt{n}$. If $n_j \neq n_i$, conservative limits, whose confidence level is marginally greater than $1-\alpha$, are given by

$$\overline{y}_j - \overline{y}_i \pm R_{\alpha,\nu,d}\, s_e \sqrt{\frac{1}{2}\left(\frac{1}{n_j} + \frac{1}{n_i}\right)}$$

In either case, calculate the appropriate confidence limits and determine whether they contain the value zero. If the confidence limits do not contain zero, we may conclude that μ_j and μ_i are significantly different. Otherwise, evidence is insufficient to say that μ_j and μ_i are different.

Example 13.4. To clarify this general process, let us apply it to the results given in Table 13.3.

1. Order the $\overline{y}_j$.

$\overline{y}_3$	$\overline{y}_1$	$\overline{y}_2$
19.217	20.043	22.826

2. $s_e^2 = 7.7299$ and $k(n-1) = 3(22) = 66$.

3. Let $\alpha = 0.05$,

d	2	3
$R_{\alpha,k(n-1),d}$	2.83	3.40

4. $s_{\overline{Y}} = s_e/\sqrt{n} = 0.5797$ and $\overline{y}_j - \overline{y}_i \pm R_{\alpha,\nu,d}\, s_e/\sqrt{n}$. Therefore,

$$\overline{y}_2 - \overline{y}_3 \pm R_{\alpha,\nu,d}\frac{s_e}{\sqrt{n}} = 22.826 - 19.217 \pm 3.40(0.5797) = 3.609 \pm 1.9710$$

$$\overline{y}_2 - \overline{y}_1 \pm R_{\alpha,\nu,d}\frac{s_e}{\sqrt{n}} = 22.826 - 20.043 \pm 2.83(0.5797) = 2.783 \pm 1.6406$$

$$\overline{y}_1 - \overline{y}_3 \pm R_{\alpha,\nu,d}\frac{s_e}{\sqrt{n}} = 20.043 - 19.217 \pm 2.83(0.5797) = 0.826 \pm 1.6406$$

These results indicate that μ_2 is significantly different from both μ_1 and μ_3 and that μ_1 is not significantly different from μ_3.

An easy way to interpret these results, particularly when several more treatment levels are present, is to start with the largest observed differences and to draw lines under those pairs of sample means that are not significantly different. In so doing, the sample means that need not be considered will be obvious in a graphical way. For example, suppose that, as shown in the following graphic, the comparisons indicated that $\mu_E > \mu_A$, that μ_E is not different from μ_B, that $\mu_D > \mu_A$, and that μ_C is not different from μ_A.

$\overline{y}_A$ $\overline{y}_B$ $\overline{y}_C$ $\overline{y}_D$ $\overline{y}_E$

Obviously, no other comparisons need to be made.

For Example 13.4, the line-drawing procedure would produce the following results.

$\overline{y}_3$	$\overline{y}_1$	$\overline{y}_2$
19.217	20.043	22.826

13.4 THE OPERATING CHARACTERISTIC CURVE FOR THE COMPLETELY RANDOMIZED SINGLE-FACTOR DESIGN

As noted earlier, in the test of any hypothesis, both the α and the β errors must be considered. As usual, the α error for the ANOVA is explicitly stated prior to the performance of the test. If the β error is ignored, the most common result is that too small a sample size is selected. If the sample size is too small, it is very unlikely that H_0 will be rejected when it is false. Once more, in determining the appropriate sample size for any hypothesis test, we must stipulate the amount of difference from the null hypothesis that should be detected and the error level that is to be tolerated at that amount of difference. In ANOVA, however, the manner in which we must specify the difference from H_0 is not as straightforward as in the hypothesis tests described in Chapters 9 and 10.

Recall that, for the *fixed-effects* model,

$$E(s_T^2) = \frac{n \sum T_j^2}{k-1} + \sigma^2$$

which implies that

$$E\left(\frac{(k-1)S_T^2}{\sigma^2}\right) = (k-1) + \frac{n \sum T_j^2}{\sigma^2} = (k-1) + \lambda \qquad (13.11)$$

We know that $[(k-1)S_T^2]/\sigma^2$ is a central χ_{k-1}^2 random variable whose expected value is $(k-1)$ if and only if H_0 is *true* ($\sum T_j^2 \equiv 0$). If H_0 is *not true,* $[(k-1)S_j^2]/\sigma^2$ is a *noncentral* chi-square random variable, $\chi_{k-1,\lambda}^2$, with noncentrality parameter, $\lambda = (n \sum T_j^2)/\sigma^2$. From these facts we may state that, if H_0 is false, F (Equation 13.9) is governed by a *noncentral F* distribution with noncentrality parameter, λ; i.e., $F \sim F_{k-1,N-k,\lambda}$. (For discussions on the noncentral χ^2 and noncentral F distributions, interested readers are referred to Hald [1965] and to Abramowitz and Stegun [1970].)

Therefore, in terms of $f = S_T^2/S_e^2$, any departure from H_0 is measured in units proportional to $\sum T_j^2$ (since n and σ^2 are both constants). Unfortunately, there is some ambiguity in this measure of departure from H_0. Because of the presence of more than two populations, there may be an uncountably infinite number of sets of the T_j's that could give rise to any particular value of λ.

Example 13.5. Suppose that $n = 10$, $\sigma^2 = 1$, and $\lambda = 60$, which implies that $\sum T_j^2 = 6$. What possible departures from H_0 might give rise to such a situation? (Recall that $\sum T_j = 0$.)

Table 13.7 gives 10 selected sets of the T_j (for $k = 3$) that would yield $\sum T_j^2 = 6$. (There are an uncountably infinite number of such sets.) Some additional sets of the T_j can easily be generated from those of Table 13.7 by (1) selecting any of the other orderings of any of the sets given in Table 13.7 (i.e., $[-2, 1, 1]$ yields $[1, -2, 1]$ and $[1, 1, -2]$ as equally valid sets of the T_j) or by (2) multiplying through any set by -1 (i.e., $[-2, 1, 1]$ yields $[2, -1, -1]$ and $[-1, 2, -1]$ *and* $[-1, -1, 2]$ as equally valid sets of the T_j).

TABLE 13.7
Selected sets of the T_j that yield $\sum T_j^2 = 6$

Set	T_1	T_2	T_3
1	−2.0	1.000	1.000
2	−1.9	0.409	1.491
3	−1.8	0.145	1.655
4	−1.7	−0.062	1.762
5	−1.5	−0.396	1.896
6	−1.2	−0.786	1.986
7	−0.9	−1.097	1.997
8	−0.6	−1.352	1.952
9	−0.3	−1.562	1.862
10	0.0	$-1.732(-\sqrt{3})$	$1.732(\sqrt{3})$

In set 1 of Table 13.7, the range of the means (treatments) is 3 units, whereas two of the treatments are the same. On the other hand, in set 10 the range is 3.464 ($2\sqrt{3}$) units and all three treatments are different. However, because all 10 sets of the T_j in Table 13.7 generate the same value of λ, each of these departures from the null hypothesis will have the same associated β error.

Although the theoretical considerations associated with the development of the o.c. curve are beyond the scope of this text, the use of the basic concepts is not. Using the program for operating characteristic curves for analysis of variance contained in your program library, you can draw the curves for single-factor ANOVA designs.

Example 13.6. Let us consider an o.c. curve for the analysis of the data of Table 13.1. After the program has been loaded, the title page appears. Once the fixed-model option is selected (as opposed to the random-model option), a request is issued for the number of treatment levels, k, the *common* number of observations for each level, n, an estimate of σ, and the level of confidence, α, for the ANOVA design. In Table 13.1, $k = 3$, $n = 23$, and $\alpha = 0.05$. Because we have no prior external estimate of σ, our best estimate of σ is the square root of the mean square error, $\sqrt{\text{MSE}} = \sqrt{7.7299} = 2.78$, from Table 13.4. Given this information, the program generates the o.c. curve and plots that curve on request. Table 13.8 and Figure 13.1 present the program's results for this example. Notice that the horizontal axis is indexed in two measures, $\sum T_j^2$ and $\phi = \sqrt{\lambda/k}$.*

*Inclusion of the modified noncentrality measure, ϕ, is done in deference to a uniformity of presentation made possible by Pearson and Hartley [1943]. The interpretation of ϕ is like that of a standard deviation of a mean; ϕ gives the average number of units of departure from the null hypothesis per treatment. By redefining ϕ for each ANOVA design, Pearson and Hartley found that they could use a single set of o.c. curves rather than requiring a separate set of curves for each design. The user of the o.c. curve program has no driving need for such uniformity. However, since the range of the o.c. curve consistently falls within the limits of $0 < \phi < 5$, ϕ provides a convenient scaling measure of the o.c. curve and ensures that values of β near $1 - \alpha$ are considered.

TABLE 13.8
The o.c. curve for the ANOVA of Table 13.1

$\sum T_j^2$	ϕ	β	$\sum T_j^2$	ϕ	β	$\sum T_j^2$	ϕ	β
0.000	0.0	0.950	1.008	1.0	0.703	4.032	2.0	0.121
0.010	0.1	0.949	1.220	1.1	0.646	4.446	2.1	0.088
0.040	0.2	0.943	1.452	1.2	0.584	4.879	2.2	0.062
0.091	0.3	0.932	1.704	1.3	0.519	5.333	2.3	0.042
0.161	0.4	0.916	1.976	1.4	0.453	5.806	2.4	0.028
0.252	0.5	0.896	2.268	1.5	0.386	6.300	2.5	0.018
0.363	0.6	0.870	2.581	1.6	0.322	6.814	2.6	0.011
0.494	0.7	0.838	2.913	1.7	0.262	7.349	2.7	0.007
0.645	0.8	0.799	3.266	1.8	0.208	7.903	2.8	0.004
0.817	0.9	0.754	3.639	1.9	0.161	8.478	2.9	0.002

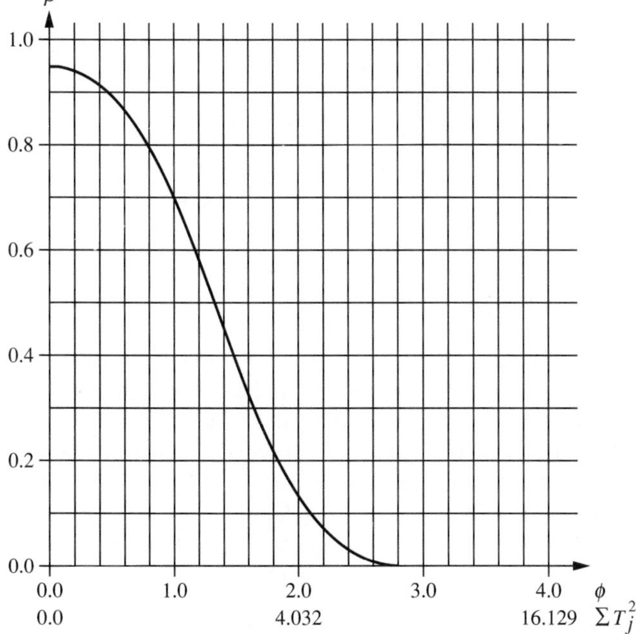

FIGURE 13-1
The o.c. curve for the ANOVA of Table13.1

Perhaps the most important information that this o.c. curve gives us for Example 13.6 is that, given the 23 observations of each of the three treatment levels, we have the reasonable probability of rejecting the null hypothesis, $1 - \beta$, of about 0.9—if $\sum T_j^2$ is approximately 4 (ϕ is approximately 2). From the data set of Table 13.1, our best estimate of $\sum T_j^2$ is derived from the results of Table 13.6; i.e., $\sum T_j^2 = (-0.652)^2 + (2.130)^2 + (-1.478)^2 = 7.146$. The o.c. curve indicates that we should detect such a difference more than 99% of the time.

As pointed out earlier, the only difference in the analysis between the fixed-effects model and the *random-effects model* for the completely randomized single-factor design is in the construction of the o.c. curve. The mathematical expression for the random-effects model is identical to that of the fixed-effects model as given in Equation 13.4, i.e.,

$$Y_{ij} = \mu + T_j + \varepsilon_{ij} \tag{13.4}$$

However, the interpretation is quite different. No longer are the treatment effects, the T_j, considered to be true but unknown constant parameters in a linear model. Rather, they are assumed to be random variables distributed as $N(0, \sigma_T^2)$, which are statistically independent of the ε_{ij}. The assumption that $E(T_j) = 0$ is analogous to the stipulation of $\sum T_j = 0$ in the fixed-effects model. Similarly, just as $H_0 : \sum T_j^2 = 0$ is an equivalent expression of the null hypothesis for the fixed-effects model, $H_0 : \sigma_T^2 = 0$ is an equivalent expression of the null hypothesis for the random-effects model. This can be given intuitive justification by noting that, if $\sigma_T^2 = 0$, the T_j can add nothing to the variability of the Y_{ij}.

Further clarification is provided by the fact that, for the random-effects model, $E(S_T^2) = \sigma^2 + n\sigma_T^2$. Clearly, F corresponding to Equation 13.9 will be distributed as a central $F_{k-1,k(n-1)}$ distribution if and only if H_0 is true ($\sigma_T^2 = 0$).

From the preceding discussion, a natural measure of departure from the null hypothesis is the magnitude of $n\sigma_T^2$. However, a more convenient and popular measure of departure is

$$\lambda = \sqrt{\frac{\sigma^2 + n\sigma_T^2}{\sigma^2}}$$

The physical interpretation of the magnitude of σ_T^2 has the same associated difficulties as does the magnitude of $\sum T_j^2$ in the fixed model. Indeed, the interpretation may be marginally more difficult since one must think in terms of a distribution of T_j's with a variance of σ_T^2. Readers interested in a more extensive discussion of this interpretation are referred to Bowker and Lieberman [1972, pp. 396–398].

The derivation of the o.c. curve for the random-effects model is beyond the scope of this book, but using the program for o.c. curves for analysis of variance will allow you to plot the o.c. curve associated with random or fixed models for a single-factor ANOVA.

Example 13.7. Table 13.9 and Figure 13.2 present the random-effects model o.c. curve for the analysis of the data of Table 13.1, where the same parameters as before—$n = 23$, $k = 3$, $\sigma = 2.78$, and $\alpha = 0.05$—have been provided to the program.

In the next section of this chapter we consider a single-factor design in which an external variable contributes to the variability of the Y_{ij}. Indeed, this added variability may be so great that it masks whatever differences there might be between the treatment levels of our factor of interest. When this external variable can be identified and controlled, direct methods exist that account for this additional variability.

TABLE 13.9
The random effects o.c. curve for the ANOVA of Table 13.1

σ_T	λ	β	σ_T	λ	β	σ_T	λ	β
0.000	1.000	0.950	2.988	5.250	0.107	5.476	9.500	0.034
0.435	1.250	0.858	3.135	5.500	0.098	5.622	9.750	0.032
0.648	1.500	0.745	3.282	5.750	0.090	5.768	10.000	0.031
0.832	1.750	0.635	3.429	6.000	0.083	5.913	10.250	0.029
1.004	2.000	0.539	3.576	6.250	0.077	6.059	10.500	0.028
1.168	2.250	0.459	3.723	6.500	0.071	6.204	10.750	0.027
1.328	2.500	0.392	3.870	6.750	0.066	6.350	11.000	0.026
1.485	2.750	0.338	4.016	7.000	0.062	6.495	11.250	0.024
1.640	3.000	0.293	4.162	7.250	0.058	6.641	11.500	0.023
1.793	3.250	0.256	4.309	7.500	0.054	6.786	11.750	0.022
1.944	3.500	0.225	4.455	7.750	0.051	6.932	12.000	0.022
2.095	3.750	0.199	4.601	8.000	0.048	7.077	12.250	0.021
2.245	4.000	0.178	4.747	8.250	0.045	7.223	12.500	0.020
2.394	4.250	0.159	4.893	8.500	0.042	7.368	12.750	0.019
2.543	4.500	0.143	5.039	8.750	0.040	7.513	13.000	0.018
2.692	4.750	0.130	5.185	9.000	0.038	7.659	13.250	0.018
2.840	5.000	0.118	5.331	9.250	0.036	7.804	13.500	0.017

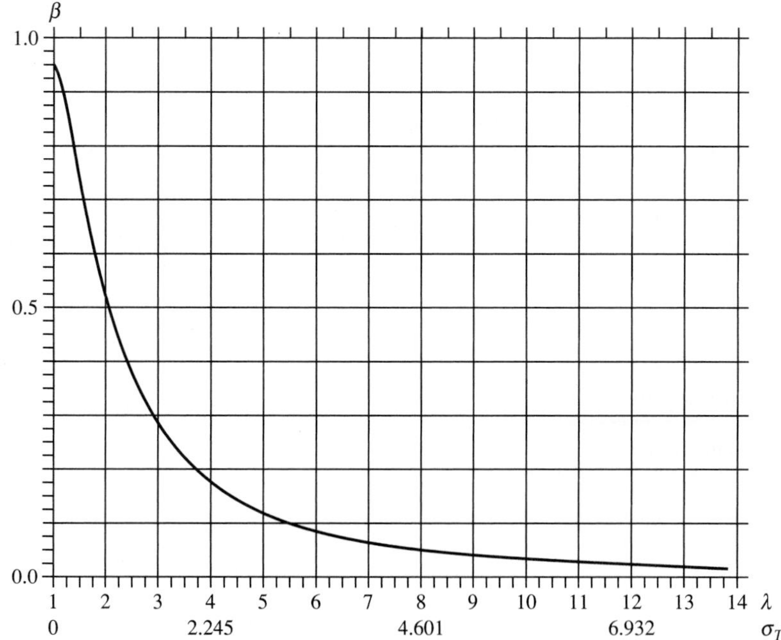

FIGURE 13-2
The random-effects model o.c. curve for Table 13.1.

13.5 THE RANDOMIZED BLOCK DESIGN: A SINGLE-FACTOR DESIGN WITH ONE RESTRICTION ON RANDOMIZATION

In many experimental situations, the experimenter is confronted with an external "nuisance" variable, which must not be ignored because of its strong effect on the Y_{ij}.

Example 13.8. Suppose that we are interested in the comparative distillation efficiency of four brands of electrically powered water stills. It is well known that the purity of the water that is to be distilled (the *feedwater*) has a major effect on the distillation efficiency of all four machines. Further, even though we could control the purity of the water used in our experiment, the results of an experiment with only one kind of feedwater would not satisfy the original intent: We want to know if there is a significant difference in the efficiencies of the four machines when they are used over the entire range of water purities likely to be found in the general consumer environment.

Thus, we want to overcome the difficulty presented by the nuisance variable (feedwater purity) and to satisfy our intent to reach a conclusion that is valid for the range of possible values of that variable. One way to do that is to select a representative set of *blocks*, i.e., specific values of the external variable, and to observe the response variable at each selected value of the external variable.

Consider the general data set as presented in Table 13.10. The mathematical model underlying this new design is

$$Y_{ij} = \mu + B_i + T_j + \varepsilon_{ij} \tag{13.12}$$

Equation 13.12 differs from the mathematical model of the completely randomized single-factor design, as given in Equation 13.4, only in the additional term, $B_i, i = 1, \ldots, n$. The B_i are the unknown constant parameters representing the effect of the "blocking factor" at level i. (We assume for the purposes of the analysis, and without loss of generality, that $\sum B_i = 0$.) Again, the relation given

TABLE 13.10
The general data layout for a randomized block single-factor design

	Factor (treatment)					
Block	**1**	**2**	$\cdots$	j	$\cdots$	k
1	y_{11}	y_{12}	$\cdots$	y_{1j}	$\cdots$	y_{1k}
2	y_{21}	y_{22}	$\cdots$	y_{2j}	$\cdots$	y_{2k}
$\vdots$	$\vdots$	$\vdots$		$\vdots$		$\vdots$
i	y_{i1}	y_{i2}	$\cdots$	y_{ij}	$\cdots$	y_{ik}
$\vdots$	$\vdots$	$\vdots$		$\vdots$		$\vdots$
n	y_{n1}	y_{n2}	$\cdots$	y_{nj}	$\cdots$	y_{nk}

in Equation 13.12 can be expanded into a more general form, similar to that of Equation 13.5, and a direct analogy between the randomized block design and its corresponding MLR model may be drawn.

Like the completely randomized design, we may impose all the assumptions of the MLR model on the model of Equation 13.12 and test whether a significant difference among the means of the levels of the treatment factor exists by testing the hypothesis of Equation 13.3:

$$\begin{aligned} &H_0 : T_1 = T_2 = \cdots = T_k = 0 \\ &H_A : T_p \neq 0 \qquad \text{for some } p \end{aligned} \tag{13.3}$$

In addition, it may be informative to test whether there is a significant difference among the levels of the blocking factor by testing the hypothesis

$$\begin{aligned} &H_0 : B_1 = B_2 = \cdots = B_n = 0 \\ &H_A : B_p \neq 0 \qquad \text{for some } p \end{aligned} \tag{13.13}$$

If we should find that the null hypothesis of Equation 13.13 cannot be rejected, this would raise the question of whether the external blocking factor should have been incorporated into the design. On the other hand, if the H_0 associated with the B_i is rejected, we are given evidence that the blocking factor's presence is appropriate, since it has a significant effect on the value of the response variable.

Table 13.11 gives the general form of the ANOVA table that the program in your software library for the randomized block single-factor design uses to perform the required analysis. It may be shown, in a manner very similar to the techniques used in Section 13.2, that s_T^2 and s_B^2 are unbiased estimates of σ^2 if and only if their respective null hypotheses are true. Further, s_e^2 is an unbiased estimate of σ^2 regardless of the truth of either of the two null hypotheses. For these reasons, we may construct the two f values given in the rightmost column of Table 13.11 and use them to test their respective hypotheses. Those readers interested in the detailed mathematical development of the contents of Table 13.11 are referred to Morrison [1983, pp. 336–343].

TABLE 13.11
ANOVA table for a randomized block single-factor design

Source of Variation	Degrees of Freedom	Sum of Squares	Mean Square	f
Between Treatments	$k-1$	$n\sum_{j=1}^{k}(\bar{y}_j - \bar{y}_\bullet)^2$	s_T^2	$\frac{s_T^2}{s_e^2}$
Between Blocks	$n-1$	$k\sum_{i=1}^{n}(\bar{y}_i - \bar{y}_\bullet)^2$	s_B^2	$\frac{s_B^2}{s_e^2}$
Error	$(k-1)(n-1)$	$\sum_{j=1}^{k}\sum_{i=1}^{n}(y_{ij} - \bar{y}_i - \bar{y}_j + \bar{y}_\bullet)^2$	s_e^2	
Totals	$kn-1$	$\sum_{j=1}^{k}\sum_{i=1}^{n}(y_{ij} - \bar{y}_\bullet)^2$		

Notice that the $n - 1$ degrees of freedom associated with the blocking effect have been "taken" from the degrees of freedom of the error term. In order for the blocking factor to be effective, the sum of squares between blocks must account for at least a proportionally equal amount of the error sum of squares. From an MLR context, we should not include the blocking factor in the ANOVA model unless it has membership credentials. If the blocking factor should not be included in the model, the β error associated with the test of hypothesis for the treatment means will be *larger* for the randomized block design than for the completely randomized design. Thus, a penalty, in the form of a lesser power of discrimination, is paid for including an improper blocking factor in the design. Often the decision on whether to include a blocking effect in the design is based on similar previous experiments in which the blocking effect was included and tested.

Example 13.9. As an example of the analysis of a randomized blocked design, let us return to the water stills of Example 13.8. Table 13.12 presents a data set that has been obtained for the purposes of this comparison, and Table 13.13 gives the results of the analysis in accordance with Table 13.11.

Based on the results given in Table 13.13, we may conclude not only that the four brands of stills are different in their average efficiencies but also that the differing levels of feedwater purity have a marked effect on efficiency. In order to determine which of the four stills (or the five levels of blocking) are significantly different, we once again apply the Tukey-Kramer procedure of Section 13.3 to the sample means of the four treatment levels (or five blocking levels).

Once again, to plan and execute the analysis of a randomized block design appropriately, we must consider the β error and the o.c. curve for the associated hypothesis test. Fortunately, the details of the o.c. curve construction for this design do not differ greatly from those of the completely randomized design. The only difference in the construction of the o.c. curve for the hypothesis test about the treatment means, for both the random and fixed-effect models, is that the error degrees of freedom have changed from $k(n - 1)$ to $(k - 1)(n - 1)$. The o.c. curve for the hypothesis test about the levels of blocking, for the fixed-effect model, is very similar in construction to that for the treatment effects. This may be seen by observing that for the block-effect hypothesis test, $\lambda = (k \sum B_i^2)/\sigma^2$ and $\phi = \sqrt{\lambda/n}$. The interpretation of λ and ϕ is also performed in the same manner as was done in completely randomized design.

TABLE 13.12
Efficiency data for water stills comparison

Water Purity Level	Brand of Still			
	1	2	3	4
1	0.54	0.52	0.37	0.39
2	0.65	0.58	0.50	0.53
3	0.74	0.64	0.57	0.61
4	0.72	0.76	0.57	0.70
5	0.73	0.70	0.67	0.72

TABLE 13.13
The ANOVA table for Table 13.12

	DF	SS	MS	f	PR> f
TREAT	3	.056	.0186	12.15	.0006
BLOCK	4	.168	.0420	27.50	.0000
ERROR	12	.018	.0015		
TOTAL	19	.242			

The random-effects model is analyzed in exactly the same way as the fixed-effects model for either hypothesis. The o.c. curve for the block-effect hypothesis test is quite similar in construction. The only differences are that the equivalent hypothesis to that of Equation 13.13 is $H_0 : \sigma_B^2 = 0$ and that the definition of the measure of departure from that hypothesis is $\lambda = \sqrt{(\sigma^2 + k\sigma_B^2)/\sigma^2}$.

As before, the o.c. curves for any of the cases mentioned in this section may be easily constructed through the use of o.c. curves for ANOVA program in your software library.

In many cases, a single observation in each "cell" (i.e., treatment level–block level combination), is insufficient, in terms of the total sample size, to give an acceptable power of discrimination (small-enough β error) to the hypothesis test. As we know, this may cause us to *fail* to reject H_0 when H_0 is false. As we have seen, the only way to increase this discrimination power is to increase the sample size. In the case of the randomized block design, the easiest way to increase the sample size is to take one or more additional observations at *every cell*. The additional observations should be taken in the form of *replications*. A replication is constructed by taking data in random fashion until every cell of the design is filled. Once this process is completed, additional replications are formed until the desired sample size is achieved.

When more than one replication is present in the data, the following mathematical model underlies the design.

$$Y_{ijm} = \mu + B_i + T_j + R_m + \varepsilon_{ijm}$$
$$i = 1, \ldots, n; j = 1, \ldots, k; m = 1, \ldots, r \qquad (13.14)$$

With $r > 1$ replications present, a new ANOVA table is required and is given in general form in Table 13.14. In addition, a sum of squares for differences in the replications can be added to the design; the replication effect is handled just like the blocking effect.

As might be inferred, the formula for the sum of squares for error, SSE, is rarely used. Rather, in practice, SSE is obtained by subtracting the other sums of squares (for treatments, blocks, and replications) from the total sum of squares.

Example 13.10. As an example of a randomized block design with more than one replication, suppose that two more replications were taken in Example 13.8, yielding the data set of Table 13.15. Using the randomized block program, which handles multiple replications just as easily as a single replication, the results given in Table 13.16 have been obtained. In comparing Table 13.16 to Table 13.13 we see that the additional number of observations in the data set has caused a marked increase in the f ratios both for the treatment effect and the block effect. This reflects the increased

TABLE 13.14
ANOVA table for a randomized block design with more than one replication

Source of Variation	Degrees of Freedom	Sum of Squares	Mean Square	f
Between Treatments	$k-1$	$rn\sum_{j=1}^{k}(\overline{y}_j-\overline{y}_\bullet)^2$	s_T^2	$\frac{s_T^2}{s_e^2}$
Between Blocks	$n-1$	$rk\sum_{i=1}^{n}(\overline{y}_i-\overline{y}_\bullet)^2$	s_B^2	$\frac{s_B^2}{s_e^2}$
Between Replications	$r-1$	$kn\sum_{m=1}^{r}(\overline{y}_m-\overline{y}_\bullet)^2$	s_R^2	$\frac{s_R^2}{s_e^2}$
Error	$rkn-k-n$ $-r+2$	$\sum_{j=1}^{k}\sum_{i=1}^{n}\sum_{m=1}^{r}(y_{ijm}-\overline{y}_i$ $-\overline{y}_j-\overline{y}_m+2\overline{y}_\bullet)^2$	s_e^2	
Totals	$rkn-1$	$\sum_{j=1}^{k}\sum_{i=1}^{n}\sum_{m=1}^{r}(y_{ijm}-\overline{y}_\bullet)^2$		

TABLE 13.15
Efficiency data for water still comparison: three replications

	Brand of Still											
	Rep 1				Rep 2				Rep 3			
Purity	**1**	**2**	**3**	**4**	**1**	**2**	**3**	**4**	**1**	**2**	**3**	**4**
1	0.54	0.52	0.37	0.39	0.54	0.51	0.43	0.43	0.54	0.52	0.37	0.39
2	0.65	0.58	0.50	0.53	0.52	0.58	0.54	0.56	0.65	0.58	0.50	0.53
3	0.74	0.64	0.57	0.61	0.67	0.64	0.51	0.58	0.74	0.64	0.57	0.61
4	0.72	0.76	0.57	0.70	0.76	0.73	0.69	0.62	0.72	0.76	0.57	0.70
5	0.73	0.70	0.67	0.72	0.69	0.80	0.64	0.55	0.73	0.70	0.67	0.62

TABLE 13.16
The ANOVA Table for Table 13.15

	DF	SS	MS	f	PR> f
REP	2	.001	.0006	0.360	.6992
TREAT	3	.146	.049	28.986	.0000
BLOCK	4	.441	.110	65.427	.0000
ERROR	50	.084	.002		
TOTAL	59	.672			

TABLE 13.17
Noncentrality parameters for randomized block ANOVA designs with multiple replications

		Treatment	Block
Fixed Model	λ	$\frac{rn\sum T_j^2}{\sigma^2}$	$\frac{rk\sum B_i^2}{\sigma^2}$
	ϕ	$\sqrt{\frac{\lambda}{k}}$	$\sqrt{\frac{\lambda}{n}}$
Random Model	λ	$\sqrt{\frac{\sigma^2 + rn\sigma_T^2}{\sigma^2}}$	$\sqrt{\frac{\sigma^2 + rk\sigma_B^2}{\sigma^2}}$

power of discrimination made possible by the increased degrees of freedom for the error term. As might have been expected, the replication effect does not have a significant f ratio.

In many cases, the sum of squares for replication is not included in the analysis. This is done only when there is a strong reason to believe, *before* the experiment, that the difference between replications will not be significant. If the replication sum of squares is not computed, Table 13.14 changes in the deletion of replication effect row and the modification of the error sum of squares to the form

$$\sum_{j=1}^{k}\sum_{i=1}^{n}\sum_{m=1}^{r}(y_{ijm} - \overline{y}_i - \overline{y}_j + \overline{y}_\bullet)^2$$

with $rkn - k - n + 1$ degrees of freedom. The decision of whether to include the replication effect follows the same logical argument that is used earlier in the chapter in the discussion of whether to include the test for the blocking factor in Table 13.11.

Again, the determination of which of the various effects levels differ may be accomplished using the Tukey-Kramer procedure. The construction of the o.c. curves is very similar to the approaches discussed earlier. You may refer to Table 13.17 to assist you in interpreting the o.c. curves that are easily constructed using the ANOVA o.c. curve program.

13.6 SOME COMMENTS ON ADDITIONAL SINGLE-FACTOR DESIGNS AND ON MISSING DATA

Often there can be two or more nuisance variables. Although their detailed consideration is beyond the scope of this text, ANOVA designs for more than one external variable merit a brief overview discussion.

Suppose that two external variables are present. In this case, one would resort to the *Latin-Square design*, so named because of the additional restriction that the second external variable imposes on the structure of the design. Two external variables imply that, in every replication, each level of the treatment factor must appear once and only once with each level of the two external variables. This fact forces the number of levels of all three factors to be the same. (Thus the design *must* be "square.") As an example, consider the following three-factor Latin-Square design for treatment levels A, B, and C.

		External Variable I		
		1	2	3
External Variable II	1	C	A	B
	2	A	B	C
	3	B	C	A

Notice that all randomization in the design has *not* been removed, since there are other data structures that also satisfy the restrictions of the Latin-Square design. However, the choices are more restricted. If an additional external variable is present, we can resort to a *Graeco-Latin-Square* design, which must also be square in the above sense. The analyses of the Latin-Square and Graeco-Latin-Square designs are simple extensions of the analyses that have just been presented. Readers interested in the details on these designs are referred to Hicks [1982].

As a final comment, it is possible and sometimes probable that one or more data values may be lost. In the completely randomized design introduced in Section 13.2, this presents no theoretical or practical problem in the analysis or interpretation of the results. Unfortunately, this statement is not true of the randomized block design. If a data value is missing in a randomized block design, the *orthogonality* property on which Tables 13.11 and 13.14 are based is destroyed. From a strictly theoretical view, this presents no problem, since we can always resort to the multiple linear regression view of analysis of variance. However, from a practical viewpoint, we know that the MLR model could be difficult to implement and interpret.

Two primary approaches can be used to solve the problems presented by missing data. The first is to replace the missing data with least squares estimates of their values and then use Table 13.12 or Table 13.14. The second is to use one of the commercially available systems for general statistical analysis, such as SPSS or SAS, that implement the MLR view of any ANOVA design.

In the next chapter, we consider some additional and more powerful designs in the realm of analysis of variance.

EXERCISES

13.1. The AKKP Class Ring Company uses four specific furnaces to heat the molds for their ring production using the "lost wax" method of casting. For best production of the molds, the furnaces need to heat quickly up to a temperature of 1200°F.

A sample of the times that the furnaces took to reach 1200°F, starting at room temperature, is given in the following table. Use analysis of variance to determine whether the furnaces take significantly different times to warm up. If the warm–up times are significantly different (at $\alpha = 0.10$), determine which furnaces differ from one another. What is the p-value at α at which the null hypothesis would be rejected?

Furnace Number			
1	**2**	**3**	**4**
15.22	15.65	15.04	14.93
15.27	14.45	15.03	15.17
14.44	15.13	14.68	15.78
13.98	14.55	15.23	15.36
15.85	14.86	15.10	15.50
15.56	14.86	15.73	14.35
14.97	15.08	15.19	15.56
15.67	16.16	15.55	14.71
13.91	15.46	15.14	14.18
14.75	14.26	14.73	14.63
15.95	15.07	15.00	15.40
14.87	14.51	13.53	15.00
15.18	14.50	15.65	14.95
15.07	14.88	16.01	15.13
15.51	14.23	16.13	15.45

13.2. Three specific brands of alkaline batteries are tested under heavy loading conditions. Given here are the times, in hours, that 10 batteries of each brand functioned before running out of power. Use analysis of variance to determine whether the battery brands take significantly different times to completely discharge. If the discharge times are significantly different (at the 0.05 level of confidence), determine which battery brands differ from one another.

Battery Type		
1	**2**	**3**
5.60	5.38	6.40
5.43	6.63	5.91
4.83	4.60	6.56
4.22	2.31	6.64
5.78	4.55	5.59
5.22	2.93	4.93
4.35	3.90	6.30
3.63	3.47	6.77
5.02	4.25	5.29
5.17	7.35	5.18

13.3. For the furnaces in Exercise 13.1, the true value of the standard deviation associated with the time to achieve 1200°F is known to be 0.5 minute. With the sample sizes given in that exercise, what is the probability of rejecting the null hypothesis of equal means if the treatment effects for the furnaces were (a) $T_1 = 0, T_2 = -0.2, T_3 = 0.4$, and $T_4 = -0.2$? (b) If the treatment effects for the furnaces were $T_1 = -0.4, T_2 = 0, T_3 = 0.4$, and $T_4 = 0$?

13.4. *Excluding* the possibility of reordering the values of the T_j in Exercise 13.3, find three other sets of T_j that will yield the same value of ϕ that is yielded in part (a).

13.5. Five different copper-silver alloys are being considered for the conducting material in large coaxial cables, for which conductivity is a very important material characteristic. Because of differing availabilities of the five kinds, it was impossible to make as many samples from alloys 2 and 3 as from other alloys. Given next are the coded conductivity measurements from samples of wire made from each of the alloys. Determine whether the alloys have significantly different conductivities. If the conductivities are significantly different (at $\alpha = 0.05$), determine which alloys differ from one another.

		Alloy		
1	**2**	**3**	**4**	**5**
60.60	58.88	62.90	60.72	57.93
58.93	59.43	63.63	60.41	59.85
58.40	59.30	62.33	59.60	61.06
58.63	56.97	63.27	59.72	57.31
60.64	59.02	61.25	59.79	61.28
59.05	58.59	62.67	62.35	59.68
59.93	60.19	61.29	60.26	57.82
60.82	57.99	60.77	60.53	59.29
58.77	59.24		58.91	58.65
59.15	57.38		58.55	61.96
61.40			61.20	57.96
59.00			59.73	59.42
			60.12	59.40
			60.49	60.30
				60.15

13.6. Three different teaching methods are being used to teach engineering statistics. Method I is the classical lecture-recitation method with three 50-minute lectures per week. Method II comprises one 75-minute lecture and two 45-minute problem sessions per week. Method III is a self-paced, or programmed, instruction approach. A random sample of students taught by each method is given a standard test, the highest score of which is 15. The results of that test are given here. Determine whether the students taught by the three methods have significantly different scores on the test. If the scores are significantly different (at $\alpha = 0.075$), determine which methods differ from one another.

Method		
I	II	III
12.31	12.10	12.54
12.54	13.00	12.64
13.18	13.09	12.81
12.56	12.55	12.06
12.08	12.04	12.78
12.28	12.50	13.00
	12.67	12.98
	13.12	11.79
	12.25	13.34
		11.86
		11.74
		13.33

13.7. Construct the o.c. curve for the treatment effect associated with the data of Table 13.12. As suggested in the chapter, the treatment effects are fixed. How well will this design detect a situation in which $T_1 = T_2 = 0$, $T_3 = 0.01$, and $T_4 = -0.01$?

13.8. For the data of Table 13.12, determine which of the still brands significantly differ from one another. (Use $\alpha = 0.05$.)

13.9. For the design given in Table 13.15, determine the probability that the following treatment effects will be detected (resulting in the rejection of H_0 at $\alpha = 0.06$): $T_1 = 0.01$, $T_2 = 0.025$, $T_3 = -0.035$, and $T_4 = 0.00$.

13.10. Three types of steel have been used to construct compression springs and three tempering methods have been used in the final hardening process. The following table presents four replications; the given data is the coded value of the stress failure point of each spring as measured by a compression test bench. Determine whether significant differences exist, in regard to spring strength, between the three types of steel and the three finishing methods. (Use $\alpha = 0.05$.) If there are significant differences, indicate what combination of steel type and finishing method yields the strongest spring.

	Replication 1			Replication 3		
	Tempering					
Steel Type	1	2	3	1	2	3
1	33.45	31.66	32.71	33.35	33.17	31.95
2	31.97	30.83	31.97	32.33	31.46	31.02
3	32.54	32.88	31.14	31.94	31.92	30.41

	Replication 2			Replication 4		
	Tempering					
Steel Type	1	2	3	1	2	3
1	33.09	32.42	32.43	33.26	33.52	32.90
2	31.60	32.23	31.66	31.37	30.25	31.30
3	31.68	31.54	31.63	32.23	32.55	32.16

13.11. Construct the o.c. curve for the hypothesis test associated with the types of steel in Exercise 13.10. Do you feel that enough observations have been taken in this experiment?

13.12. Use ANOVA to analyze the data of Table 10.2. When compared with the paired two–sample t test, what observations can you make about the similarities and differences between the two methods of analysis?

13.13. Four types of epoxy resin mixes containing steel filings and granules are tested to determine their average bonding strength in plumbing pipe repair. The data given are the scaled loads that caused the bonds to fail in each of 35 bonds for the four types of epoxy resin mixtures. Determine if there are significant differences in the average strengths of the four mixtures. (Use $\alpha = 0.06$.) If there are significant differences, determine which mixtures differ from one another. What is the probability, using a data set of the size given, that we will be able to detect a maximum difference of 0.5 between the four mean strengths?

Mixture Type							
1	**2**	**3**	**4**	**1**	**2**	**3**	**4**
9.24	10.43	9.93	11.77	9.32	8.74	10.01	12.32
8.12	9.53	10.33	11.93	8.84	8.07	10.85	11.17
9.63	9.76	9.97	11.57	9.19	9.80	10.14	11.21
10.22	10.04	9.79	12.11	9.33	10.27	9.83	11.38
9.27	10.31	9.20	11.03	8.70	9.46	9.84	12.63
8.78	9.02	10.17	11.71	8.71	9.79	9.63	11.86
8.67	9.09	9.12	11.14	8.82	10.73	9.61	11.56
9.36	8.97	9.50	12.21	9.46	9.16	9.23	12.14
9.30	9.58	9.74	11.90	9.07	9.06	9.94	12.04
9.44	9.38	9.77	11.13	9.09	9.86	9.66	11.21
8.71	9.80	10.03	11.33	8.96	9.00	9.33	11.56
8.53	9.77	9.97	11.94	8.60	8.92	10.89	11.54
8.85	9.24	10.52	11.91	8.47	9.54	9.79	12.10
9.11	10.13	10.12	11.18	8.49	10.31	10.44	11.17
8.26	9.39	10.70	11.16	8.60	9.18	9.31	12.18
8.21	8.96	10.20	12.31	8.25	9.56	9.32	12.08
9.40	10.27	9.57	11.56	8.86	9.90	9.90	11.30
9.29	9.26	10.22	11.35				

13.14. Five milling machines are used to cut keyways in a critical component during the production of large metal lathes. The amount of "chatter" that the cutting head experiences during the actual cutting stroke is a major cause of problems in the keyway process. The data give a dimensionless measure of chatter for each of 12 measurements taken for each of the five milling machines. Based on this data set, may we say that there is a significant difference in the amount of chatter that the machines experience? If so, which machines differ from one another? Do you feel that the sample size is adequate to detect a maximum difference of 0.9 between the mean chatter measures of the five machines?

Milling Machine Number				
1	2	3	4	5
25.66	26.39	24.40	25.91	25.17
24.17	25.01	25.73	24.18	24.75
25.80	25.64	24.25	23.36	25.97
24.80	25.94	24.88	25.46	24.80
25.59	25.45	24.71	25.06	26.52
26.10	24.68	25.42	25.99	25.86
25.04	24.81	24.90	26.72	23.51
24.59	24.54	24.44	25.26	26.39
24.97	24.28	24.75	25.82	24.27
25.73	24.44	25.04	25.00	24.09
23.98	24.45	24.23	27.34	24.11
25.39	25.53	23.70	24.75	25.77

13.15. Under a grant from the Navy, the Center for Electro-Magnatronifics has developed a rail gun that uses stored electrical power instead of chemical explosives to fire projectiles from shipboard 8-inch guns. Five variations on the basic design have been manufactured in prototype form and have been tested. The given data present measures of the relative accuracy of the five prototypes. The experiment was replicated over 4 days and the same six gun operators were used on each day. Determine if there is a statistically significant difference in the accuracy of the five different prototype guns. If so, determine which of the guns is the most accurate. Is it necessary to consider explicitly the effect of the days and operators? Why or why not?

	Day									
	1					2				
	Gun Number									
Operator	1	2	3	4	5	1	2	3	4	5
1	.35	.49	.29	.35	.36	−.25	−.1	.06	−.28	−.09
2	.14	.16	.13	.04	.34	−.19	−.34	−.18	−.23	−.15
3	.16	.1	.23	.24	.14	−.14	−.27	−.13	−.27	−.28
4	.39	.23	.15	.43	.36	−.08	−.15	−.02	−.09	−.19
5	.04	.16	.11	.07	.09	−.4	−.43	−.29	−.39	−.25
6	.09	.21	.16	.32	.22	−.07	−.18	−.26	−.3	−.04

	Day									
	3					4				
	Gun Number									
Operator	1	2	3	4	5	1	2	3	4	5
1	.27	.29	.17	−.09	.37	−.27	−.07	−.15	−.21	−.19
2	.27	.28	.35	.46	.14	−.19	−.16	−.11	−.13	.13
3	−.08	.03	.06	.35	.02	−.29	−.23	−.41	−.4	−.16
4	.19	.31	.29	.18	.15	−.27	−.18	−.13	−.27	−.26
5	.29	.39	.22	.18	.48	−.09	−.13	−.15	0	−.02
6	.19	.03	−.05	.08	.24	−.37	−.38	−.36	−.44	−.16

13.16. Concrete support beams are constructed out of three different mixes by two different mixing machines on 6 different days. The following data set gives the modulus of rupture, in pounds per square inch, for each beam in the experiment. Determine if there is a statistically significant difference in the average strengths of the beams made from the three concrete mixes. If so, determine which of the mixes differ from one another. Is it necessary to consider explicitly the effect of the days and mixers? Why or why not? How likely are we to detect an actual range of 40 pounds per square inch in the mean moduli of rupture among the different types of beams?

	Replication								
	1			2			3		
	Concrete Type								
Mixer Type	**1**	**2**	**3**	**1**	**2**	**3**	**1**	**2**	**3**
1	614.54	662.10	672.27	615.85	637.61	644.79	605.22	657.81	639.91
2	606.33	638.22	692.15	598.42	567.86	708.84	591.70	650.05	685.24

	Replication								
	4			5			6		
	Concrete Type								
Mixer Type	**1**	**2**	**3**	**1**	**2**	**3**	**1**	**2**	**3**
1	603.90	627.71	682.60	616.97	630.58	663.22	653.63	646.41	666.41
2	629.60	638.18	705.91	598.16	620.01	685.71	637.46	593.19	656.02

CHAPTER 14

FACTORIAL ANALYSIS OF VARIANCE

Given in this list are the major ideas and concepts of this chapter.

- Understand the differences between a single-factor ANOVA design with one or more blocking variables and a *multifactor factorial* ANOVA design
- Understand what is meant by an *interaction effect* and how to interpret such an effect when it is present in the model
- With the use of the factorial ANOVA programs, be able to perform either a two- or three-factor factorial ANOVA and interpret the results
- Be able to construct, using the appropriate program, the o.c. curves for the hypothesis tests regarding the A, B, and AB effects in a two-factor factorial design; be able to use those curves to assist you in sizing the number of observations in the data set so that appropriate α and β errors are achieved
- Understand that higher-order factorial ANOVA designs are just extensions of the two-factor factorial ANOVA designs

14.1 INTRODUCTION

In Chapter 13, our interest was directed primarily to the question of whether the levels of a *single factor* have a significant effect on the response variable. This was true even when another "extraneous" variable had a significant effect on the response variable, as in the randomized block design. The reason for inclusion of the blocking factor is *to remove* its contribution to the variability of the response variable. (This removal of explainable variability is very similar to what is accomplished in regression analysis when a new variable with good "membership credentials" is added to the regression model; see Figure 11.6.) When this variability is removed, possible differences between the treatment levels are no longer masked by the nuisance factor's presence. In many cases, however, there may be two or more *factors* of equal interest that are not external to the experiment.

When two or more factors are present in an experiment, we are interested both in the effect of each independent factor *level*, and in the possibility of a *joint* effect due to a simultaneous *interaction* between two, three, or up to as many as all the factors in the model. As a starting point for the discussion, let us consider a factorial ANOVA design with two factors.

14.2 A TWO-FACTOR FACTORIAL ANOVA DESIGN

The mathematical model for a factorial ANOVA design with two factors is

$$\begin{aligned} Y_{ijm} &= \mu + A_i + B_j + AB_{ij} + R_m + \varepsilon_{ijm} \\ i &= 1, \ldots, n; \; j = 1, \ldots, k; \; m = 1, \ldots, r \end{aligned} \tag{14.1}$$

where Y_{ijm} is the response variable in replication m at level i of factor A and level j of factor B. In addition, we will assume without loss of generality (and to ensure a unique solution to the normal equations associated with Equation 14.1) that

$$\sum_{i=1}^{n} A_i = \sum_{j=1}^{k} B_j = \sum_{\substack{i=1 \\ (j=1,\ldots,k)}}^{n} AB_{ij} = \sum_{\substack{j=1 \\ (i=1,\ldots,n)}}^{k} AB_{ij} = \sum_{m=1}^{r} R_m \equiv 0 \tag{14.2}$$

As before, ε_{ijm} is the inherent unremovable error and is distributed NID(0, σ^2). Finally, the term AB_{ij} represents the joint, or interaction, effect as a result of the simultaneous presence of factor A at level i and factor B at level j.

Thus, AB may be viewed as an additional additive factor in the model that is present at kn unique levels. The effect of a nonzero interaction can best be visualized by considering the simplest meaningful version of the model of Equation 14.1. Let $n = 2$, $k = 2$, and $m = 1$; i.e., we have a single replication of a model in which each independent factor occurs at two levels and, therefore, the interaction factor occurs at four levels.

In Figure 14.1*a* we see a schematic representation in which, excepting μ, all terms in the model (*including* the ε_{ijm}) have a value of zero. In this case, the model contour is a level plane, everywhere equal to μ. In Figure 14.1*b*, we add the effect of factor A. Since Equation 14.2 dictates that $A_1 + A_2 \equiv 0$ (or $A_2 = -A_1$) and, for this example, $A_2 > 0$, the model contour is still a plane. However, this plane is sloping upward from left to right.

The effect of adding factor B to the model of Figure 14.1*b* is seen in Figure 14.1*c*, where, similarly, $B_1 + B_2 \equiv 0$ (or $B_1 = -B_2$). Since, in this case, B_1 is

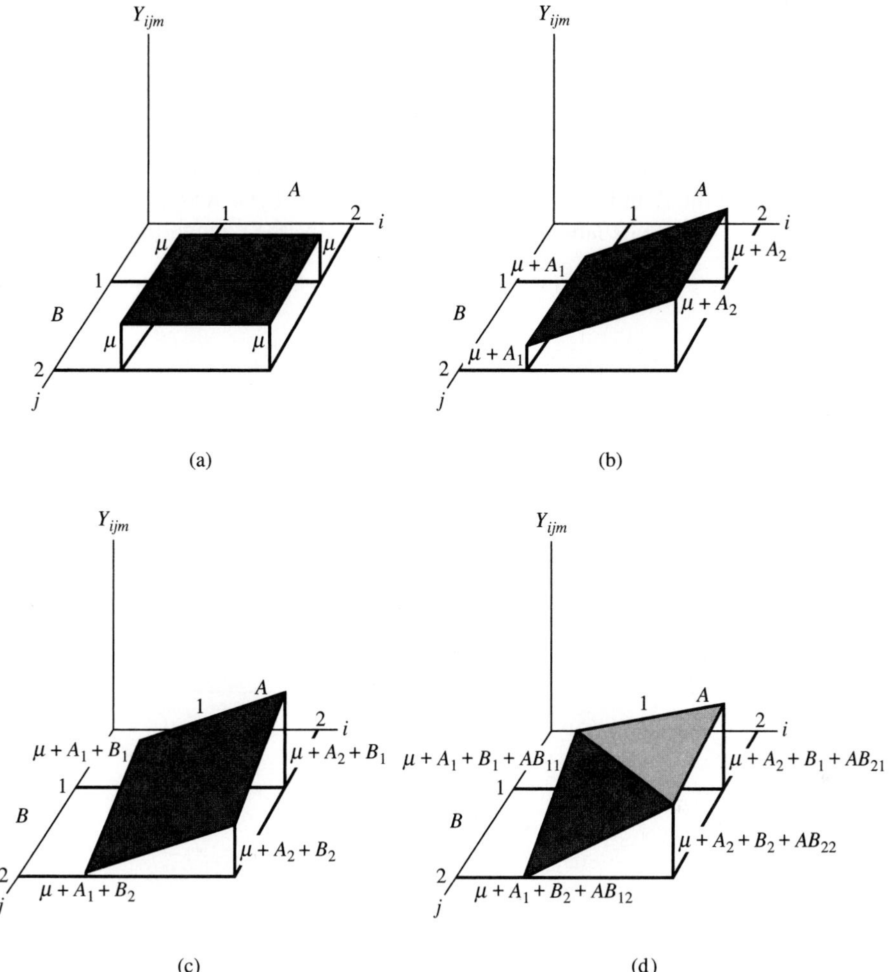

FIGURE 14-1
An illustration of the two-factor factorial model.

greater than zero, B_2 must be an equal amount less than zero. These facts cause the back line of the plane in Figure 14.1*b* to be uniformly raised, and the front line of that plane is uniformly dropped by the same amount. This uniform change in the contour of the model simply gives another plane with a slightly different orientation.

Figure 14.1*d* shows what adding the interaction effect does to the representation in Figure 14.1*c*. Equation 14.2 and some algebraic manipulation yield $AB_{11} = AB_{22} = -AB_{12} = -AB_{21}$. In Figure 14.1*d*, $AB_{11} = AB_{22} > 0$ and $AB_{12} = AB_{21} < 0$. Imposing these terms on the representation of Figure 14.1*c* destroys the planar contour of the model. This effect will always be present when a nonzero interaction effect is present in the ANOVA model.

We have not added the error term to the graphical representation of the model in Figure 14.1 because its effect is, by definition, unpredictable. Indeed, if the error variance, σ^2, is large enough, the error term can very well overwhelm and thus mask the effects of any of the factors in the model.

Figure 14.2 presents an alternate view of the contents of Figure 14.1, where the front and back lines of the four parts of Figure 14.1 are projected onto the plane formed by the *A* and *Y* axes. Figure 14.2*c* and *d* are *particularly informative*. In Figure 14.2*c*, the lines labeled B_1 and B_2 represent the change in the response

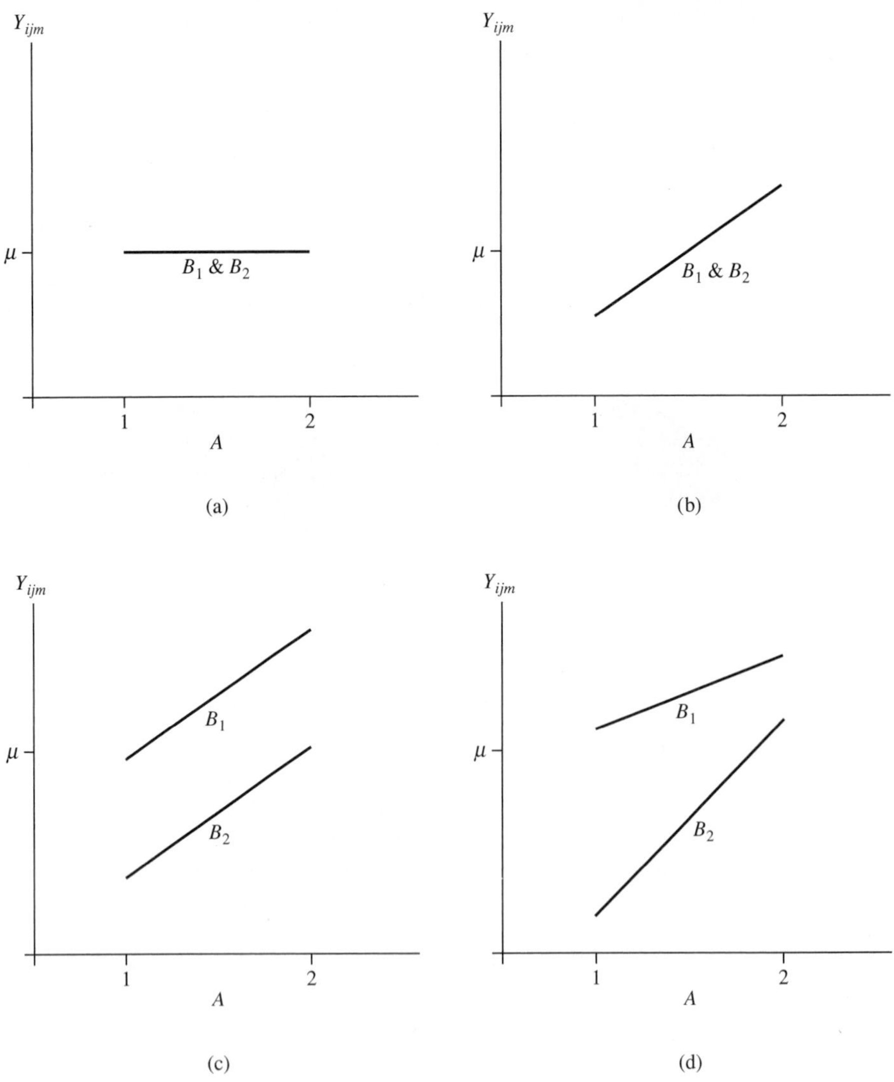

FIGURE 14-2
A Projection View of Figure 14.1

variable when factor A changes levels—while factor B is held at level 1 or 2, respectively. Since the lines are parallel, the amount of change in the response variable is the same *regardless* of the level of B. This is a characteristic of any pair of factors in any ANOVA model in which no interaction between the pair is present.

The corresponding lines in Figure 14.2*d*, where the interaction effect *is* present, are not parallel. This means that the change in Y when factor A moves to a new level is dependent on the level of B. Anytime that this is true for any pair of factors in any ANOVA model, we know that there is an interaction effect present between the two factors.

However, we must remember that when random error is present in the model, an interaction may appear to be present when none exists. In that case, we may perform a statistical hypothesis test, like the ones developed in Chapter 13 for other factors, to assist us in deciding if an interaction effect is actually present.

As in the models discussed in Chapter 13, an ANOVA table for the two-factor factorial design may also be constructed. We limit our discussion in this chapter to fixed-effect models. The consideration of a random-effect model for factorial analysis of variance, both in the required analysis and in the construction of the associated o.c. curves, introduces additional complexities. Readers interested in the details of this more advanced topic are referred to Hicks [1982], Morrison [1983], and Neter, Wasserman, and Kutner [1990].

The general ANOVA table for this design, assuming a fixed model, is given in Table 14.1. Each of the first four items in the mean square column are unbiased estimators of σ^2 if and only if their respective null hypotheses of "no significant effect" are true. Thus, the f values contained in the rightmost column may be used to test each of the associated hypotheses. When the interaction effect is included in the model, a minimum of two replications is required. A single replication

TABLE 14.1
ANOVA table for a two-factor factorial design

Source of Variation	Degrees of Freedom	Sum of Squares	Mean Square	f
Factor A	$n-1$	$rk\sum_{i=1}^{n}(\bar{y}_i - \bar{y}_\bullet)^2$	s_A^2	$\frac{s_A^2}{s_e^2}$
Factor B	$k-1$	$rn\sum_{j=1}^{k}(\bar{y}_j - \bar{y}_\bullet)^2$	s_B^2	$\frac{s_B^2}{s_e^2}$
Factor AB	$(k-1)(n-1)$	$r\sum_{j=1}^{k}\sum_{i=1}^{n}(\bar{y}_{ij} - \bar{y}_i - \bar{y}_j + \bar{y}_\bullet)^2$	s_{AB}^2	$\frac{s_{AB}^2}{s_e^2}$
Between Replications	$r-1$	$kn\sum_{m=1}^{r}(\bar{y}_m - \bar{y}_\bullet)^2$	s_R^2	$\frac{s_R^2}{s_e^2}$
Error	$(nk-1)(r-1)$	By subtraction	s_e^2	
Totals	$rkn-1$	$\sum_{j=1}^{k}\sum_{i=1}^{n}\sum_{m=1}^{r}(y_{ijm} - \bar{y}_\bullet)^2$		

results in zero degrees of freedom for the error term, which causes the sum of squares for error to be equal to zero.

Just as in the randomized block ANOVA, it is often appropriate to exclude the sum of squares for replications from the analysis. If this is done for a two-factor factorial ANOVA, the row in Table 14.1 for replications is deleted and the replication degrees of freedom and sum of squares are added to the error degrees of freedom and the error sum of squares.

Let us now consider a specific example of the analysis of a fixed-model two-factor factorial analysis of variance.

Example 14.1. Suppose that a study has been conducted to determine if there are significant differences in the ductility of cold-rolled steel produced by three types of machines. It is well known that the properties of the raw steel that serves as feedstock can have influences on the product from the rolling machines. A little research shows that there are three major suppliers of the feedstock for the rolling machines. Clearly, we are also interested both in determining whether significant differences exist between the feedstock provided by the three suppliers and in determining if there is an interaction effect; i.e., we want to know whether any of the machines produces markedly different results when working one particular supplier's raw steel.

The data set in Table 14.2 has been gathered to assist in obtaining answers to these questions. The ANOVA program in your software library for the two-factor factorial design yields the information given in Table 14.3 for the data set of Table 14.2.

As may be observed, all of the effects, with the exception of the effect of factor B, are highly significant. Once more, to determine which of the levels are statistically different in their mean values, we use the Tukey-Kramer procedure. Figure 14.3 is a plot of the cell means and gives a dramatic illustration of the interaction effect between the levels of factors A and B.

The construction of the o.c. curves associated with the hypotheses about factors A and B for the two-factor factorial design is very similar to the o.c. curves that are discussed in Chapter 13. Just as before, the o.c. curves are easily generated using the program for o.c. curves for analysis of variance. The definitions consistent with that program for our current design are given in Table 14.4.

TABLE 14.2
Scaled ductility data for sampled specimens of cold-rolled steel

	Replication								
	1			2			3		
	Machine (A)								
Supplier (B)	**1**	**2**	**3**	**1**	**2**	**3**	**1**	**2**	**3**
1	8.65	8.26	8.17	8.29	6.86	8.52	8.55	7.62	7.91
2	7.53	8.07	7.76	7.05	7.70	7.43	8.00	8.43	8.83
3	7.26	8.46	9.64	6.05	7.82	8.78	7.97	8.10	9.04

TABLE 14.3
ANOVA table and summary information for Table 14.2

```
         ANALYSIS OF VARIANCE TABLE
      DF     SS      MS       f     PR> f
REP    2   2.367   1.184   6.030   .0112
A      2   2.660   1.330   6.777   .0074
B      2   0.355   0.178   0.905   .4244
AB     4   5.543   1.386   7.061   .0018
ERR   16   3.140   0.196
TOT   26  14.066

                Sample Means                   Cell Means
Level          1       2       3                   A
         A  7.706   7.924   8.453          8.497   7.580   8.200
Factor   B  8.092   7.867   8.124       B  7.526   8.067   8.007
       Rep  8.200   7.611   8.272          7.093   8.127   9.153
Grand Mean  =  8.028
```

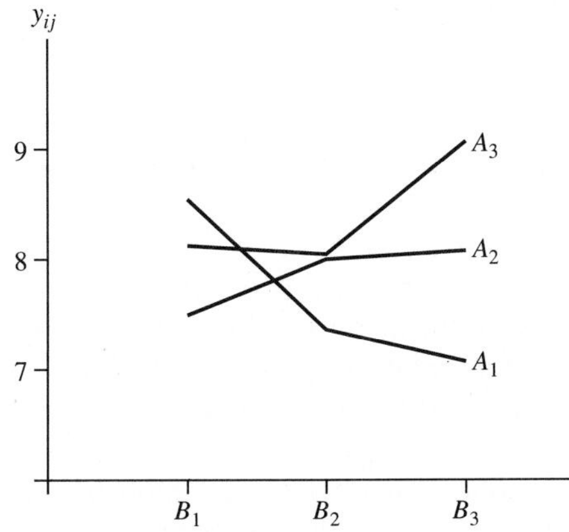

FIGURE 14-3
A plot of the interaction effect between factors *A* and *B* for Example 14.1.

TABLE 14.4
The definitions of λ and ϕ for the two-factor factorial ANOVA

		A	*B*	*AB*
Fixed Model	λ	$\frac{rn\sum A_i^2}{\sigma^2}$	$\frac{rk\sum B_j^2}{\sigma^2}$	$\frac{r\sum\sum AB_{ij}^2}{\sigma^2}$
	ϕ	$\sqrt{\frac{\lambda}{k}}$	$\sqrt{\frac{\lambda}{n}}$	$\sqrt{\frac{\lambda}{(k-1)(n-1)+1}}$

14.3 HIGHER-ORDER MULTIFACTOR FACTORIAL ANOVA DESIGNS

There are many practical experiments in which three or more factors naturally arise, factors that are of equal importance to an investigator. The analysis of such situations, through somewhat cumbersome in the presence of a large number of factors, is merely an extension of the methods presented earlier. As an indication of such analyses, let us consider the major aspects of a three-factor factorial design.

The mathematical model for a three-factor factorial design is given by

$$Y_{ijum} = \mu + A_i + B_j + AB_{ij} + C_u + AC_{iu} + BC_{ju} + ABC_{iju} + R_m + \varepsilon_{ijum} \quad (14.3)$$

$$(i = 1, \ldots, n;\ j = 1, \ldots, k;\ u = 1, \ldots, v;\ m = 1, \ldots, r)$$

where Y_{ijum} is the value of the response variable in replication m at level i of factor A, level j of factor B, and level u of factor C. The other terms in the model have an interpretation similar to the terms in Equation 14.1. In Equation 14.3, we have three two-factor interaction terms and a single three-factor interaction term. In general, a multifactor factorial ANOVA design will have one interaction term for every possible subset combination of the factors in the model.

To ensure a unique solution to the normal equations, we must impose a set of restrictions like those of Equation 14.2. This is true for any multifactor factorial ANOVA design. However, as we have seen earlier, explicit imposition of those restrictions is not necessary to construct the ANOVA table and to perform the associated F tests.

The construction of o.c. curves and the interpretation of the results for higher-order factorial designs is performed in the same way as for the two-factor factorial design. Those interested in the details of the analysis of higher-order factorial designs are referred to Hicks [1982], Morrison [1983], and Neter, Wasserman, and Kutner [1990]. In addition, a program for the analysis of any three-factor factorial ANOVA design is included in your software library.

The next chapter introduces another application of the basic probabilistic ideas presented in the earlier parts of the text—statistical quality assurance.

EXERCISES

14.1. Show, in general, that when only a single replication is present in a two-factor factorial ANOVA design in which the interaction effect is explicitly included, the degrees of freedom for the error term is zero.

14.2. For the data and the analysis contained in Tables 14.2 and 14.3, use the Tukey-Kramer procedure to determine which levels differ from one another for the A factor, and the interaction factor, AB.

14.3. Estimate the values of each of the A_i, B_j, and AB_{ij} effects for the two-factor factorial ANOVA model underlying the analysis given in Table 14.3.

14.4. Construct the o.c. curve for the interaction effect for the data of Table 14.2. What is the probability that the following departure from the null hypothesis of no interaction effect will be detected?

$$AB_{11} = -0.5 \quad AB_{12} = 0 \quad AB_{13} = 0.5$$
$$AB_{21} = 0.2 \quad AB_{22} = -0.1 \quad AB_{23} = -0.1$$
$$AB_{31} = 0.3 \quad AB_{32} = 0.1 \quad AB_{33} = -0.4$$

14.5. In Exercise 13.10, the two-way blocked ANOVA embodies the implicit assumption that no interaction effect is present in the data. Suppose that an interaction effect is possible, and reanalyze the data using the two-factor factorial ANOVA. If a significant interaction effect is found, determine which of the AB_{ij} differ from one another and give a practical interpretation of your findings.

14.6. Using the two-factor factorial ANOVA program, reanalyze (a) Exercise 13.15 and (b) Exercise 13.16. In each new analysis, summarize the differences and similarities in the results given by the two-way blocked ANOVA program and the two-factor factorial ANOVA program.

14.7. A new plastic sealant is being checked for its susceptibility to extremes of temperature and humidity. The data given here are the experimental results that show the sealant's capability after exposure to three temperature settings and two humidity settings. To increase the sample size, three replications were performed. Is there evidence of a significant effect on the sealant as a result of the different temperature or humidity settings or to their joint effect? If so, summarize your findings, detailing where the significant effects occur.

Sealing Capability Indices

	Replications					
	1		**2**		**3**	
Temperature Settings	**Humidity Settings**					
	1	**2**	**1**	**2**	**1**	**2**
1	73.38	76.96	72.91	76.74	72.57	72.24
2	74.36	76.04	72.75	75.57	76.19	74.83
3	76.73	74.39	76.76	76.76	77.35	76.41

Is the sample size adequate if a difference of 1.5 in the measure of sealing capability should be detected by our test?

14.8. Two fertilizer additives are being studied for their effect in the growth of a new blight-resistant barley. Four plots of ground are selected, where 12 identical seedlings are planted in a 3-by-4 grid. Additive A is administered in three strengths, 1, 2, and 3, along the columns of the grids, and additive B is administered in four strengths, 1, 2, 3, and 4, along the rows of the grids. The schematic layout of the four plots of ground, with the resultant growth data, are given next.

	Plot							**Plot**					
	1			**2**				**3**			**4**		
	A							**A**					
B	**1**	**2**	**3**	**1**	**2**	**3**	**B**	**1**	**2**	**3**	**1**	**2**	**3**
1	35.75	35.69	36.79	35.67	35.21	36.97	1	35.94	35.61	36.76	35.64	35.49	36.77
2	35.55	35.56	36.85	35.43	35.63	37.05	2	35.62	35.57	37.15	35.73	35.46	36.90
3	35.50	35.66	36.82	35.49	35.51	37.04	3	35.47	35.48	37.05	35.53	35.54	36.94
4	31.31	31.57	33.29	31.36	31.53	33.14	4	31.46	31.34	33.11	31.36	31.43	33.27

Determine if either additive, separately or together with the other additive, has a significant effect upon the growth of the barley.

14.9. The wearability of a certain epoxy based marine deck paint is being studied. Three factors have been identified as important to this particular paint: propellant pressure, surface roughness, and humidity level. The following data set gives the result of an experiment with three replications at three selected levels of propellant pressure, three selected levels of surface roughness, and four levels of ambient humidity. Determine if there is a significant difference in the wearability index due to the different levels of the three factors addressed in this study.

Wearability Index/Replication 1

	C											
	1			2			3			4		
	A											
B	**1**	**2**	**3**	**1**	**2**	**3**	**1**	**2**	**3**	**1**	**2**	**3**
1	73.48	75.86	74.86	78.07	79.66	80.86	76.95	78.14	79.38	77.23	77.66	79.55
2	73.54	74.63	73.49	77.67	76.14	81.69	77.60	78.28	78.41	75.66	76.75	78.54
3	73.01	75.64	73.25	77.33	80.25	80.51	79.43	79.07	78.57	77.62	75.41	77.05

Wearability Index/Replication 2

	C											
	1			2			3			4		
	A											
B	**1**	**2**	**3**	**1**	**2**	**3**	**1**	**2**	**3**	**1**	**2**	**3**
1	75.39	76.92	77.69	81.58	81.25	83.73	79.22	81.12	82.71	80.28	80.61	81.16
2	78.25	76.56	77.04	80.33	81.18	81.25	78.94	80.00	83.42	78.47	80.79	81.15
3	76.34	77.42	77.43	80.27	81.65	85.45	80.92	80.31	80.47	78.12	79.87	82.08

Wearability Index/Replication 3

	C											
	1			2			3			4		
	A											
B	**1**	**2**	**3**	**1**	**2**	**3**	**1**	**2**	**3**	**1**	**2**	**3**
1	79.88	79.77	80.54	84.59	85.06	85.66	82.94	84.88	84.21	82.82	81.94	84.08
2	78.26	80.18	81.71	83.91	84.01	84.66	81.84	85.79	85.09	81.58	84.20	84.23
3	78.32	79.41	80.42	84.11	84.07	86.54	82.71	83.48	86.03	83.25	83.23	83.35

CHAPTER

15

AN INTRODUCTION TO STATISTICAL QUALITY CONTROL

The following list presents the important procedures and concepts explored in this chapter.

- Be able to describe what kinds of *quality characteristics* may be studied using the $\overline{x}$-R chart and be able to give examples of such quality characteristics
- Be able to state the five things that must be done prior to using an $\overline{x}$-R control chart
- Understand the concept of *rational subgroups* and how they are used to aid in sampling from a process being studied by means of an $\overline{x}$-R chart
- Understand how an $\overline{x}$-R chart is constructed and know how to construct one using the software program
- Know how to interpret the results of the completed $\overline{x}$-R chart and how to determine if there are indications that the process is out of control
- Understand what is meant by an *assignable cause* and what the presence of an assignable cause tells us about a production process
- Understand that the primary purpose of the $\overline{x}$-R control chart is to *prevent* defective units of production
- Understand the difference between *engineering tolerance limits*, i.e., engineering specifications, and the *control limits* that are generated as part of the construction of an $\overline{x}$-R control chart
- Know how to use a completed $\overline{x}$-R chart to determine how well a production process will satisfy engineering specifications

- Know when an $\overline{x}$-R chart should be applied and when an attribute chart should be applied
- Understand why an attribute chart is less efficient than an $\overline{x}$-R chart
- Know the four major kinds of attribute charts and what situations are appropriate for each
- Know how to use a p data sheet and a p chart in the analysis of production data
- Understand that the same indications of assignable causes that are used for an $\overline{x}$-R chart are also used for a p chart, an np chart, a c chart, and a $\overline{c}$ chart
- Understand how to construct each of the four attribute charts discussed in this chapter
- Know how to analyze and interpret the results from an attribute chart for determining whether a process is under control
- Understand the purposes and the construction of scientific acceptance sampling plans
- Understand the meanings of the terms *acceptable quality level, rejectable quality level*, and *average outgoing quality*
- Understand the differences between *single*, *double*, *multiple*, and *sequential* sampling plans

15.1 INTRODUCTION

In statistical quality control, *quality* is defined as "conformance to requirements." One of the most widespread applications of probability and statistical analysis is in ensuring that the products of industrial manufacturing meet their requirements in a consistent and uniform manner. The methods of statistical quality control are not primarily directed at finding the reasons for "bad" products and counting the numbers of such bad products. Rather, the goal of statistical quality control is to prevent bad products by removing their causes and by detecting the onset of problems prior to bad production.

The most widely used tool of statistical quality control is the *control chart*. In essence, control charts are based on the well-known fact that any process contains a certain amount of inherent unremovable variability. Control charts assist us in determining the magnitude of that inherent variability and, thereby, give us a reference against which to measure the total variability observed in a process at any point in time. If the observed variability exceeds or shows signs that it will exceed the inherent variability, we know that an *assignable cause* is active in the process. In this chapter, we study two kinds of control charts—one for variables and one for attributes.

15.2 A CONTROL CHART FOR VARIABLES: THE $\overline{x}$-R CHART

Often called the $\overline{x}$-R chart, the control chart for variables is used when the level of the quality characteristic to be studied can be expressed as a number. Since many things are measured relative to a numerical scale, the field of application for

$\overline{x}$-R charts is very broad. The following list is intended only as an illustration of some of the many quality characteristics to which the $\overline{x}$-R chart might be applied.

1. Volumes of containers
2. Weekly sales
3. Strengths of circuit boards
4. Daily numbers of customer complaints
5. Diameters of bearing shafts
6. Resistivity of nonconducting materials
7. Number of typographical errors per page
8. Purity of metallic conductors
9. Sizes of formed plastic parts
10. Days lateness for arrival of purchased lots
11. Weights of assembled products

The following paragraphs describe what must be done before you can use the $\overline{x}$-R chart to assess any production system or any process of interest.

First, you must decide what quality characteristic is going to be studied. This seemingly obvious point is often overlooked. Many products have two or more characteristics that are quite important to an overall quality assessment. For example, the manufacturer of ball bearings is interested in the diameter, the sphericity, the hardness, and the surface smoothness of the bearings that she produces. It is very important to decide which characteristics are important and to set up a separate control chart for each one of those characteristics.

Second, you must decide when and how often the production system should be sampled to provide data for the control chart. This question generally reduces to one of economics. At one extreme, we might decide to look at every unit produced, i.e., to perform an "exhaustive inspection." This is almost always prohibitively expensive and in some cases, such as when destructive testing is required, completely inappropriate. At the other extreme, we could decide to test very few units (or none). This strategy will usually generate costs, in terms of product returns and disappointed customers, far in excess of a more reasonable approach that forms a compromise between the two extremes.

There is no general method for determining how frequently the process should be sampled. This decision must be based on experience with the process and made with an awareness of the resources available. In any event, the data should be obtained in such a way that the groups of data are taken from rational subgroups that represent homogeneous behavior from the process under study. In order to detect differences in the process that are due to some cause rather than random variation, the data set must include observations from the whole spectrum of activity. In particular, observations should be taken from different periods during which we have logical reasons to believe that differences in the process might exist.

One way of obtaining homogeneous *rational subgroups* of data is to analyze the process and determine where changes in process behavior are most likely to occur. Examples of two such cases are seen in the change of a work shift or the change of a machine operator. When those situations exist, you should take data from each shift or from the production of each operator.

Third, you must design a random sampling method to be used to obtain the units from the production process.

Fourth, we must answer the question, How many units should be measured at each sampling time? There are many highly mathematical and cumbersome techniques to determine the theoretically correct number. However, from a practical viewpoint, it is almost always sufficient to extract five or six units at each process sampling time and to adjust the frequency of the checks, if necessary, to maintain appropriate control data.

Fifth, you must obtain, calibrate, and make available, at the selected sampling times, the appropriate equipment and measuring devices to obtain reliable data about the process. Failure to do a thorough job here will very likely make all your efforts in the other phases useless. There is a popular saying in the computer industry: Garbage in, garbage out! Nowhere is that idea more valid than in securing data for an $\overline{x}$-R chart. This is even more important when one realizes that the user of faulty data is often unaware that it is faulty. It is not hard to imagine a completely wrong and costly decision being made on the basis of "bad" data, the source of which was faulty or uncalibrated measuring devices.

In summary, prior to using an $\overline{x}$-R chart you need to

1. Stipulate the quality characteristic or characteristics to be studied.
2. Decide when and how often to check the process.
3. Design a random sampling method with which to obtain the data at the sampling times.
4. Decide how many units to measure at each sampling time.
5. Fully prepare the measuring equipment for every use in gathering data for the control chart.

The $\overline{x}$-R data sheet presented in Table 15.1 is typical of forms used in data gathering for the construction of $\overline{x}$-R control charts. Notice that some effort has been put forth to ensure complete identification of the source of the data. This kind of information is often essential if an accurate and intelligent interpretation is to be made. The leftmost column, under the column title *No.*, gives the *time-ordered index* assigned to each sample of five units. Thus we observe that a total of 20 samples of size five are included in the data set presented in Table 15.1, and we observe that the *engineering tolerance limits* or *specifications* of the brass shaft are 0.500 ± 0.030 inch with the nominal value set at 0.500 inch.

Let us proceed to the $\overline{x}$-R chart that is exemplified in Table 15.2 and Figure 15.1, where $\overline{x}$ is the arithmetic average of $x_1, x_2, x_3, x_4,$ and x_5 and R is the sample range for the five observations in the sample. The $\overline{x}$'s, which give insight into the average quality, and the R's, which present information about the variability of the quality characteristic, should both be plotted and analyzed. It is important that the $\overline{x}$'s and R's are both plotted in the *time order* that they were taken. If this is not done, most of the power of this analysis method will be rendered ineffective.

TABLE 15.1
An example of an $\bar{x}$-R data sheet

$\bar{x}$-R DATA SHEET

Date 6/23/93
Part Name Brass Bushing Shaft Part No. OS121
Operation No. & Desc. 9—Final Lathe
Machine No. 27 Dept. No. 4
Engineering Specs. 0.500 ± 0.030 Nominal 0.500
Parts per Hour 350 Subgroup Size 175 Sample Size 5

No.	x_1	x_2	x_3	x_4	x_5
1	0.5110	0.5038	0.5090	0.5072	0.4793
2	0.4943	0.5093	0.5163	0.5041	0.4985
3	0.4890	0.5080	0.5033	0.4960	0.5106
4	0.4913	0.4847	0.5127	0.4972	0.4731
5	0.5114	0.5058	0.4925	0.4979	0.5126
6	0.4955	0.5009	0.5067	0.5235	0.4968
7	0.5043	0.5169	0.4929	0.5026	0.4782
8	0.5132	0.4949	0.4877	0.5053	0.4929
9	0.4927	0.5074	0.4949	0.4891	0.4865
10	0.4965	0.4888	0.5067	0.4855	0.5196
11	0.5190	0.5055	0.4921	0.5120	0.4796
12	0.4950	0.4967	0.5112	0.4973	0.4942
13	0.4625	0.5040	0.5169	0.5012	0.4940
14	0.5140	0.5037	0.4940	0.5049	0.5030
15	0.5119	0.5012	0.5065	0.4911	0.5015
16	0.5015	0.5121	0.5066	0.4994	0.5055
17	0.4957	0.5151	0.5103	0.4885	0.5145
18	0.5130	0.5135	0.5272	0.5031	0.4981
19	0.5058	0.5102	0.5048	0.4901	0.4933
20	0.5043	0.5138	0.5080	0.4915	0.4831

After the $\bar{x}$-R program plots all of the $\bar{x}$'s and R's, it obtains the grand averages, $\bar{\bar{x}}$ and $\bar{R}$, and the upper and lower *control limits,* $\text{UCL}_{\bar{x}}$, $\text{LCL}_{\bar{x}}$, UCL_R, and LCL_R, for the $\bar{x}$ and R charts, respectively. The calculation of those quantities is performed using the following equations, 15.1 through 15.4.

$$\bar{\bar{x}} = \frac{\sum \bar{x}}{k} \tag{15.1}$$

where $\bar{\bar{x}}$ is the grand average and k is the number of $\bar{x}$'s

$$\bar{R} = \frac{\sum R}{k} \tag{15.2}$$

where R is the average range for the k values of R

$$\text{UCL}_{\bar{x}} = \bar{\bar{x}} + A_2\bar{R} \qquad \text{LCL}_{\bar{x}} = \bar{\bar{x}} - A_2\bar{R} \tag{15.3}$$

where $\text{UCL}_{\bar{x}}$ and $\text{LCL}_{\bar{x}}$ are the upper and lower control limits for the $\bar{x}$ chart and A_2 is a multiplier that may be shown to be the theoretically correct value to ensure that the control limits are set at approximately $3\sigma_{\bar{X}}$ away from the centerline value,

TABLE 15.2
The $\bar{x}$-R chart for Table 15.1

Capability Study for Brass Shaft

Analyst: J. Doe — Date: 6/23/93

Number of Samples = 20 Sample Size = 5

R Chart Parameters		$\bar{x}$ Chart Parameters	
Lower Control Limit	= 0.000	Lower Control Limit	= 0.485
Centerline	= 0.0276	Centerline	= 0.501
Upper Control Limit	= 0.0585	Upper Control Limit	= 0.517

Sample Number	Sample Range	Sample Mean
1	0.0317	0.502
2	0.0220	0.505
3	0.0216	0.501
4	0.0396	0.492
5	0.0201	0.504
6	0.0280	0.505
7	0.0387	0.499
8	0.0255	0.499
9	0.0209	0.494
10	0.0341	0.499
11	0.0394	0.502
12	0.0170	0.499
13	0.0544	0.496
14	0.0200	0.504
15	0.0208	0.502
16	0.0127	0.505
17	0.0266	0.505
18	0.0291	0.511
19	0.0201	0.501
20	0.0307	0.500

$\bar{\bar{x}}$. This fact implies that $A_2\bar{R}$ estimates $3\sigma_{\bar{X}} = 3\sigma/\sqrt{n}$, where σ is the population standard deviation and n is the number of measurements in each group of data.

$$\text{UCL}_R = D_4\bar{R} \qquad \text{LCL}_R = D_3\bar{R} \tag{15.4}$$

where UCL_R and LCL_R are the upper and lower control limits for the R chart and D_4 and D_3 are theoretically derived multipliers that give control limits for R at approximately three standard deviations about the center value of $\bar{R}$. This means that $D_4\bar{R}$ estimates $\bar{R} + 3\sigma_R$ and that $D_3\bar{R}$ estimates $\bar{R} - 3\sigma_R$. Of course, $\bar{R} - 3\sigma_R$ is valid only if the lower limit is greater than zero. For small values of $\bar{R}$, the lower limit can yield a negative value. Since a negative range has no practical meaning, D_3 is set to zero in such cases.

Referring to Figure 15.1, we see that $\bar{\bar{x}}$ and $\bar{R}$ are plotted as solid lines on the $\bar{x}$ and R charts, respectively. The upper and lower control limits are plotted as dashed lines on their respective control charts.

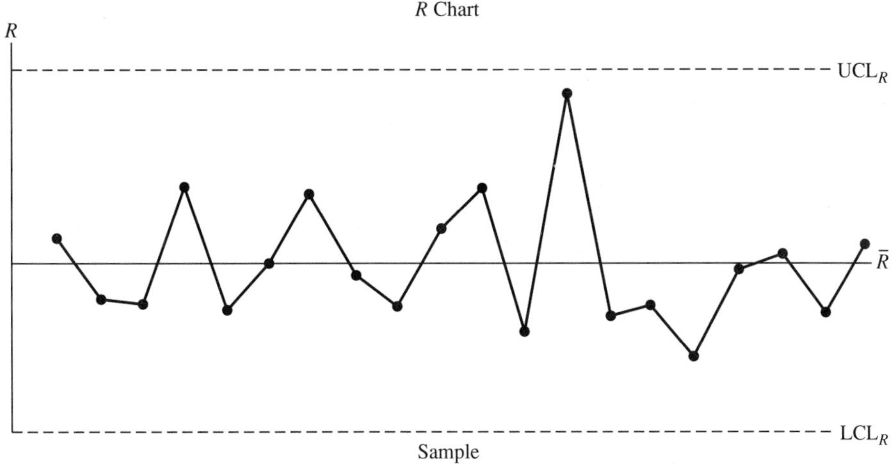

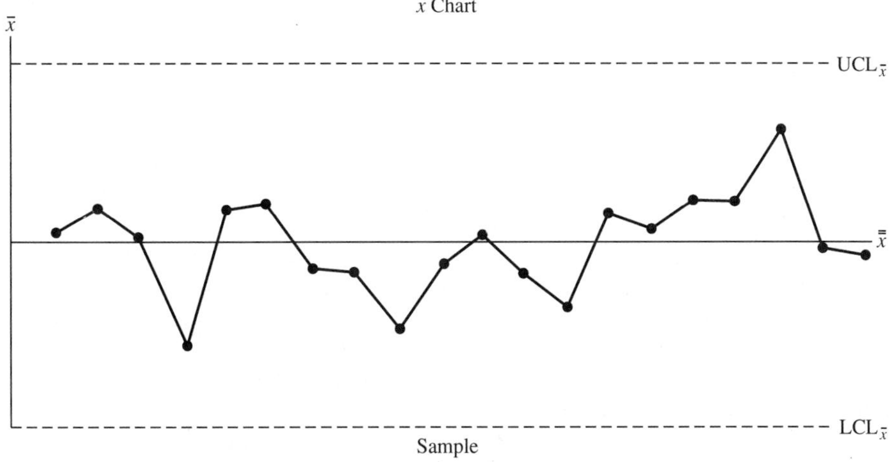

FIGURE 15-1
The $\overline{x}$-R chart for Table 15.1.

Although the derivation of the values of A_2, D_3, and D_4 is beyond the scope of this book, it is important that we have the values available for our use. Table 15.3 contains those values and shows that they are dependent on the sample size, n. The d_2 factor is used in the comparison of the process with engineering specifications, and its discussion will be deferred until that topic is presented.

TABLE 15.3
Control limit coefficients for the $\bar{x}$-R chart

Sample Size, n	Averages, A_2	Ranges D_3	Ranges D_4	Standard Deviation, d_2	Sample Size, n
2	1.880	0.0	3.268	1.128	2
3	1.023	0.0	2.574	1.693	3
4	0.729	0.0	2.282	2.059	4
5	0.577	0.0	2.115	2.326	5
6	0.483	0.0	2.004	2.534	6
7	0.419	0.076	1.924	2.704	7

Using the data of Table 15.1 and Equations 15.1 through 15.4, the values of the parameters are $\bar{\bar{x}} = 0.5012$, $\text{UCL}_{\bar{x}} = 0.5172$, $\text{LCL}_{\bar{x}} = 0.4853$, $\bar{R} = 0.027649$, $\text{UCL}_R = 0.0585$, and $\text{LCL}_R = 0$.

Having considered how to construct a control chart, let us discuss how to interpret the results that we obtain. We already know that $\bar{\bar{x}}$ and $\bar{R}$ give us both the best measure of the average size of the unit and a feeling for how widely the values in such samples will range. In our example, the value of $\bar{\bar{x}} = 0.5012$ is greater than the engineering specification, or nominal size, of 0.500. If the grand average is very far from the nominal value (in the middle of the specifications), the symmetric variation about the average that is a characteristic of the normal distribution would make it more likely that manufactured parts will fall outside of the engineering specifications or tolerance limits. For this reason, we would like the grand average to be precisely equal to the nominal size. However, it is very improbable that such an unusual event will occur. Whether $\bar{\bar{x}}$, in this case, is far enough from the nominal value to have a significant effect on the number of units falling outside the tolerance limits remains to be seen.

Because the range is a measure of the variability associated (in our example) with the manufacture of the part, the best of all possible values is zero—also an improbable event. Acceptable values of the range can be quantified by studying the process capability. This is discussed later in this section.

In order to understand how and why the control chart helps in detecting whether the production process is "in control," i.e., behaving in an ordinary, "good" fashion, we must know how much variability is expected to be present when the process is in control. There are, in general, only two sources of variation in any production process, the inherent random variability discussed above and variability that can be traced to an assignable cause. If no assignable cause is present, the process is operating as well as can be expected and is said to be *in control*. If an assignable cause is present, there will be a change in the pattern of variation that would not ordinarily be present when the process is in control. Figure 15.2 shows some typical control chart indications of the presence of an assignable cause.

It is important to realize that the presence of an assignable cause can mean that defective units are produced. However, an assignable cause will *not necessar-*

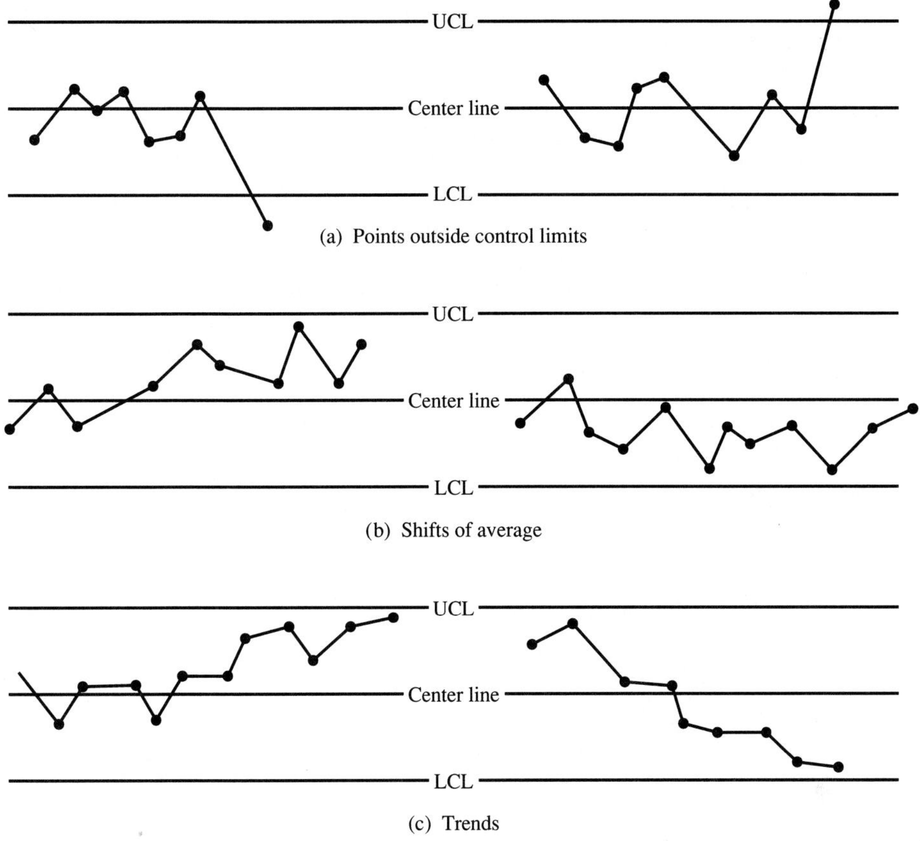

FIGURE 15-2
Indications of assignable causes in a control chart.

ily cause defective units. Indeed, one of the more important things that a control chart can do is to give advance notice of the possibility of bad units and allow the individuals who are in charge of the process to take corrective action prior to the production of any defective units.

In Figure 15.2*a*, we see the case where one or more units cause a plotting to fall outside the control limits. This is always a sufficient reason to conduct an investigation into whether corrective action is required. A point outside the control limits is almost always the result of an assignable cause.

In Figure 15.2*b* and Figure 15.2*c*, we see examples in which the process average has shifted or in which a trend is present. Various rules have been suggested to help in recognizing whether a shift has occurred or a trend is present. Among them are the following.

- 2 out of 3 successive plottings between two standard deviations and the control limit on the same side of the center line

- 4 out of 5 successive plottings between one standard deviation and the control limit on the same side of the center line
- 8 successive plottings on the same side of the center line

From a practical viewpoint, however, if the pattern present in the control chart is enough to raise a question in your mind, that is a sufficient reason to investigate for the presence of an assignable cause. Figure 15.3 presents examples of each of these circumstances to clarify the exact meanings of these statements.

It should be emphasized that the control limits are *not* the tolerance limits that correspond to the engineering specifications. This adds impact to the earlier

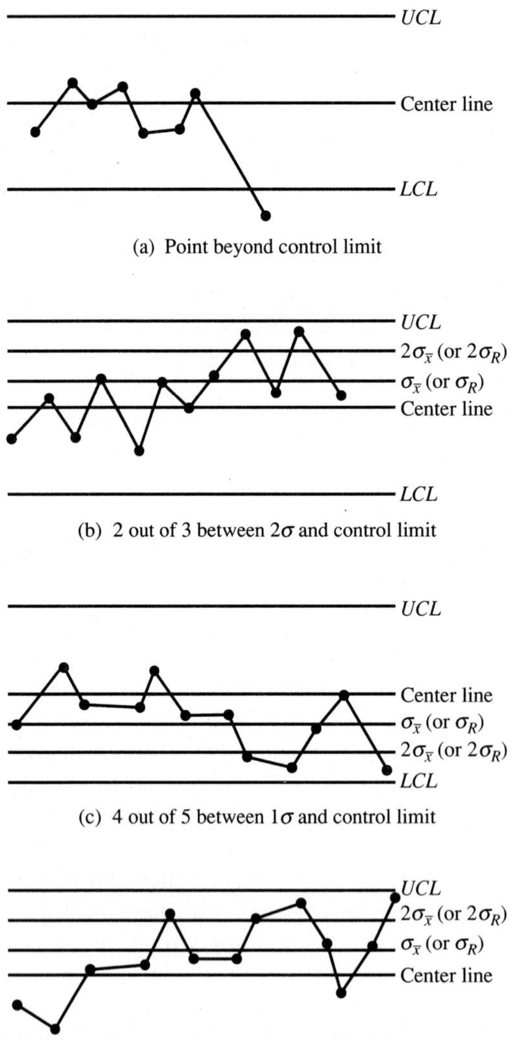

FIGURE 15-3
Specific examples of the "rules" for detecting assignable causes.

statement that a process out of control does not necessarily imply that defective parts are being made. Further, a process in control is only operating within the "natural" limits of the process; i.e., no assignable causes are present. It is possible that a process may have control limits that are not in accordance with engineering specifications. Indeed, as we have said, it may be impossible for the process to satisfy the engineering specifications. At that point, either the specifications, the process, or both must be modified.

After a process has been brought under control and sufficient data have been taken to construct the $\overline{x}$ and R control charts, it is a relatively easy task to determine whether the process will produce units that satisfy the engineering specifications. Furthermore, such an analysis will indicate the changes that need to be made to bring the control limits and the specifications into conformance with one another. Indeed, in the initial use of a control chart on a process, it may be necessary to revise the control limits one or more times before the process is finally brought into control and the control limits finally reflect the actual process capability.

In performing a comparison of the process characteristics against the specifications, we are primarily interested in the answers to the following four questions.

1. Are parts likely to be produced that are greater than the upper specification?
2. Are parts likely to be produced that are less than the lower specification?
3. How does the process average compare with the middle specification?
4. How does the process spread compare with the total tolerance?

The engineering specifications for the brass shaft in our example are

Upper specification	=	0.530
Middle specification	=	0.500
Lower specification	=	0.470
Total tolerance	=	0.060

How do these specifications compare to the natural limits of the process as obtained earlier? The control limits derived above are in terms of $\overline{x}$, from a sample of size 5. What we need are the natural limits in terms of a single unit of product. They are easily found through the use of the d_2 factor in Table 15.3. It can be shown that $3\overline{R}/d_2$ approximates the value of the quantity 3σ, where σ is the standard deviation of the distribution of the quality characteristic under study.

Using this information, the natural limits of the process, $\text{UNL}_{\overline{x}}$ and $\text{LNL}_{\overline{x}}$, in terms of a single unit are obtained by computing

$$\begin{aligned} \text{UNL}_{\overline{x}} &= \overline{\overline{x}} + \frac{3\overline{R}}{d_2} \\ \text{LNL}_{\overline{x}} &= \overline{\overline{x}} - \frac{3\overline{R}}{d_2} \end{aligned} \tag{15.5}$$

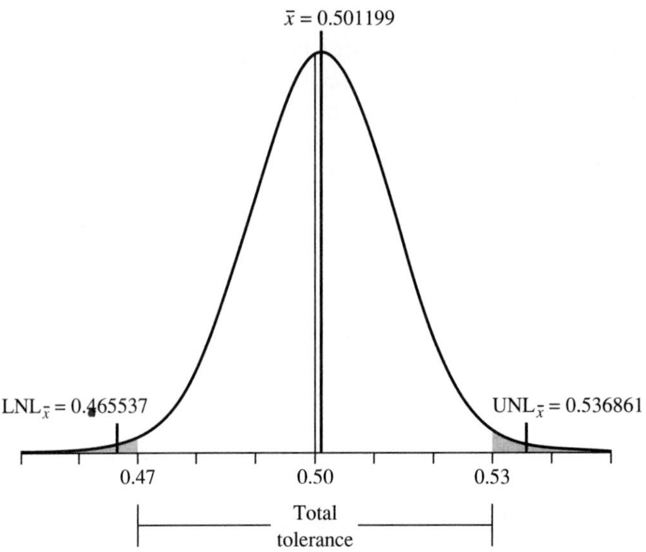

FIGURE 15-4
The relationship between engineering tolerances and natural limits.

Use of the formulas in Equation 15.5 for our example yields

Upper natural limit	=	0.536861
Process average	=	0.501199
Lower natural limit	=	0.465537
Process spread	=	0.071324

In comparing the natural limits in our example against the specifications, the process spread is greater than the total tolerance, and both the upper and lower natural limits fall outside their specification. Figure 15.4 pictures the implications of these facts.

The probability that a unit produced by this process will be greater than the upper tolerance limit is determined by evaluating the probability that a normal random variable with a mean of $\overline{x} = 0.501199$ and a standard deviation of $\overline{R}/d_2 = 0.119$ will exceed the upper specification of 0.53. As illustrated in Figure 15.4, the probability that a unit will exceed the upper specification is about 0.0084, or 0.84%. Similarly, the probability that a unit will be less than the lower specification is about 0.0040 or 0.40%. Thus, on the average, we can expect a total of about 1.24% of the units produced to be either too small or too large. If this is an unacceptable state of affairs, we have no choice but to tighten the process spread or to widen the specifications.

15.3 CONTROL CHARTS FOR ATTRIBUTES

The $\overline{x}$-R chart can be applied only to quality characteristics that can be measured, i.e., described in terms of a number or a numerical scale. Many quality characteristics do not require such measurements. For example, we often do not care if support members are too strong; we are concerned that they are strong enough. If such members break at less than a specified minimum stress level, they are bad. If they do not fail, they are good.

Some characteristics are not measurable in the sense of an $\overline{x}$-R chart. For example, how does one measure the beauty of a rose? Finally, quality characteristics that could be measured are sometimes either too difficult or too expensive to measure. For example, a single part may have many small characteristics that could be measured. However, the construction of an $\overline{x}$-R chart on every characteristic would be prohibitively expensive. Often a single attribute chart can be used in the place of numerous $\overline{x}$-R charts at much less cost and with an acceptable level of protection for a production process.

Attribute charts are never as efficient as $\overline{x}$-R charts. Classifying a part as good or bad simply does not contain as much information as an actual measurement that permits comparison with the engineering tolerance limits. We overcome this comparative lack of efficiency by using appreciably larger sample sizes in attribute charts than are used in $\overline{x}$-R charts.

When obtaining a measurement is infeasible, we must resort to other techniques. One common quality rating method is to classify the unit as either "good" or "bad." In the use of control charts for attributes, specific definitions of the terms *defect* and *defective* are used. A part can possess a defect, meaning that some quality characteristic does not meet the engineering specifications, without being classified as a "defective" part. A defective is a part that has more than an allowable number of defects. (Often no defects of a specified type are allowed. If so, a single defect will be sufficient cause to declare the part defective.)

The following list gives the four major kinds of attribute charts.

- p charts—used to record the fraction or proportion of defective parts in a sample
- np charts—used to record the actual number of defectives in a sample (used when defectives are relatively rare)
- c charts—used to record the number of defects on each unit inspected
- $\overline{c}$ charts—used to record the average number of defects per unit inspected (used when the average number of defects per unit is small); $\overline{c}$ charts are also known as *u charts*

The most popular attribute chart is the p chart, or *fraction defective chart*. Therefore, we consider it in some detail prior to a shorter look at the other three types of attribute charts.

Table 15.4 and Figure 15.5 present typical examples of a p data sheet and a p chart. Note that all the relevant information concerning the part under investigation is contained in the heading material of the p data sheet. As can be observed, there can be several reasons why a part might be classified as a defective.

Once a set of samples has been inspected (typically 20 to 30 groups) and the numbers of defects have been recorded, this information is used to construct the p chart. First, the p chart program computes the proportion defective for each sample. In the example of Table 15.4, we have 25 samples of 50 angle

TABLE 15.4
A typical *p* data sheet

p DATA SHEET — Date 2/12/93

Dept No. 22
Machine No. 3 — Part Name Angle Iron Support — Part No. 37-2
Subgroup Size 500 — Oper. Desc. Bevel and Drill Bolt Attach
Sample Size 50

No.	Number Defective	Fraction Defective	Reasons for Reject
1	7	0.14	
2	4	0.08	Bad taper, bent
3	4	0.08	
4	4	0.08	
5	5	0.10	Hole mispositioned
6	2	0.04	
7	3	0.06	
8	1	0.02	
9	6	0.12	Remill needed
10	6	0.12	
11	4	0.08	Material defect
12	3	0.06	
13	4	0.08	
14	5	0.10	
15	8	0.16	
16	8	0.16	
17	2	0.04	
18	3	0.06	
19	9	0.18	Incorrect sheet metal thickness
20	5	0.10	
21	7	0.14	
22	7	0.14	
23	6	0.12	
24	5	0.10	
25	3	0.06	

iron supports. After computing the proportions defective for all 25 samples, the program plots these proportions as on the p chart of Figure 15.5. Next, both the center line and the control limits are plotted to help us in interpreting what information the data contains. The center line or average proportion defective, symbolized by $\overline{p}$, may be obtained by computing the arithmetic average of the sample proportions defective or by forming the ratio of the total number defective to the total number inspected. For our specific example, $\overline{p} = 121/1250 = 0.0968$.

If $\overline{p}$ is large enough, the upper and lower control limits are easily found by noting that $\bar{P}$ is approximately normally distributed with a standard deviation $\sigma_{\overline{P}}$ estimated by $s_{\overline{P}} = \sqrt{[\overline{p}(1-\overline{p})]/n}$, where n is the sample size for each sample. As before, we want the control limits to be placed at the plus and minus three standard deviations away from the mean. Therefore,

Total Number Inspected = 1250
Total Number Rejected = 121
Observed Value of $\overline{p}$ = 0.0968

UCL = 0.2222
LCL = 0
There are 0 points above the upper control limit.

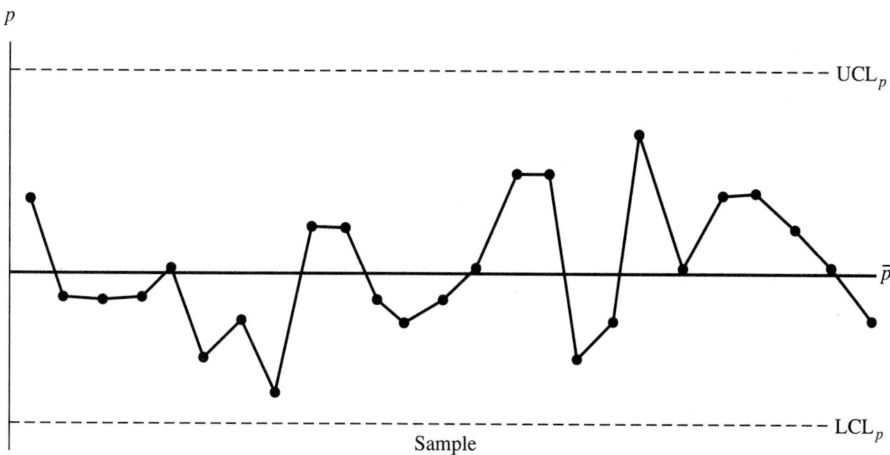

FIGURE 15-5
p chart for Table 15.4

$$\begin{aligned} UCL_p &= \overline{p} + 3\sqrt{\frac{\overline{p}(1-\overline{p})}{n}} \\ LCL_p &= \overline{p} - 3\sqrt{\frac{\overline{p}(1-\overline{p})}{n}} \end{aligned} \tag{15.7}$$

For our example, $UCL_p = 0.2222$ and $LCL_p = -0.02475$, which implies that $LCL_p = 0$.

When $\overline{p}$ is too small, LCL_p will be negative. If this occurs, LCL_p should be reset to a value of zero. In our current example, the computed value of -0.02475 is only marginally negative. Thus the normal approximation is sufficiently close to use for this data set.

What can we conclude about the process? In using p charts, we look for the same chart indications of assignable causes that we look for when we use $\bar{x}$-R charts—i.e., shifts, trends, points outside the control limits, and so on—the unnatural patterns presented in the previous section. The finished p chart for our example gives no indication that the process is out of control, and, because it does appear to be in control, we can use the current values of $\overline{p}$, UCL_p, and LCL_p for further quality assessment.

Since no general rules exist for the setting of an acceptable quality level, whether $\overline{p}$ is acceptable remains a managerial decision that must be reached on a case-by-case basis.

Earlier, we studied the normal approximation to the binomial distribution. There we saw that the approximation is good when $np \geq 5$ and $p \leq 0.5$ or when $n(1-p) \geq 5$ and $p \geq 0.5$. If these criteria are not at least approximately satisfied, the values for UCL_p and LCL_p as computed earlier may not be sufficiently accurate. If this is the case, the control limits must be computed directly from the binomial distribution. With an adequate sample size, this will rarely be necessary.

Let us briefly consider the other three kinds of attribute charts mentioned earlier. The first is the *np* chart, which serves the same function as the *p* chart. The *np* chart is used in preference to the *p* chart when the number of defectives rather than the proportion of defectives is the important measure of quality. The major difference between the two charts is that the vertical scale for the *np* chart is calibrated in terms of number of defectives rather than proportion defective.

The relations given in Equation 15.8 are used to compute the center line and the control limits for the *np* chart:

$$
\begin{aligned}
n\overline{p} &= \frac{\text{Total number of defectives}}{\text{Total number of samples}} \\
\text{UCL}_{np} &= n\overline{p} + 3\sqrt{n\overline{p}(1-\overline{p})} \qquad (15.8) \\
\text{LCL}_{np} &= n\overline{p} - 3\sqrt{n\overline{p}(1-\overline{p})}
\end{aligned}
$$

The analysis procedure for the *np* chart exactly parallels that of the *p* and $\overline{x}$-R charts.

The *c* chart is used when we want to study the number of defects, not defectives. For example, consider a large support frame made up of welded steel pipe. It is almost certain that on such an assembly there will be one or more welding defects. However, only a few welding defects would probably not make the assembly defective.

A complete study of the quality characteristics of such a large assembly would probably be prohibitive in cost, even if attribute charts were used for each weld point. Consider instead a chart that records the total number of defects on each assembly that is inspected. Although it does not give the wealth of information available from other, more detailed charts, such a chart would give good indications of when an overall change in production quality had taken place.

The center line and the control limits for the *c* chart are computed using Equation 15.9.

$$
\begin{aligned}
\overline{c} &= \frac{\text{Total number of defects}}{\text{Total number of samples}} \\
\text{UCL}_c &= \overline{c} + 3\sqrt{\overline{c}} \qquad (15.9) \\
\text{LCL}_c &= \overline{c} - 3\sqrt{\overline{c}}
\end{aligned}
$$

The *c* chart is usually used for cases in which the sample size for each data point is a single unit (although this is not required).

Suppose that the expected number of defects per unit is so small that most units have zero defects and yet we still want to study the average number of defects per unit. In that case, we would use the $\overline{c}$ chart (which is also known as a u chart). In constructing a $\overline{c}$ chart, we are forced to look at several units per sample in order to raise the expected number of defects to a value that would ensure that most samples yield one or more defects.

The formulas for the center line and control limits are given in Equation 15.10.

$$\overline{\overline{c}} = \frac{\text{Total number of defects}}{\text{Total number of units inspected}}$$

$$\text{UCL}_{\overline{c}} = \overline{\overline{c}} + 3\sqrt{\frac{\overline{\overline{c}}}{n}} \qquad (15.10)$$

$$\text{LCL}_{\overline{c}} = \overline{\overline{c}} - 3\sqrt{\frac{\overline{\overline{c}}}{n}}$$

In both the c and $\overline{c}$ charts, care should be taken to ensure that the sample size or inspection unit is a constant value for all data that are taken for subsequent use on the selected kind of control chart. If this is not done, each point and the distribution that governs it could be sufficiently different to cause the control chart to be confusing, at best, and to cause incorrect conclusions, at worst.

In the next and final section of the chapter, we present a brief introduction to the subject of acceptance sampling.

15.4 ACCEPTANCE SAMPLING: CONSTRUCTION OF SAMPLING PLANS AND THEIR USES

In many production processes, the producer must rely on outside sources for both raw material and components. In that situation, one problem is regularly confronted: Should the lot of material that has been provided be accepted and used? Clearly, if the material is substandard and is used, an excellent chance of decreasing the quality of the product exists. This statement is true both when the supplier is a group inside the company or an outside vendor. Some techniques currently used in industry to attempt to solve the problem are listed here.

1. No inspection
2. 100% inspection
3. Spot-checking
4. Constant percentage sampling
5. Scientific sampling

If no inspection is conducted, the material is sent directly to inventory. We find the material to be substandard only when we really need it for production. The approach of not inspecting is prohibitively expensive and can also cause bad relations to develop between the company, its suppliers, and its customers.

Inspecting every part, 100% inspection, is also quite expensive. The sheer effort required is often overwhelming—causing "inspection fatigue"—and is an inefficient use of inspection personnel. Of course, this approach is clearly infeasible if the inspection procedure is destructive.

Spot-checking makes it possible to let some lots through without any checking. This is contrary to a common-sense approach to the problem.

Constant percentage sampling is probably the most common of all approaches currently used. Unfortunately, such an approach disregards the fact that a constant percentage sample of 10% from a lot of 50 items, for example, provides a *much less stringent* statistical measure of quality than does a 10% sample from a lot of size 500.

Of course, a scientific acceptance sampling plan is the best of the five approaches listed. It forms a compromise between no sampling and 100% sampling and does it in such a way that both the supplier and the manufacturer know precisely the risks of an incorrect decision. The following preliminary steps should be taken prior to starting a scientific acceptance sampling plan.

1. Identify the characteristic(s) to be checked.
2. Obtain reliable and accurate measuring instruments.
3. Provide a method of random sampling.

An acceptance sampling plan that is based in sound statistical concepts is nothing more than the application of a test of hypothesis, just like those discussed in Chapter 9. An acceptance sampling plan possesses all the properties and characteristics of a hypothesis test, including α and β errors, and an operating characteristic curve. For our current application, the null and alternate hypotheses are the following.

$$H_0 : \text{The material is up to standards.}$$
$$H_A : \text{The material is substandard.}$$

The associated errors are $\alpha = P(H_0 \text{ rejected when true}) =$ producer's risk and $\beta = P(H_0 \text{ accepted when false}) =$ consumer's risk. The o.c. curve of the hypothesis test is a plot of the P(Accepting the lot) against the true proportion of defects in the lot.

How do we construct an o.c. curve for an acceptance sampling plan? Let us assume that we have a lot of size N, with an actual proportion defective, p, where we have randomly sampled and inspected n items taken from the lot. Logically, our decision on whether to accept the lot will be based on the number of defectives found in the n sampled items. We will accept the lot if no more than c defectives are found in our sample. Knowing the values N, n, and c, this is a perfect application of the hypergeometric distribution. Letting $L(p)$ be the probability of accepting a lot that has a proportion of defective units equal to p, we have

$$L(p) = \sum_{x=0}^{c} \frac{{}_{Np}C_x \; {}_{N-Np}C_{n-x}}{{}_{N}C_n} \tag{15.11}$$

If $N = 100$, $n = 10$, and $c = 1$, the probability of accepting the lot is the probability of finding 0 or 1 defective in the sample. Therefore,

$$L(p) = \frac{{}_{100p}C_0 \; {}_{100(1-p)}C_{10} + {}_{100p}C_1 \; {}_{100(1-p)}C_9}{{}_{100}C_{10}}$$

In most cases, we use the binomial approximation of the hypergeometric distribution to compute the o.c. curve. (Recall that this approximation is appropriate when $N/10 \geq n$.) Since the lot size is 10 times larger than the sample size in our current example, we may approximate $L(p)$ with the binomial distribution using the distribution evaluator program in your software library. Table 15.5 and Figure 15.6 present the o.c. curve for $N = 100$, $n = 10$, and $c = 1$. From Figure 15.6, we can observe that

- If there are no defectives in the lot, we are certain to accept the lot.
- We will accept a lot with as many as 40% defective items about 5% of the time! (This kind of error will be intolerable in most cases.)

TABLE 15.5
The o.c. curve for $N = 100$, $n = 10$, and $c = 1$

p	$L(p)$	p	$L(p)$	p	$L(p)$
0.01	0.9957	0.12	0.6583	0.30	0.1493
0.03	0.9655	0.15	0.5443	0.35	0.0860
0.05	0.9139	0.17	0.4730	0.40	0.0464
0.07	0.8482	0.20	0.3758	0.45	0.0232
0.10	0.7361	0.25	0.2440	0.50	0.0107

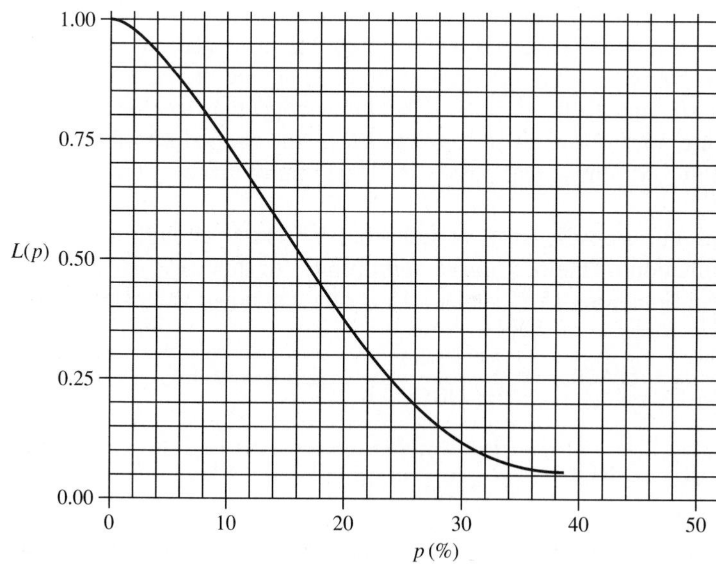

FIGURE 15-6
Plot of the o.c. curve for $N = 100$, $n = 10$, and $c = 1$.

A natural question to ask at this point is, What minimum percentage of defective items in the lots should we require from the producer? in other words, What is an acceptable quality level (AQL)? First, for practical purposes, 0% defective is merely a dream. One way to determine an AQL is to review past practices and determine what percentage of defective items, in the long term, is reasonable from the viewpoint of overall cost and quality. The setting of the AQL is usually a high-level managerial decision (made in conjunction with representatives of the "producer's" company), which is, in general, beyond the scope of this book. However, once the AQL is set, we need a sampling plan that will accept nearly all the lots that meet the AQL and reject nearly all the lots that do not meet the AQL. As we have seen, given N, n, and c, it is simple to construct the o.c. curve using either the binomial or hypergeometric distribution.

This procedure does allow both the "producer" and "consumer" to specify the risks and associated quality levels that are acceptable to them. Therefore, in order to custom-tailor an acceptance sampling plan, we must first set an AQL and the associated α error that are acceptable to the producer of the lots to be inspected. Next, the consumer, or user, of the items contained in the inspected lots must provide a β error and an associated *rejectable quality level* (RQL). The RQL (also known as the *lot tolerance percentage defective,* or LTPD) is the level of percentage defective that would do economic damage to the consumer's business if such lots are passed. Hence, we must reject nearly all the lots that have defects at the RQL level or greater.

Just as in the earlier tests of hypotheses, the greater discrimination that we require from our test of hypothesis, the larger the sample that will be required. In our current test, the closer the AQL and RQL values, the greater the required sample size to discriminate between them with acceptably small values of the α and β errors.

The setting of α, AQL, β, and RQL are sufficient to place two points on the o.c. curve. Suppose that the following have been agreed on: $\alpha = 0.05$, AQL $= 0.03$, $\beta = 0.10$, and RQL $= 0.10$. The knowledge that we have so far is presented in Figure 15.7. Now all we have to do is find the values of N, n, and c that will yield an o.c. curve that passes (at least approximately) through those two points.

For our present purposes, let us assume that N is large enough to make the binomial approximation of the o.c. curve appropriate. In this case, all that needs to be done is solve Equations 15.12 and 15.13 for the unknown values of n and c.

$$1 - \alpha = L(\text{AQL}) = \sum_{i=0}^{c} \frac{n!}{i!(n-i)!}\text{AQL}^i(1 - \text{AQL})^{n-i} \tag{15.12}$$

$$\beta = L(\text{RQL}) = \sum_{i=0}^{c} \frac{n!}{i!(n-i)!}\text{RQL}^i(1 - \text{RQL})^{n-i} \tag{15.13}$$

As may be observed, not only are the equations nonlinear but they also require integer values of the unknowns. These kinds of equations do not offer an easy analytical solution. Fortunately, the nomograph presented in Figure 15.8 [Johnson and Kotz, 1969] allows us to find an *approximate* graphical solution very easily.

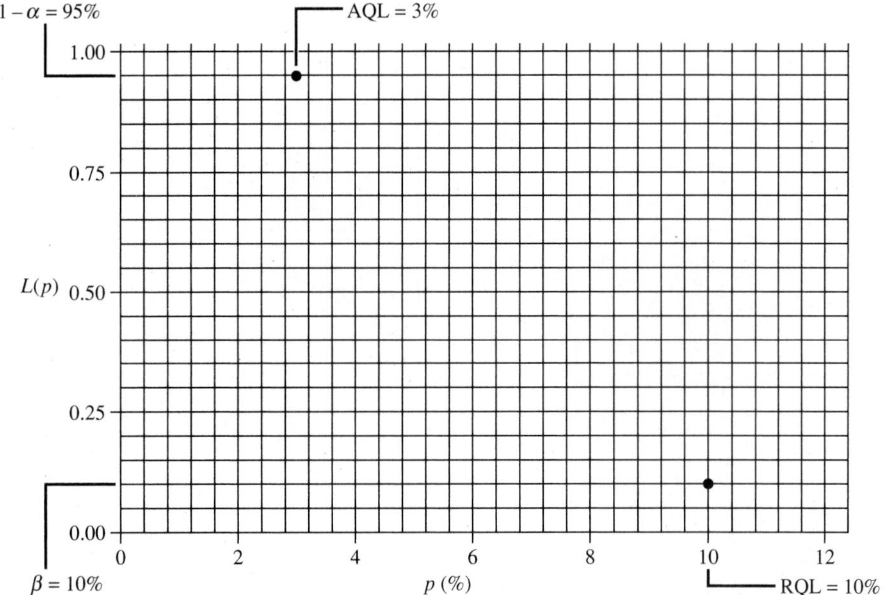

FIGURE 15-7
Two prespecified points on the o.c. curve.

Using Figure 15.8 and sharpening the nomograph's approximate results with the o.c. curve program, we find, for $\alpha = 0.05$, AQL $= 0.03$, $\beta = 0.10$, and RQL $= 0.10$ that $n = 90$ and $c = 5$. Figure 15.9 presents the o.c. curve.

Now that we can construct a tailor-made acceptance sampling plan, a natural question is, How much does the use of such a plan actually help? In other words, What is the quality of the accepted lots, the *average outgoing quality,* or AOQ?

The AOQ is dependent on (1) the acceptance sampling plan, (2) the quality of the incoming lots, and (3) the disposition of the rejected lots. A common method of handling rejected lots is to require that the supplier perform 100% inspection and replace all defective units with good units. The "refurbished" lot is then passed back to the customer. Assuming that all rejected lots are transformed into lots with no defects, we may express the AOQ as a function of the incoming quality, p, and the probability of acceptance at that incoming quality, $L(p)$. It follows that

$$\text{AOQ}(p) = pL(p) + 0(1 - L(p)) = pL(p) \qquad (15.14)$$

Thus, once the o.c. curve is obtained, the AOQ(p) curve can be produced with only one additional step. The AOQ curve associated with the o.c. curve of Figure 15.9 is given in Figure 15.10. Notice that the worst value on the AOQ curve, the AOQ limit of AOQL $= 0.03534$, occurs at an incoming quality level of $p = 0.0475$. At any other value of p the AOQ will be less. The key point is that we are assured that the quality of the product that is *actually used* can have, on the average, no greater percentage defective than the AOQL *regardless* of the incoming quality level.

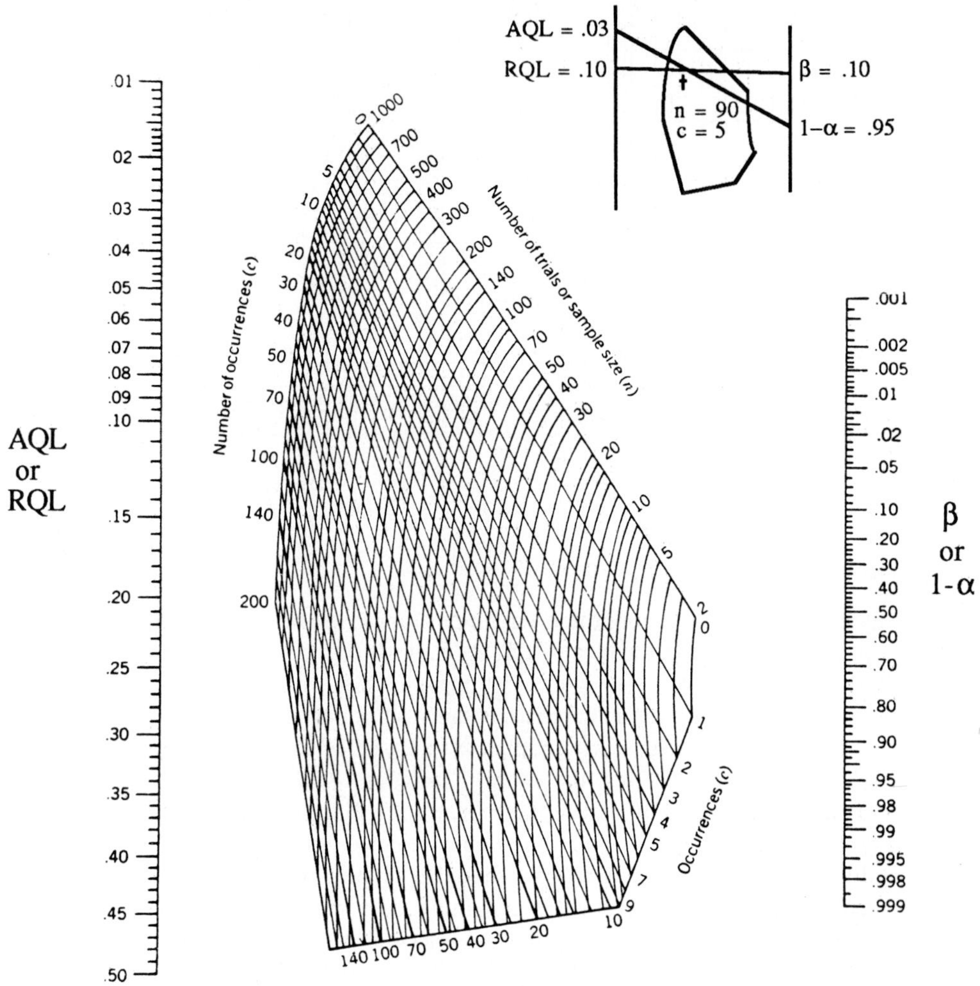

FIGURE 15-8
A nomograph for obtaining n and c for an acceptance sampling plan (Adapted with permission from Johnson and Kotz, *Discrete Distributions,* Wiley Interscience, 1969.

Up to this point we have considered acceptance sampling plans that base the decision on one sample taken from the lot—thus the name *single sampling plans*. There are other kinds of sampling plans that use two or more samples and are known as *double sampling plans* and *multiple sampling plans*. A double sampling plan is carried out as follows.

If the first sample of size n_1 yields no more than c_1 defectives, accept the lot and terminate testing.

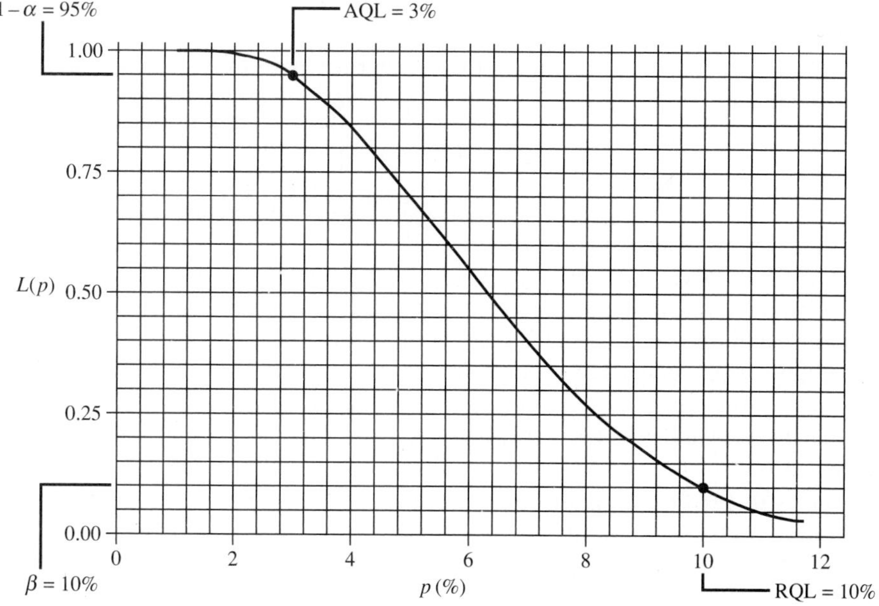

FIGURE 15-9
The o.c. curve for $n = 90$ and $c = 5$.

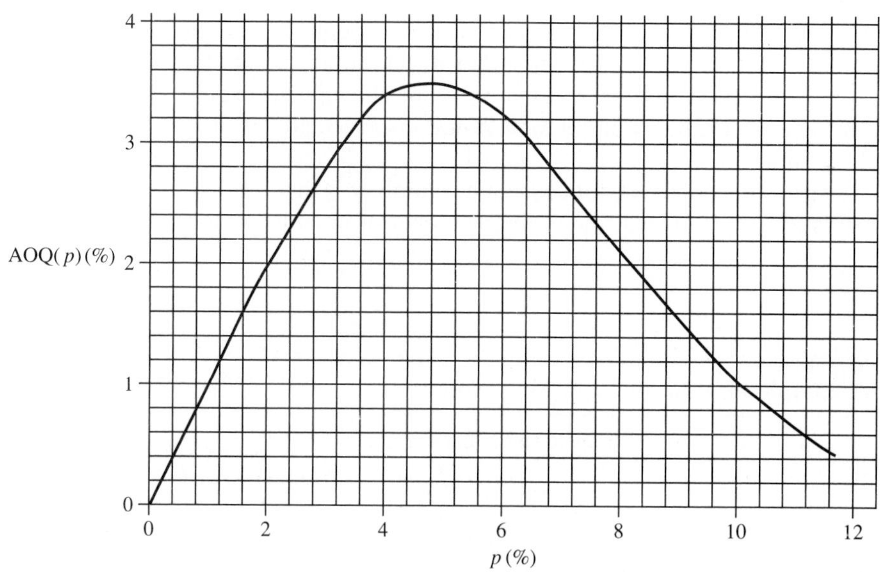

FIGURE 15-10
The AOQ curve for the o.c. curve of Figure 15.9.

If the first sample yields more than c_2 defectives, reject the lot and terminate testing.

If the first sample yields more than c_1 but no more than c_2 defectives, take a second sample of size n_2. If the total number of defectives from both samples is no greater than c_3, accept the lot; otherwise, reject the lot.

A multiple sampling plan is conducted in a similar fashion, but there is a possibility that three or more samples will be required.

It can be shown that, by adjusting the sample sizes and acceptance numbers, it is possible to match the o.c. curve of any single sampling plan with a double or multiple sampling plan. The advantage of these more complex inspection schemes is that they lessen the overall average number of units that must be inspected while maintaining the same level of protection for the producer and the consumer. In particular, for very bad or very good lots, the number of units that must be inspected is dramatically lessened. The disadvantage is that a lot of "intermediate quality" will possibly require markedly more units to be inspected than in the equivalent single sampling plan.

Consider the double sampling plan of Table 15.6. If incoming quality is $p = 0.05$, the probability that a second sample is taken is equal to the probability that 2 or 3 defectives are found in the first sample. Assuming that the lot size, N, is large, the probability of this occurring is approximately 0.087839. The average number of units that will be inspected is $50 + 0.087839(100) = 58.784$. This compares very well with the 75 units that would have to be inspected in the equivalent single sampling plan where $n = 75$ and $c = 2$.

The idea of multiple sampling is carried to its logical extreme by the method of *sequential sampling*, which has, in theory, no definite termination condition. In sequential sampling, we decide to either reject the lot, accept the lot, or continue sampling after each unit is inspected. It can be shown that, in general, the numbers for acceptance and rejection for the nth unit inspected are given in Equations 15.15 and 15.16.

Again, the advantage of sequential sampling is that the amount of inspection effort is lessened. Indeed, the average number of units inspected is reduced about

TABLE 15.6
An example of a double sampling plan

		Combined Samples		
Sample	**Sample Size**	**Size**	**Accept Number**	**Reject Number**
1	50	50	1	4
2	100	150	3	4

50% over single sampling plans. In addition, in the practical use of sequential sampling plans there is always a point at which the sampling is truncated. The exact methods of doing this are cumbersome and are not discussed here.

$$a_n = \left(\log \frac{\beta}{1-\alpha} + n \log \frac{1-\text{AQL}}{1-\text{RQL}}\right) \Big/ \left(\log \frac{\text{RQL}}{\text{AQL}} - \log \frac{1-\text{RQL}}{1-\text{AQL}}\right) \quad (15.15)$$

$$r_n = \left(\log \frac{1-\beta}{\alpha} + n \log \frac{1-\text{AQL}}{1-\text{RQL}}\right) \Big/ \left(\log \frac{\text{RQL}}{\text{AQL}} - \log \frac{1-\text{RQL}}{1-\text{AQL}}\right) \quad (15.16)$$

where α = producer's risk, β = consumer's risk, and log is a logarithm of base 10, or *common logarithm*. If a_n is not an integer, use the largest integer not larger than a_n (truncate a_n). If r_n is not an integer, use the smallest integer larger than r_n (round r_n).

In addition to the computer-based methods just described, many excellent tables, provided by private companies and by governmental agencies, exist for acceptance sampling and inspection. Each table differs in various ways: Some emphasize selecting the plans by acceptable quality level and others by average outgoing quality limit. Most sampling tables provide different inspection levels; single, double, and multiple types of sampling plans; and normal, tightened, and reduced degrees of inspection. Many of these tables include plots of the o.c. curves and AOQ curves for the plans they provide. If a person keeps these curves in mind and recognizes their value, any of the tables may be used efficiently and satisfactorily.

In the next and final chapter of this book, we consider some additional techniques of statistical analysis that apply to situations that engineers encounter less frequently than the situations described in the earlier chapters. In the first part of Chapter 16, we discuss a method for assessing "goodness of fit," i.e., how well a data set "fits" or agrees with a proposed underlying probability distribution function. In the remainder of Chapter 16 we consider alternate methods to the t test, the paired two-sample t test, the two-sample t-test, and the completely randomized single factor ANOVA. All of these alternate methods are characterized by the fact that they do not require stringent prior assumptions about the distribution that governs the phenomenon or phenomena being studied. As such, the latter part of Chapter 16 comprises an introduction to "distribution-free" or *nonparametric* statistical analysis.

EXERCISES

15.1. Analyze the given $\overline{x}$-R data set. In this chapter, the importance of preserving the order in which the data are obtained and plotting the data in that same order has been emphasized. While preserving the contents of each group of five measurements, randomize the order in which the groups of data appear and then again perform the $\overline{x}$-R analysis.

Obs						Obs					
1	4.96	4.86	5.00	4.85	5.11	14	5.27	5.14	5.07	5.21	5.02
2	5.03	5.05	5.09	4.81	4.98	15	5.18	5.12	5.15	4.99	5.17
3	4.80	5.04	5.16	5.11	5.15	16	5.26	5.11	5.12	5.31	5.14
4	5.05	4.98	5.00	5.11	4.94	17	5.20	5.23	5.03	5.22	5.21
5	5.08	4.81	5.07	5.11	4.97	18	5.18	4.99	5.01	5.03	5.18
6	5.02	4.85	4.98	5.08	4.97	19	5.26	5.23	5.24	5.14	5.30
7	4.99	5.26	5.05	5.11	5.04	20	5.16	5.24	5.15	5.25	5.13
8	5.08	5.20	5.06	4.87	5.16	21	5.46	5.42	5.14	5.11	5.21
9	5.12	5.07	5.00	5.05	5.10	22	5.18	5.25	5.32	5.27	5.29
10	5.08	5.14	4.95	4.99	5.14	23	5.31	5.44	5.32	5.16	5.14
11	5.16	5.12	4.98	5.14	5.11	24	5.30	5.12	5.35	5.29	5.21
12	4.92	5.05	4.94	5.13	5.07	25	5.11	5.27	5.22	5.27	5.26
13	5.19	5.09	5.14	4.94	5.04						

Compare the two analyses, and comment on the differences and similarities that you observe.

15.2. Magnetron, Inc., produces a line of small electromagnets. The given data are measurements of the magnetic field strength for sampled magnets from the Magnetron production line. Based on this data set, may we conclude that the production process is under control?

Obs						Obs					
1	7.77	7.60	7.72	7.68	6.98	13	6.56	7.60	7.92	7.53	7.35
2	7.36	7.73	7.91	7.60	7.46	14	7.85	7.59	7.35	7.62	7.58
3	7.23	7.70	7.58	7.40	7.77	15	7.78	7.53	7.66	7.28	7.54
4	7.28	7.12	7.82	7.43	6.83	16	7.54	7.80	7.67	7.48	7.64
5	7.78	7.63	7.31	7.45	7.82	17	7.39	7.88	7.76	7.21	7.86
6	7.39	7.52	7.67	8.09	7.42	18	7.82	7.84	8.18	7.58	7.45
7	7.61	7.92	7.32	7.57	6.95	19	7.65	7.76	7.62	7.25	7.33
8	7.83	7.37	7.19	7.63	7.32	20	7.61	7.85	7.70	7.29	7.08
9	7.32	7.69	7.37	7.23	7.16	21	7.74	7.40	7.42	7.42	7.53
10	7.41	7.22	7.67	7.14	7.99	22	7.14	7.31	7.00	7.59	7.63
11	8.12	7.64	7.30	7.80	6.74	23	7.35	7.29	7.32	7.49	7.90
12	7.37	7.42	7.78	7.43	7.36						

15.3. In the production of aluminum cylindrical supports for hardware part bins, the indention in the end of the support determines how well the support will function in holding up the bin shelf. Twenty-five samples of size 5 were sampled from the output of the casting process that produces the supports. The following data gives the indention measurements of the sampled supports. Does the process appear to be in control?

Obs						Obs					
1	26.10	25.38	25.90	25.72	22.93	2	24.43	25.93	26.63	25.41	24.85
3	23.90	25.80	25.33	24.60	26.06	4	24.13	25.52	26.27	24.72	22.31
5	26.14	25.52	24.25	24.79	26.28	6	24.55	25.09	25.67	27.35	24.68
7	25.43	26.69	24.29	25.26	22.82	8	26.32	24.49	23.77	25.53	24.29
9	24.27	25.74	24.49	23.91	23.65	10	24.65	23.88	25.67	23.55	26.96
11	26.90	25.55	24.21	26.20	22.96	12	24.50	24.67	26.12	24.73	24.42
13	21.25	25.40	26.69	25.12	24.40	14	26.40	25.37	24.40	25.49	25.30
15	26.12	25.12	25.65	24.12	25.15	16	25.15	26.21	25.66	24.94	25.55
17	24.57	26.51	26.03	23.85	26.45	18	26.30	26.35	27.72	25.31	24.81
19	25.58	26.02	25.48	24.01	24.33	20	25.43	26.38	25.80	24.15	23.31
21	25.95	24.60	24.68	24.69	25.13	22	23.55	24.33	23.00	25.35	25.52
23	24.41	24.17	24.30	24.97	26.59	24	26.19	23.58	23.93	27.03	24.80
25	26.07	26.23	25.28	25.02	24.16						

15.4. Figure 15.2 gives four unique indications of the presence of an assignable cause. Determine the probability associated with the occurrence of each indication and comment on the meanings, if any, of the similarities and differences among those probabilities.

15.5. Every 8-hour shift, the air around a spray-painting booth is measured at four randomly selected times to determine the amount of suspended paint "mist" that has escaped the booth's enclosure. The following data set gives the measured amounts of mist that were present in the last 20 shifts. Do the precautionary measures for preventing paint escape from the booth appear to be consistent; i.e., is the amount of mist under control?

Obs					Obs				
1	2.42	2.16	2.19	2.50	11	2.16	2.31	2.27	2.42
2	2.28	2.41	2.42	2.33	12	2.33	2.31	2.27	2.42
3	2.30	2.22	2.29	2.34	13	2.25	2.25	2.29	2.47
4	2.37	2.31	2.45	2.38	14	2.25	2.45	2.41	2.44
5	2.34	2.34	2.29	2.21	15	2.18	2.28	2.31	2.37
6	2.26	2.21	2.33	2.30	16	2.29	2.13	2.18	2.08
7	2.28	2.37	2.30	2.31	17	2.30	2.22	2.11	2.23
8	2.36	2.29	2.39	2.17	18	2.25	2.28	2.23	2.34
9	2.37	2.24	2.42	2.23	19	2.28	2.20	2.32	2.12
10	2.23	2.18	2.19	2.38	20	2.29	2.19	2.25	2.18

15.6. Suppose that the engineering specifications on the indentions of the aluminum supports of Exercise 15.3 are 25.00 ± 2.50. Is the process capable of meeting the engineering specifications?

15.7. The Moka-Kola bottle-manufacturing plant produces six sizes of bottles. Fifty bottles are randomly sampled from every 2 hours of production from the new and experimental 12-ounce "line." Data from the last 50 hours of production are given here. Is the 12-ounce design in control?

#	Insp	Rej	#	Insp	Rej	#	Insp	Rej	#	Insp	Rej
1	50	10	2	50	8	3	50	10	4	50	9
5	50	2	6	50	6	7	50	10	8	50	12
9	50	9	10	50	7	11	50	5	12	50	10
13	50	8	14	50	6	15	50	10	16	50	5
17	50	4	18	50	11	19	50	7	20	50	1
21	50	10	22	50	9	23	50	6	24	50	7
25	50	11									

15.8. Plocter and Ramble Soap, Inc., produce their Satin Skin facial bar only in the "personal" size. Data from the last 30 hours of production are given in the following table; defects are nicks that occur during the bar-casting step. Is the process in control?

#	Insp	Rej	#	Insp	Rej	#	Insp	Rej	#	Insp	Rej
1	60	10	2	60	8	3	60	10	4	60	9
5	60	2	6	60	6	7	60	10	8	60	12
9	60	9	10	60	7	11	60	5	12	60	10
13	60	8	14	60	6	15	60	10	16	60	5
17	60	4	18	60	11	19	60	7	20	60	1
21	60	10	22	60	9	23	60	8	24	60	9
25	60	11	26	60	12	27	60	8	28	60	9
29	60	14	30	60	11						

15.9. In Exercise 15.7, management and engineering have set up a schedule of development for the new bottle production process. If the design goal, at this point in time, is 20%, is the experimental 12-ounce design meeting expectations?

15.10. Suppose that in the construction of a particular p chart, the proportion defective was too small for the normal approximation to the distribution of p to be appropriate. Outline in detail how you would approach the setting of the upper and lower control limits for that situation.

15.11. In Section 15.4, it is stated that constant percentage sampling is much more stringent for large lots than for small lots. Construct and plot the o.c. curves for the following sampling plans: (N, n, c) = (50,5,0), (100,10,0), (250,25,0), (50,10,0), (∞, 10,0). Discuss how the preceding statement is supported by these o.c. curves, and, further, show that in this example a constant *number* of items sampled is much more consistent than a constant percentage sampled.

15.12. An agreement has been reached between your company and one of its larger suppliers. As part of this agreement, it was decided that a certain part would undergo a single sample acceptance sampling plan with AQL = 2%, $\alpha = 0.05$, RQL = 4%, and $\beta = 0.05$. Find the values of N, n, and c that match these values of AQL, α, β, and RQL.

15.13. Compute the AOQ(p) curve for the sampling plan of Exercise 15.12, and determine the AOQL.

15.14. Suppose that a sequential random sample of size 15 has been obtained and has yielded the following results.

Number Inspected	Number Defective
1	0
2	1
3	2
4	3
5	3
6	4

Use Equations 15.15 and 15.16 to determine the values of a_n and b_n, $n = 1, 2, \ldots, 6$, for AQL $= 0.04$, $\alpha = 0.1$, $\beta = 0.1$, and RQL $= 0.15$. Can the lot be either rejected or accepted on the basis of this sample?

CHAPTER 16

SOME ADDITIONAL METHODS OF DATA ANALYSIS

The following are the more important procedures and concepts of this chapter.

- Understand and be able to apply the χ^2 *goodness of fit test* to determine whether it is reasonable to infer that a given data set was drawn from a given completely specified probability distribution function
- Understand and be able to apply the *Wilcoxon signed rank test* as a *distribution-free* alternative to the one-sample t test
- Be able to generate the distribution of the null hypothesis, for small samples, for the Wilcoxon signed rank test
- Understand the use of the p-value in determining whether to reject the null hypothesis for the Wilcoxon signed rank test, the *Wilcoxon rank sum test*, and the *Kruskal-Wallis test*
- Understand and be able to apply the Wilcoxon rank sum test as a distribution-free alternative to the two-sample t test
- Be able to generate the distribution of the null hypothesis, for small samples, for the Wilcoxon rank sum test
- Understand and be able to apply the Kruskal-Wallis test as a distribution-free alternative to the completely randomized single-factor ANOVA

16.1 INTRODUCTION

In Section 16.2, we consider a method for assessing *goodness of fit*, i.e., how well a data set "fits," or agrees with, a proposed underlying probability distribution function. In the remainder of this chapter, we consider methods that may be used instead of several of the hypothesis tests presented in Chapters 9, 10, and 13. These alternate methods do not require stringent prior assumptions about the associated probability distribution function. Hence, the latter part of Chapter 16 is an introduction to *distribution-free*, or *nonparametric*, statistical analysis.

16.2 The χ^2 Test for Goodness of Fit

In Exercise 6.2, three data sets are given. In addition to a typical analysis, you are asked if the data are governed by one of the named distributions that had been introduced in Chapters 3 and 4. Given the information provided by the book to that point, all that you could have been expected to do was to study the shape of the frequency polygon and to determine if the estimates of μ, σ, β_1, and β_2 were in reasonable agreement with the theoretical values for the proposed underlying distribution. Fortunately, methods have been designed for this problem that allow the user to perform a formal hypothesis test about a proposed underlying probability distribution function. The most versatile of these methods [Cramer, 1946] is based on the normal approximation of the binomial distribution joined with the fact that squared standardized normal random variables form χ^2 random variables when they are summed (as discussed in Chapter 4).

Suppose that a data set is believed to be governed by a completely specified finite discrete probability distribution:

x	x_1	x_2	$\cdots$	x_k
$p_0(x)$	p_{01}	p_{02}	$\cdots$	p_{0k}

In other words, the null hypothesis,

$$H_0\text{: The distribution governing the data is } p_0(x).$$

is to be tested against the alternate hypothesis,

$$H_A\text{: Some other distribution governs the data.}$$

Suppose that random sampling yields a sample of size n and that n_1 of the data values were equal to x_1, n_2 of the data values were equal to $x_2, \ldots,$ and n_k of the data values were equal to x_k. If H_0 is true, then we would *expect* that np_{01} of the data would be equal to x_1, np_{02} of the data would be equal to $x_2, \ldots,$ and np_{0k} of the data would be equal to x_k. Further, if H_0 is true, it may be shown that

$$\chi^2 = \sum_{i=1}^{k} \frac{(n_i - n\,p_{0i})^2}{n\,p_{0i}} = \sum_{\text{all possibilities}} \frac{(\text{observed} - \text{expected})^2}{\text{expected}} \tag{16.1}$$

is the realization of an associated random variable that is approximately distributed as a χ^2_{k-1} distribution. To preserve the integrity of the approximation of the χ^2_{k-1} distribution, it is preferable that $np_{0i} \geq 5$ for all $i = 1, 2, \ldots, k$. However, when $k \geq 3$ and no np_{0i} is less than 1, as many as 20% of the np_{0i} can be less than 5 without a serious distortion of the test.

To test the preceding H_0, we compute χ^2 according to Equation 16.1 and compare it against the value of the χ^2_{k-1} distribution at the preselected value of α, the type I error for the hypothesis test. If $\chi^2 > \chi^2_{k-1,\alpha}$, we reject the null hypothesis and conclude that the proposed distribution does not govern the data set. Equivalently, we can determine the p-value (the right-hand-tail probability) associated with χ^2—i.e., $P(\chi^2_{k-1} > \chi^2)$—and compare the p-value with α. If $P(\chi^2_{k-1} > \chi^2) < \alpha$, we reject the null hypothesis.

Example 16.1. Suppose that we have a good reason to believe that a data set containing 100 values is governed by a binomial distribution with $n = 10$ and $p = 0.3$. The observed n_i and the theoretical $p_0(x)$ are

x	$p_0(x)$	n_i
0	0.02825	3
1	0.12106	12
2	0.23347	19
3	0.26683	29
4	0.20012	21
5	0.10292	6
6	0.03676	9
7	0.00900	0
8	0.00145	0
9	0.00014	1
10	0.00000	0

In order to improve the approximation of the χ^2_{k-1} distribution, we aggregate the last five values of x into one group for the purpose of our goodness of fit test. Therefore, the input information to the goodness of fit program is

x	$p_0(x)$	n_i
0	0.02825	3
1	0.12106	12
2	0.23347	19
3	0.26683	29
4	0.20012	21
5	0.10292	6
6–10	0.04735	10

This leaves the first and last cells for $x = 0$ and x ranging from 6 to 10 with $np_{0i} < 5$. However, because neither np_{0i} is extremely small and k (the number of categories, or "cells") is equal to 7, there is no serious damage to the test.

Passing this $p_0(x)$ and the associated n_i to the goodness of fit program, we obtain $\chi^2 = 8.715$ and the fact that $P(\chi^2_{k-1} > \chi^2) = P(\chi^2_6 > 8.715) = 0.1902$. Hence, for any value of α less than 0.1902, there is insufficient evidence to reject the null hypothesis that these data were drawn from a binomial distribution with $n = 10$ and $p = 0.3$. (We recall, however, that this does *not* constitute a proof of H_0.)

In the preceding discussion, we assumed that the underlying probability distribution was discrete. As long as we have explicit knowledge of $f(x)$, very little difficulty is added if the distribution is assumed to be continuous. In the case of a continuous probability distribution, the cells and their associated p_{0i} are formed by integrating $f(x)$ over a selected set of subranges that cover the entire range of x.

Example 16.2. Suppose that a data set of 75 values is believed to be governed by a continuous uniform distribution, $f(x) = 1$, for $0 \le x \le 1$.

Range	p_{0i}	n_i
0.0–0.1	0.1	5
0.1–0.2	0.1	8
0.2–0.3	0.1	3
0.3–0.4	0.1	11
0.4–0.5	0.1	4
0.5–0.6	0.1	5
0.6–0.7	0.1	4
0.7–0.8	0.1	14
0.8–0.9	0.1	13
0.9–1.0	0.1	8

One acceptable selection of subranges for integration would be 0.0–0.1, 0.1–0.2, . . . , and 0.9–1.0. This selection of subranges and the 75 data values yield the above input data for the goodness of fit program.

Although not required, this selection does provide the popular characteristic that all np_{0i} are *equal* at a value of 7.5. (Additionally, they all exceed 5.)

Passing this information to the goodness of fit program, we obtain $\chi^2 = 19.0$ and the fact that $P(\chi^2_{k-1} > \chi^2) = P(\chi^2_9 > 19.0) = 0.025$. Hence, for any value of α greater than 0.025, there is sufficient evidence to reject the null hypothesis that this data was drawn from a uniform distribution over the range $0 \le x \le 1$.

From Examples 16.1 and 16.2, we see that performing a χ^2 goodness of fit test is reasonably straightforward. Unfortunately, obtaining the operating characteristic curve for such a test is much more difficult and is not considered in this book. The primary reason for the difficulty is the great number of possible specific alternate hypothesis distributions that might govern the data when H_0 is false. Although the difficulty encountered here is much greater and multifaceted, it resembles the problem encountered in Section 10.5 in the computation of the β error for the comparison of two proportions.

In the preceding discussion, we have assumed that we had *explicit* knowledge of either $p(x)$ or $f(x)$. The technique illustrated in Examples 16.1 and 16.2 can also be extended to problems in which we have knowledge only of the distributional form; i.e., we might know that the data's underlying distribution is a gamma distribution but have no knowledge of the values of λ and r. Using *maximum likelihood estimation* methods to obtain estimates of λ and r, we could proceed in much the same fashion as above to perform goodness of fit tests. Additional applications of the χ^2 statistic, similar to the ones illustrated in this section, are found in the testing for statistical independence or for homogeneous structure. Readers interested in these other uses of the χ^2 statistic are referred to Bowker and Lieberman [1972], Hald [1965], Mosteller and Rourke [1973], and Gibbons [1985].

16.3 A DISTRIBUTION-FREE ALTERNATIVE TO THE *t* TEST: THE WILCOXON SIGNED RANK TEST

Chapter 9 shows that the t distribution can be used to test conjectures about a mean when the sample size is small and the value of σ is unknown. However, the application of the t distribution is correct only if the probability distribution governing the possible values of the data is normal. In some cases, the normality assumption will not be supportable.

If the normality assumption is not supportable, a good alternative technique is the *Wilcoxon signed rank test*. Suppose that the population's underlying distribution is not normal but is known to be *continuous and symmetric* and that we have randomly sampled n items from that population. Once more, we desire to test the null hypothesis, $H_0 : \mu = \mu_0$, against the usual triad of alternate hypotheses, $H_A : \mu > \mu_0$, $\mu < \mu_0$, or $\mu \neq \mu_0$.

The Wilcoxon signed rank test is performed in the following three steps.

1. Randomly sample n observations, $x_1, x_2, \ldots x_n$, from the population under study. Subtract μ_0 from each x_i , forming the $d_i = x_i - \mu_0$. Place the observations in the data set in ascending order by the magnitudes of the *absolute values* of the d_i, marking those observations that are greater than μ_0. (Discard any values that are equal to μ_0.)
2. Define the *rank* of any observation to be its index in the ordered data set of step 1, and compute s^+, the observed value of random variable S^+. s^+ is the sum of the *positive ranks,* i.e., the ranks of the observations that are greater than μ_0. (If any subset of two or more observations have the same absolute magnitude, redefine each of their ranks to be the average rank of the observations in the subset.)
3. Compute the p-value(s), $p^+ = P(S^+ \geq s^+ \mid H_0 \text{ is true})$ and/or $p^- = P(S^+ \leq s^+ \mid H_0 \text{ is true})$. Use Table 16.1 to decide whether to reject H_0. (Note that in the two-tailed H_A, the p-value is actually twice the minimum of p^+ or p^-.)

TABLE 16.1
Decision table for the Wilcoxon signed rank test

Alternate Hypothesis	Reject H_0 if
$H_A : \mu > \mu_0$	$p^+ \leq \alpha$
$H_A : \mu < \mu_0$	$p^- \leq \alpha$
$H_A : \mu \neq \mu_0$	$p^+ \leq \alpha/2$ or $p^- \leq \alpha/2$

Example 16.3. Suppose that the following six data are thought to be drawn from a beta distribution with $\alpha = 0.5$ and $\beta = 0.5$ (a symmetric, "bowl-shaped" continuous distribution, see Exercise 4.6).

0.2719, 0.7889, 0.1961, 0.2514, 0.0453, 0.5106

If the conjecture is true, the appropriate null hypothesis for the mean of the distribution is $H_0 : \mu = \mu_0 = 0.5$. Further, let $H_A : \mu > \mu_0$.

When the computer program for the Wilcoxon signed rank test performs step 1 on these data, it obtains

Rank	+1	−2	−3	+4	−5	−6
	0.0106	0.2281	0.2486	0.2889	0.3039	0.4547

In step 2 for this example, the program finds that the sum of the positive ranks is $s^+ = 1 + 4 = 5$. In step 3, we must assess whether this s^+ is consistent with $H_0 : \mu = \mu_0 = 0.5$. In order to perform this assessment, we must have available sufficient information about the probability distribution of S^+.

In this example, we have 6 observations. Excluding the theoretically impossible happening that an observation may be *exactly* equal to μ_0, each observation may be either larger than μ_0 (has a *positive rank*) or smaller than μ_0 (has a *negative rank*). Since we have assumed that the distribution is symmetric, the prior probability of either a positive or a negative rank for any observation is 0.5. Hence, any of the $2^6 = 64$ possible signed rank orderings has a probability of occurrence of $(0.5)^6 = \frac{1}{64}$. Those possibilities are detailed in Table 16.2.

TABLE 16.2
The 64 possible signed rank orderings when $n = 6$

Outcome	s^+	Outcome	s^+	Outcome	s^+	Outcome	s^+
−1 − 2 − 3 − 4 − 5 − 6	0	−1 − 2 − 3 − 4 + 5 − 6	5	−1 − 2 − 3 − 4 − 5 + 6	6	−1 − 2 − 3 − 4 + 5 + 6	11
+1 − 2 − 3 − 4 − 5 − 6	1	+1 − 2 − 3 − 4 + 5 − 6	6	+1 − 2 − 3 − 4 − 5 + 6	7	+1 − 2 − 3 − 4 + 5 + 6	12
−1 + 2 − 3 − 4 − 5 − 6	2	−1 + 2 − 3 − 4 + 5 − 6	7	−1 + 2 − 3 − 4 − 5 + 6	8	−1 + 2 − 3 − 4 + 5 + 6	13
+1 + 2 − 3 − 4 − 5 − 6	3	+1 + 2 − 3 − 4 + 5 − 6	8	+1 + 2 − 3 − 4 − 5 + 6	9	+1 + 2 − 3 − 4 + 5 + 6	14
−1 − 2 + 3 − 4 − 5 − 6	3	−1 − 2 + 3 − 4 + 5 − 6	8	−1 − 2 + 3 − 4 − 5 + 6	9	−1 − 2 + 3 − 4 + 5 + 6	14
+1 − 2 + 3 − 4 − 5 − 6	4	+1 − 2 + 3 − 4 + 5 − 6	9	+1 − 2 + 3 − 4 − 5 + 6	10	+1 − 2 + 3 − 4 + 5 + 6	15
−1 + 2 + 3 − 4 − 5 − 6	5	−1 + 2 + 3 − 4 + 5 − 6	10	−1 + 2 + 3 − 4 − 5 + 6	11	−1 + 2 + 3 − 4 + 5 + 6	16
+1 + 2 + 3 − 4 − 5 − 6	6	+1 + 2 + 3 − 4 + 5 − 6	11	+1 + 2 + 3 − 4 − 5 + 6	12	+1 + 2 + 3 − 4 + 5 + 6	17
−1 − 2 − 3 + 4 − 5 − 6	4	−1 − 2 − 3 + 4 + 5 − 6	9	−1 − 2 − 3 + 4 − 5 + 6	10	−1 − 2 − 3 + 4 + 5 + 6	15
+1 − 2 − 3 + 4 − 5 − 6	5	+1 − 2 − 3 + 4 + 5 − 6	10	+1 − 2 − 3 + 4 − 5 + 6	11	+1 − 2 − 3 + 4 + 5 + 6	16
−1 + 2 − 3 + 4 − 5 − 6	6	−1 + 2 − 3 + 4 + 5 − 6	11	−1 + 2 − 3 + 4 − 5 + 6	12	−1 + 2 − 3 + 4 + 5 + 6	17
+1 + 2 − 3 + 4 − 5 − 6	7	+1 + 2 − 3 + 4 + 5 − 6	12	+1 + 2 − 3 + 4 − 5 + 6	13	+1 + 2 − 3 + 4 + 5 + 6	18
−1 − 2 + 3 + 4 − 5 − 6	7	−1 − 2 + 3 + 4 + 5 − 6	12	−1 − 2 + 3 + 4 − 5 + 6	13	−1 − 2 + 3 + 4 + 5 + 6	18
+1 − 2 + 3 + 4 − 5 − 6	8	+1 − 2 + 3 + 4 + 5 − 6	13	+1 − 2 + 3 + 4 − 5 + 6	14	+1 − 2 + 3 + 4 + 5 + 6	19
−1 + 2 + 3 + 4 − 5 − 6	9	−1 + 2 + 3 + 4 + 5 − 6	14	−1 + 2 + 3 + 4 − 5 + 6	15	−1 + 2 + 3 + 4 + 5 + 6	20
+1 + 2 + 3 + 4 − 5 − 6	10	+1 + 2 + 3 + 4 + 5 − 6	15	+1 + 2 + 3 + 4 − 5 + 6	16	+1 + 2 + 3 + 4 + 5 + 6	21

When the rank orderings with the same values of s^+ are aggregated, the probability distribution of $\mathbb{S}^+$ for $n = 6$ is obtained.

s^+	0	1	2	3	4	5	6	7	8	9	10	11	12	13	14	15	16	17	18	19	20	21
$p(s^+)$	$\frac{1}{64}$	$\frac{1}{64}$	$\frac{1}{64}$	$\frac{1}{32}$	$\frac{1}{32}$	$\frac{3}{64}$	$\frac{1}{16}$	$\frac{1}{16}$	$\frac{1}{16}$	$\frac{5}{64}$	$\frac{5}{64}$	$\frac{5}{64}$	$\frac{5}{64}$	$\frac{1}{16}$	$\frac{1}{16}$	$\frac{1}{16}$	$\frac{3}{64}$	$\frac{1}{32}$	$\frac{1}{32}$	$\frac{1}{64}$	$\frac{1}{64}$	$\frac{1}{64}$

From this probability distribution of $\mathbb{S}^+$, we can conclude that, for Example 16.3, $p^+ = P(\mathbb{S}^+ \geq s^+ \mid H_0 \text{ is true}) = P(\mathbb{S}^+ \geq 5 \mid H_0 \text{ is true}) = 1 - \frac{7}{64} = 0.890625$. Therefore, for any reasonable value of α ($\alpha < 0.890625$), we do not have sufficient reason to reject the null hypothesis. (In hindsight, $H_A : \mu < \mu_0$ might have been a more appropriate alternate hypothesis.)

Notice that, if α is set to a value of 0.05, s^+ would have to be at least 19 before H_0 could be rejected. By referring to Table 16.2, we see that only the last three most positive rank orderings fulfill that condition. In addition, both the number of observations that exceed μ_0 and the *relative* amounts by which they exceed μ_0 are important. For example, the outcome +1, −2, +3, +4, +5, +6 has 5 observations greater than μ_0 and has $s^+ = 19$. However, 4 of the 6 outcomes that have 5 observations greater than μ_0 have s^+ values less than 19 and would not cause H_0 to be rejected if $\alpha = 0.05$.

An easy way to obtain the probability distribution of $\mathbb{S}^+$ is to use the discrete convolution program to form the probability distribution of the sum of a set of *scaled* Bernoulli random variables. In general, to obtain the distribution of $\mathbb{S}^+$ for a sample of size n, form the n probability distributions,

$$p(x_i, i) = \begin{cases} 0.5 & \text{for } x_i = 0 \\ 0.5 & \text{for } x_i = i \end{cases} \qquad i = 1, 2, \ldots, n$$

and then obtain the summation convolution of the X_i, $\mathbb{S}^+ = \sum_{i=1}^{n} X_i$. This method was used to form the distributions for $\mathbb{S}^+$ (for $n = 1, 2, \ldots, 20$) for the Wilcoxon signed rank Test program that is part of your software library.

Perhaps the most frequent use of the Wilcoxon signed rank test is in the comparison of paired samples (see Section 10.3) in which neither population is governed by a normal distribution. If the two populations are governed by continuous probability distributions that differ only in the value of their respective means, it may be shown that the probability distribution of the differences will be a symmetric continuous distribution.

Example 16.4. Suppose that the friction associated with each of 10 bushings is measured before and after an antirust coating is applied and that the experiment yielded the following coded differences: −4, −4, −3, −2, −2, 0, 1, −2, −3, 3. Since we are interested only in whether there is a significant difference in the means of the samples, $H_0 : \mu = \mu_0 = 0$ and $H_A : \mu \neq \mu_0 = 0$.

Following the procedure for the Wilcoxon signed rank test, we find, after steps 1 and 2 (discarding the single value of 0)

Rank	1	−3	−3	−3	−6	−6	6	−8.5	−8.5
Difference	1	−2	−2	−2	−3	−3	3	−4	−4

Since there are ties in the absolute values of differences, the "ranks" assigned are the averages of the indices of the members of each tied group. For example, the second, third, and fourth differences in the ordered data all have absolute values of 2. Hence, each is assigned a rank of $(2 + 3 + 4)/3$. Since $s^+ = 7$, $p^+ = P(S^+ \geq s^+ \mid H_0$ is true$) = 0.9727$ and $p = P(S^+ \leq s^+ \mid H_0$ is true$) = 0.0371$. Therefore, the p-value for this test is $2(0.0371) = 0.0742$. The null hypothesis will be rejected only if α is set greater than or equal to 0.0742.

Because of the dramatic increase in the computational effort to evaluate explicitly the probability distribution of S^+ as the number of observations, n, becomes large, it is indeed fortunate that a large-sample approximation of that distribution exists. It may be shown that for $n > 20$, S^+ is distributed approximately as

$$N\left(\frac{n(n+1)}{4}, \sqrt{\frac{n(n+1)(2n+1)}{24}}\right)$$

This result follows directly from the generalization of the central limit theorem, as discussed at the end of Section 7.2; i.e., S^+ is the result of summing n nonidentical random variables.

Indeed, as pictured in Figure 16.1, the distribution of S^+, where $n = 20$, approximates a normal distribution very well. Therefore, to perform the Wilcoxon

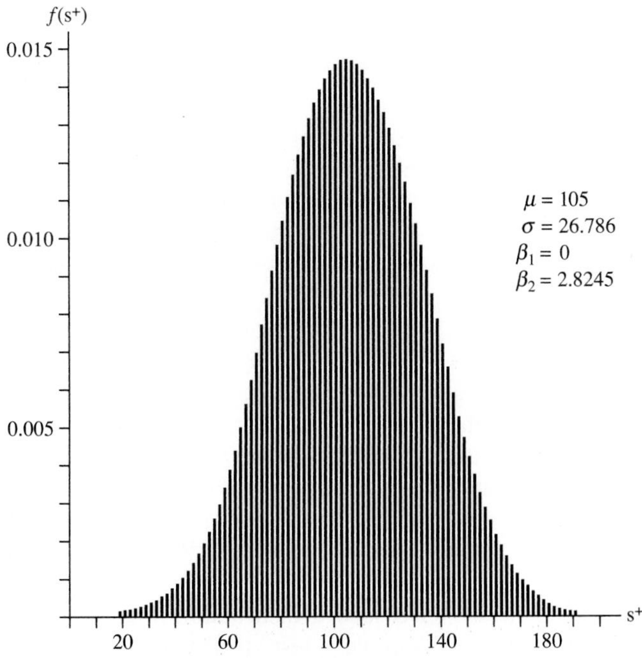

FIGURE 16-1
The probability distribution of S^+ when $n = 20$.

TABLE 16.3
Decision table for the large-sample Wilcoxon signed rank test

$H_A : \mu > \mu_0$	$p^+ \leq \alpha \quad (z \geq z_\alpha)$
$H_A : \mu < \mu_0$	$p^- \leq \alpha \quad (z \leq z_\alpha)$
$H_A : \mu \neq \mu_0$	$p^+ \leq \frac{\alpha}{2}$ or $p^- \leq \frac{\alpha}{2}$ $(z \geq z_{\alpha/2}$ or $z \leq -z_{\alpha/2})$

signed rank test when $n > 20$, we compute

$$z = \frac{s^+ - \frac{n(n+1)}{4}}{\sqrt{\frac{n(n+1)(2n+1)}{24}}} \tag{16.2}$$

and evaluate the appropriate tail probabilities for the selected alternate hypothesis. Table 16.3 details the appropriate decisions to be made in this case.

Some authors recommend that a correction factor for ties be applied to Equation 16.2. Fortunately, unless there are a *great many* ties, the correction factor will cause no practical difference in the results and thus it will not be considered here.

Example 16.5. To evaluate this approximation for small values of n, consider the large-sample approximation for Example 16.4. Substituting $s^+ = 7$ and $n = 9$ into Equation 16.2, we obtain $z = 1.836$. This value of z (with a right-hand-tail probability of 0.0332) implies an approximate p-value of 0.664. In view of the small sample, this is in very good agreement with the exact p-value, 0.0742.

For reasons similar to those associated with the test of the last section, the computation of the β error is not considered here. However, certain general comments can be made. When the assumptions associated with the t test are satisfied, the t test is the best test to use for either application discussed here. However, even when the assumptions are satisfied, the Wilcoxon signed rank test has an *asymptotic relative efficiency*, or ARE, equal to 0.955. This means that, when sample sizes are *large*, the t test and the Wilcoxon signed rank test have the same value of the β error if the ratio of the sample size for the t test divided by the sample size for the Wilcoxon signed rank test is equal to 0.955. In other words, when sample sizes are large, the Wilcoxon signed rank test sample must be $1/0.955 = 1.047$ times as large as the sample for the t test to achieve the same level of error. No matter what the governing distribution, the ARE for the Wilcoxon signed rank test will never be less than 0.864, and, for distributions markedly different from the normal distribution, the ARE can be much greater than 1.

In this section of the chapter, we have seen a distribution-free alternative to the single-sample t test and the paired-sample t test. In the next section we discuss a similar distribution-free alternative that may be used for situations in which two statistically independent samples need to be compared.

16.4 A DISTRIBUTION-FREE ALTERNATIVE TO THE TWO-SAMPLE *t* TEST: THE WILCOXON RANK SUM TEST

In Section 10.2, the two-sample t test requires the assumptions that the two populations are governed by normal probability distributions and that $\sigma_1 = \sigma_2$. The Wilcoxon rank sum test requires neither of these assumptions.

Suppose that we have two statistically independent random samples taken from populations governed by continuous probability distributions that have the same shape and differ only in the values of their respective means. We desire to test the null hypothesis, $H_0 : \mu_1 - \mu_2 = d_0 (H_0 : \mu_1 - d_0 = \mu_2)$ against the usual triad of alternate hypotheses, $H_A : \mu_1 - \mu_2 > d_0$, $\mu_1 - \mu_2 < d_0$, or $\mu_1 - \mu_2 \neq d_0$. Without loss of generality, let the sample size from population 1 be less than or equal to the sample size from population 2; i.e., $n_1 \leq n_2$.

Example 16.6. Let us set $d_0 = 0$, and select the one-tailed alternate hypothesis, $H_A : \mu_1 - \mu_2 < 0$. Suppose that random sampling has obtained the following from the two populations, where $n_1 = 4$ and $n_2 = 5$.

$$x_{11} \quad x_{12} \quad x_{13} \quad x_{14} \qquad \text{and} \qquad x_{21} \quad x_{22} \quad x_{23} \quad x_{24} \quad x_{25}$$

Now suppose that the two samples are combined and the observations are placed in ascending order by their values. If H_0 is true and we care only about whether an observation belongs to sample 1 or sample 2, any of the possible ${}_9C_4 = 126$ combinations of the two samples are equally likely. Let us define w_n to be equal to the sum of the ranks of the observations from sample 1 as they appear in the ordered combined data set. One possible ordering is

Rank	1	2	3	4	5	6	7	8	9
Observation	x_{12}	x_{14}	x_{23}	x_{11}	x_{21}	x_{22}	x_{13}	x_{24}	x_{25}

In this realization, $w_n = 1 + 2 + 4 + 7 = 14$. What can we conclude based on this result?

As in the Wilcoxon signed rank test, we must determine the probability associated with such a value of w_n given that H_0 is true. By exhaustively enumerating the 126 combinations, evaluating w_n for each one of them, and aggregating the combinations with the same w_n, the following probability and cumulative probability distributions may be obtained for W_n, the random variable associated with w_n.

w_n	10	11	12	13	14	15	16	17	18	19	20
$p(w_n)$	0.008	0.008	0.016	0.024	0.039	0.048	0.063	0.072	0.087	0.087	0.096
$F(w_n)$	0.008	0.016	0.032	0.056	0.095	0.143	0.206	0.278	0.365	0.452	0.548

w_n	30	29	28	27	26	25	24	23	22	21
$p(w_n)$	0.008	0.008	0.016	0.024	0.039	0.048	0.063	0.072	0.087	0.087
$F(w_n)$	1.000	0.992	0.984	0.968	0.944	0.905	0.857	0.794	0.722	0.635

Using the information contained in the cumulative distribution, we observe that $w_n = 14$ is a somewhat rare event. Indeed, values as small or smaller than 14 occur with a probability of only 0.095 when H_0 is true. Thus the p-value for $w_n = 14$ when $H_A : \mu_1 - \mu_2 < 0$ is 0.095; i.e., a value of $w_n = 14$ would be cause for rejection of the null hypothesis at any a $\alpha \geq 0.095$.

The program in your software package that performs the Wilcoxon rank sum test automatically references the appropriate W_n distribution and reports the p-value associated with the w_n from your data set. If the reported p-value is less than the selected value of α, there is sufficient evidence to reject the null hypothesis.

Thus, the rank sum test is performed in three steps.

1. Subtract d_0 from each data value in sample 1, and then combine the two samples into one data set. Place the observations in the combined data set in ascending order by value, marking the n_1 observations that belong to sample 1.
2. Define the *rank* of any observation to be its index in the ordered data set of step 1, and compute the observed value of W_n, w_n—the sum of the ranks of the observations from sample 1. (If the members of any subset of two or more observations have the same magnitude, redefine each of their ranks to be the average rank of the observations in the subset.)
3. Compute the p-value(s), $p^+ = P(W_n \geq w_n \mid H_0 \text{ is true})$ and/or $p^- = P(W_n \leq w_n \mid H_0 \text{ is true})$. Use Table 16.4 to decide whether to reject H_0.

If either of the samples has more than 10 observations, it may be shown that W_n is distributed approximately as a normal distribution with

$$\mu = \frac{n_1(n_1 + n_2 + 1)}{2} \qquad \text{and} \qquad \sigma^2 = \frac{n_1 n_2(n_1 + n_2 + 1)}{12}$$

This means that we may form

$$z = \frac{w_n - \frac{n_1(n_1+n_2+1)}{2}}{\sqrt{\frac{n_1 n_2(n_1+n_2+1)}{12}}} \tag{16.3}$$

and know that its associated random variable, Z, will be distributed as $N(0, 1)$ if H_0 is true. Table 16.5 details the appropriate decisions to be made in this case.

TABLE 16.4
Decision table for the Wilcoxon rank sum test

Alternative Hypothesis	Reject H_0 if
$H_A : d > d_0$	$p^+ \leq \alpha$
$H_A : d < d_0$	$p^- \leq \alpha$
$H_A : d \neq d_0$	$p^+ \leq \frac{\alpha}{2}$ or $p^- \leq \frac{\alpha}{2}$

TABLE 16.5
Decision table for the large-sample Wilcoxon rank sum test

Alternate Hypothesis	Reject H_0 if
$H_A : d > d_0$	$p^+ \leq \alpha \quad (z \geq z_\alpha)$
$H_A : d < d_0$	$p^- \leq \alpha \quad (z \leq z_\alpha)$
$H_A : d \neq d_0$	$p^+ \leq \frac{\alpha}{2}$ or $p^- \leq \frac{\alpha}{2}$ $(z \geq z_{\alpha/2}$ or $z \leq -z_{\alpha/2})$

Example 16.7. A vendor supplies two similar kinds of high-intensity light bulbs. The more expensive bulb is claimed to last an average of *more than* 200 hours longer that the cheaper bulb. Further, the assumption of normality is not supportable. Seven of the expensive bulbs and 12 of the cheap bulbs are burned continuously until they fail. The following times to failure (in hours) are recorded during the experiment.

Cheap	6,591	2,215	4,180	19,148	642	7,311	31,377
	334	2,182	18,536	8,065	3,268		
Expensive	1,238	42,351	6,719	12,421	19,138	3,477	2,869

Is it reasonable to conclude that the expensive bulbs last an average of 200 hours longer than the cheap bulbs?

In this example, $H_0 : \mu_1 - \mu_2 = 200$ is tested against the alternate hypothesis, $H_A : \mu_1 - \mu_2 > 200$. Before the samples are combined and ranked, each of the x_{1i}'s is replaced by $x_{1i} - 200$ to reflect the fact that $d_0 = 200$. The samples are then combined, yielding

Rank	1	2	3	4	5	6	7	8	9	10
Observation	334	642	<u>1,038</u>	2,182	2,215	<u>2,669</u>	3,268	<u>3,277</u>	4,180	<u>6,519</u>

Rank	11	12	13	14	15	16	17	18	19
Observation	6,591	7,311	8,065	<u>12,221</u>	18,536	<u>18,938</u>	19,148	31,377	<u>42,151</u>

Summing the ranks of the underlined observations, we find that $w_n = 76$. Since sample 2 (from the cheap bulbs) has 12 observations, we may use the large-sample approximation to determine the p-value for this test. Substituting the values of w_n, n_1, and n_2 into Equation 16.3 yields $z = 0.507$, which implies that the p-value is $p^+ = 0.306$.

For reasons similar to those cited in the previous sections of this chapter, no discussion of the computation of the β error is given here. However, when compared with the two-sample t test, the Wilcoxon rank sum test enjoys the same ARE characteristics as the Wilcoxon signed rank test possesses relative to the one-sample t test. In the next and concluding section of the book, we consider a distribution-free alternative to the comparison of three or more population means.

16.5 A DISTRIBUTION-FREE ALTERNATIVE TO THE COMPLETELY RANDOMIZED SINGLE-FACTOR ANOVA: THE KRUSKAL-WALLIS TEST

As demonstrated in Section 13.2, the proper use of any analysis of variance (ANOVA) design requires that all the assumptions associated with multivariate regression analysis be satisfied. Principal among these assumptions are that the populations being compared are distributed according to a normal distribution and that they share the same value of the variance. Just as in Section 16.4, there may be occasions when we need to compare three or more populations and these assumptions cannot be supported.

The Kruskal-Wallis test [Kruskal and Wallis, 1952] requires only that the distributions governing the ε_{ij} be the same for all the populations that are to be compared. As in the similar ANOVA design, the null hypothesis is

$$H_0\text{: All the population means are the same}$$

and the alternate hypothesis is

$$H_A\text{: At least two of the means differ}$$

The data set for this test of hypothesis is schematically shown in Table 16.6.

Let $n = \sum_{j=1}^{k} n_j$, the total number of values in the data set. Suppose that the data from all k of the populations are combined into a single data set and the data are rank-ordered from the smallest to largest value. If H_0 is true, all the possible $n!/(n_1!\ n_2!\ \cdots\ n_k!)$ rank assignments of the n data from the k populations are equally likely. If H_0 is false, orderings in which more of the lower ranks are assigned to the populations with the lower means become more likely.

Therefore, one way to assess whether a particular ordering of all the data is consistent with H_0 is to compute the r_j, the sum of the ranks of the observations belonging to each population, $j = 1, \ldots, k$. By exhaustively enumerating the possible rank assignments and determining the probability associated with such

TABLE 16.6
The general data layout for the Kruskal-Wallis test

		Population				
1	2	$\cdots$	j	$\cdots$	k
y_{11}	y_{12}	$\cdots$	y_{1j}	$\cdots$	y_{1k}
y_{21}	y_{22}	$\cdots$	y_{2j}	$\cdots$	y_{2k}
$\vdots$	$\vdots$		$\vdots$		$\vdots$
y_{i1}	y_{i2}	$\cdots$	y_{ij}	$\cdots$	y_{ik}
$\vdots$	$\vdots$		$\vdots$		$\vdots$
	$y_{n_2 2}$				$y_{n_k k}$
$y_{n_1 1}$			$y_{n_3 j}$		

a collection of $r_j, j = 1, \ldots, k$, we could then perform that assessment. This process is made easier by defining

$$h = \frac{12}{n(n+1)} \sum_{j=1}^{k} \frac{r_j^2}{n_j} - 3(n+1) \tag{16.4}$$

Since the expected rank of any observation, given that H_0 is true, is $(n+1)/2$, the expected average rank of any population, given that H_0 is true, is also $(n+1)/2$. Equation 16.4 measures the total amount of difference of all of the compared populations relative to that expected average rank. If the value of h is too large, H_0 is rejected.

If $k = 3$ and all the $n_j > 5$ or if $k > 3$ and all the $n_j \geq 5$, it may be shown that the random variable, H, realized in Equation 16.4, is distributed approximately as χ^2_{k-1}. If $k = 3$ and one or more of the $n_j \leq 5$, the χ^2_{k-1} approximation is not sufficiently accurate, and we must resort to the explicit values of the distribution of the null hypothesis.

Even though the values of the explicit distributions may be evaluated, as evidenced by the existence of tables of such distributions [Siegel, 1956], their generation and use is cumbersome. In addition, with such small samples, the β error can be inflated beyond acceptable values. For these reasons, consideration in this book and in the program that implements the Kruskal-Wallis test is limited to the cases where the χ^2_{k-1} approximation is acceptable.

Thus, the Kruskal-Wallis Test is performed in three steps.

1. Combine the k samples into one data set. Place the observations in the combined data set in ascending order by value, marking the n_j observations that belong to sample $j, j = 1, \ldots, k$.
2. Define the *rank* of any observation to be its index in the ordered data set of step 1, and compute h. (If the members of any subset of two or more observations have the same magnitude, redefine each of their ranks to be the average rank of the observations in the subset.)
3. Compute the p-value, $p = P(\chi^2_{k-1} \geq h \mid H_0 \text{ is true})$. If $p < \alpha$, or, equivalently, if $h > \chi^2_{k-1,\alpha}$, reject H_0.

Example 16.8. Suppose that once a particular type of electronic component survives past the early burnout period (when defective items fail), its practical lifetime is governed by a uniform probability distribution function. Three different vendors supply this component to your company. After the early burnout failures were discarded, the data given in Table 16.7 were collected on components from each vendor; the overall rank of each datum is given in parentheses.

Is there a significant difference in the means of the time-to-failure distributions of the components provided by the three vendors? To answer this question, the Kruskal-Wallis program performs steps 1 and 2, described above, using Equation 16.4 to compute h:

$$h = \frac{12}{30(30+1)} \left(\frac{127^2}{10} + \frac{164^2}{9} + \frac{174^2}{11} \right) - 3(30+1) = 1.8866$$

TABLE 16.7
The data for example 16.8

Vendor		
I	II	III
643 (20)	868 (30)	683 (22)
526 (17)	269 (6)	448 (16)
373 (11)	233 (2)	267 (5)
246 (3)	725 (23)	614 (19)
425 (14)	859 (29)	294 (7)
321 (9)	659 (21)	802 (28)
396 (12)	261 (4)	414 (13)
438 (15)	727 (24)	224 (1)
302 (8)	764 (25)	792 (26)
565 (18)		794 (27)
		348 (10)

The program then finds that the p-value for this Kruskal-Wallis test is equal to $P(\chi_2^2 \geq 1.8886) = 0.389$. This means that α would have to be set to a value of 0.389 or larger before the null hypothesis could be rejected on the basis of the given data.

As indicated at the start of the chapter, Sections 16.3, 16.4, and 16.5 merely scratch the surface of the great store of techniques collectively known as *non-parametric statistics*. Two of the better books devoted to the subject are those by Gibbons [1985] and Siegel [1956], the former being appropriate for more mathematically inclined readers.

EXERCISES

16.1. Use the χ^2 goodness of fit test to determine whether it is reasonable to state that data set A of Exercise 6.2 has been drawn from a uniform distribution bounded between the limits of 0 and 1.

16.2. Use the χ^2 goodness of fit test to determine whether it is reasonable to state that data set B of Exercise 6.2 has been drawn from an exponential distribution with $\lambda = 2$.

16.3. Use the χ^2 goodness of fit test to answer the question posed in Exercise C2.2.

16.4. Suppose that, in Example 10.5, the data of Table 10.2 is found not to be distributed according to a normal distribution. Reanalyze the data using the Wilcoxon signed rank test. Does your conclusion differ from that of Example 10.5?

16.5. Use the convolution program to show that the distribution of S^+ when $n = 5$ is

s^+	0	1	2	3	4	5	6	7	8	9	10	11	12	13	14	15
$p(s^+)$	$\frac{1}{32}$	$\frac{1}{32}$	$\frac{1}{32}$	$\frac{1}{16}$	$\frac{1}{16}$	$\frac{3}{32}$	$\frac{3}{32}$	$\frac{3}{32}$	$\frac{3}{32}$	$\frac{3}{32}$	$\frac{3}{32}$	$\frac{1}{16}$	$\frac{1}{16}$	$\frac{1}{32}$	$\frac{1}{32}$	$\frac{1}{32}$

16.6. The following data have been drawn from a population known to be distributed according to a continuous symmetric distribution that is not normal. Test whether the mean of the distribution is less than 5.1.

4.818 4.751 4.241 4.724 4.262 4.666 5.750

3.335 2.739 5.248 5.108 6.087 5.469 3.302

16.7. The following data are the coded measures of the elastic breaking point from randomly sampled waistbands of one brand of men's underwear. Is the manufacturer's claim that the average breaking point is at least 6.0 justified? (Because of the nonlinear nature of the elastic material, the assumption of normally distributed breaking points cannot be supported without a good deal more study than is possible at this time.)

5.8 6.1 6.1 6.4 5.7 6.0 6.0 5.5 6.7 6.3 4.6 6.1 5.8 5.7 5.3 6.0 5.7

6.1 5.5 5.3 5.2 6.5 4.7 6.4 5.6 7.0 6.3 6.1 6.2 5.9 5.1 5.8

16.8. Determine the distribution of W_n for the Wilcoxon rank sum test, assuming H_0 is true, if we are interested in comparing the means from two populations where the sample sizes are $n_1 = 3$ and $n_2 = 4$.

16.9. Suppose that it has been determined that the viscosity data of Exercise 10.8 cannot reasonably be assumed to follow a normal distribution. Use the Wilcoxon signed rank test to determine if a significant difference exists in the average viscosity of the two types of oil after they have been used in the demanding operating environment of the Borchah racing team.

16.10. Random sampling the amounts of fluoride (in parts per million) in the well water from two counties in central Texas yielded the following data.

County 1	15.5	20.3	32.0	32.9	29.7	21.6	22.6	29.4	
County 2	36.9	21.7	26.7	20.3	25.2	36.3	21.1	33.8	26.4

Is there a significant difference in the average amount of fluoride in the wells within the two counties? (Owing to geologic reasons, we may not assume that the fluoride concentrations follow a normal distribution.)

16.11. Suppose that, after some additional study, it has been conclusively shown that the variability present in the data of Table 13.1 is not distributed according to a normal distribution. Use the Kruskal-Wallis test to determine if a significant difference exists in the average amounts of the chemical by-product from the three laboratories.

16.12. Recent research has raised some questions about the validity of the normality assumption that must be made to apply a single-factor ANOVA to the data of Exercise 13.6. In order to substantiate the conclusion that was reached using ANOVA, reanalyze the data using the Kruskal-Wallis test. Should we change our earlier conclusions? Why or why not?

16.13. Determine the exact probability distribution and cumulative probability distribution of H for the Kruskal-Wallis test, if $k = 3$ and the sample sizes from the three populations are 2, 2, and 3.

16.14. Suppose that the conductivities of the five materials discussed in Exercise 13.5 are not governed by a normal distribution. Use the Kruskal-Wallis test to determine if a significant difference between the conductivities exists.

REFERENCES AND SUGGESTED READINGS

Abramowitz, M. and I. Stegun, *Handbook of Mathematical Functions,* National Bureau of Standards, U.S. Government Printing Office, Washington, DC, 1970.

Berry, D. A., and B. W. Lindgren, *Statistics: Theory and Methods*, Brooks/Cole Pub. C., Pacific Grove, CA, 1990.

Bowker, A. H., and G. J. Lieberman, *Engineering Statistics,* Prentice Hall, Englewood Cliffs, NJ, 1972.

Cramer, H., *Mathematical Methods of Statistics,* Princeton University Press, Princeton, NJ, 1946.

Daniel, C., and F. S. Wood, *Fitting Equations to Data,* Wiley Interscience, New York, 1971.

Draper, N. R., and H. Smith, *Applied Regression Analysis,* John Wiley & Sons, New York, 1981.

Gibbons, J. D., *Nonparametric Methods for Quantitative Analysis,* American Sciences Press, New York, 1985.

Hald, A., *Statistical Theory with Engineering Applications,* John Wiley & Sons, New York, 1965.

Hayter, A. J., "A Proof of the Conjecture That the Tukey-Kramer Multiple Comparisons Procedure Is Conservative," *The Annals of Statistics,* 1984, vol. 12, no. 1, pp. 61–75.

Hicks, C. R., *Fundamental Concepts in the Design of Experiments,* Holt, Rinehart & Winston, New York, 1982.

Hillier, F. S., and G. J. Lieberman, *Operations Research,* Holden-Day, San Francisco, 1974.

IMSL, International Mathematical and Statistical Library, 9th ed., IMSL, Inc., Houston, 1984.

Johnson, N., and S. Kotz, *Discrete Distributions,* Wiley Interscience, New York, 1969.

Koopmans, L. H., *An Introduction to Contemporary Statistics,* Duxbury, Boston, 1981.

Kramer, C. Y., "Extensions of Multiple Range Tests to Group Means with Unequal Numbers of Replications," *Biometrics,* 1956, vol. 12, pp. 307–310.

Kruskal, W. H., and W. A. Wallis, "Use of Ranks in One-Criterion Variance Analysis," *Journal of the American Statistical Association*, 1952, vol. 47, pp. 583–621.

Mood, A. M., and F. A. Graybill, *Introduction to the Theory of Statistics,* 2d ed. McGraw-Hill, New York, 1963.

Morrison, D. F., *Applied Linear Statistical Methods,* Prentice Hall, Englewood Cliffs, NJ 1983.

Mosteller, F., and R. Rourke, *Sturdy Statistics,* Addison-Wesley, Reading, MA, 1973.

Neter, J., W. Wasserman, and M. Kutner, *Applied Linear Statistical Models,* 3rd ed., Richard D. Irwin, Homewood, IL, 1990.

Nie, N. H. et al., *SPSS, Statistical Package for the Social Sciences,* 2d ed., McGraw-Hill, New York, 1984.

Pearson, E. S., and H. O. Hartley "Charts of the Power Function of the Analysis of Variance Tests, Derived from the Non-Central F Distribution," *Biometrika,* 1943, vol. 38, pp. 112–130.

SAS, Statistical Analysis System, SAS Institute, Cary, NC, 1979.

Siegel, S., *Nonparametric Statistics for the Behavioral Sciences,* McGraw-Hill, New York, 1956.

Tufte, Edward R., *The Visual Display of Quantitative Data,* Graphics Press, Cheshire, Connecticut, 1983.

Tukey, J. W., "The Problem of Multiple Comparisons," unpublished report, Princeton University, Princeton, NJ, 1953.

Vellman, P., and D. Hoaglin, *Applications, Basics, and Computing of Exploratory Data Analysis,* Duxbury, Boston, 1981.

Appendix

STATISTICAL TABLES

TABLE A.1

A cumulative binomial distribution table

$F(x) = \sum_{k=0}^{x} {}_nC_x\, p^x (1-p)^{n-x}$

		p																				
	x	0.01	0.05	0.10	0.15	0.20	0.25	0.30	0.35	0.40	0.45	0.50	0.55	0.60	0.65	0.70	0.75	0.80	0.85	0.90	0.95	0.99
$n = 2$	0	.980	.902	.810	.723	.640	.562	.490	.423	.360	.303	.250	.203	.160	.123	.090	.063	.040	.023	.010	.002	.000
	1	1.00	.997	.990	.977	.960	.937	.910	.878	.840	.798	.750	.698	.640	.578	.510	.437	.360	.278	.190	.098	.020
$n = 3$	0	.970	.857	.729	.614	.512	.422	.343	.275	.216	.166	.125	.091	.064	.043	.027	.016	.008	.003	.001	.000	.000
	1	1.00	.993	.972	.939	.896	.844	.784	.718	.648	.575	.500	.425	.352	.282	.216	.156	.104	.061	.028	.007	.000
	2	1.00	1.00	.999	.997	.992	.984	.973	.957	.936	.909	.875	.834	.784	.725	.657	.578	.488	.386	.271	.143	.030
$n = 4$	0	.961	.815	.656	.522	.410	.316	.240	.179	.130	.092	.063	.041	.026	.015	.008	.004	.002	.001	.000	.000	.000
	1	.999	.986	.948	.890	.819	.738	.652	.563	.475	.391	.313	.241	.179	.126	.084	.051	.027	.012	.004	.000	.000
	2	1.00	1.00	.996	.988	.973	.949	.916	.874	.821	.759	.688	.609	.525	.437	.348	.262	.181	.110	.052	.014	.001
	3	1.00	1.00	1.00	.999	.998	.996	.992	.985	.974	.959	.938	.908	.870	.821	.760	.684	.590	.478	.344	.185	.039
$n = 5$	0	.951	.774	.590	.444	.328	.237	.168	.116	.078	.050	.031	.018	.010	.005	.002	.001	.000	.000	.000	.000	.000
	1	.999	.977	.919	.835	.737	.633	.528	.428	.337	.256	.188	.131	.087	.054	.031	.016	.007	.002	.000	.000	.000
	2	1.00	.999	.991	.973	.942	.896	.837	.765	.683	.593	.500	.407	.317	.235	.163	.104	.058	.027	.009	.001	.000
	3	1.00	1.00	1.00	.998	.993	.984	.969	.946	.913	.869	.813	.744	.663	.572	.472	.367	.263	.165	.081	.023	.001
	4	1.00	1.00	1.00	1.00	1.00	.999	.998	.995	.990	.982	.969	.950	.922	.884	.832	.763	.672	.556	.410	.226	.049
$n = 6$	0	.941	.735	.531	.377	.262	.178	.118	.075	.047	.028	.016	.008	.004	.002	.001	.000	.000	.000	.000	.000	.000
	1	.999	.967	.886	.776	.655	.534	.420	.319	.233	.164	.109	.069	.041	.022	.011	.005	.002	.000	.000	.000	.000
	2	1.00	.998	.984	.953	.901	.831	.744	.647	.544	.442	.344	.255	.179	.117	.070	.038	.017	.006	.001	.000	.000
	3	1.00	1.00	.999	.994	.983	.962	.930	.883	.821	.745	.656	.558	.456	.353	.256	.169	.099	.047	.016	.002	.000
	4	1.00	1.00	1.00	1.00	.998	.995	.989	.978	.959	.931	.891	.836	.767	.681	.580	.466	.345	.224	.114	.033	.001
	5	1.00	1.00	1.00	1.00	1.00	1.00	.999	.998	.996	.992	.984	.972	.953	.925	.882	.822	.738	.623	.469	.265	.059
$n = 7$	0	.932	.698	.478	.321	.210	.133	.082	.049	.028	.015	.008	.004	.002	.001	.000	.000	.000	.000	.000	.000	.000
	1	.998	.956	.850	.717	.577	.445	.329	.234	.159	.102	.063	.036	.019	.009	.004	.001	.000	.000	.000	.000	.000
	2	1.00	.996	.974	.926	.852	.756	.647	.532	.420	.316	.227	.153	.096	.056	.029	.013	.005	.001	.000	.000	.000
	3	1.00	1.00	.997	.988	.967	.929	.874	.800	.710	.608	.500	.392	.290	.200	.126	.071	.033	.012	.003	.000	.000
	4	1.00	1.00	1.00	.999	.995	.987	.971	.944	.904	.847	.773	.684	.580	.468	.353	.244	.148	.074	.026	.004	.000
	5	1.00	1.00	1.00	1.00	1.00	.999	.996	.991	.981	.964	.938	.898	.841	.766	.671	.555	.423	.283	.150	.044	.002
	6	1.00	1.00	1.00	1.00	1.00	1.00	1.00	.999	.998	.996	.992	.985	.972	.951	.918	.867	.790	.679	.522	.302	.068
$n = 8$	0	.923	.663	.430	.272	.168	.100	.058	.032	.017	.008	.004	.002	.001	.000	.000	.000	.000	.000	.000	.000	.000
	1	.997	.943	.813	.657	.503	.367	.255	.169	.106	.063	.035	.018	.009	.004	.001	.000	.000	.000	.000	.000	.000
	2	1.00	.994	.962	.895	.797	.679	.552	.428	.315	.220	.145	.088	.050	.025	.011	.004	.001	.000	.000	.000	.000
	3	1.00	1.00	.995	.979	.944	.886	.806	.706	.594	.477	.363	.260	.174	.106	.058	.027	.010	.003	.000	.000	.000
	4	1.00	1.00	1.00	.997	.990	.973	.942	.894	.826	.740	.637	.523	.406	.294	.194	.114	.056	.021	.005	.000	.000
	5	1.00	1.00	1.00	1.00	.999	.996	.989	.975	.950	.912	.855	.780	.685	.572	.448	.321	.203	.105	.038	.006	.000
	6	1.00	1.00	1.00	1.00	1.00	1.00	.999	.996	.991	.982	.965	.937	.894	.831	.745	.633	.497	.343	.187	.057	.003
	7	1.00	1.00	1.00	1.00	1.00	1.00	1.00	1.00	.999	.998	.996	.992	.983	.968	.942	.900	.832	.728	.570	.337	.077

$n = 9$	0	.914	.630	.387	.232	.134	.075	.040	.021	.010	.005	.002	.001	.000	.000	.000	.000	.000	.000	.000	.000	.000
	1	.997	.929	.775	.599	.436	.300	.196	.121	.071	.039	.020	.009	.004	.001	.000	.000	.000	.000	.000	.000	.000
	2	1.00	.992	.947	.859	.738	.601	.463	.337	.232	.150	.090	.050	.025	.011	.004	.001	.000	.000	.000	.000	.000
	3	1.00	.999	.992	.966	.914	.834	.730	.609	.483	.361	.254	.166	.099	.054	.025	.010	.003	.001	.000	.000	.000
	4	1.00	1.00	.999	.994	.980	.951	.901	.828	.733	.621	.500	.379	.267	.172	.099	.049	.020	.006	.001	.000	.000
	5	1.00	1.00	1.00	.999	.997	.990	.975	.946	.901	.834	.746	.639	.517	.391	.270	.166	.086	.034	.008	.001	.000
	6	1.00	1.00	1.00	1.00	1.00	.999	.996	.989	.975	.950	.910	.850	.768	.663	.537	.399	.262	.141	.053	.008	.000
	7	1.00	1.00	1.00	1.00	1.00	1.00	1.00	.999	.996	.991	.980	.961	.929	.879	.804	.700	.564	.401	.225	.071	.003
	8	1.00	1.00	1.00	1.00	1.00	1.00	1.00	1.00	1.00	.999	.998	.995	.990	.979	.960	.925	.866	.768	.613	.370	.086
$n = 10$	0	.904	.599	.349	.197	.107	.056	.028	.013	.006	.003	.001	.000	.000	.000	.000	.000	.000	.000	.000	.000	.000
	1	.996	.914	.736	.544	.376	.244	.149	.086	.046	.023	.011	.005	.002	.001	.000	.000	.000	.000	.000	.000	.000
	2	1.00	.988	.930	.820	.678	.526	.383	.262	.167	.100	.055	.027	.012	.005	.002	.000	.000	.000	.000	.000	.000
	3	1.00	.999	.987	.950	.879	.776	.650	.514	.382	.266	.172	.102	.055	.026	.011	.004	.001	.000	.000	.000	.000
	4	1.00	1.00	.998	.990	.967	.922	.850	.751	.633	.504	.377	.262	.166	.095	.047	.020	.006	.001	.000	.000	.000
	5	1.00	1.00	1.00	.999	.994	.980	.953	.905	.834	.738	.623	.496	.367	.249	.150	.078	.033	.010	.002	.000	.000
	6	1.00	1.00	1.00	1.00	.999	.996	.989	.974	.945	.898	.828	.734	.618	.486	.350	.224	.121	.050	.013	.001	.000
	7	1.00	1.00	1.00	1.00	1.00	1.00	.998	.995	.988	.973	.945	.900	.833	.738	.617	.474	.322	.180	.070	.012	.000
	8	1.00	1.00	1.00	1.00	1.00	1.00	1.00	.999	.998	.995	.989	.977	.954	.914	.851	.756	.624	.456	.264	.086	.004
	9	1.00	1.00	1.00	1.00	1.00	1.00	1.00	1.00	1.00	1.00	.999	.997	.994	.987	.972	.944	.893	.803	.651	.401	.096
$n = 11$	0	.895	.569	.314	.167	.086	.042	.020	.009	.004	.001	.000	.000	.000	.000	.000	.000	.000	.000	.000	.000	.000
	1	.995	.898	.697	.492	.322	.197	.113	.061	.030	.014	.006	.002	.001	.000	.000	.000	.000	.000	.000	.000	.000
	2	1.00	.985	.910	.779	.617	.455	.313	.200	.119	.065	.033	.015	.006	.002	.001	.000	.000	.000	.000	.000	.000
	3	1.00	.998	.981	.931	.839	.713	.570	.426	.296	.191	.113	.061	.029	.012	.004	.001	.000	.000	.000	.000	.000
	4	1.00	1.00	.997	.984	.950	.885	.790	.668	.533	.397	.274	.174	.099	.050	.022	.008	.002	.000	.000	.000	.000
	5	1.00	1.00	1.00	.997	.988	.966	.922	.851	.753	.633	.500	.367	.247	.149	.078	.034	.012	.003	.000	.000	.000
	6	1.00	1.00	1.00	1.00	.998	.992	.978	.950	.901	.826	.726	.603	.467	.332	.210	.115	.050	.016	.003	.000	.000
	7	1.00	1.00	1.00	1.00	1.00	.999	.996	.988	.971	.939	.887	.809	.704	.574	.430	.287	.161	.069	.019	.002	.000
	8	1.00	1.00	1.00	1.00	1.00	1.00	.999	.998	.994	.985	.967	.935	.881	.800	.687	.545	.383	.221	.090	.015	.000
	9	1.00	1.00	1.00	1.00	1.00	1.00	1.00	1.00	.999	.998	.994	.986	.970	.939	.887	.803	.678	.508	.303	.102	.005
	10	1.00	1.00	1.00	1.00	1.00	1.00	1.00	1.00	1.00	1.00	1.00	.999	.996	.991	.980	.958	.914	.833	.686	.431	.105
$n = 12$	0	.886	.540	.282	.142	.069	.032	.014	.006	.002	.001	.000	.000	.000	.000	.000	.000	.000	.000	.000	.000	.000
	1	.994	.882	.659	.443	.275	.158	.085	.042	.020	.008	.003	.001	.000	.000	.000	.000	.000	.000	.000	.000	.000
	2	1.00	.980	.889	.736	.558	.391	.253	.151	.083	.042	.019	.008	.003	.001	.000	.000	.000	.000	.000	.000	.000
	3	1.00	.998	.974	.908	.795	.649	.493	.347	.225	.134	.073	.036	.015	.006	.002	.000	.000	.000	.000	.000	.000
	4	1.00	1.00	.996	.976	.927	.842	.724	.583	.438	.304	.194	.112	.057	.026	.009	.003	.001	.000	.000	.000	.000
	5	1.00	1.00	.999	.995	.981	.946	.882	.787	.665	.527	.387	.261	.158	.085	.039	.014	.004	.001	.000	.000	.000
	6	1.00	1.00	1.00	.999	.996	.986	.961	.915	.842	.739	.613	.473	.335	.213	.118	.054	.019	.005	.001	.000	.000
	7	1.00	1.00	1.00	1.00	.999	.997	.991	.974	.943	.888	.806	.696	.562	.417	.276	.158	.073	.024	.004	.000	.000
	8	1.00	1.00	1.00	1.00	1.00	1.00	.998	.994	.985	.964	.927	.866	.775	.653	.507	.351	.205	.092	.026	.002	.000
	9	1.00	1.00	1.00	1.00	1.00	1.00	1.00	.999	.997	.992	.981	.958	.917	.849	.747	.609	.442	.264	.111	.020	.000
	10	1.00	1.00	1.00	1.00	1.00	1.00	1.00	1.00	1.00	.999	.997	.992	.980	.958	.915	.842	.725	.557	.341	.118	.006
	11	1.00	1.00	1.00	1.00	1.00	1.00	1.00	1.00	1.00	1.00	1.00	.999	.998	.994	.986	.968	.931	.858	.718	.460	.114

TABLE A.1
(Continued)

		p																				
	x	0.01	0.05	0.10	0.15	0.20	0.25	0.30	0.35	0.40	0.45	0.50	0.55	0.60	0.65	0.70	0.75	0.80	0.85	0.90	0.95	0.99
$n = 13$	0	.878	.513	.254	.121	.055	.024	.010	.004	.001	.000	.000	.000	.000	.000	.000	.000	.000	.000	.000	.000	.000
	1	.993	.865	.621	.398	.234	.127	.064	.030	.013	.005	.002	.001	.000	.000	.000	.000	.000	.000	.000	.000	.000
	2	1.00	.975	.866	.692	.502	.333	.202	.113	.058	.027	.011	.004	.001	.000	.000	.000	.000	.000	.000	.000	.000
	3	1.00	.997	.966	.882	.747	.584	.421	.278	.169	.093	.046	.020	.008	.003	.001	.000	.000	.000	.000	.000	.000
	4	1.00	1.00	.994	.966	.901	.794	.654	.501	.353	.228	.133	.070	.032	.013	.004	.001	.000	.000	.000	.000	.000
	5	1.00	1.00	.999	.992	.970	.920	.835	.716	.574	.427	.291	.179	.098	.046	.018	.006	.001	.000	.000	.000	.000
	6	1.00	1.00	1.00	.999	.993	.976	.938	.871	.771	.644	.500	.356	.229	.129	.062	.024	.007	.001	.000	.000	.000
	7	1.00	1.00	1.00	1.00	.999	.994	.982	.954	.902	.821	.709	.573	.426	.284	.165	.080	.030	.008	.001	.000	.000
	8	1.00	1.00	1.00	1.00	1.00	.999	.996	.987	.968	.930	.867	.772	.647	.499	.346	.206	.099	.034	.006	.000	.000
	9	1.00	1.00	1.00	1.00	1.00	1.00	.999	.997	.992	.980	.954	.907	.831	.722	.579	.416	.253	.118	.034	.003	.000
	10	1.00	1.00	1.00	1.00	1.00	1.00	1.00	1.00	.999	.996	.989	.973	.942	.887	.798	.667	.498	.308	.134	.025	.000
	11	1.00	1.00	1.00	1.00	1.00	1.00	1.00	1.00	1.00	.999	.998	.995	.987	.970	.936	.873	.766	.602	.379	.135	.007
	12	1.00	1.00	1.00	1.00	1.00	1.00	1.00	1.00	1.00	1.00	1.00	1.00	.999	.996	.990	.976	.945	.879	.746	.487	.122
$n = 14$	0	.869	.488	.229	.103	.044	.018	.007	.002	.001	.000	.000	.000	.000	.000	.000	.000	.000	.000	.000	.000	.000
	1	.992	.847	.585	.357	.198	.101	.047	.021	.008	.003	.001	.000	.000	.000	.000	.000	.000	.000	.000	.000	.000
	2	1.00	.970	.842	.648	.448	.281	.161	.084	.040	.017	.006	.002	.001	.000	.000	.000	.000	.000	.000	.000	.000
	3	1.00	.996	.956	.853	.698	.521	.355	.220	.124	.063	.029	.011	.004	.001	.000	.000	.000	.000	.000	.000	.000
	4	1.00	1.00	.991	.953	.870	.742	.584	.423	.279	.167	.090	.043	.018	.006	.002	.000	.000	.000	.000	.000	.000
	5	1.00	1.00	.999	.988	.956	.888	.781	.641	.486	.337	.212	.119	.058	.024	.008	.002	.000	.000	.000	.000	.000
	6	1.00	1.00	1.00	.998	.988	.962	.907	.816	.692	.546	.395	.259	.150	.075	.031	.010	.002	.000	.000	.000	.000
	7	1.00	1.00	1.00	1.00	.998	.990	.969	.925	.850	.741	.605	.454	.308	.184	.093	.038	.012	.002	.000	.000	.000
	8	1.00	1.00	1.00	1.00	1.00	.998	.992	.976	.942	.881	.788	.663	.514	.359	.219	.112	.044	.012	.001	.000	.000
	9	1.00	1.00	1.00	1.00	1.00	1.00	.998	.994	.982	.957	.910	.833	.721	.577	.416	.258	.130	.047	.009	.000	.000
	10	1.00	1.00	1.00	1.00	1.00	1.00	1.00	.999	.996	.989	.971	.937	.876	.780	.645	.479	.302	.147	.044	.004	.000
	11	1.00	1.00	1.00	1.00	1.00	1.00	1.00	1.00	.999	.998	.994	.983	.960	.916	.839	.719	.552	.352	.158	.030	.000
	12	1.00	1.00	1.00	1.00	1.00	1.00	1.00	1.00	1.00	1.00	.999	.997	.992	.979	.953	.899	.802	.643	.415	.153	.008
	13	1.00	1.00	1.00	1.00	1.00	1.00	1.00	1.00	1.00	1.00	1.00	1.00	.999	.998	.993	.982	.956	.897	.771	.512	.131
$n = 15$	0	.860	.463	.206	.087	.035	.013	.005	.002	.000	.000	.000	.000	.000	.000	.000	.000	.000	.000	.000	.000	.000
	1	.990	.829	.549	.319	.167	.080	.035	.014	.005	.002	.000	.000	.000	.000	.000	.000	.000	.000	.000	.000	.000
	2	1.00	.964	.816	.604	.398	.236	.127	.062	.027	.011	.004	.001	.000	.000	.000	.000	.000	.000	.000	.000	.000
	3	1.00	.995	.944	.823	.648	.461	.297	.173	.091	.042	.018	.006	.002	.000	.000	.000	.000	.000	.000	.000	.000
	4	1.00	.999	.987	.938	.836	.686	.515	.352	.217	.120	.059	.025	.009	.003	.001	.000	.000	.000	.000	.000	.000
	5	1.00	1.00	.998	.983	.939	.852	.722	.564	.403	.261	.151	.077	.034	.012	.004	.001	.000	.000	.000	.000	.000
	6	1.00	1.00	1.00	.996	.982	.943	.869	.755	.610	.452	.304	.182	.095	.042	.015	.004	.001	.000	.000	.000	.000
	7	1.00	1.00	1.00	.999	.996	.983	.950	.887	.787	.654	.500	.346	.213	.113	.050	.017	.004	.001	.000	.000	.000
	8	1.00	1.00	1.00	1.00	.999	.996	.985	.958	.905	.818	.696	.548	.390	.245	.131	.057	.018	.004	.000	.000	.000
	9	1.00	1.00	1.00	1.00	1.00	.999	.996	.988	.966	.923	.849	.739	.597	.436	.278	.148	.061	.017	.002	.000	.000

	10	1.00	1.00	1.00	1.00	1.00	1.00	.999	.997	.991	.975	.941	.880	.783	.648	.485	.314	.164	.062	.013	.001	.000
	11	1.00	1.00	1.00	1.00	1.00	1.00	1.00	1.00	.998	.994	.982	.958	.909	.827	.703	.539	.352	.177	.056	.005	.000
	12	1.00	1.00	1.00	1.00	1.00	1.00	1.00	1.00	1.00	.999	.996	.989	.973	.938	.873	.764	.602	.396	.184	.036	.000
	13	1.00	1.00	1.00	1.00	1.00	1.00	1.00	1.00	1.00	1.00	1.00	.998	.995	.986	.965	.920	.833	.681	.451	.171	.010
	14	1.00	1.00	1.00	1.00	1.00	1.00	1.00	1.00	1.00	1.00	1.00	1.00	1.00	.998	.995	.987	.965	.913	.794	.537	.140
$n = 16$	0	.851	.440	.185	.074	.028	.010	.003	.001	.000	.000	.000	.000	.000	.000	.000	.000	.000	.000	.000	.000	.000
	1	.989	.811	.515	.284	.141	.063	.026	.010	.003	.001	.000	.000	.000	.000	.000	.000	.000	.000	.000	.000	.000
	2	.999	.957	.789	.561	.352	.197	.099	.045	.018	.007	.002	.001	.000	.000	.000	.000	.000	.000	.000	.000	.000
	3	1.00	.993	.932	.790	.598	.405	.246	.134	.065	.028	.011	.003	.001	.000	.000	.000	.000	.000	.000	.000	.000
	4	1.00	.999	.983	.921	.798	.630	.450	.289	.167	.085	.038	.015	.005	.001	.000	.000	.000	.000	.000	.000	.000
	5	1.00	1.00	.997	.976	.918	.810	.660	.490	.329	.198	.105	.049	.019	.006	.002	.000	.000	.000	.000	.000	.000
	6	1.00	1.00	.999	.994	.973	.920	.825	.688	.527	.366	.227	.124	.058	.023	.007	.002	.000	.000	.000	.000	.000
	7	1.00	1.00	1.00	.999	.993	.973	.926	.841	.716	.563	.402	.256	.142	.067	.026	.007	.001	.000	.000	.000	.000
	8	1.00	1.00	1.00	1.00	.999	.993	.974	.933	.858	.744	.598	.437	.284	.159	.074	.027	.007	.001	.000	.000	.000
	9	1.00	1.00	1.00	1.00	1.00	.998	.993	.977	.942	.876	.773	.634	.473	.312	.175	.080	.027	.006	.001	.000	.000
	10	1.00	1.00	1.00	1.00	1.00	1.00	.998	.994	.981	.951	.895	.802	.671	.510	.340	.190	.082	.024	.003	.000	.000
	11	1.00	1.00	1.00	1.00	1.00	1.00	1.00	.999	.995	.985	.962	.915	.833	.711	.550	.370	.202	.079	.017	.001	.000
	12	1.00	1.00	1.00	1.00	1.00	1.00	1.00	1.00	.999	.997	.989	.972	.935	.866	.754	.595	.402	.210	.068	.007	.000
	13	1.00	1.00	1.00	1.00	1.00	1.00	1.00	1.00	1.00	.999	.998	.993	.982	.955	.901	.803	.648	.439	.211	.043	.001
	14	1.00	1.00	1.00	1.00	1.00	1.00	1.00	1.00	1.00	1.00	1.00	.999	.997	.990	.974	.937	.859	.716	.485	.189	.011
	15	1.00	1.00	1.00	1.00	1.00	1.00	1.00	1.00	1.00	1.00	1.00	1.00	1.00	.999	.997	.990	.972	.926	.815	.560	.149
$n = 17$	0	.843	.418	.167	.063	.023	.008	.002	.001	.000	.000	.000	.000	.000	.000	.000	.000	.000	.000	.000	.000	.000
	1	.988	.792	.482	.252	.118	.050	.019	.007	.002	.001	.000	.000	.000	.000	.000	.000	.000	.000	.000	.000	.000
	2	.999	.950	.762	.520	.310	.164	.077	.033	.012	.004	.001	.000	.000	.000	.000	.000	.000	.000	.000	.000	.000
	3	1.00	.991	.917	.756	.549	.353	.202	.103	.046	.018	.006	.002	.000	.000	.000	.000	.000	.000	.000	.000	.000
	4	1.00	.999	.978	.901	.758	.574	.389	.235	.126	.060	.025	.009	.003	.001	.000	.000	.000	.000	.000	.000	.000
	5	1.00	1.00	.995	.968	.894	.765	.597	.420	.264	.147	.072	.030	.011	.003	.001	.000	.000	.000	.000	.000	.000
	6	1.00	1.00	.999	.992	.962	.893	.775	.619	.448	.290	.166	.083	.035	.012	.003	.001	.000	.000	.000	.000	.000
	7	1.00	1.00	1.00	.998	.989	.960	.895	.787	.641	.474	.315	.183	.092	.038	.013	.003	.000	.000	.000	.000	.000
	8	1.00	1.00	1.00	1.00	.997	.988	.960	.901	.801	.663	.500	.337	.199	.099	.040	.012	.003	.000	.000	.000	.000
	9	1.00	1.00	1.00	1.00	1.00	.997	.987	.962	.908	.817	.685	.526	.359	.213	.105	.040	.011	.002	.000	.000	.000
	10	1.00	1.00	1.00	1.00	1.00	.999	.997	.988	.965	.917	.834	.710	.552	.381	.225	.107	.038	.008	.001	.000	.000
	11	1.00	1.00	1.00	1.00	1.00	1.00	.999	.997	.989	.970	.928	.853	.736	.580	.403	.235	.106	.032	.005	.000	.000
	12	1.00	1.00	1.00	1.00	1.00	1.00	1.00	.999	.997	.991	.975	.940	.874	.765	.611	.426	.242	.099	.022	.001	.000
	13	1.00	1.00	1.00	1.00	1.00	1.00	1.00	1.00	1.00	.998	.994	.982	.954	.897	.798	.647	.451	.244	.083	.009	.000
	14	1.00	1.00	1.00	1.00	1.00	1.00	1.00	1.00	1.00	1.00	.999	.996	.988	.967	.923	.836	.690	.480	.238	.050	.001
	15	1.00	1.00	1.00	1.00	1.00	1.00	1.00	1.00	1.00	1.00	1.00	.999	.998	.993	.981	.950	.882	.748	.518	.208	.012
	16	1.00	1.00	1.00	1.00	1.00	1.00	1.00	1.00	1.00	1.00	1.00	1.00	1.00	.999	.998	.992	.977	.937	.833	.582	.157
$n = 18$	0	.835	.397	.150	.054	.018	.006	.002	.000	.000	.000	.000	.000	.000	.000	.000	.000	.000	.000	.000	.000	.000
	1	.986	.774	.450	.224	.099	.039	.014	.005	.001	.000	.000	.000	.000	.000	.000	.000	.000	.000	.000	.000	.000
	2	.999	.942	.734	.480	.271	.135	.060	.024	.008	.003	.001	.000	.000	.000	.000	.000	.000	.000	.000	.000	.000
	3	1.00	.989	.902	.720	.501	.306	.165	.078	.033	.012	.004	.001	.000	.000	.000	.000	.000	.000	.000	.000	.000
	4	1.00	.998	.972	.879	.716	.519	.333	.189	.094	.041	.015	.005	.001	.000	.000	.000	.000	.000	.000	.000	.000

TABLE A.1
(Continued)

		p																				
	x	0.01	0.05	0.10	0.15	0.20	0.25	0.30	0.35	0.40	0.45	0.50	0.55	0.60	0.65	0.70	0.75	0.80	0.85	0.90	0.95	0.99
	5	1.00	1.00	.994	.958	.867	.717	.534	.355	.209	.108	.048	.018	.006	.001	.000	.000	.000	.000	000	.000	.000
$n = 18$	6	1.00	1.00	.999	.988	.949	.861	.722	.549	.374	.226	.119	.054	.020	.006	.001	.000	.000	.000	.000	.000	.000
	7	1.00	1.00	1.00	.997	.984	.943	.859	.728	.563	.391	.240	.128	.058	.021	.006	.001	.000	.000	.000	.000	.000
	8	1.00	1.00	1.00	.999	.996	.981	.940	.861	.737	.578	.407	.253	.135	.060	.021	.005	.001	.000	.000	.000	.000
	9	1.00	1.00	1.00	1.00	.999	.995	.979	.940	.865	.747	.593	.422	.263	.139	.060	.019	.004	.001	.000	.000	.000
	10	1.00	1.00	1.00	1.00	1.00	.999	.994	.979	.942	.872	.760	.609	.437	.272	.141	.057	.016	.003	.000	.000	.000
	11	1.00	1.00	1.00	1.00	1.00	1.00	.999	.994	.980	.946	.881	.774	.626	.451	.278	.139	.051	.012	.001	.000	.000
	12	1.00	1.00	1.00	1.00	1.00	1.00	1.00	.999	.994	.982	.952	.892	.791	.645	.466	.283	.133	.042	.006	.000	.000
	13	1.00	1.00	1.00	1.00	1.00	1.00	1.00	1.00	.999	.995	.985	.959	.906	.811	.667	.481	.284	.121	.028	.002	.000
	14	1.00	1.00	1.00	1.00	1.00	1.00	1.00	1.00	1.00	.999	.996	.988	.967	.922	.835	.694	.499	.280	.098	.011	.000
	15	1.00	1.00	1.00	1.00	1.00	1.00	1.00	1.00	1.00	1.00	.999	.997	.992	.976	.940	.865	.729	.520	.266	.058	.001
	16	1.00	1.00	1.00	1.00	1.00	1.00	1.00	1.00	1.00	1.00	1.00	1.00	.999	.995	.986	.961	.901	.776	.550	.226	.014
	17	1.00	1.00	1.00	1.00	1.00	1.00	1.00	1.00	1.00	1.00	1.00	1.00	1.00	1.00	.998	.994	.982	.946	.850	.603	.165
$n = 19$	0	.826	.377	.135	.046	.014	.004	.001	.000	.000	.000	.000	.000	.000	.000	.000	.000	.000	.000	.000	.000	.000
	1	.985	.755	.420	.198	.083	.031	.010	.003	.001	.000	.000	.000	.000	.000	.000	.000	.000	.000	.000	.000	.000
	2	.999	.933	.705	.441	.237	.111	.046	.017	.005	.002	.000	.000	.000	.000	.000	.000	.000	.000	.000	.000	.000
	3	1.00	.987	.885	.684	.455	.263	.133	.059	.023	.008	.002	.001	.000	.000	.000	.000	.000	.000	.000	.000	.000
	4	1.00	.998	.965	.856	.673	.465	.282	.150	.070	.028	.010	.003	.001	.000	.000	.000	.000	.000	.000	.000	.000
	5	1.00	1.00	.991	.946	.837	.668	.474	.297	.163	.078	.032	.011	.003	.001	.000	.000	.000	.000	.000	.000	.000
	6	1.00	1.00	.998	.984	.932	.825	.666	.481	.308	.173	.084	.034	.012	.003	.001	.000	.000	.000	.000	.000	.000
	7	1.00	1.00	1.00	.996	.977	.923	.818	.666	.488	.317	.180	.087	.035	.011	.003	.000	.000	.000	.000	.000	.000
	8	1.00	1.00	1.00	.999	.993	.971	.916	.815	.667	.494	.324	.184	.088	.035	.011	.002	.000	.000	.000	.000	.000
	9	1.00	1.00	1.00	1.00	.998	.991	.967	.913	.814	.671	.500	.329	.186	.087	.033	.009	.002	.000	.000	.000	.000
	10	1.00	1.00	1.00	1.00	1.00	.998	.989	.965	.912	.816	.676	.506	.333	.185	.084	.029	.007	.001	.000	.000	.000
	11	1.00	1.00	1.00	1.00	1.00	1.00	.997	.989	.965	.913	.820	.683	.512	.334	.182	.077	.023	.004	.000	.000	.000
	12	1.00	1.00	1.00	1.00	1.00	1.00	.999	.997	.988	.966	.916	.827	.692	.519	.334	.175	.068	.016	.002	.000	.000
	13	1.00	1.00	1.00	1.00	1.00	1.00	1.00	.999	.997	.989	.968	.922	.837	.703	.526	.332	.163	.054	.009	.000	.000
	14	1.00	1.00	1.00	1.00	1.00	1.00	1.00	1.00	.999	.997	.990	.972	.930	.850	.718	.535	.327	.144	.035	.002	.000
	15	1.00	1.00	1.00	1.00	1.00	1.00	1.00	1.00	1.00	.999	.998	.992	.977	.941	.867	.737	.545	.316	.115	.013	.000
	16	1.00	1.00	1.00	1.00	1.00	1.00	1.00	1.00	1.00	1.00	1.00	.998	.995	.983	.954	.889	.763	.559	.295	.067	.001
	17	1.00	1.00	1.00	1.00	1.00	1.00	1.00	1.00	1.00	1.00	1.00	1.00	.999	.997	.990	.969	.917	.802	.580	.245	.015
	18	1.00	1.00	1.00	1.00	1.00	1.00	1.00	1.00	1.00	1.00	1.00	1.00	1.00	1.00	.999	.996	.986	.954	.865	.623	.174
$n = 20$	0	.818	.358	.122	.039	.012	.003	.001	.000	.000	.000	.000	.000	.000	.000	.000	.000	.000	.000	.000	.000	.000
	1	.983	.736	.392	.176	.069	.024	.008	.002	.001	.000	.000	.000	.000	.000	.000	.000	.000	.000	.000	.000	.000
	2	.999	.925	.677	.405	.206	.091	.035	.012	.004	.001	.000	.000	.000	.000	.000	.000	.000	.000	.000	.000	.000
	3	1.00	.984	.867	.648	.411	.225	.107	.044	.016	.005	.001	.000	.000	.000	.000	.000	.000	.000	.000	.000	.000
	4	1.00	.997	.957	.830	.630	.415	.238	.118	.051	.019	.006	.002	.000	.000	.000	.000	.000	.000	.000	.000	.000
	5	1.00	1.00	.989	.933	.804	.617	.416	.245	.126	.055	.021	.006	.002	.000	.000	.000	.000	.000	.000	.000	.000

n	k																					
$n = 20$	6	1.00	1.00	.998	.978	.913	.786	.608	.417	.250	.130	.058	.021	.006	.002	.000	.000	.000	.000	.000	.000	.000
	7	1.00	1.00	1.00	.994	.968	.898	.772	.601	.416	.252	.132	.058	.021	.006	.001	.000	.000	.000	.000	.000	.000
	8	1.00	1.00	1.00	.999	.990	.959	.887	.762	.596	.414	.252	.131	.057	.020	.005	.001	.000	.000	.000	.000	.000
	9	1.00	1.00	1.00	1.00	.997	.986	.952	.878	.755	.591	.412	.249	.128	.053	.017	.004	.001	.000	.000	.000	.000
	10	1.00	1.00	1.00	1.00	.999	.996	.983	.947	.872	.751	.588	.409	.245	.122	.048	.014	.003	.000	.000	.000	.000
	11	1.00	1.00	1.00	1.00	1.00	.999	.995	.980	.943	.869	.748	.586	.404	.238	.113	.041	.010	.001	.000	.000	.000
	12	1.00	1.00	1.00	1.00	1.00	1.00	.999	.994	.979	.942	.868	.748	.584	.399	.228	.102	.032	.006	.000	.000	.000
	13	1.00	1.00	1.00	1.00	1.00	1.00	1.00	.998	.994	.979	.942	.870	.750	.583	.392	.214	.087	.022	.002	.000	.000
	14	1.00	1.00	1.00	1.00	1.00	1.00	1.00	1.00	.998	.994	.979	.945	.874	.755	.584	.383	.196	.067	.011	.000	.000
	15	1.00	1.00	1.00	1.00	1.00	1.00	1.00	1.00	1.00	.998	.994	.981	.949	.882	.762	.585	.370	.170	.043	.003	.000
	16	1.00	1.00	1.00	1.00	1.00	1.00	1.00	1.00	1.00	1.00	.999	.995	.984	.956	.893	.775	.589	.352	.133	.016	.000
	17	1.00	1.00	1.00	1.00	1.00	1.00	1.00	1.00	1.00	1.00	1.00	.999	.996	.988	.965	.909	.794	.595	.323	.075	.001
	18	1.00	1.00	1.00	1.00	1.00	1.00	1.00	1.00	1.00	1.00	1.00	1.00	.999	.998	.992	.976	.931	.824	.608	.264	.017
	19	1.00	1.00	1.00	1.00	1.00	1.00	1.00	1.00	1.00	1.00	1.00	1.00	1.00	1.00	.999	.997	.988	.961	.878	.642	.182
$n = 21$	0	.810	.341	.109	.033	.009	.002	.001	.000	.000	.000	.000	.000	.000	.000	.000	.000	.000	.000	.000	.000	.000
	1	.981	.717	.365	.155	.058	.019	.006	.001	.000	.000	.000	.000	.000	.000	.000	.000	.000	.000	.000	.000	.000
	2	.999	.915	.648	.370	.179	.075	.027	.009	.002	.001	.000	.000	.000	.000	.000	.000	.000	.000	.000	.000	.000
	3	1.00	.981	.848	.611	.370	.192	.086	.033	.011	.003	.001	.000	.000	.000	.000	.000	.000	.000	.000	.000	.000
	4	1.00	.997	.948	.803	.586	.367	.198	.092	.037	.013	.004	.001	.000	.000	.000	.000	.000	.000	.000	.000	.000
	5	1.00	1.00	.986	.917	.769	.567	.363	.201	.096	.039	.013	.004	.001	.000	.000	.000	.000	.000	.000	.000	.000
	6	1.00	1.00	.997	.971	.891	.744	.551	.357	.200	.096	.039	.013	.004	.001	.000	.000	.000	.000	.000	.000	.000
	7	1.00	1.00	.999	.992	.957	.870	.723	.536	.350	.197	.095	.038	.012	.003	.001	.000	.000	.000	.000	.000	.000
	8	1.00	1.00	1.00	.998	.986	.944	.852	.706	.524	.341	.192	.091	.035	.011	.002	.000	.000	.000	.000	.000	.000
	9	1.00	1.00	1.00	1.00	.996	.979	.932	.838	.691	.512	.332	.184	.085	.031	.009	.002	.000	.000	.000	.000	.000
	10	1.00	1.00	1.00	1.00	.999	.994	.974	.923	.826	.679	.500	.321	.174	.077	.026	.006	.001	.000	.000	.000	.000
	11	1.00	1.00	1.00	1.00	1.00	.998	.991	.969	.915	.816	.668	.488	.309	.162	.068	.021	.004	.000	.000	.000	.000
	12	1.00	1.00	1.00	1.00	1.00	1.00	.998	.989	.965	.909	.808	.659	.476	.294	.148	.056	.014	.002	.000	.000	.000
	13	1.00	1.00	1.00	1.00	1.00	1.00	.999	.997	.988	.962	.905	.803	.650	.464	.277	.130	.043	.008	.001	.000	.000
	14	1.00	1.00	1.00	1.00	1.00	1.00	1.00	.999	.996	.987	.961	.904	.800	.643	.449	.256	.109	.029	.003	.000	.000
	15	1.00	1.00	1.00	1.00	1.00	1.00	1.00	1.00	.999	.996	.987	.961	.904	.799	.637	.433	.231	.083	.014	.000	.000
	16	1.00	1.00	1.00	1.00	1.00	1.00	1.00	1.00	1.00	.999	.996	.987	.963	.908	.802	.633	.414	.197	.052	.003	.000
	17	1.00	1.00	1.00	1.00	1.00	1.00	1.00	1.00	1.00	1.00	.999	.997	.989	.967	.914	.808	.630	.389	.152	.019	.000
	18	1.00	1.00	1.00	1.00	1.00	1.00	1.00	1.00	1.00	1.00	1.00	.999	.998	.991	.973	.925	.821	.630	.352	.085	.001
	19	1.00	1.00	1.00	1.00	1.00	1.00	1.00	1.00	1.00	1.00	1.00	1.00	1.00	.999	.994	.981	.942	.845	.635	.283	.019
	20	1.00	1.00	1.00	1.00	1.00	1.00	1.00	1.00	1.00	1.00	1.00	1.00	1.00	1.00	.999	.998	.991	.967	.891	.659	.190
$n = 22$	0	.802	.324	.098	.028	.007	.002	.000	.000	.000	.000	.000	.000	.000	.000	.000	.000	.000	.000	.000	.000	.000
	1	.980	.698	.339	.137	.048	.015	.004	.001	.000	.000	.000	.000	.000	.000	.000	.000	.000	.000	.000	.000	.000
	2	.999	.905	.620	.338	.154	.061	.021	.006	.002	.000	.000	.000	.000	.000	.000	.000	.000	.000	.000	.000	.000
	3	1.00	.978	.828	.575	.332	.162	.068	.025	.008	.002	.000	.000	.000	.000	.000	.000	.000	.000	.000	.000	.000
	4	1.00	.996	.938	.774	.543	.323	.165	.072	.027	.008	.002	.000	.000	.000	.000	.000	.000	.000	.000	.000	.000
	5	1.00	.999	.982	.900	.733	.517	.313	.163	.072	.027	.008	.002	.000	.000	.000	.000	.000	.000	.000	.000	.000
	6	1.00	1.00	.996	.963	.867	.699	.494	.302	.158	.071	.026	.008	.002	.000	.000	.000	.000	.000	.000	.000	.000
	7	1.00	1.00	.999	.989	.944	.838	.671	.474	.290	.152	.067	.024	.007	.002	.000	.000	.000	.000	.000	.000	.000

TABLE A.1
(concluded)

		p																				
	x	0.01	0.05	0.10	0.15	0.20	0.25	0.30	0.35	0.40	0.45	0.50	0.55	0.60	0.65	0.70	0.75	0.80	0.85	0.90	0.95	0.99
n = 22	8	1.00	1.00	1.00	.997	.980	.925	.814	.647	.454	.276	.143	.062	.021	.006	.001	.000	.000	.000	.000	.000	.000
	9	1.00	1.00	1.00	.999	.994	.970	.908	.792	.624	.435	.262	.133	.055	.018	.004	.001	.000	.000	.000	.000	.000
	10	1.00	1.00	1.00	1.00	.998	.990	.961	.893	.772	.604	.416	.246	.121	.047	.014	.003	.000	.000	.000	.000	.000
	11	1.00	1.00	1.00	1.00	1.00	.997	.986	.953	.879	.754	.584	.396	.228	.107	.039	.010	.002	.000	.000	.000	.000
	12	1.00	1.00	1.00	1.00	1.00	.999	.996	.982	.945	.867	.738	.565	.376	.208	.092	.030	.006	.001	.000	.000	.000
	13	1.00	1.00	1.00	1.00	1.00	1.00	.999	.994	.979	.938	.857	.724	.546	.353	.186	.075	.020	.003	.000	.000	.000
	14	1.00	1.00	1.00	1.00	1.00	1.00	1.00	.998	.993	.976	.933	.848	.710	.526	.329	.162	.056	.011	.001	.000	.000
	15	1.00	1.00	1.00	1.00	1.00	1.00	1.00	1.00	.998	.992	.974	.929	.842	.698	.506	.301	.133	.037	.004	.000	.000
	16	1.00	1.00	1.00	1.00	1.00	1.00	1.00	1.00	1.00	.998	.992	.973	.928	.837	.687	.483	.267	.100	.018	.001	.000
	17	1.00	1.00	1.00	1.00	1.00	1.00	1.00	1.00	1.00	1.00	.998	.992	.973	.928	.835	.677	.457	.226	.062	.004	.000
	18	1.00	1.00	1.00	1.00	1.00	1.00	1.00	1.00	1.00	1.00	1.00	.998	.992	.975	.932	.838	.668	.425	.172	.022	.000
	19	1.00	1.00	1.00	1.00	1.00	1.00	1.00	1.00	1.00	1.00	1.00	1.00	.998	.994	.979	.939	.846	.662	.380	.095	.001
	20	1.00	1.00	1.00	1.00	1.00	1.00	1.00	1.00	1.00	1.00	1.00	1.00	1.00	.999	.996	.985	.952	.863	.661	.302	.020
	21	1.00	1.00	1.00	1.00	1.00	1.00	1.00	1.00	1.00	1.00	1.00	1.00	1.00	1.00	1.00	.998	.993	.972	.902	.676	.198
n = 23	0	.794	.307	.089	.024	.006	.001	.000	.000	.000	.000	.000	.000	.000	.000	.000	.000	.000	.000	.000	.000	.000
	1	.978	.679	.315	.120	.040	.012	.003	.001	.000	.000	.000	.000	.000	.000	.000	.000	.000	.000	.000	.000	.000
	2	.998	.895	.592	.308	.133	.049	.016	.004	.001	.000	.000	.000	.000	.000	.000	.000	.000	.000	.000	.000	.000
	3	1.00	.974	.807	.540	.297	.137	.054	.018	.005	.001	.000	.000	.000	.000	.000	.000	.000	.000	.000	.000	.000
	4	1.00	.995	.927	.744	.501	.283	.136	.055	.019	.005	.001	.000	.000	.000	.000	.000	.000	.000	.000	.000	.000
	5	1.00	.999	.977	.881	.695	.468	.269	.131	.054	.019	.005	.001	.000	.000	.000	.000	.000	.000	.000	.000	.000
	6	1.00	1.00	.994	.954	.840	.654	.440	.253	.124	.051	.017	.005	.001	.000	.000	.000	.000	.000	.000	.000	.000
	7	1.00	1.00	.999	.985	.928	.804	.618	.414	.237	.115	.047	.015	.004	.001	.000	.000	.000	.000	.000	.000	.000
	8	1.00	1.00	1.00	.996	.973	.904	.771	.586	.388	.220	.105	.041	.013	.003	.001	.000	.000	.000	.000	.000	.000
	9	1.00	1.00	1.00	.999	.991	.959	.880	.741	.556	.364	.202	.094	.035	.010	.002	.000	.000	.000	.000	.000	.000
	10	1.00	1.00	1.00	1.00	.997	.985	.945	.858	.713	.528	.339	.184	.081	.028	.007	.001	.000	.000	.000	.000	.000
	11	1.00	1.00	1.00	1.00	.999	.995	.979	.932	.836	.687	.500	.313	.164	.068	.021	.005	.001	.000	.000	.000	.000
	12	1.00	1.00	1.00	1.00	1.00	.999	.993	.972	.919	.816	.661	.472	.287	.142	.055	.015	.003	.000	.000	.000	.000
	13	1.00	1.00	1.00	1.00	1.00	1.00	.998	.990	.965	.906	.798	.636	.444	.259	.120	.041	.009	.001	.000	.000	.000
	14	1.00	1.00	1.00	1.00	1.00	1.00	.999	.997	.987	.959	.895	.780	.612	.414	.229	.096	.027	.004	.000	.000	.000
	15	1.00	1.00	1.00	1.00	1.00	1.00	1.00	.999	.996	.985	.953	.885	.763	.586	.382	.196	.072	.015	.001	.000	.000
	16	1.00	1.00	1.00	1.00	1.00	1.00	1.00	1.00	.999	.995	.983	.949	.876	.747	.560	.346	.160	.046	.006	.000	.000
	17	1.00	1.00	1.00	1.00	1.00	1.00	1.00	1.00	1.00	.999	.995	.981	.946	.869	.731	.532	.305	.119	.023	.001	.000
	18	1.00	1.00	1.00	1.00	1.00	1.00	1.00	1.00	1.00	1.00	.999	.995	.981	.945	.864	.717	.499	.256	.073	.005	.000
	19	1.00	1.00	1.00	1.00	1.00	1.00	1.00	1.00	1.00	1.00	1.00	.999	.995	.982	.946	.863	.703	.460	.193	.026	.000
	20	1.00	1.00	1.00	1.00	1.00	1.00	1.00	1.00	1.00	1.00	1.00	1.00	.999	.996	.984	.951	.867	.692	.408	.105	.002
	21	1.00	1.00	1.00	1.00	1.00	1.00	1.00	1.00	1.00	1.00	1.00	1.00	1.00	.999	.997	.988	.960	.880	.685	.321	.022
	22	1.00	1.00	1.00	1.00	1.00	1.00	1.00	1.00	1.00	1.00	1.00	1.00	1.00	1.00	1.00	.999	.994	.976	.911	.693	.206
n = 24	0	.786	.292	.080	.020	.005	.001	.000	.000	.000	.000	.000	.000	.000	.000	.000	.000	.000	.000	.000	.000	.000
	1	.976	.661	.292	.106	.033	.009	.002	.000	.000	.000	.000	.000	.000	.000	.000	.000	.000	.000	.000	.000	.000
	2	.998	.884	.564	.280	.115	.040	.012	.003	.001	.000	.000	.000	.000	.000	.000	.000	.000	.000	.000	.000	.000

	3	1.00	.970	.786	.505	.264	.115	.042	.013	.004	.001	.000	.000	.000	.000	.000	.000	.000	.000	.000	.000	.000
	4	1.00	.994	.915	.713	.460	.247	.111	.042	.013	.004	.001	.000	.000	.000	.000	.000	.000	.000	.000	.000	.000
	5	1.00	.999	.972	.861	.656	.422	.229	.104	.040	.013	.003	.001	.000	.000	.000	.000	.000	.000	.000	.000	.000
	6	1.00	1.00	.993	.943	.811	.607	.389	.211	.096	.036	.011	.003	.001	.000	.000	.000	.000	.000	.000	.000	.000
	7	1.00	1.00	.998	.980	.911	.766	.565	.358	.192	.086	.032	.010	.002	.000	.000	.000	.000	.000	.000	.000	.000
	8	1.00	1.00	1.00	.994	.964	.879	.725	.526	.328	.173	.076	.027	.008	.002	.000	.000	.000	.000	.000	.000	.000
	9	1.00	1.00	1.00	.999	.987	.945	.847	.687	.489	.299	.154	.065	.022	.005	.001	.000	.000	.000	.000	.000	.000
	10	1.00	1.00	1.00	1.00	.996	.979	.926	.817	.650	.454	.271	.134	.053	.016	.004	.001	.000	.000	.000	.000	.000
	11	1.00	1.00	1.00	1.00	.999	.993	.969	.906	.787	.615	.419	.242	.114	.042	.012	.002	.000	.000	.000	.000	.000
	12	1.00	1.00	1.00	1.00	1.00	.998	.988	.958	.886	.758	.581	.385	.213	.094	.031	.007	.001	.000	.000	.000	.000
	13	1.00	1.00	1.00	1.00	1.00	.999	.996	.984	.947	.866	.729	.546	.350	.183	.074	.021	.004	.000	.000	.000	.000
	14	1.00	1.00	1.00	1.00	1.00	1.00	.999	.995	.978	.935	.846	.701	.511	.313	.153	.055	.013	.001	.000	.000	.000
	15	1.00	1.00	1.00	1.00	1.00	1.00	1.00	.998	.992	.973	.924	.827	.672	.474	.275	.121	.036	.006	.000	.000	.000
	16	1.00	1.00	1.00	1.00	1.00	1.00	1.00	1.00	.998	.990	.968	.914	.808	.642	.435	.234	.089	.020	.002	.000	.000
	17	1.00	1.00	1.00	1.00	1.00	1.00	1.00	1.00	.999	.997	.989	.964	.904	.789	.611	.393	.189	.057	.007	.000	.000
	18	1.00	1.00	1.00	1.00	1.00	1.00	1.00	1.00	1.00	.999	.997	.987	.960	.896	.771	.578	.344	.139	.028	.001	.000
	19	1.00	1.00	1.00	1.00	1.00	1.00	1.00	1.00	1.00	1.00	.999	.996	.987	.958	.889	.753	.540	.287	.085	.006	.000
	20	1.00	1.00	1.00	1.00	1.00	1.00	1.00	1.00	1.00	1.00	1.00	.999	.996	.987	.958	.885	.736	.495	.214	.030	.000
	21	1.00	1.00	1.00	1.00	1.00	1.00	1.00	1.00	1.00	1.00	1.00	1.00	.999	.997	.988	.960	.885	.720	.436	.116	.002
	22	1.00	1.00	1.00	1.00	1.00	1.00	1.00	1.00	1.00	1.00	1.00	1.00	1.00	1.00	.998	.991	.967	.894	.708	.339	.024
	23	1.00	1.00	1.00	1.00	1.00	1.00	1.00	1.00	1.00	1.00	1.00	1.00	1.00	1.00	1.00	.999	.995	.980	.920	.708	.214
$n = 25$	0	.778	.277	.072	.017	.004	.001	.000	.000	.000	.000	.000	.000	.000	.000	.000	.000	.000	.000	.000	.000	.000
	1	.974	.642	.271	.093	.027	.007	.002	.000	.000	.000	.000	.000	.000	.000	.000	.000	.000	.000	.000	.000	.000
	2	.998	.873	.537	.254	.098	.032	.009	.002	.000	.000	.000	.000	.000	.000	.000	.000	.000	.000	.000	.000	.000
	3	1.00	.966	.764	.471	.234	.096	.033	.010	.002	.000	.000	.000	.000	.000	.000	.000	.000	.000	.000	.000	.000
	4	1.00	.993	.902	.682	.421	.214	.090	.032	.009	.002	.000	.000	.000	.000	.000	.000	.000	.000	.000	.000	.000
	5	1.00	.999	.967	.838	.617	.378	.193	.083	.029	.009	.002	.000	.000	.000	.000	.000	.000	.000	.000	.000	.000
	6	1.00	1.00	.991	.930	.780	.561	.341	.173	.074	.026	.007	.002	.000	.000	.000	.000	.000	.000	.000	.000	.000
	7	1.00	1.00	.998	.975	.891	.727	.512	.306	.154	.064	.022	.006	.001	.000	.000	.000	.000	.000	.000	.000	.000
	8	1.00	1.00	1.00	.992	.953	.851	.677	.467	.274	.134	.054	.017	.004	.001	.000	.000	.000	.000	.000	.000	.000
	9	1.00	1.00	1.00	.998	.983	.929	.811	.630	.425	.242	.115	.044	.013	.003	.000	.000	.000	.000	.000	.000	.000
	10	1.00	1.00	1.00	1.00	.994	.970	.902	.771	.586	.384	.212	.096	.034	.009	.002	.000	.000	.000	.000	.000	.000
	11	1.00	1.00	1.00	1.00	.998	.989	.956	.875	.732	.543	.345	.183	.078	.025	.006	.001	.000	.000	.000	.000	.000
	12	1.00	1.00	1.00	1.00	1.00	.997	.983	.940	.846	.694	.500	.306	.154	.060	.017	.003	.000	.000	.000	.000	.000
	13	1.00	1.00	1.00	1.00	1.00	.999	.994	.975	.922	.817	.655	.457	.268	.125	.044	.011	.002	.000	.000	.000	.000
	14	1.00	1.00	1.00	1.00	1.00	1.00	.998	.991	.966	.904	.788	.616	.414	.229	.098	.030	.006	.000	.000	.000	.000
	15	1.00	1.00	1.00	1.00	1.00	1.00	1.00	.997	.987	.956	.885	.758	.575	.370	.189	.071	.017	.002	.000	.000	.000
	16	1.00	1.00	1.00	1.00	1.00	1.00	1.00	.999	.996	.983	.946	.866	.726	.533	.323	.149	.047	.008	.000	.000	.000
	17	1.00	1.00	1.00	1.00	1.00	1.00	1.00	1.00	.999	.994	.978	.936	.846	.694	.488	.273	.109	.025	.002	.000	.000
	18	1.00	1.00	1.00	1.00	1.00	1.00	1.00	1.00	1.00	.998	.993	.974	.926	.827	.659	.439	.220	.070	.009	.000	.000
	19	1.00	1.00	1.00	1.00	1.00	1.00	1.00	1.00	1.00	1.00	.998	.991	.971	.917	.807	.622	.383	.162	.033	.001	.000
	20	1.00	1.00	1.00	1.00	1.00	1.00	1.00	1.00	1.00	1.00	1.00	.998	.991	.968	.910	.786	.579	.318	.098	.007	.000
	21	1.00	1.00	1.00	1.00	1.00	1.00	1.00	1.00	1.00	1.00	1.00	1.00	.998	.990	.967	.904	.766	.529	.236	.034	.000
	22	1.00	1.00	1.00	1.00	1.00	1.00	1.00	1.00	1.00	1.00	1.00	1.00	1.00	.998	.991	.968	.902	.746	.463	.127	.002
	23	1.00	1.00	1.00	1.00	1.00	1.00	1.00	1.00	1.00	1.00	1.00	1.00	1.00	1.00	.998	.993	.973	.907	.729	.358	.026
	24	1.00	1.00	1.00	1.00	1.00	1.00	1.00	1.00	1.00	1.00	1.00	1.00	1.00	1.00	1.00	.999	.996	.983	.928	.723	.222

TABLE A.2

A cumulative poisson distribution table

$F(x) = \sum_{k=0}^{x}(\lambda^k e^{-\lambda}/k!)$

	x										
λ	**0**	**1**	**2**	**3**	**4**	**5**	**6**	**7**	**8**	**9**	**10**
.02	.980	1.00									
.04	.961	.999	1.00								
.06	.942	.998	1.00								
.08	.923	.997	1.00								
.10	.905	.995	1.00								
.15	.861	.990	.999	1.00							
.20	.819	.982	.999	1.00							
.25	.779	.974	.998	1.00							
.30	.741	.963	.996	1.00							
.35	.705	.951	.994	1.00							
.40	.670	.938	.992	.999	1.00						
.45	.638	.925	.989	.999	1.00						
.50	.607	.910	.986	.998	1.00						
.55	.577	.894	.982	.998	1.00						
.60	.549	.878	.977	.997	1.00						
.65	.522	.861	.972	.996	.999	1.00					
.70	.497	.844	.966	.994	.999	1.00					
.75	.472	.827	.959	.993	.999	1.00					
.80	.449	.809	.953	.991	.999	1.00					
.85	.427	.791	.945	.989	.998	1.00					
.90	.407	.772	.937	.987	.998	1.00					
.95	.387	.754	.929	.984	.997	1.00					
1.00	.368	.736	.920	.981	.996	.999	1.00				
1.10	.333	.699	.900	.974	.995	.999	1.00				
1.20	.301	.663	.879	.966	.992	.998	1.00				
1.30	.273	.627	.857	.957	.989	.998	1.00				
1.40	.247	.592	.833	.946	.986	.997	.999	1.00			
1.50	.223	.558	.809	.934	.981	.996	.999	1.00			
1.60	.202	.525	.783	.921	.976	.994	.999	1.00			
1.70	.183	.493	.757	.907	.970	.992	.998	1.00			
1.80	.165	.463	.731	.891	.964	.990	.997	.999	1.00		
1.90	.150	.434	.704	.875	.956	.987	.997	.999	1.00		
2.00	.135	.406	.677	.857	.947	.983	.995	.999	1.00		
2.20	.111	.355	.623	.819	.928	.975	.993	.998	1.00		
2.40	.091	.308	.570	.779	.904	.964	.988	.997	.999	1.00	
2.60	.074	.267	.518	.736	.877	.951	.983	.995	.999	1.00	
2.80	.061	.231	.469	.692	.848	.935	.976	.992	.998	.999	1.00
3.00	.050	.199	.423	.647	.815	.916	.966	.988	.996	.999	1.00
3.20	.041	.171	.380	.603	.781	.895	.955	.983	.994	.998	1.00

TABLE A.2
(Continued)

					x				
λ	**0**	**1**	**2**	**3**	**4**	**5**	**6**	**7**	**8**
3.40	.033	.147	.340	.558	.744	.871	.942	.977	.992
3.60	.027	.126	.303	.515	.706	.844	.927	.969	.988
3.80	.022	.107	.269	.473	.668	.816	.909	.960	.984
4.00	.018	.092	.238	.433	.629	.785	.889	.949	.979
4.20	.015	.078	.210	.395	.590	.753	.867	.936	.972
4.40	.012	.066	.185	.359	.551	.720	.844	.921	.964
4.60	.010	.056	.163	.326	.513	.686	.818	.905	.955
4.80	.008	.048	.143	.294	.476	.651	.791	.887	.944
5.00	.007	.040	.125	.265	.440	.616	.762	.867	.932
5.40	.005	.029	.095	.213	.373	.546	.702	.822	.903
5.80	.003	.021	.072	.170	.313	.478	.638	.771	.867
6.00	.002	.017	.062	.151	.285	.446	.606	.744	.847
6.20	.002	.015	.054	.134	.259	.414	.574	.716	.826
6.40	.002	.012	.046	.119	.235	.384	.542	.687	.803
6.60	.001	.010	.040	.105	.213	.355	.511	.658	.780
6.80	.001	.009	.034	.093	.192	.327	.480	.628	.755
7.00	.001	.007	.030	.082	.173	.301	.450	.599	.729
7.20	.001	.006	.025	.072	.156	.276	.420	.569	.703

					x				
λ	**9**	**10**	**11**	**12**	**13**	**14**	**15**	**16**	**17**
3.40	.997	.999	1.00						
3.60	.996	.999	1.00						
3.80	.994	.998	.999	1.00					
4.00	.992	.997	.999	1.00					
4.20	.989	.996	.999	1.00					
4.40	.985	.994	.998	.999	1.00				
4.60	.980	.992	.997	.999	1.00				
4.80	.975	.990	.996	.999	1.00				
5.00	.968	.986	.995	.998	.999	1.00			
5.40	.951	.977	.990	.996	.999	1.00			
5.80	.929	.965	.984	.993	.997	.999	1.00		
6.00	.916	.957	.980	.991	.996	.999	.999	1.00	
6.20	.902	.949	.975	.989	.995	.998	.999	1.00	
6.40	.886	.939	.969	.986	.994	.997	.999	1.00	
6.60	.869	.927	.963	.982	.992	.997	.999	.999	1.00
6.80	.850	.915	.955	.978	.990	.996	.998	.999	1.00
7.00	.830	.901	.947	.973	.987	.994	.998	.999	1.00
7.20	.810	.887	.937	.967	.984	.993	.997	.999	1.00

TABLE A.2
(Continued)

	x													
λ	0	1	2	3	4	5	6	7	8	9	10	11	12	13
7.40	.001	.005	.022	.063	.140	.253	.392	.539	.676	.788	.871	.926	.961	.980
7.60	.001	.004	.019	.055	.125	.231	.365	.510	.648	.765	.854	.915	.954	.976
7.80	.000	.004	.016	.048	.112	.210	.338	.481	.620	.741	.835	.902	.945	.971
8.00	.000	.003	.014	.042	.100	.191	.313	.453	.593	.717	.816	.888	.936	.966
8.50	.000	.002	.009	.030	.074	.150	.256	.386	.523	.653	.763	.849	.909	.949
9.00	.000	.001	.006	.021	.055	.116	.207	.324	.456	.587	.706	.803	.876	.926
9.50	.000	.001	.004	.015	.040	.089	.165	.269	.392	.522	.645	.752	.836	.898
10.0	.000	.000	.003	.010	.029	.067	.130	.220	.333	.458	.583	.697	.792	.864
10.5	.000	.000	.002	.007	.021	.050	.102	.179	.279	.397	.521	.639	.742	.825
11.0	.000	.000	.001	.005	.015	.038	.079	.143	.232	.341	.460	.579	.689	.781
11.5	.000	.000	.001	.003	.011	.028	.060	.114	.191	.289	.402	.520	.633	.733
12.0	.000	.000	.001	.002	.008	.020	.046	.090	.155	.242	.347	.462	.576	.682
12.5	.000	.000	.000	.002	.005	.015	.035	.070	.125	.201	.297	.406	.519	.628
13.0	.000	.000	.000	.001	.004	.011	.026	.054	.100	.166	.252	.353	.463	.573
13.5	.000	.000	.000	.001	.003	.008	.019	.041	.079	.135	.211	.304	.409	.518
14.0	.000	.000	.000	.000	.002	.006	.014	.032	.062	.109	.176	.260	.358	.464
14.5	.000	.000	.000	.000	.001	.004	.010	.024	.048	.088	.145	.220	.311	.413
15.0	.000	.000	.000	.000	.001	.003	.008	.018	.037	.070	.118	.185	.268	.363

λ	4	5	6	7	8	9	10	11	12	13	14	15	16	17
16	.000	.001	.004	.010	.022	.043	.077	.127	.193	.275	.368	.467	.566	.659
17	.000	.001	.002	.005	.013	.026	.049	.085	.135	.201	.281	.371	.468	.564
18	.000	.000	.001	.003	.007	.015	.030	.055	.092	.143	.208	.287	.375	.469
19	.000	.000	.001	.002	.004	.009	.018	.035	.061	.098	.150	.215	.292	.378
20	.000	.000	.000	.001	.002	.005	.011	.021	.039	.066	.105	.157	.221	.297
21	.000	.000	.000	.000	.001	.003	.006	.013	.025	.043	.072	.111	.163	.227
22	.000	.000	.000	.000	.001	.002	.004	.008	.015	.028	.048	.077	.117	.169
23	.000	.000	.000	.000	.000	.001	.002	.004	.009	.017	.031	.052	.082	.123
24	.000	.000	.000	.000	.000	.000	.001	.003	.005	.011	.020	.034	.056	.087
25	.000	.000	.000	.000	.000	.000	.001	.001	.003	.006	.012	.022	.038	.060

λ	34	35	36	37	38	39	40	41	42	43
19	.999	1.00								
20	.999	.999	1.00							
21	.997	.998	.999	.999	1.00					
22	.994	.996	.998	.999	.999	1.00				
23	.988	.993	.996	.997	.999	.999	1.00			
24	.979	.987	.992	.995	.997	.998	.999	.999	1.00	
25	.966	.978	.985	.991	.994	.997	.998	.999	.999	1.00

TABLE A.2
(Concluded)

x															
14	**15**	**16**	**17**	**18**	**19**	**20**	**21**	**22**	**23**	**24**	**25**	**26**	**27**	**28**	**29**
.991	.996	.998	.999	1.00											
.989	.995	.998	.999	1.00											
.986	.993	.997	.999	1.00											
.983	.992	.996	.998	.999	1.00										
.973	.986	.993	.997	.999	.999	1.00									
.959	.978	.989	.995	.998	.999	1.00									
.940	.967	.982	.991	.996	.998	.999	1.00								
.917	.951	.973	.986	.993	.997	.998	.999	1.00							
.888	.932	.960	.978	.988	.994	.997	.999	.999	1.00						
.854	.907	.944	.968	.982	.991	.995	.998	.999	1.00						
.815	.878	.924	.954	.974	.986	.992	.996	.998	.999	1.00					
.772	.844	.899	.937	.963	.979	.988	.994	.997	.999	.999	1.00				
.725	.806	.869	.916	.948	.969	.983	.991	.995	.998	.999	.999	1.00			
.675	.764	.835	.890	.930	.957	.975	.986	.992	.996	.998	.999	1.00			
.623	.718	.798	.861	.908	.942	.965	.980	.989	.994	.997	.998	.999	1.00		
.570	.669	.756	.827	.883	.923	.952	.971	.983	.991	.995	.997	.999	.999	1.00	
.518	.619	.711	.790	.853	.901	.936	.960	.976	.986	.992	.996	.998	.999	.999	1.00
.466	.568	.664	.749	.819	.875	.917	.947	.967	.981	.989	.994	.997	.998	.999	1.00
18	**19**	**20**	**21**	**22**	**23**	**24**	**25**	**26**	**27**	**28**	**29**	**30**	**31**	**32**	**33**
.742	.812	.868	.911	.942	.963	.978	.987	.993	.996	.998	.999	.999	1.00		
.655	.736	.805	.861	.905	.937	.959	.975	.985	.991	.995	.997	.999	.999	1.00	
.562	.651	.731	.799	.855	.899	.932	.955	.972	.983	.990	.994	.997	.998	.999	1.00
.469	.561	.647	.725	.793	.849	.893	.927	.951	.969	.980	.988	.993	.996	.998	.999
.381	.470	.559	.644	.721	.787	.843	.888	.922	.948	.966	.978	.987	.992	.995	.997
.302	.384	.471	.558	.640	.716	.782	.838	.883	.917	.944	.963	.976	.985	.991	.994
.232	.306	.387	.472	.556	.637	.712	.777	.832	.877	.913	.940	.959	.973	.983	.989
.175	.238	.310	.389	.472	.555	.635	.708	.772	.827	.873	.908	.936	.956	.971	.981
.128	.180	.243	.314	.392	.473	.554	.632	.704	.768	.823	.868	.904	.932	.953	.969
.092	.134	.185	.247	.318	.394	.473	.553	.629	.700	.763	.818	.863	.900	.929	.950

TABLE A.3

A cumulative standard normal distribution table

$P(Z \le z_\alpha) = \int_{-\infty}^{z_\alpha}(1/\sqrt{2\pi})\exp -(x^2/2)\,dx = 1-\alpha$

z	**0.09**	**0.08**	**0.07**	**0.06**	**0.05**	**0.04**	**0.03**	**0.02**	**0.01**	**0.00**
−3.5	.00017	.00017	.00018	.00019	.00019	.00020	.00021	.00022	.00022	.00023
−3.4	.00024	.00025	.00026	.00027	.00028	.00029	.00030	.00031	.00032	.00034
−3.3	.00035	.00036	.00038	.00039	.00040	.00042	.00043	.00045	.00047	.00048
−3.2	.00050	.00052	.00054	.00056	.00058	.00060	.00062	.00064	.00066	.00069
−3.1	.00071	.00074	.00076	.00079	.00082	.00084	.00087	.00090	.00094	.00097
−3.0	.00100	.00104	.00107	.00111	.00114	.00118	.00122	.00126	.00131	.00135
−2.9	.00139	.00144	.00149	.00154	.00159	.00164	.00169	.00175	.00181	.00187
−2.8	.00193	.00199	.00205	.00212	.00219	.00226	.00233	.00240	.00248	.00256
−2.7	.00264	.00272	.00280	.00289	.00298	.00307	.00317	.00326	.00336	.00347
−2.6	.00357	.00368	.00379	.00391	.00402	.00415	.00427	.00440	.00453	.00466
−2.5	.00480	.00494	.00508	.00523	.00539	.00554	.00570	.00587	.00604	.00621
−2.4	.00639	.00657	.00676	.00695	.00714	.00734	.00755	.00776	.00798	.00820
−2.3	.00842	.00866	.00889	.00914	.00939	.00964	.00990	.01017	.01044	.01072
−2.2	.01101	.01130	.01160	.01191	.01222	.01255	.01287	.01321	.01355	.01390
−2.1	.01426	.01463	.01500	.01539	.01578	.01618	.01659	.01700	.01743	.01786
−2.0	.01831	.01876	.01923	.01970	.02018	.02068	.02118	.02169	.02222	.02275
−1.9	.02330	.02385	.02442	.02500	.02559	.02619	.02680	.02743	.02807	.02872
−1.8	.02938	.03005	.03074	.03144	.03216	.03288	.03362	.03438	.03515	.03593
−1.7	.03673	.03754	.03836	.03920	.04006	.04093	.04182	.04272	.04363	.04457
−1.6	.04551	.04648	.04746	.04846	.04947	.05050	.05155	.05262	.05370	.05480
−1.5	.05592	.05705	.05821	.05938	.06057	.06178	.06301	.06426	.06552	.06681
−1.4	.06811	.06944	.07078	.07215	.07353	.07493	.07636	.07780	.07927	.08076
−1.3	.08226	.08379	.08534	.08692	.08851	.09012	.09176	.09342	.09510	.09680
−1.2	.09853	.10027	.10204	.10383	.10565	.10749	.10935	.11123	.11314	.11507
−1.1	.11702	.11900	.12100	.12302	.12507	.12714	.12924	.13136	.13350	.13567
−1.0	.13786	.14007	.14231	.14457	.14686	.14917	.15151	.15386	.15625	.15866
−0.9	.16109	.16354	.16602	.16853	.17106	.17361	.17619	.17879	.18141	.18406
−0.8	.18673	.18943	.19215	.19489	.19766	.20045	.20327	.20611	.20897	.21186
−0.7	.21476	.21770	.22065	.22363	.22663	.22965	.23270	.23576	.23885	.24196
−0.6	.24510	.24825	.25143	.25463	.25785	.26109	.26435	.26763	.27093	.27425
−0.5	.27760	.28096	.28434	.28774	.29116	.29460	.29806	.30153	.30503	.30854
−0.4	.31207	.31561	.31918	.32276	.32636	.32997	.33360	.33724	.34090	.34458
−0.3	.34827	.35197	.35569	.35942	.36317	.36693	.37070	.37448	.37828	.38209
−0.2	.38591	.38974	.39358	.39743	.40129	.40517	.40905	.41294	.41683	.42074
−0.1	.42465	.42858	.43251	.43644	.44038	.44433	.44828	.45224	.45620	.46017
−0.0	.46414	.46812	.47210	.47608	.48006	.48405	.48803	.49202	.49601	.50000

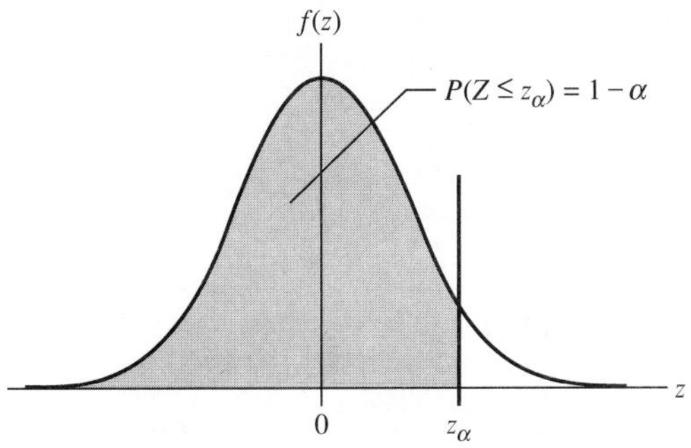

z	**0.00**	**0.01**	**0.02**	**0.03**	**0.04**	**0.05**	**0.06**	**0.07**	**0.08**	**0.09**
+0.0	.50000	.50399	.50798	.51197	.51595	.51994	.52392	.52790	.53188	.53586
+0.1	.53983	.54380	.54776	.55172	.55567	.55962	.56356	.56749	.57142	.57535
+0.2	.57926	.58317	.58706	.59095	.59483	.59871	.60257	.60642	.61026	.61409
+0.3	.61791	.62172	.62552	.62930	.63307	.63683	.64058	.64431	.64803	.65173
+0.4	.65542	.65910	.66276	.66640	.67003	.67364	.67724	.68082	.68439	.68793
+0.5	.69146	.69497	.69847	.70194	.70540	.70884	.71226	.71566	.71904	.72240
+0.6	.72575	.72907	.73237	.73565	.73891	.74215	.74537	.74857	.75175	.75490
+0.7	.75804	.76115	.76424	.76730	.77035	.77337	.77637	.77935	.78230	.78524
+0.8	.78814	.79103	.79389	.79673	.79955	.80234	.80511	.80785	.81057	.81327
+0.9	.81594	.81859	.82121	.82381	.82639	.82894	.83147	.83398	.83646	.83891
+1.0	.84134	.84375	.84614	.84849	.85083	.85314	.85543	.85769	.85993	.86214
+1.1	.86433	.86650	.86864	.87076	.87286	.87493	.87698	.87900	.88100	.88298
+1.2	.88493	.88686	.88877	.89065	.89251	.89435	.89617	.89796	.89973	.90147
+1.3	.90320	.90490	.90658	.90824	.90988	.91149	.91308	.91466	.91621	.91774
+1.4	.91924	.92073	.92220	.92364	.92507	.92647	.92785	.92922	.93056	.93189
+1.5	.93319	.93448	.93574	.93699	.93822	.93943	.94062	.94179	.94295	.94408
+1.6	.94520	.94630	.94738	.94845	.94950	.95053	.95154	.95254	.95352	.95449
+1.7	.95543	.95637	.95728	.95818	.95907	.95994	.96080	.96164	.96246	.96327
+1.8	.96407	.96485	.96562	.96638	.96712	.96784	.96856	.96926	.96995	.97062
+1.9	.97128	.97193	.97257	.97320	.97381	.97441	.97500	.97558	.97615	.97670
+2.0	.97725	.97778	.97831	.97882	.97932	.97982	.98030	.98077	.98124	.98169
+2.1	.98214	.98257	.98300	.98341	.98382	.98422	.98461	.98500	.98537	.98574
+2.2	.98610	.98645	.98679	.98713	.98745	.98778	.98809	.98840	.98870	.98899
+2.3	.98928	.98956	.98983	.99010	.99036	.99061	.99086	.99111	.99134	.99158
+2.4	.99180	.99202	.99224	.99245	.99266	.99286	.99305	.99324	.99343	.99361
+2.5	.99379	.99396	.99413	.99430	.99446	.99461	.99477	.99492	.99506	.99520
+2.6	.99534	.99547	.99560	.99573	.99585	.99598	.99609	.99621	.99632	.99643
+2.7	.99653	.99664	.99674	.99683	.99693	.99702	.99711	.99720	.99728	.99736
+2.8	.99744	.99752	.99760	.99767	.99774	.99781	.99788	.99795	.99801	.99807
+2.9	.99813	.99819	.99825	.99831	.99836	.99841	.99846	.99851	.99856	.99861
+3.0	.99865	.99869	.99874	.99878	.99882	.99886	.99889	.99893	.99896	.99900
+3.1	.99903	.99906	.99910	.99913	.99916	.99918	.99921	.99924	.99926	.99929
+3.2	.99931	.99934	.99936	.99938	.99940	.99942	.99944	.99946	.99948	.99950
+3.3	.99952	.99953	.99955	.99957	.99958	.99960	.99961	.99962	.99964	.99965
+3.4	.99966	.99968	.99969	.99970	.99971	.99972	.99973	.99974	.99975	.99976
+3.5	.99977	.99978	.99978	.99979	.99980	.99981	.99981	.99982	.99983	.99983

TABLE A.4
A cumulative chi-square distribution table
Values of χ^2_α

	α							
ν	**0.995**	**0.99**	**0.975**	**0.95**	**0.05**	**0.025**	**0.01**	**0.005**
1	0.0000393	0.000157	0.000982	0.00393	3.841	5.024	6.635	7.879
2	0.0100	0.0201	0.0506	0.103	5.991	7.378	9.210	10.597
3	0.0717	0.115	0.216	0.352	7.815	9.348	11.345	12.838
4	0.207	0.297	0.484	0.711	9.488	11.143	13.277	14.860
5	0.412	0.554	0.831	1.145	11.071	12.833	15.086	16.749
6	0.676	0.872	1.237	1.635	12.592	14.449	16.812	18.548
7	0.989	1.239	1.690	2.167	14.067	16.013	18.475	20.278
8	1.344	1.646	2.180	2.733	15.507	17.535	20.090	21.955
9	1.735	2.088	2.700	3.325	16.919	19.023	21.666	23.590
10	2.156	2.558	3.247	3.940	18.307	20.483	23.209	25.188
11	2.603	3.053	3.816	4.575	19.675	21.920	24.725	26.758
12	3.074	3.571	4.404	5.226	21.026	23.337	26.217	28.299
13	3.565	4.107	5.009	5.892	22.362	24.736	27.688	29.820
14	4.075	4.660	5.629	6.571	23.685	26.119	29.142	31.320
15	4.601	5.229	6.262	7.261	24.996	27.489	30.578	32.801
16	5.142	5.812	6.908	7.962	26.296	28.845	32.000	34.268
17	5.697	6.408	7.564	8.672	27.587	30.191	33.409	35.717
18	6.265	7.015	8.231	9.390	28.869	31.526	34.805	37.156
19	6.844	7.633	8.907	10.117	30.144	32.853	36.191	38.581
20	7.434	8.260	9.591	10.851	31.410	34.170	37.566	39.997
21	8.034	8.897	10.283	11.591	32.671	35.479	38.932	41.400
22	8.643	9.542	10.982	12.338	33.926	36.781	40.289	42.796
23	9.260	10.196	11.689	13.091	35.172	38.075	41.638	44.184
24	9.886	10.856	12.401	13.848	36.415	39.364	42.980	45.559
25	10.520	11.524	13.120	14.611	37.652	40.646	44.314	46.930
26	11.160	12.198	13.844	15.379	38.885	41.924	45.643	48.290
27	11.808	12.878	14.573	16.151	40.113	43.195	46.963	49.647
28	12.461	13.565	15.308	16.928	41.337	44.461	48.278	50.994
29	13.121	14.256	16.047	17.708	42.557	45.722	49.588	52.338
30	13.787	14.953	16.791	18.493	43.773	46.979	50.892	53.673

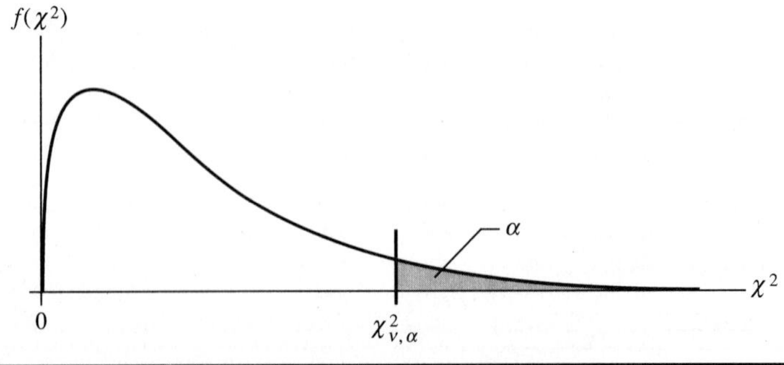

TABLE A.5
A cumulative t distribution table
Values of t_α

	α				
ν	**0.10**	**0.05**	**0.025**	**0.01**	**0.005**
1	3.078	6.314	12.706	31.820	63.656
2	1.886	2.920	4.303	6.965	9.925
3	1.638	2.353	3.182	4.541	5.841
4	1.533	2.132	2.776	3.747	4.604
5	1.476	2.015	2.571	3.365	4.032
6	1.440	1.943	2.447	3.143	3.707
7	1.415	1.895	2.365	2.998	3.499
8	1.397	1.860	2.306	2.896	3.355
9	1.383	1.833	2.262	2.821	3.250
10	1.372	1.812	2.228	2.764	3.169
11	1.363	1.796	2.201	2.718	3.106
12	1.356	1.782	2.179	2.681	3.055
13	1.350	1.771	2.160	2.650	3.012
14	1.345	1.761	2.145	2.624	2.977
15	1.341	1.753	2.131	2.602	2.947
16	1.337	1.746	2.120	2.583	2.921
17	1.333	1.740	2.110	2.567	2.898
18	1.330	1.734	2.101	2.552	2.878
19	1.328	1.729	2.093	2.539	2.861
20	1.325	1.725	2.086	2.528	2.845
21	1.323	1.721	2.080	2.518	2.831
22	1.321	1.717	2.074	2.508	2.819
23	1.319	1.714	2.069	2.500	2.807
24	1.318	1.711	2.064	2.492	2.797
25	1.316	1.708	2.060	2.485	2.787
26	1.315	1.706	2.056	2.479	2.779
27	1.314	1.703	2.052	2.473	2.771
28	1.313	1.701	2.048	2.467	2.763
29	1.311	1.699	2.045	2.462	2.756
∞	1.282	1.645	1.960	2.326	2.576

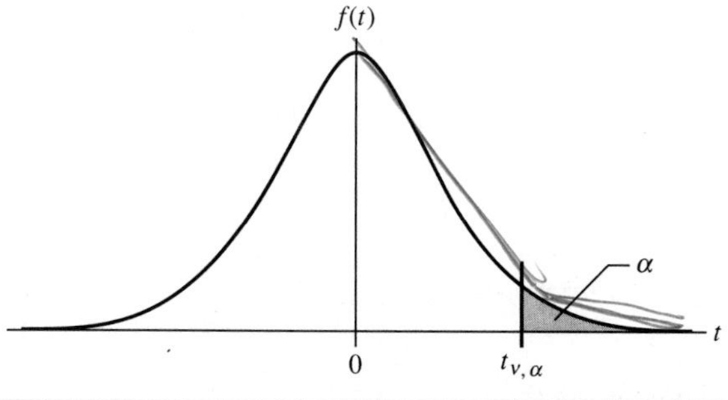

TABLE A.6*a*
A cumulative F distribution table
Values of $F_{\nu_1,\nu_2,.05}$

	ν_1								
ν_2	1	2	3	4	5	6	7	8	9
1	161.14	199.5	215.7	224.6	230.2	234.0	236.8	238.9	240.5
2	18.51	19.00	19.16	19.25	19.30	19.33	19.35	19.37	19.38
3	10.13	9.55	9.28	9.12	9.01	8.94	8.89	8.85	8.81
4	7.71	6.94	6.59	6.39	6.26	6.16	6.09	6.04	6.00
5	6.61	5.79	5.41	5.19	5.05	4.95	4.88	4.82	4.77
6	5.99	5.14	4.76	4.53	4.39	4.28	4.21	4.15	4.10
7	5.59	4.74	4.35	4.12	3.97	3.87	3.79	3.73	3.68
8	5.32	4.46	4.07	3.84	3.69	3.58	3.50	3.44	3.39
9	5.12	4.26	3.86	3.63	3.48	3.37	3.29	3.23	3.18
10	4.96	4.10	3.71	3.48	3.33	3.22	3.14	3.07	3.02
11	4.84	3.98	3.59	3.36	3.20	3.09	3.01	2.95	2.90
12	4.75	3.89	3.49	3.26	3.11	3.00	2.91	2.85	2.80
13	4.67	3.81	3.41	3.18	3.03	2.92	2.83	2.77	2.71
14	4.60	3.74	3.34	3.11	2.96	2.85	2.76	2.70	2.65
15	4.54	3.68	3.29	3.06	2.90	2.79	2.71	2.64	2.59
16	4.49	3.63	3.24	3.01	2.85	2.74	2.66	2.59	2.54
17	4.45	3.59	3.20	2.96	2.81	2.70	2.61	2.55	2.49
18	4.41	3.55	3.16	2.93	2.77	2.66	2.58	2.51	2.46
19	4.38	3.52	3.13	2.90	2.74	2.63	2.54	2.48	2.42
20	4.35	3.49	3.10	2.87	2.71	2.60	2.51	2.45	2.39
21	4.32	3.47	3.07	2.84	2.68	2.57	2.49	2.42	2.37
22	4.30	3.44	3.05	2.82	2.66	2.55	2.46	2.40	2.34
23	4.28	3.42	3.03	2.80	2.64	2.53	2.44	2.37	2.32
24	4.26	3.40	3.01	2.78	2.62	2.51	2.42	2.36	2.30
25	4.24	3.39	2.99	2.76	2.60	2.49	2.40	2.34	2.28
30	4.17	3.32	2.92	2.69	2.53	2.42	2.33	2.27	2.21
40	4.08	3.23	2.84	2.61	2.45	2.34	2.25	2.18	2.12
60	4.00	3.15	2.76	2.53	2.37	2.25	2.17	2.10	2.04
120	3.92	3.07	2.68	2.45	2.29	2.17	2.09	2.02	1.96
∞	3.84	3.00	2.60	2.37	2.21	2.10	2.01	1.94	1.88

TABLE A.6*a*
(Continued)

ν_1									
10	**12**	**15**	**20**	**24**	**30**	**40**	**60**	**120**	∞
241.9	243.9	245.9	248.0	249.1	250.1	251.1	252.2	253.3	254.3
19.40	19.41	19.43	19.45	19.45	19.46	19.47	19.48	19.49	19.50
8.79	8.74	8.70	8.66	8.64	8.62	8.59	8.57	8.55	8.53
5.96	5.91	5.86	5.80	5.77	5.75	5.72	5.69	5.66	5.63
4.74	4.68	4.62	4.56	4.53	4.50	4.46	4.43	4.40	4.36
4.06	4.00	3.94	3.87	3.84	3.81	3.77	3.74	3.70	3.67
3.65	3.57	3.51	3.44	3.41	3.38	3.34	3.30	3.27	3.23
3.14	3.07	3.01	2.94	2.90	2.86	2.83	2.79	2.75	2.71
2.98	2.91	2.85	2.77	2.74	2.70	2.66	2.62	2.58	2.54
2.85	2.79	2.72	2.65	2.61	2.57	2.53	2.49	2.45	2.40
2.75	2.69	2.62	2.54	2.51	2.47	2.43	2.38	2.34	2.30
2.67	2.60	2.53	2.46	2.42	2.38	2.34	2.30	2.25	2.21
2.60	2.53	2.46	2.39	2.35	2.31	2.34	2.30	2.25	2.21
2.54	2.48	2.40	2.33	2.29	2.25	2.20	2.16	2.11	2.07
2.49	2.42	2.35	2.28	2.24	2.19	2.15	2.10	2.06	2.02
2.45	2.38	2.31	2.23	2.19	2.15	2.10	2.06	2.01	1.96
2.41	2.34	2.27	2.19	2.15	2.11	2.06	2.02	1.97	1.92
2.38	2.31	2.23	2.16	2.11	2.07	2.03	1.98	1.93	1.88
2.35	2.28	2.20	2.12	2.08	2.04	1.99	1.95	1.90	1.84
2.32	2.25	2.18	2.10	2.05	2.01	1.96	1.92	1.87	1.81
2.30	2.23	2.15	2.07	2.03	1.98	1.94	1.89	1.84	1.78
2.27	2.20	2.13	2.05	2.01	1.96	1.91	1.86	1.81	1.76
2.25	2.18	2.11	2.03	1.98	1.94	1.89	1.84	1.79	1.73
2.24	2.16	2.09	2.01	1.96	1.92	1.87	1.82	1.77	1.71
2.16	2.09	2.01	1.93	1.89	1.84	1.79	1.74	1.68	1.62
2.08	2.00	1.92	1.84	1.79	1.74	1.69	1.64	1.58	1.51
1.99	1.92	1.84	1.75	1.70	1.65	1.59	1.53	1.47	1.39
1.91	1.83	1.75	1.66	1.61	1.55	1.50	1.43	1.35	1.25
1.83	1.75	1.67	1.57	1.52	1.46	1.39	1.32	1.22	1.00

TABLE A.6*b*

A cumulative *F* distribution table

Values of $F_{\nu_1,\nu_2,.01}$

	ν_1								
ν_2	1	2	3	4	5	6	7	8	9
1	4052	4999.5	5403	5625	5764	5859	5928	5981	6022
2	98.50	99.00	99.17	99.25	99.30	99.33	99.36	99.37	99.39
3	34.12	30.82	29.46	28.71	28.24	27.91	27.67	27.49	27.35
4	21.20	18.00	16.69	15.98	15.52	15.21	14.98	14.80	14.66
5	16.26	13.27	12.06	11.39	10.97	10.67	10.46	10.29	10.16
6	13.75	10.92	9.78	9.15	8.75	8.47	8.26	8.10	7.98
7	12.25	9.55	8.45	7.85	7.46	7.19	6.99	6.84	6.72
8	11.26	8.65	7.59	7.01	6.63	6.37	6.18	6.03	5.91
9	10.56	8.02	6.99	6.42	6.06	5.80	5.61	5.47	5.35
10	10.04	7.56	6.55	5.99	5.64	5.39	5.20	5.06	4.94
11	9.65	7.21	6.22	5.67	5.32	5.07	4.89	4.74	4.63
12	9.33	6.93	5.95	5.41	5.06	4.82	4.64	4.50	4.39
13	9.07	6.70	5.74	5.21	4.86	4.62	4.41	4.30	4.19
14	8.86	6.51	5.56	5.04	4.69	4.46	4.28	4.14	4.03
15	8.68	6.36	5.42	4.89	4.56	4.32	4.14	4.00	3.89
16	8.53	6.23	5.29	4.77	4.44	4.20	4.03	3.89	3.78
17	8.40	6.11	5.18	4.67	4.34	4.10	3.93	3.79	3.68
18	8.29	6.01	5.09	4.58	4.25	4.01	3.84	3.71	3.60
19	8.18	5.93	5.01	4.50	4.17	3.94	3.77	3.63	3.52
20	8.10	5.85	4.94	4.43	4.10	3.87	3.70	3.56	3.46
21	8.02	5.78	4.87	4.37	4.04	3.81	3.64	3.51	3.40
22	7.95	5.72	4.82	4.31	3.99	3.76	3.59	3.45	3.35
23	7.88	5.66	4.76	4.26	3.94	3.71	3.54	3.41	3.30
24	7.82	5.61	4.72	4.22	3.90	3.67	3.50	3.36	3.26
25	7.77	5.57	4.68	4.18	3.85	3.63	3.46	3.32	3.22
30	7.56	5.39	4.51	4.02	3.70	3.47	3.30	3.17	3.07
40	7.31	5.18	4.31	3.83	3.51	3.29	3.12	2.99	2.89
60	7.08	4.98	4.13	3.65	3.34	3.12	2.95	2.82	2.72
120	6.85	4.79	3.95	3.48	3.17	2.96	2.79	2.66	2.56
∞	6.63	4.61	3.78	3.32	3.02	2.80	2.64	2.51	2.41

TABLEA.6*b*
(Continued)

ν_1									
10	**12**	**15**	**20**	**24**	**30**	**40**	**60**	**120**	∞
6056	6106	6157	6209	6235	6261	6287	6313	6339	6366
99.40	99.42	99.43	99.45	99.46	99.47	99.47	99.48	99.49	99.50
27.23	27.05	26.87	26.69	26.60	26.50	26.41	26.32	26.22	26.13
14.55	14.37	14.20	14.02	13.93	13.84	13.75	13.65	13.56	13.46
10.05	9.89	9.72	9.55	9.47	9.38	9.29	9.20	9.11	9.02
7.87	7.72	7.56	7.40	7.31	7.23	7.14	7.06	6.97	6.88
6.62	6.47	6.31	6.16	6.07	5.99	5.91	5.82	5.74	5.65
5.81	5.67	5.52	5.36	5.28	5.20	5.12	5.03	4.95	4.86
5.26	5.11	4.96	4.81	4.73	4.65	4.57	4.48	4.40	4.31
4.85	4.71	4.56	4.41	4.33	4.25	4.17	4.08	4.00	3.91
4.54	4.40	4.25	4.10	4.02	3.94	3.86	3.78	3.69	3.60
4.30	4.16	4.01	3.86	3.78	3.70	3.62	3.54	3.45	3.36
4.10	3.96	3.82	3.66	3.59	3.51	3.43	3.34	3.25	3.17
3.94	3.80	3.66	3.51	3.43	3.35	3.27	3.18	3.09	3.00
3.80	3.67	3.52	3.37	3.29	3.21	3.13	3.05	2.96	2.87
3.69	3.55	3.41	3.26	3.18	3.10	3.02	2.93	2.84	2.75
3.59	3.46	3.31	3.16	3.08	3.00	2.92	2.83	2.75	2.65
3.51	3.37	3.23	3.08	3.00	2.92	2.84	2.75	2.66	2.57
3.43	3.30	3.15	3.00	2.92	2.84	2.76	2.67	2.58	2.49
3.37	3.23	3.09	2.94	2.86	2.78	2.69	2.61	2.52	2.42
3.31	3.17	3.03	2.88	2.80	2.72	2.64	2.55	2.46	2.36
3.26	3.12	2.98	2.83	2.75	2.67	2.58	2.50	2.40	2.31
3.21	3.07	2.93	2.78	2.70	2.62	2.54	2.45	2.35	2.26
3.17	3.03	2.89	2.74	2.66	2.58	2.49	2.40	2.31	2.21
3.13	2.99	2.85	2.70	2.62	2.54	2.45	2.36	2.27	2.17
2.98	2.84	2.70	2.55	2.47	2.39	2.30	2.21	2.11	2.01
2.80	2.66	2.52	2.37	2.29	2.20	2.11	2.02	1.92	1.80
2.63	2.50	2.35	2.20	2.12	2.03	1.94	1.84	1.73	1.60
2.47	2.34	2.19	2.03	1.95	1.86	1.76	1.66	1.53	1.38
2.32	2.18	2.04	1.88	1.79	1.70	1.59	1.47	1.32	1.00

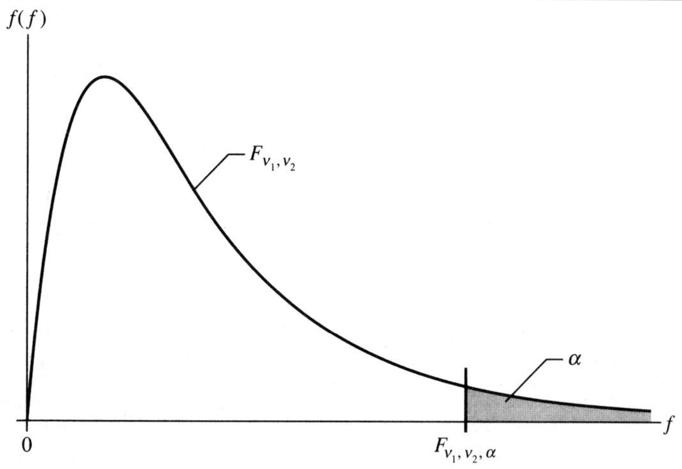

TABLE A.7
A studentized range distribution table
Values of $R_{\alpha,\nu,d}$

		ν_1									
ν_2	α	2	3	4	5	6	7	8	9	10	11
5	.05	3.64	4.60	5.22	5.67	6.03	6.33	6.58	6.80	6.99	7.17
	.01	5.70	6.98	7.80	8.42	8.91	9.32	9.67	9.97	10.24	10.48
6	.05	3.46	4.34	4.90	5.30	5.63	5.90	6.12	6.32	6.49	6.65
	.01	5.24	6.33	7.03	7.56	7.97	8.32	8.61	8.87	9.10	9.30
7	.05	3.34	4.16	4.68	5.06	5.36	5.61	5.82	6.00	6.16	6.30
	.01	4.91	5.92	6.54	7.01	7.37	7.68	7.94	8.17	8.37	8.55
8	.05	3.26	4.04	4.53	4.89	5.17	5.40	5.60	5.77	5.92	6.05
	.01	4.75	5.64	6.20	6.62	6.96	7.24	7.47	7.68	7.86	8.03
9	.05	3.20	3.95	4.41	4.76	5.02	5.24	5.43	5.59	5.74	5.87
	.01	4.60	5.43	5.96	6.35	6.66	6.91	7.13	7.33	7.49	7.65
10	.05	3.15	3.88	4.33	4.65	4.91	5.12	5.30	5.46	5.60	5.72
	.01	4.48	5.27	5.77	6.14	6.43	6.67	6.87	7.05	7.21	7.36
11	.05	3.11	3.82	4.26	4.57	4.82	5.03	5.20	5.35	5.49	5.61
	.01	4.39	5.15	5.62	5.97	6.25	6.48	6.67	6.84	6.99	7.13
12	.05	3.08	3.77	4.20	4.51	4.75	4.95	5.12	5.27	5.39	5.51
	.01	4.32	5.05	5.50	5.84	6.10	6.32	6.51	6.67	6.81	6.94
13	.05	3.06	3.73	4.15	4.45	4.69	4.88	5.05	5.19	5.32	5.43
	.01	4.26	4.96	5.40	5.73	5.98	6.19	6.37	6.53	6.67	6.79
14	.05	3.03	3.70	4.11	4.41	4.64	4.83	4.99	5.13	5.25	5.36
	.01	4.21	4.89	5.32	5.63	5.88	6.08	6.26	6.41	6.54	6.66
15	.05	3.01	3.67	4.08	4.37	4.59	4.78	4.94	5.08	5.20	5.31
	.01	4.17	4.84	5.25	5.56	5.80	5.99	6.16	6.31	6.44	6.55
16	.05	3.00	3.65	4.05	4.33	4.56	4.74	4.90	5.03	5.15	5.26
	.01	4.13	4.79	5.19	5.49	5.72	5.92	6.08	6.22	6.35	6.46
17	.05	2.98	3.63	4.02	4.30	4.52	4.70	4.86	4.99	5.11	5.21
	.01	4.10	4.74	5.14	5.43	5.66	5.85	6.01	6.15	6.27	6.38
18	.05	2.97	3.61	4.00	4.28	4.49	4.67	4.82	4.96	5.07	5.17
	.01	4.07	4.70	5.09	5.38	5.60	5.79	5.94	6.08	6.20	6.31
19	.05	2.96	3.59	3.98	4.25	4.47	4.65	4.79	4.92	5.04	5.14
	.01	4.05	4.67	5.05	5.33	5.55	5.73	5.89	6.02	6.14	6.25
20	.05	2.95	3.58	3.96	4.23	4.45	4.62	4.77	4.90	5.01	5.11
	.01	4.02	4.64	5.02	5.29	5.51	5.69	5.84	5.97	6.09	6.19
24	.05	2.92	3.53	3.90	4.17	4.37	4.54	4.68	4.81	4.92	5.01
	.01	3.96	4.55	4.91	5.17	5.37	5.54	5.69	5.81	5.92	6.02
30	.05	2.89	3.49	3.85	4.10	4.30	4.46	4.60	4.72	4.82	4.92
	.01	3.89	4.45	4.80	5.05	5.24	5.40	5.54	5.65	5.76	5.85
40	.05	2.86	3.44	3.79	4.04	4.23	4.39	4.52	4.63	4.73	4.82
	.01	3.82	4.37	4.70	4.93	5.11	5.26	5.39	5.50	5.60	5.69
60	.05	2.83	3.40	3.74	3.98	4.16	4.31	4.44	4.55	4.65	4.73
	.01	3.76	4.28	4.59	4.82	4.99	5.13	5.25	5.36	5.45	5.53
120	.05	2.80	3.36	3.68	3.92	4.10	4.24	4.36	4.47	4.56	4.64
	.01	3.70	4.20	4.50	4.71	4.87	5.01	5.12	5.21	5.30	5.37
∞	.05	2.77	3.31	3.63	3.86	4.03	4.17	4.29	4.39	4.47	4.55
	.01	3.64	4.12	4.40	4.60	4.76	4.88	4.99	5.08	5.16	5.23

Source: This table is abridged from Table 29 in *Biometrica Tables for Statisticians,* vol. 1, 3rd ed., by E. S. Pearson and H. O. Hanley (eds.). Reproduced with the kind permission of the Trustees of *Biometrica,* 1966.

TABLE A.7
(Continued)

				ν_1						
12	**13**	**14**	**15**	**16**	**17**	**18**	**19**	**20**	α	∞
7.32	7.47	7.60	7.72	7.83	7.93	8.03	8.12	8.21	.05	5
10.70	10.89	11.08	11.24	11.40	11.55	11.68	11.81	11.93	.01	
6.79	6.92	7.03	7.14	7.24	7.34	7.43	7.51	7.59	.05	6
9.48	9.65	8.91	9.95	10.08	10.21	10.31	10.43	10.54	.01	
6.43	6.55	6.66	6.76	6.85	6.94	7.02	7.10	7.17	.05	7
8.71	8.86	9.00	9.12	9.24	9.35	9.46	9.55	9.65	.01	
6.18	6.29	6.39	6.48	6.57	6.65	6.73	6.80	6.87	.05	8
8.18	8.31	8.44	8.55	8.66	8.76	8.85	8.94	9.03	.01	
5.98	6.09	6.19	6.28	6.36	6.44	6.51	6.58	6.64	.05	9
7.78	7.91	8.03	8.13	8.23	8.33	8.41	8.49	8.57	.01	
5.83	5.93	6.03	6.11	6.19	6.27	6.34	6.40	6.47	.05	10
7.49	7.60	7.71	7.81	7.91	7.99	8.08	8.15	8.23	.01	
5.71	5.81	5.90	5.98	6.06	6.13	6.20	6.27	6.33	.05	11
7.25	7.36	7.46	7.56	7.65	7.73	7.81	7.88	7.95	.01	
5.61	5.71	5.80	5.88	5.95	6.02	6.09	6.15	6.21	.05	12
7.06	7.17	7.26	7.36	7.44	7.52	7.59	7.66	7.73	.01	
5.53	5.63	5.71	5.79	5.86	5.93	5.99	6.05	6.11	.05	13
6.90	7.01	7.10	7.19	7.27	7.35	7.42	7.48	7.55	.01	
5.46	5.55	5.64	5.71	5.79	5.85	5.91	5.97	6.03	.05	14
6.77	6.87	6.96	7.05	7.13	7.20	7.27	7.33	7.39	.01	
5.40	5.49	5.57	5.65	5.72	5.78	5.85	5.90	5.96	.05	15
6.66	6.76	6.84	6.93	7.00	7.07	7.14	7.20	7.26	.01	
5.35	5.44	5.52	5.59	5.66	5.73	5.79	5.84	5.90	.05	17
6.57	6.66	6.74	6.82	6.90	6.97	7.03	7.09	7.15	.01	
5.31	5.39	5.47	5.54	5.61	5.67	5.73	5.79	5.84	.05	17
6.48	6.57	6.66	6.73	6.81	6.87	6.94	7.00	7.05	.01	
5.27	5.35	5.43	5.50	5.57	5.63	5.69	5.74	5.79	.05	18
6.41	6.50	6.58	6.65	6.73	6.79	6.85	6.91	6.97	.01	
5.23	5.31	5.39	5.46	5.53	5.59	5.65	5.70	5.75	.05	19
6.34	6.43	6.51	6.58	6.65	6.72	6.78	6.84	6.89	.01	
5.20	5.28	5.36	5.43	5.49	5.55	5.61	5.66	5.71	.05	20
6.28	6.37	6.45	6.52	6.59	6.65	6.71	6.77	6.82	.01	
5.10	5.18	5.25	5.32	5.38	5.44	5.49	5.55	5.59	.05	24
6.11	6.19	6.26	6.33	6.39	6.45	6.51	6.56	6.61	.01	
5.00	5.08	5.15	5.21	5.27	5.33	5.38	5.43	5.47	.05	30
5.93	6.01	6.08	6.14	6.20	6.26	6.31	6.36	6.41	.01	
4.90	4.98	5.04	5.11	5.16	5.22	5.27	5.31	5.36	.05	40
5.76	5.83	5.90	5.96	6.02	6.07	6.12	6.16	6.21	.01	
4.81	4.88	4.94	5.00	5.06	5.11	5.15	5.20	5.24	.05	60
5.60	5.67	5.73	5.78	5.84	5.89	5.93	5.97	6.01	.01	
4.71	4.78	4.84	4.90	4.95	5.00	5.04	5.09	5.13	.05	120
5.44	5.50	5.56	5.61	5.66	5.71	5.75	5.79	5.83	.01	
4.62	4.68	4.74	4.80	4.85	4.89	4.93	4.97	5.01	.05	∞
5.29	5.35	5.40	5.45	5.49	5.54	5.57	5.61	5.65	.01	

ANSWERS TO SELECTED EXERCISES

Chapter 2

2.3.a. .18 **b.** $\frac{151}{200}$ **2.4.** $\frac{35}{36}$ **2.5.a.** $\frac{4}{52}$ **b.** $\frac{1}{13}$ **c.** $\frac{2}{13}$

2.6. $\frac{1}{11}$ **2.8.a** $\frac{1}{2}$ **b.** $\frac{1}{2}$ **c.** $\frac{1}{2}$ **d.** $\frac{3}{4}$ **e.** 1 **f.** $\frac{3}{4}$ **g.** $\frac{1}{4}$

h. 0 **i.** $\frac{1}{4}$ **2.12.** $P(A) = \frac{2}{3}.P(A \cap B) = \frac{7}{60}.P(B|A) = \frac{7}{40}$ **2.15.** $\frac{2}{10}$

2.16. .00108 **2.17.a.** .32 **b.** .12 **c.** 0.56 **2.18.a.** Yes. **b.** $\frac{3}{4}$

c. A and B are not mutually exclusive events. **2.20.** $R!$ **2.21.** ${}_nC_r$ ways

2.24. $(5/6)^{10} = .1615$ **2.25.a.** $\frac{11}{60}$ **b.** $\frac{17}{24}$ **c.** $\frac{11}{60}$

d. A and B are not independent events. **2.26.** $\frac{8}{9}$ **2.27.** $\frac{5}{11}$ **2.29.a.** 500

b. 96 **2.32.** 210 **2.34.** 54

Chapter 3

3.1. $P(X = 2) = \frac{1}{15}, P(X = 3) = \frac{2}{15}, P(X = 4) = \frac{3}{15}, P(X = 5) = \frac{4}{15}, P(X = 6) = \frac{1}{3}$

3.2. $P(X = 1) = \frac{9}{25}, P(X = 2) = \frac{7}{25}, P(X = 3) = \frac{5}{25}, P(X = 4) = \frac{4}{25}$

3.3.

x	0	1	2	3	4	5	6
$p(x)$	$\frac{36}{60}$	$\frac{2}{60}$	$\frac{8}{60}$	$\frac{6}{60}$	$\frac{2}{60}$	0	$\frac{6}{60}$

3.4. Binomial with $n = 4$, $p = \frac{1}{18}$ **3.6.** 0.5178 **3.7.** 0.6229

3.8. 0.137 **3.9.** 0.782 **3.10** 0.79 **3.12** Geometric distribution ($p = 0.6$); Four switches are needed.

3.13.a. $k = 0.3591$ **b.** 0.8479 **3.19** 0.00345 **3.22.** ≈ 0

3.26. $F(11) = 0.19404$ **3.29.a.** $p(3) = 0.012255$

b. $P(X \leq 1) = 0.8284$ **3.32.a.** Yes. **b.** 0.00 c. 0.2857

3.33a. The probability of the intersection of any column and row is the product of the probabilities of that column and row. In addition, the rows and columns are proportional multiples of one another. For example, row 2 is $\frac{1}{2}$ of row one and column 3 is $\frac{2}{3}$ of column 2.

Chapter 4

4.1.a. $\frac{1}{4}$ **b.** $\frac{1}{2}$ **c.** $\frac{15}{16}$ **4.3a.** $k = \frac{1}{4}$ **b.** $e^{-1/2} - e^{-2}$ **c.** $e^{-7/4}$

d. $1 - e^{-1/4}$ **e.** $4\ln(20) \approx 11.98$ **f.** $c \approx .1$ **4.4.** $-7.5 = a$

4.5.a. .0043 **b.** .0427 **c.** .8997 **d.** .9343 **e.** $C = 2.34$

f. $d = 1.33$ **g.** $a = -2.575$ $b = 2.575$

4.7. .9962 **4.9.** .0283 **4.12.** 0.3536

4.14. 3.993 seconds **4.16.** $t^* = 1.126$ **4.18.** 0.5509

4.20. The needle will touch or cross the lines 240° out of the 360° total, hence the probability is $\frac{2}{3}$. The distribution is a continuous uniform pdf bounded between 0° and 360°.

4.22a. $k = 100$ **b.** $P(X \geq 500) = 0.2$ **4.24.** .01322 **4.26.** .8358

4.27. .0132 **4.29.** 0.1641 **4.30.a.** $k = \frac{6}{5}$

b. $f_{X_1}(x_1) = \frac{6}{5}\left(x_1 + \frac{1}{3}\right)$, $f_{X_1}(x_2) = \frac{6}{5}\left(\frac{1}{2} + x_2^2\right)$

c. They are not statistically independent. **d.** $1.327x_1 + .3363$

4.33. $f_{X_1}(x_1) = (\sqrt{.5}/\sqrt{\pi})e^{-.5x}$, $f_{X_1}(x_2) = (\sqrt{.5}/\sqrt{\pi})e^{-.5x}$ **a.** Yes.

b. $(\sqrt{.5}/\sqrt{\pi})e^{-.5x}$

Chapter 5

5.2.a. $\mu = 5$, $\sigma^2 = 12.74$; median is anywhere in the range $5 < x < 7$; mode is at $x = 7$

b. $\beta_1 = 0$, $\beta_2 = 1.8$ **c.** 43.76

d.

y	1.75	31	57	133
$p(y)$	.333	.167	.375	.125

e. $\mu_y = 43.76$, $\sigma_y^2 = 1644.98$;
the median is anywhere in the range $31 < x < 57$.

f. For Y, $\beta_1 = 0.88(+)$ and $\beta_2 = 3.197$.

5.3.a. $k = 1/24 = 0.041$ **b.** $\mu = 6.2222$, $\sigma^2 = 1.284$;
mode is at $x = 8$.
The median is $a = 6.3246$. **c.** $\beta_1 = .05269(-)$, $\beta_2 = 1.8936$.

d. $E(2x) = 12.44$

5.4a.& b. $\mu = 3$, $\sigma^2 = 6$ $(\sigma = \sqrt{6})$, $\beta_1 = \frac{8}{3}$, $\beta_2 = 7$,
$a =$ median $= 2.3661$, the mode $x = 1$.

5.4c. $E(X^{-.5}) = \frac{2}{\sqrt{2\pi}}$

5.11.a. *(Exercise 3.32)* $\text{Cov}(X_1, X_2) = .1416$, $\text{Corr}(X_1, X_2) = 0.2899$

b. *(Exercise 3.33)* $\text{Corr}(X_1, X_2) = 0$ **5.11.f.** *(Exercise 4.30)* $\text{Cov}(X_1, X_2) = -0.01$, $\text{Corr}(X_1, X_2) = -0.1306$ **5.11.g.** *(Exercise 4.31)* $\text{Cov}(X_1, X_2) = -0.0027$, $\text{Corr}(X_1, X_2) = -0.0171$

5.11.j. *(Exercise 4.34)* $\text{Cov}(X_2, X_2) = 0.0178$, $\text{Corr}(X_1, X_2) = 0.4924$

5.13. $P\{-18 \leq X \leq 27\} > 0.99$

5.18.a. *(Exercise 4.1)* $\mu = \frac{2}{3}, \sigma^2 = \frac{1}{18}$ **b.** *(Exercise 4.2)* $\mu = 7$, $\sigma^2 = \frac{1}{6}$

e. *(Exercise 4.21)* $\mu = 2.4, \sigma^2 = 6.4475$ **f.** *(Exercise 4.22)* μ is undefined; therefore σ^2 is undefined. **k.** *(Exercise 3.13)* $\mu = 0.09, \sigma^2 = 2.51$
o. *(Exercise 4.33)* $\mu = 0.7979, \sigma^2 = 0.3634$

5.19. for $n = 1, \sigma^2$ is undefined; for $n = 2, \sigma^2$ is undefined.

5.20.a. All values are modes. **b.** $(r - 1)/\lambda$ **c.** $n - 2(n > 1)$

5.21.c. $\mu = \Psi\eta/(\Psi - 1), \sigma^2 = \Psi\eta^2/[(\Psi - 2)(\Psi - 1)^2]$

5.24. $L = 1/e, \mu = 1, \sigma^2 = 1$

Chapter 6

6.2a. Data Set A: range $= .9418$, median $= .4825$, $x = .519$,

$s^2 = .0786, b_1 = .015(+), b_2 = 1.72$; with a starting value of 0.05 and a class interval of 0.2, we obtain a good frequency polygon. **b.** This data set was generated by simulation from a uniform distribution over the limits from 0 to 1. Sampling error obscures this fact. However, the values of the sample estimators and the theoretical values for all the uniform distribution are in good agreement; i.e.,

Data: $x = .519, s^2 = .0786, b_1 = 0.015(+), b_2 = 1.72.$

Theoretical: $\mu = .5, \sigma^2 = .0833, \beta_1 = 0, \beta_2 = 1.8.$

6.4. Problem 6.4's data set is the combination of 50 data generated from an $N(10, 0.5)$ distribution and 75 data from an $N(8, 1)$ distribution. This will be hidden from the analyst before a frequency polygon or histogram is constructed via the data analysis routine. The bimodal nature of the data will then be encountered.

6.7. This data is drawn (using the Monte Carlo simulation program) from a binomial distribution with $n = 15$ and $p = 0.04$. From the frequency table, it appears that it is unlikely that any sample of 15 devices will have more than 2 defectives. The best guess of the proportion of defectives is $\frac{0.5}{15} = 0.0333$.

6.10. The sample average is 0.7467, implying an average proportion of failures of $0.7467/8 = 0.0933375$. It certainly appears that the average proportion defectives is well above the required level of 7%. In the presentation of acceptance sampling methods in Chapter 15 we will study specific methods designed to answer the kind of question presented in this exercise.

6.12. The best data estimate from this data set of the proportion of nonmatching pairs is $0.625/10 = 0.0625$. This is not very different from 0.08, which was the estimate from the data set of Exercise 6.11. However, we do not have any dependable method of performing a scientific comparison of these proportions. A method of comparing two such proportions in a statistically valid manner is presented in Chapter 10.

Chapter 7

7.1. For $N = 40000$, we may ignore the finite population correction factor: the new sample is 1600. If $N = 5000$, the finite population correction factor cannot be ignored. Required sample size is about 1200. **7.2.** 0.0668 **7.3.** No.

7.4. The probability that $\sigma^2 \le 12$ when $s^2 \ge 21.96$ is 0.0577.

7.6.

$\overline{x}$	2.00	2.33	2.66	3.00	3.33	3.66	4.00	4.33	4.66	5.00
$p(\overline{x})$	0.05	0.05	0.10	0.15	0.15	0.15	0.15	0.10	0.05	0.05

7.8. 0.9606 **7.10.** 0.0223 **7.20.** 0.022 **7.21.** 0.47

Chapter 8

8.2. Confidence interval with $\alpha = 0.05$ is [.2022, .4978].

8.3. 2 **8.4.** 80% confidence interval is [5.0446, 5.0754].

8.5. 90% confidence interval is [0.00031, 0.00057].

8.6. 90% confidence interval is [0.2799, 0.6292].

8.8a. The symmetric confidence limits are $[.2631s^2, 1.742s^2]$.

b. The minimum-width 95% confidence interval limits are $[0.313s^2, 2.95s^2]$.

8.9. The 95% confidence limits are [0.01235, 0.317].

8.10. The 95% equal-tail confidence interval on λ is [0.08074, 1.85722]

8.13.a. $s^2 = 2.69$ **b.** [0.96, 4.70]

8.15. [11.541, 12.498]

8.17.a. [4.07, 4.50] **b.** [0.0648, 0.4563]

8.20. On μ, [0.343, 0.657]; on σ, [0.618, 0.848].

8.24. Yes.

Chapter 9

9.1. $\beta = 0.2043$ **9.2.** $n = 7$; reject $H_0 : \mu = 15$. **9.3.** Do not reject H_0 *(barely)*.

9.7. If $\frac{1}{3}$ of 1% is sufficient accuracy, then $n = 60$ is an acceptable value.

9.9. The probability of not rejecting H_0: $\sigma^2 = 25$ when σ^2 is *actually* equal to 35 is 0.75. A larger sample is required to bring β to an acceptable size.

9.10. Reject H_0.

9.11. Foghorn will almost certainly win the election.

9.14. $\alpha = 0.1, C = 30.5711, \overline{x} = 34.5 > 30.5711$. Null hypothesis is rejected.

9.17. $\alpha = 0.1, C = 6.388, \overline{x} = 6.3 < 6.388$. Reject the null hypothesis. Klone's claim is incorrect.

9.19. $\sigma = 0.01, n = 10, s = 0.018, \alpha = 0.05 \rightarrow C = 0.0002$. H_0 is rejected. Sample size is adequate.

Chapter 10

10.1. *Reject* H_0: $\mu_1 - \mu_2 = 0$. **10.2.** Reject H_0; there is a significant difference.

10.3. Cannot reject H_0; no significant difference implied.

10.4. $n_1 = n_2 = n = 198$. Now we can use the O.C. curve program for hypothesis test of 2 means.

10.6. $\hat{P}_{1,A} = 0.598$ is a value of $\hat{P}_{1,A}$ that minimizes $\hat{\sigma}^2_A$. ($\hat{P}_{2,A} = 0.848$.)

10.9. Reject the null hypothesis. The sample size is adequate.

10.12. Using $\alpha = 0.1$, do not reject the null hypothesis.

10.15. $\alpha = 0.1$; do not reject the null hypothesis.

Chapter 11

11.1. The multiple correlation coefficient squared is .996. Clearly, there is a strong linear relationship between x and y, with $r = .998$.

The model is $Y = B_0 + B_1X = 4.98 + 2X$.

The fit, the confidence band, and the ECDF look very good. Unfortunately, the plot of the residuals versus y shows clearly that the variance is not constant. These results cannot be used. We must resort to a weighted least squares approach, a subject beyond the scope of this text.

11.3. A hyperbolic transformation of the data yields an excellent fit.

11.4.b. Applying the inverse transform $e^{Y'} = \exp(b_0' + b_1X' + \varepsilon)$, one obtains $Y'' = b_0x^{b_1}e^{\varepsilon}$. The error structure in the untransformed space is thus multiplicative and exponential-normal.

11.6. For the untransformed model, the correlation coefficient is $r = .988$, and the multiple correlation coefficient squared is .976144.

The model $Y = \beta_0 + \beta_1 x + \epsilon$ accounts for most of the variability in the data, although there is a significant nonlinear effect missing from the model.There is a strong likelihood that all that is required is the addition of a term in x^2; i.e., fit $Y = \beta_0 + \beta_1 x + \beta_2 x^2 + \varepsilon$ to the data.

11.10. Although there is a significant linear relation between Y and x, much of the variability remains unexplained by the SLR model(r^2 is only 0.714). In the residual plot where e_i is plotted against y_i, there is an indication that another factor with a linear relationship to Y has not been included in the model.

Chapter 12

12.2. On β_0, [47.48, 57.66]; on β_1, [1.194, 1.742]; on β_2, [0.5579, 0.7665]. The box is *not* reliable because of the high correlation of -0.87534 between $\hat{\beta}_0$ and $\hat{\beta}_2$. We would have to use the confidence ellipsoid described in Section 12.4.

12.3. The full model, $Y = \beta_0 + \beta_1 x_1 + \beta_2 x_2 + \beta_3 x_3 + \beta_4 x_4 + \varepsilon$, has an acceptable fit with no clear indications of lack of fit from the residual plots. It is also superior to the subset models in values of r^2, f, s_e, and C_p. We will just have to live with $r^2 = 0.87$.

12.5. The confidence interval about the mean with $\alpha = 0.05$ is [86.876, 90.204]; for a future observation, [82.926, 94.153].

12.6. The full model is $Y = \beta_0 + \sum_{i=1}^{4} \beta_i X_i + \varepsilon$. The table below gives summary results for each possible submodel.

Variables in Model	r^2	MSE	C_p
*1,2,3,4	0.9824	6.00	5.0
1	0.5339	115.1	202.5
2	0.6663	82.4	142.5
*1,2	0.9787	5.83	2.7
3	0.2859	176.3	315.2
1,3	0.5482	122.7	198.1
2,3	0.8470	41.5	62.4
*1,2,3	0.9823	5.31	3.0
4	0.6745	80.4	138.7
*1,4	0.9725	7.47	5.5
2,4	0.6801	86.8	138.2
*1,2,4	0.9823	5.31	3.0
3,4	0.9353	17.6	22.4
*1,3,4	0.9813	5.6	3.5
2,3,4	0.9728	8.2	7.3

* The asterisks indicate the competing models. The new models help, but not much.

12.8. It appears that a single point may be destroying the goodness of the fit. First, eliminate that datum and rerun the model as is. If the goodness of fit is corrected, report the results with and without the questionable point. If the goodness of fit is not improved, investigate subset models, first dropping x_4 and then x_3 from the model because of their very low t values relative to the other x_j. Then consider adding terms to the model that are functions of x_1, x_2, and x_3. What terms to add would depend on the residual indications from the subset models cited above.

Chapter 13

13.1. The p value of α is 0.6195. H_0 is not rejected.

13.2. Since p value of α is 0.0165, the F value is significant at the 0.05 level of confidence. Brand 3 is significantly different from both brands 1 and 2.

13.5. We easily reject H_0 at the $\alpha = 0.05$ level. $\overline{\mu}_2$ and $\overline{\mu}_5$ are not significantly different. $\overline{\mu}_5$, $\overline{\mu}_1$, and $\overline{\mu}_4$ are not significantly different. All other pairs are significantly different.

13.6. Do not reject H_0.

13.8. Only [$\overline{\mu}_1$ and $\overline{\mu}_2$] and [$\overline{\mu}_2$ and $\overline{\mu}_4$] are *not* significantly different.

13.10. Only the "blocks" or hardening processes exhibit a statistically significant difference at $\alpha = 0.05$. Finishing process 1 may be considered superior when used with any of the three types of steel.

13.13. $f = 208$; the mixtures are different. All pairs of mixtures are significantly different. The probability is about 0.025.

13.15. There are no significant differences between guns or operators. The day (replication) effect is very significant, so days must be considered in the analysis.

Chapter 14

14.3. From Table 14.1 we see that the effects for A, B, and AB may be estimated by $A_j = (\overline{y}_j - \overline{y}_\bullet)$, $B_i = (\overline{y}_i - \overline{y}_\bullet)$ and $AB_{ij} = (\overline{y}_{ij} - \overline{y}_i - \overline{y}_j + \overline{y}_\bullet)$. The grand mean is $\overline{y}_\bullet = 8.027$. Therefore,

$$A_1 = -0.322 \qquad A_2 = -0.107 \qquad A_3 = +0.426$$
$$B_1 = +0.063 \qquad B_2 = -0.160 \qquad B_3 = +0.097$$

$$\begin{array}{llllll} AB_{11} & = 0.7266 & AB_{12} & = -0.4088 & AB_{13} & = -0.3177 \\ AB_{21} & = -0.0177 & AB_{22} & = 0.3033 & AB_{23} & = -0.2855 \\ AB_{31} & = -0.7088 & AB_{32} & = 0.1055 & AB_{33} & = 0.6033 \end{array}$$

Notice that the interaction effect AB_{11} is actually greater than AB_{33} even though $\bar{y}_{33}$ is greater than $\bar{y}_{11}$. This fact is accounted for by the effects of the major factors being so different at levels 1 and 3.

14.7. Factor B is significant. No, the sample is not large enough. The β error associated with this difference is more than 50% for both the A and B effects.

14.9. Only the A and C effects are significant. All other effects are not significant.

Chapter 15

15.2. From observation of the R chart, we see an indication of lack of control in data points 11 and 13, which satisfy the condition of Figure 15.2*b*. Thus, the variability of the process should be investigated further.

15.3. The charts give no indication that the process is out of control.

15.4.a. 0.0026 **b.** 0.0027 **c.** 0.00516 **d.** 0.0078

15.5. Since the last 5 points in the $\bar{x}$ chart satisfy the condition of Figure 15.2*c*, we may conclude that the amount of mist is *not* in control. In addition, and perhaps most important, there appears to be a trend towards lesser values of escaping paint mist. We need to find out what caused this and make sure we continue to do it!

15.6. The process will produce about 3% of the components with identions outside of the engineering specifications. If 3% defective components is unacceptable, the process will be deemed incapable of meeting the specifications and the process will have to be modified.

15.7. Since there are no indications of an assignable cause, the process is in control.

15.9. No. **15.12.** $n = 785$, $C = 22$, N is *large*.

15.13. The AOQP is about 0.01969 and is located at a value of 0.022.

15.14. Reject the lot at the third sample.

Chapter 16

16.1. Using the ranges and p_{0i} of Example 16.2 yields a χ^2 value of 7.4, which implies a p-value of .5955. H_0 is not rejected.

16.6. Use the Wilcoxon signed rank test $s^+ = 27$ which has a p-value of 0.0595. Reject H_0 for any value of α greater than 0.0595.

16.10. Wilcoxon rank sum test $s^+ = 65.5$ which has a p-value of 0.542. H_0 may not be rejected for any value of α less than 0.542.

16.14. *(Exercise 13.5)* using the Kruskal-Wallis test we obtain $h = 24.18$, which yields a p-value of 0.0001. Reject H_0 for any $\alpha > 0.0001$. Essentially, it is the same result as from ANOVA.

INDEX

A

B

C

D

E

I

J

K

Q

R

S

T

U

V

W

X

Z